THE MAGELLANIC SYSTEM:
STARS, GAS, AND GALAXIES

IAU SYMPOSIUM No. 256

COVER ILLUSTRATION: THE TECHNICOLOR MAGELLANIC CLOUD

This picture shows the Large Magellanic Cloud in its multi-colour splendour. It is a combination of images covering wavelengths from optical to radio, reflecting the different "eyes" astronomers have created for themselves over the past half century. In purple through to blue are shown light emitted in the Hα, [S II] and [O III] atomic transitions (data from the Magellanic Clouds Emission-Line Survey, MCELS). This mainly pinpoints the sites of star formation activity and stellar death. Yellow-green colours show infrared emission at 3.6, 5.8 and 24 μm from the *Spitzer* images obtained by the SAGE Legacy Team, and orange-yellow colours show older infrared data at 12, 25, 60 and 100 μm from *IRAS*. These hues reflect the glow of dust, which plays an important rôle in the evolution of galaxies. Finally, red colours show the 21-cm emission from neutral hydrogen (data from *ATNF/ATCA*), revealing the extent also of the colder interstellar gas.

Multi-wavelength views such as this one, of both Magellanic Clouds, reveal the intricate connections between the cycle of life and death of stars, and the processes that sculpt and transform the diffuse matter pervading these magnificent galaxies. The quasi-3-dimensional appearance of these composites stimulates us to think beyond the boundaries of individual data sets, to be imaginative in applying the laws of physics to combine the microscopic with the macroscopic.

These psychedelic pictures are also incredibly humbling, and make us conscious that the Universe is a truly remarkable place in which to seek beauty and truth.

Composite image courtesy of Iain McDonald, Keele University.

INTERNATIONAL ASTRONOMICAL UNION

UNION ASTRONOMIQUE INTERNATIONALE

International Astronomical Union

THE MAGELLANIC SYSTEM: STARS, GAS, AND GALAXIES

PROCEEDINGS OF THE 256th SYMPOSIUM OF THE INTERNATIONAL ASTRONOMICAL UNION HELD AT KEELE UNIVERSITY, UNITED KINGDOM JULY 28–AUGUST 1, 2008

Edited by

JACCO TH. VAN LOON
Keele University, United Kingdom

and

JOANA M. OLIVEIRA
Keele University, United Kingdom

CAMBRIDGE
UNIVERSITY PRESS

CAMBRIDGE UNIVERSITY PRESS
The Edinburgh Building, Cambridge CB2 8RU, United Kingdom
32 Avenue of the Americas, New York, NY 10013-2473, USA
477 Williamstown Road, Port Melbourne, VIC 3207, Australia
Ruiz de Alarcón 13, 28014 Madrid, Spain
Dock House, The Waterfront, Cape Town 8001, South Africa

First published 2009

Printed in the United Kingdom at the University Press, Cambridge

Typeset in System LaTeX 2_ε

A catalogue record for this book is available from the British Library

Library of Congress Cataloguing in Publication data

ISBN 9780521889872 hardback
ISSN 1743-9213

Table of Contents

Session I. Surveys of the Magellanic Clouds

Session II. The structure and dynamics of the Magellanic System

Session III. The properties of the interstellar medium

Session IV. The star formation process

Session V. The star formation history and chemical evolution

Session VI. The Magellanic Clouds as laboratories of stellar astrophysics

Session VII. The final stages of stellar evolution and feedback

Session VIII. Magellanic type systems as a class

Summary

List of posters

Posters are available on-line at http://www.astro.keele.ac.uk/iaus256/proceedings, either as fully edited/indexed manuscripts, numbered P-xx, or PDF files, numbered PDF-xx.

Preface

When five centuries ago Portuguese explorer Fernão de Magalhães sailed underneath the splendid skies of the Southern hemisphere, his name was given to the two dimly glowing clouds that appear separated from the so prominent Milky Way. The larger of these was already known to the Persians in the 3rd century after Mohammad, from reports by seafarers who traded along the East-African coasts. It was documented by the great astronomer Al Sufi (the Wise One) and called "Al Bakr", the White Ox. We now know these clouds are small gas-rich galaxies like many in the Universe, much like the Milky Way is a typical spiral galaxy. We also know that these galaxies are not as separated as they first seemed; that they move under each other's influence, disfigure, and trade gas and possibly stars. Like a true ox, the Large Magellanic Cloud ploughs through the Galactic Halo, its steamy breath leaving behind a gaseous trail.

Some fifty millenia earlier, other adventurous tribes crossed the Indonesian archipelago and settled the vast territories of Australia. These were highly imaginative peoples, who readily adopted the "Magellanic" Clouds into their dreamworld. Their stories about the relationships between these clouds have survived to the present day. So have some rather abstract pieces of artwork, depicting the clouds as tortoise shells perhaps to embody their perpetuity — the tortoise has an important place in certain North-American peoples' creation myth, and some modern astronomers indeed believe gas-rich haloes like the Magellanic Clouds to be the building blocks of larger galaxies in the Universe.

The Magellanic Clouds may have been known to Meso-Americans, as evidenced by petroglyphs and architectural alignment. They must certainly have been known to the great civilizations of the central Andes and coastal desert plains who, in the absence of a written language but in spite of their hugely expressive decorative art (pottery, textiles, petroglyphs and murals) left no imagery of these constellations. In their cosmogeny, the dark patches seen in silhouette against the Milky Way are much more important than the scattered dots of light (and probably more reliable, given the supernovae and other star-like erratic variations they must have witnessed). Dark matter within the Milky Way and Magellanic Clouds is a hot topic in contemporary cosmology.

The creativity and imagination of our ancestors, and their thirst for adventure and discovery should be an inspiration to us now, when we explore the nature and behaviour of the Magellanic Clouds and all that is within and surrounds them. Having seen them once on a clear, Moon-less night away from the noise and light and other waste emanating from wherever modern people conglomerate, one can not be left untouched by the magical beauty of these celestial bodies. In recent years stunning multi-colour images have painted the clouds as we will never see them with our own eyes. Now more than ever do they do justice to the constellations in which they appear, a golden fish (Doradus) and a most splendid — if a little freaky — bird (Tucana).

Having watched the Clouds, they don't easily let go. As graduate students, hugely impressed at the previous IAU Symposium "New Views" on the Magellanic Clouds, in a place sharing its name with Magalhães' ship the "Victoria", we are still watching, marvelling, and learning from them, seeking beauty and truth. The past decennium has seen a wealth of new information emerge, on the constitution of the Magellanic Clouds and the gas that once belonged to them, on their trajectories through space, and the many (g)astrophysical processes that operate and for which the Magellanic Clouds offer us the ideal laboratory (or, perhaps more accurately, a zoological garden to study). It was therefore obviously time for the next IAU Symposium devoted to the Magellanic System, at which we were in particular looking for synergies between disparate fields of research.

The organizing committee gathered under the hot Sun in the romantic city of Prague, and plans were made. The peaceful, green university campus of Keele was to host the meeting, as well as display a cross-section of typical british weather. As Joss and Jay point out in their conference analysis (far beyond a plain summary!) we *know* the quirks of weather and how they affect us, but weather is not *understood*. Let alone climate. To understand the Magellanic System requires weather maps, charting the dynamics of the different phases of the interstellar medium, the nucleation of stellar objects, and their feedback, and climate patterns, to be read in the star formation, chemical evolution and kineto-morphological histories.

Thus convened 152 participants from 27 countries and 30 nationalities, for a full five days of science presentations the products of which are documented in these proceedings. Besides the scientific programme, a cultural programme was put together by the local organizing committee. It started with a performance of the Shakespeare drama "Othello" by Anvil Productions, in Keele Hall (once the thunderstorms had arrived) and its gardens (before lightning and rain had become too heavy to ignore). This was followed the next day by a public lecture by acclaimed archaeo-astronomer Professor Clive Ruggles, entitled "Astronomy before History", which was attended by an audience of around 200. The programme ended on Wednesday evening with a classical concert by London Concertante, in the University Chapel, including also works by the famous German-British astronomer William Herschel (who never saw the Magellanic Clouds, but his brother did). Other public events included a teacher's event, a public display of two large stands showcasing the Spitzer surveys of the Magellanic Clouds, and a large number of postcards which had been designed to show a multi-wavelength image of the Large Magellanic Cloud and historical pictures of Magalhães and his ship. Nearly all symposium participants joined in the conference dinner, at the delightful early-17th-century Wrenbury Hall in the Cheshire countryside. A beer had been produced by a local brewery specially for the occasion, the "Magellanic Ale".

The organisers are grateful to the Royal Astronomical Society, which helped fund some of the cultural activities and supported the attendance of some young researchers. The International Astronomical Union, through its grant, made it possible for many more to attend. We also thank Cambridge University Press and Wiley-Blackwell Publishing for their generosity in making available prizes for the best posters (which are included in the printed edition of these proceedings) and for young researchers. Special thanks go to Nye Evans for chairing the local organization, Iain McDonald for much of the artistic designs, Rob Jeffries for making Othello meet the end, and Michael Feast and Despina Hatzidimitriou for judging all of the posters on top of their other contributions to the scientific organization. And last but certainly no least we would like to express our warmest thanks to all our colleagues and friends for having come and made the symposium worth while.

In 2008 we remembered Henrietta Leavitt who, a century before, discovered the Cepheid period–luminosity relationship through measurements in the Magellanic Clouds, paving the way to measure the size of the Universe. A separate conference dedicated to her achievements was held elsewhere that year. The Magellanic Clouds meeting, however, was dedicated to the memory of Bengt Westerlund. A keen and influential observer of the Clouds, he sadly passed away less than two months before the symposium.

Jacco van Loon and Joana Oliveira
Keele University, United Kingdom, November 20, 2008

THE ORGANIZING COMMITTEE

Scientific

B. Barbuy (Brazil)

Y.-H. Chu (USA)

G. Da Costa (Australia)

M. Feast (South Africa)

Y. Fukui (Japan)

E. Grebel (Germany)

D. Hatzidimitriou (Greece)

M. Heydari-Malayeri (France)

B. Moore (Switzerland)

W. Pietsch (Germany)

M. Rubio (Chile)

S. Stanimirović (USA)

J.Th. van Loon (chair, UK)

N. Walborn (USA)

Ł. Wyrzykowski (Poland)

Local

A. Evans (chair)

S.L. Faulke

R. D. Jeffries

I. McDonald

J.M. Oliveira

S.J. Rooney

B. Smalley

J.Th. van Loon

Acknowledgements

The symposium is sponsored and supported by the IAU Divisions IV (Stars), VI (Interstellar Matter) and X (Radio Astronomy); by the IAU Commissions No. 27 (Variable Stars), No. 28 (Galaxies), No. 33 (Structure & Dynamics of the Galactic System), No. 34 (Interstellar Matter), No. 40 (Radio Astronomy) and No. 45 (Stellar Classification); and by the IAU Working Groups on Massive Stars, on Abundances in Red Giants, on Active B Stars, on Ap and Related Stars, and on Variable & Binary Stars in Galaxies.

The Local Organizing Committee operated under the auspices of
Keele University.

Funding by the
International Astronomical Union,
Royal Astronomical Society,
University of Keele,
its Research Inst. for Environment, Physical Sciences & Applied Mathematics,
and its School of Physical and Geographical Sciences,
Cambridge University Press,
and
Wiley-Blackwell,
is gratefully acknowledged.

CONFERENCE PHOTOGRAPH

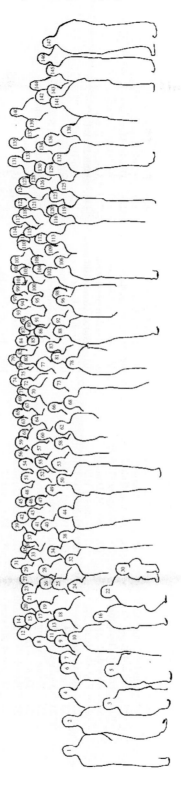

1. Dimitrios A. Gouliermis
2. Maria-Rosa Cioni
3. Alceste Bonanos
4. Norikazu Mizuno
5. Yoji Mizuno
6. Erik Muller
7. Vallia Antoniou
8. Ingrid Meschin
9. Leandro Kerber
10. Paola Marigo
11. Carme Gallart
12. Annie Hughes
13. Alessio Mucciarelli
14. Joss Bland-Hawthorn
15. Tony Wong
16. Anita
17. Gisella Clementini
18. Jason Harris
19. François Boulanger
20. Ricardo Carrera
21. Warren Reid
22. Léo Girardi
23. Patrick Müller
24. Nidia Morrell
25. Nolan Walborn
26. Thomas Lebzelter
27. Tetsuhiro Minamidani
28. Margaret Meixner
29. Christophe Martayan
30. Alessio
31. Richard de Grijs
32. Martin Cohen
33. Walter J. Maciel
34. Els van Aarle
35. Raphael Hirschi
36. João F. C. Santos Jr.
37. Marcelo Borges Fernandes
38. Tom Lloyd Evans
39. Frank Ripple
40. Lee Townsend
41. Beatríz Barbuy
42. Guillermo Bosch
43. Andrea Ahumada
44. Yasuo Fukui
45. Remy Indebetouw
46. Noelia Noël
47. Nitya Kallivayalil
48. Manfred W. Pakull
49. Mónica Alejandra Oddone
50. Sundar Srinivasan
51. Despina Hatzidimitriou
52. Uma Vijh
53. Cinthya Herrera Contreras
54. Michael Feast
55. Michaela Kraus
56. Igor Soszyński
57. Clare Worley
58. Mónica Rubio
59. Wolfgang Gieren
60. Iain McDonald
61. Vanessa Hill
62. Mohammad Heydari-Malayeri
63. Dougal Mackey
64. Stefan Keller
65. Chris Evans
66. Martha Boyer
67. Roald Guandalini
68. Elena Sabbi
69. Basílio Santiago
70. Richard Shaw
71. Andrew Cole
72. Florian Schiller
73. Tabitha C. Bush
74. Eduardo Balbinot
75. Malcolm Coe
76. Andrew Marble
77. Letizia Stanghellini
78. Rosa Williams
79. Hilding Neilson
80. Adam Leroy
81. Sui Ann Mao
82. Kenji Bekki
83. Sandro Villanova
84. Toshikazu Onishi
85. Patricia Whitelock
86. Eric Wilcots
87. Andrew Gosling
88. Taghi Mirtorabi
89. Akiko Kawamura
90. Brandon Lawton
91. Snežana Stanimirović
92. Elena D'Onghia
93. Christian Henkel
94. Itzhak Goldman
95. Jay Gallagher
96. Vali Huseynov
97. Sacha Hony
98. Gurtina Besla
99. Geoffrey Clayton
100. Takashi Shimonishi
101. Aaron Grocholski
102. Toshiya Ueta
103. Gregory C. Sloan
104. Linda Smith
105. Roeland van der Marel
106. António Mário Magalhães
107. Vincenzo Ripepi
108. Keiichi Ohnaka
109. Noriyuki Matsunaga
110. Frédéric Royer
111. Alexandra Carrick
112. Takashi Onaka
113. Bruno Dias
114. Steven Majewski
115. Yoshifusa Ita
116. Karin Sandstrom
117. Alex Fullerton
118. Pierre North
119. Joana Oliveira
120. Min Wang
121. Arunas Kučinskas
122. Stefano Rubele
123. Suzanne Madden
124. Chiara Mastropietro
125. Caroline Bot
126. Daisuke Kato
127. Robin Corbet
128. Min-Young Lee
129. Karl Gordon
130. Luciana Pompéia
131. Miroslav Filipović
132. Gary Da Costa
133. Jeffrey L. Payne
134. Martin Groenewegen
135. Paul Woods
136. Eric Lagadec
137. Frank Haberl
138. Mikako Matsuura
139. Wolfgang Pietsch
140. Andreas Zezas
141. John Dickel
142. You-Hua Chu
143. Eva Grebel
144. Annapurni Subramaniam
145. Jacco van Loon
146. Nye Evans
147. Barry Smalley

Absent from the photograph:

Nate Bastian
Jessica Bush
Rasmiyya Gasimova
Scott Gaudi
James Green
Rob Jeffries
Yaël Nazé
Quentin Parker
Adam Růžička

Participants

Andrea **Ahumada**, University of Córdoba, Argentina / ESO, Chile   andreav.ahumada@gmail.com
Vallia **Antoniou**, University of Crete, Greece / Harvard-Smithsonian, USA   vallia@physics.uoc.gr
Eduardo **Balbinot**, Universidad Federal do Rio Grande do Sul, Brazil   eduardo.balbinot@gmail.com
Beatriz **Barbuy**, University of São Paulo, Brazil   barbuy@astro.iag.usp.br
Nate **Bastian**, IoA Cambridge, UK   bastian@ast.cam.ac.uk
Kenji **Bekki**, University of New South Wales, Australia   bekki@phys.unsw.edu.au
Gurtina **Besla**, Harvard-Smithsonian, USA   gbesla@cfa.harvard.edu
Joss **Bland-Hawthorn**, University of Sydney, Australia   jbh@physics.usyd.edu.au
Alceste **Bonanos**, Space Telescope Science Institute, USA   bonanos@stsci.edu
Marcelo **Borges Fernandes**, Observatoire de la Côte d'Azur, France   marcelo.borges@obs-azur.fr
Guillermo **Bosch**, FCAG-IALP, Brazil   guille@fcaglp.unlp.edu.ar
Caroline **Bot**, Observatoire Astronomíque de Strasbourg, France   bot@astro.u-strasbg.fr
François **Boulanger**, Institut d'Astrophysique Spatiale, France   francois.boulanger@ias.u-psud.fr
Martha **Boyer**, University of Minnesota, USA   mboyer@astro.umn.edu
Jessica **Bush**,   tabitha.bush@gmail.com
Tabitha **Bush**,   tabitha.bush@gmail.com
Ricardo **Carrera**, Osservatorio Astronomico di Bologna, Italy   ricardo.carrera@bo.astro.it
Alexandra **Carrick**, Wiley, UK   alcarric@wiley.co.uk
You-Hua **Chu**, University of Illinois Urbana-Champaign, USA   chu@astro.uiuc.edu
Maria-Rosa **Cioni**, University of Hertfordshire, UK   M.Cioni@herts.ac.uk
Geoffrey **Clayton**, Louisiana State University, USA   gclayton@fenway.phys.lsu.edu
Gisella **Clementini**, University of Bologna, Italy   gisella.clementini@oabo.inaf.it
Malcolm **Coe**, University of Southampton, UK   mjcoe@soton.ac.uk
Martin **Cohen**, University of California at Berkeley, USA   mcohen@astro.berkeley.edu
Andrew **Cole**, University of Tasmania, Australia   andrew.cole@utas.edu.au
Robin **Corbet**, UMBC/NASA GSFC, USA   corbet@umbc.edu
Elena **D'Onghia**, University of Zurich, Switzerland   elena@physik.unizh.ch
Gary **Da Costa**, Australia National University, Australia   gdc@mso.anu.edu.au
Richard **de Grijs**, University of Sheffield, UK   R.deGrijs@sheffield.ac.uk
Bruno **Dias**, University of São Paulo, Brazil   bdias@astro.iag.usp.br
John **Dickel**, University of New Mexico, USA   johnd@phys.unm.edu
Chris **Evans**, Royal Observatory Edinburgh / ATC, Scotland   cje@roe.ac.uk
Aneurin (Nye) **Evans**, Keele University, UK   ae@astro.keele.ac.uk
Michael **Feast**, University of Cape Town, Republic of South Africa   mwf@artemisia.ast.uct.ac.za
Miroslav **Filipović**, University of Western Sydney, Australia   m.filipovic@uws.edu.au
Yasuo **Fukui**, Nagoya University, Japan   fukui@a.phys.nagoya-u.ac.jp
Alex **Fullerton**, Space Telescope Science Institute, USA   fullerton@stsci.edu
John (Jay) **Gallagher**, University of Wisconsin at Madison, USA   jsg@astro.wisc.edu
Carme **Gallart**, Instituto de Astrofísica de Canarias, Spain   carme@iac.es
Rasmiyya **Gasimova**, Nakchivan State University, Azerbaijan   gasimovar@yahoo.co.uk
Scott **Gaudi**, Ohio State University, USA   gaudi@astronomy.ohio-state.edu
Wolfgang **Gieren**, University of Concepción, Chile   wgieren@astro-udec.cl
Léo **Girardi**, Osservatorio Astronomico di Padova, Italy   leo.girardi@oapd.inaf.it
Itzhak **Goldman**, Afeka College / Tel Aviv University, Israel   goldman@wise.tau.ac.il
Karl **Gordon**, Space Telescope Science Institute, USA   kgordon@stsci.edu
Andrew **Gosling**, University of Oulu, Finland   ajgosling@gmail.com
Dimitrios **Gouliermis**, MPIA Heidelberg, Germany   dgoulier@mpia.de
Eva **Grebel**, University of Heidelberg, Germany   grebel@ari.uni-heidelberg.de
James **Green**, CSIRO Australia Telescope National Facility, Australia   james.green@csiro.au
Aaron **Grocholski**, Space Telescope Science Institute, USA   aarong@stsci.edu
Martin **Groenewegen**, Royal Observatory of Belgium, Belgium   marting@oma.be
Roald **Guandalini**, University of Perugia, Italy   guandalini@fisica.unipg.it
Frank **Haberl**, MPE Garching, Germany   fwh@mpe.mpg.de
Jason **Harris**, NOAO, USA   jharris@noao.edu
Despina **Hatzidimitriou**, University of Crete, Greece   dh@physics.uoc.gr
Christian **Henkel**, MPIfR Bonn, Germany   chenkel@mpifr-bonn.mpg.de
Cinthya **Herrera Contreras**, University of Chile, Chile   cnherrer@das.uchile.cl
Mohammad **Heydari-Malayeri**, Observatoire de Paris Meudon, France   m.heydari@obspm.fr
Vanessa **Hill**, Observatoire de Paris Meudon, France   Vanessa.Hill@obspm.fr
Raphael **Hirschi**, Keele University, UK   hirschi@astro.keele.ac.uk
Sacha **Hony**, CEA Saclay, France   sacha.hony@cea.fr
Annie **Hughes**, Swinburne University / ATNF, Australia   ahughes@astro.swin.edu.au
Vali **Huseynov**, Nakchivan State University, Azerbaijan   vgusseinov@yahoo.com
Remy **Indebetouw**, University of Virginia / NRAO, USA   remy@virginia.edu
Yoshifusa **Ita**, National Astronomical Observatory, Japan   yoshifusa.ita@nao.ac.jp
Rob **Jeffries**, Keele University, UK   rdj@astro.keele.ac.uk
Nitya **Kallivayalil**, Massachusetts Institute of Technology, USA   nitya@mit.edu
Daisuke **Kato**, University of Tokyo, Japan   kato@astron.s.u-tokyo.ac.jp
Akiko **Kawamura**, Nagoya University, Japan   kawamura@a.phys.nagoya-u.ac.jp
Stefan **Keller**, Australia National University, Australia   stefan@mso.anu.edu.au
Leandro **Kerber**, University of São Paulo, Brazil   kerber@astro.iag.usp.br
Michaela **Kraus**, Ondrejov, Czech Republic   kraus@sunstel.asu.cas.cz
Arunas **Kučinskas**, Institute for Theoretical Physics & Astronomy, Lithuania   arunaskc@itpa.lt
Eric **Lagadec**, Jodrell Bank Centre for Astrophysics, Manchester, UK   eric.lagadec@manchester.ac.uk
Brandon **Lawton**, New Mexico State University, USA   blawton@nmsu.edu
Thomas **Lebzelter**, University of Vienna, Austria   lebzelter@astro.univie.ac.at
Min-Young **Lee**, University of Wisconsin at Madison, USA   lee@astro.wisc.edu
Adam **Leroy**, MPIA Heidelberg, Germany   leroy@mpia.de
Thomas **Lloyd Evans**, University of St. Andrews, Scotland   thhle@st-andrews.ac.uk
Walter J. **Maciel**, University of São Paulo, Brazil   maciel@astro.iag.usp.br
Alasdair D. **Mackey**, Royal Observatory Edinburgh, Scotland   dmy@roe.ac.uk
Suzanne **Madden**, CEA Saclay, France   smadden@cea.fr
António Mário **Magalhães**, University of São Paulo, Brazil   mario@astro.iag.usp.br
Steven **Majewski**, University of Virginia, USA   srm4n@virginia.edu
Sui Ann **Mao**, Harvard-Smithsonian, USA   samao@cfa.harvard.edu
Andrew **Marble**, University of Arizona, USA   amarble@as.arizona.edu
Paola **Marigo**, University of Padova, Italy   paola.marigo@unipd.it
Christophe **Martayan**, Royal Observatory of Belgium, Belgium   martayan@oma.be
Chiara **Mastropietro**, Observatoire de Paris Meudon, France   chiara.mastropietro@obspm.fr
Noriyuki **Matsunaga**, Kyoto University, Japan   matsunaga@kusastro.kyoto-u.ac.jp
Mikako **Matsuura**, National Astronomical Observatory, Japan   mikako@optik.mtk.nao.ac.jp
Iain **McDonald**, Keele University, UK   iain@astro.keele.ac.uk

Margaret **Meixner**, Space Telescope Science Institute, USA — meixner@stsci.edu
Ingrid **Meschin**, Instituto de Astrofísica de Canarias, Spain — imeschin@iac.es
Tetsuhiro **Minamidani**, Hokkaido University, Japan — tetsu@astro1.sci.hokudai.ac.jp
Taghi **Mirtorabi**, Azzahra University, Tehran, Iran — torabi@mail.ipm.ir
Norikazu **Mizuno**, Nagoya University, Japan — norikazu@a.phys.nagoya-u.ac.jp
Yoji **Mizuno**, Nagoya University, Japan — y_mizuno@a.phys.nagoya-u.ac.jp
Nidia **Morrell**, Las Campanas Observatory, Chile — nmorrell@lco.cl
Alessio **Mucciarelli**, University of Bologna, Italy — alessio.mucciarelli@studio.unibo.it
Erik **Muller**, Nagoya University, Japan — Erik.Muller@csiro.au
Yaël **Nazé**, University of Liège, Belgium — naze@astro.ulg.ac.be
Hilding **Neilson**, University of Toronto, Canada — neilson@astro.utoronto.ca
Noelia **Noël**, Instituto de Astrofísica de Canarias, Spain — noelia@iac.es
Pierre **North**, EPFL Lausanne, Switzerland — pierre.north@epfl.ch
Mónica Alejandra **Oddone**, University of Córdoba, Argentina — mao@mail.oac.uncor.edu
Keiichi **Ohnaka**, MPIfR Bonn, Germany — kohnaka@mpifr-bonn.mpg.de
Joana **Oliveira**, Keele University, UK — joana@astro.keele.ac.uk
Takashi **Onaka**, University of Tokyo, Japan — onaka@astron.s.u-tokyo.ac.jp
Toshikazu **Onishi**, Nagoya University, Japan — ohnishi@a.phys.nagoya-u.ac.jp
Manfred **Pakull**, Observatoire Astronomíque de Strasbourg, France — pakull@astro.u-strasbg.fr
Quentin **Parker**, Macquarie University, Australia — qap@ics.mq.edu.au
Jeffrey **Payne**, James Cook University, Australia — jeffrey.payne@jcu.edu.au
Wolfgang **Pietsch**, MPE Garching, Germany — wnp@mpe.mpg.de
Luciana **Pompéia**, University of Vale do Paraíba, Brazil — pompeia@univap.br
Warren **Reid**, Macquarie University / Anglo-Australian Observatory, Australia — warren@ics.mq.edu.au
Vincenzo **Ripepi**, Osservatorio Astronomico di Capodimonte, Italy — ripepi@oacn.inaf.it
Frank **Ripple**, State University of New York at Oswego, USA — fripple@oswego.edu
Frédéric **Royer**, Observatoire de Paris Meudon, France — frederic.royer@obspm.fr
Stefano **Rubele**, Osservatorio Astronomico di Padova, Italy — stefano.rubele@oapd.inaf.it
Mónica **Rubio**, University of Chile, Chile — mrubio@das.uchile.cl
Adam **Růžička**, Astronomical Institute of Prague, Czech Republic — adam.ruzicka@gmail.com
Elena **Sabbi**, Space Telescope Science Institute, USA — sabbi@stsci.edu
Karin **Sandstrom**, University of California at Berkeley, USA — karin@astro.berkeley.edu
Basílio **Santiago**, Universidad Federal do Rio Grande do Sul, Brazil — santiago@if.ufrgs.br
João F. C. **Santos** Jr., Universidad Federal de Minas Gerais, Brazil — jsantos@fisica.ufmg.br
Florian **Schiller**, University of Erlangen, Germany — Florian.Schiller@sternwarte.uni-erlangen.de
Richard **Shaw**, NOAO, USA — shaw@noao.edu
Takashi **Shimonishi**, University of Tokyo, Japan — shimonishi@astron.s.u-tokyo.ac.jp
Gregory **Sloan**, Cornell University, USA — gcs22@cornell.edu
Linda **Smith**, Space Telescope Science Institute, USA — lsmith@stsci.edu
Igor **Soszyński**, Warsaw University, Poland — soszynsk@astrouw.edu.pl
Sundar **Srinivasan**, Johns Hopkins University, USA — sundar@pha.jhu.edu
Letizia **Stanghellini**, NOAO, USA — letizia@noao.edu
Snežana **Stanimirović**, University of Wisconsin at Madison, USA — sstanimi@astro.wisc.edu
Annapurni **Subramaniam**, Indian Institute of Astrophysics, India — purni@iiap.res.in
Lee **Townsend**, University of Southampton, UK — l.j.townsend@phys.soton.ac.uk
Toshiya **Ueta**, Denver University, USA — tueta@du.edu
Els **van Aarle**, Katholieke Universiteit Leuven, Belgium — els.vanaarle@ster.kuleuven.be
Roeland **van der Marel**, Space Telescope Science Institute, USA — marel@stsci.edu
Jacco **van Loon**, Keele University, UK — jacco@astro.keele.ac.uk
Uma **Vijh**, University of Toledo, USA — Uma.Vijh@utoledo.edu
Sandro **Villanóva**, University of Concepcíon, Chile — svillanova@astro-udec.cl
Nolan **Walborn**, Space Telescope Science Institute, USA — walborn@stsci.edu
Min **Wang**, Purple Mountain Observatory, China — mwang@pmo.ac.cn
Patricia **Whitelock**, South African Astronomical Observatory, Republic of South Africa — paw@saao.ac.za
Eric **Wilcots**, University of Wisconsin at Madison, USA — ewilcots@astro.wisc.edu
Rosa **Williams**, Columbus State University, USA — rosanina@ccssc.org
Tony **Wong**, University of Illinois at Urbana-Champaign, USA — wongt@astro.uiuc.edu
Paul **Woods**, Jodrell Bank Centre for Astrophysics, Manchester, UK — dr.paul.woods@gmail.com
Clare **Worley**, University of Canterbury, New Zealand — clare.worley@pg.canterbury.ac.nz
Andreas **Zezas**, Harvard-Smithsonian, USA — azezas@cfa.harvard.edu

Dedicated to the memory of Bengt E. Westerlund (1921–2008)

Bengt Westerlund, a stellar astronomer and a leading expert in the study of the Magellanic Clouds, has passed away in June 2008 after a short period of illness. He was born in 1921 in Gävle, a city about 110 km north of Stockholm, Sweden. He studied at Uppsala University and got his Ph.D. there in 1953 with a thesis on "Luminosity effects and colour-equivalents as measured in short stellar spectra". He took up an assistant professorship at Uppsala, continued research and teaching, and also worked as a guest at l'Observatoire de Haute Provence together with Daniel Chalonge. After that he became the first astronomer at the new "Uppsala Southern Station" at Mount Stromlo Observatory. Here, the Uppsala astronomers, in collaboration with the Australian colleagues had established a Schmidt telescope with 52 cm aperture, also equipped with an objective prism. This instrument, in spite of its moderate size, played a very significant rôle in the mapping of the Southern Sky, and Bengt Westerlund was leading this work. Here, he started his systematic study of the stellar content of the Southern Milky Way and the Magellanic Clouds which made him internationally well known. He later became a member of the ordinary staff at Mount Stromlo.

In the late 1960s he moved to Steward Observatory in Arizona. Then, in 1970 he became local director of ESO in Chile, where he played an important rôle in a critical phase of the building up of this observatory. In 1975 he was asked to take up the old professorship in astronomy at Uppsala university and the directorship of its Observatory, the position which Anders Celsius carried in the 18th century and which later belonged to Gunnar Malmquist and Erik Holmberg. Westerlund retired in 1987 but continued actively his research, and published no less than 60 papers after 1987. he was member of several academies, including the Royal Swedish Academy of Sciences.

Bengt Westerlund was a generous and enthusiastic teacher. He was the leading and often cited expert on the stellar content in the Magellanic Clouds. He summarized the knowledge about the Clouds in a much appreciated monograph, "The Magellanic Clouds", Cambridge University Press, in 1997. Westerlund also made a number of interesting discoveries. In 1961 he found a star cluster in Ara in the Southern Milky Way behind about 10 magnitudes of extinction which got his name, Westerlund I, and which has turned out to be the richest young cluster known in our Galaxy. Its estimated age is 3 to 5 Myr, and it contains a great number of very bright stars, including about 50 stars with masses above 30 solar masses, among which are numerous Wolf Rayet stars. There is also an active pulsar present in the cluster, suggesting a relatively recent supernova. This cluster is a beautiful symbol of a brilliant research life achievement.

In a poem, "The star day", the Swedish author and Nobel prize winner in literature, Harry Martinson describes how he, when visiting the Mount Stromlo Observatory, is shown the sight of another rich cluster, ω Centauri, through the telescope by a "star man". In the poem Martinson speculates how the sky looks for those who live there.

> *Difficult to imagine, but still it seems*
> *that such a space so rich in suns*
> *must be suitable as paradise for shining persons*

The star man, presenting the cluster to Martinson, was Bengt Westerlund. With his kind and friendly enthusiasm, his encyclopedic knowledge, and his humble wisdom he has spread light for many of us.

Bengt Gustafsson

Session I

Surveys of the Magellanic Clouds

The Magellanic System: Stars, Gas, and Galaxies
Proceedings IAU Symposium No. 256, 2008
Jacco Th. van Loon & Joana M. Oliveira, eds.

Measuring the lifecycle of baryonic matter in the Large Magellanic Cloud with the *Spitzer* SAGE-LMC survey

Margaret Meixner[1], Jean-Philippe Bernard[2], Robert D. Blum[3], Remy Indebetouw[4], William Reach[5], Sundar Srinivasan[6], Marta Sewilo[1] and Barbara A. Whitney[7]

[1] Space Telescope Science Institute, 3700 San Martin Dr., Baltimore, MD 21093, USA
email: meixner@stsci.edu, mmsewilo@stsci.edu

[2] Direction de la Recherche, Centre dEtude Spatiale des Rayonnements, 18Avenue Edouard Belin, Toulouse, Cedex F-31055, France
email: jean-philippe.bernard@cesr.fr

[3] NOAO, PO Box 26732, Tucson AZ 85726-6732, USA
email: rblum@noao.edu

[4] Depart. of Astronomy, University of Virginia, PO Box 3818, Charlottesville, VA 22903, USA
email: remy@virginia.edu

[5] Spitzer Science Center, California Institute of Technology, 220-6, Pasadena, CA, 91125, USA
email: reach@ipac.caltech.edu

[6] Department of Physics and Astronomy, The Johns Hopkins University, 3400 North Charles St., Baltimore, MD 21218, USA
email: sundar@pha.jhu.edu

[7] Space Science Institute, 4750 Walnut St. Suite 205, Boulder, CO 80301, USA
email: bwhitney@spacescience.org

Abstract. The recycling of matter between the interstellar medium (ISM) and stars are key evolutionary drivers of a galaxy's baryonic matter. The Spitzer wavelengths provide a sensitive probe of circumstellar and interstellar dust and hence, allow us to study the physical processes of the ISM, the formation of new stars and the injection of mass by evolved stars and their relationships on the galaxy-wide scale of the LMC. Due to its proximity, favorable viewing angle, multi-wavelength information, and measured tidal interactions with the Small Magellanic Cloud (SMC), the LMC is uniquely suited for surveying the agents of a galaxy's evolution (SAGE), the ISM and stars. The SAGE-LMC project is measuring these key transition points in the life cycle of baryonic matter in the LMC. Here we present a connective view of the preliminary quantities estimated from SAGE-LMC for the total mass of the ISM, the galaxy wide star formation rate and the current stellar mass loss return. For context, we compare these numbers to the LMC's stellar mass.

Keywords. surveys, stars: AGB and post-AGB, circumstellar matter, stars: formation, stars: mass loss, dust, extinction, ISM: evolution, galaxies: evolution, galaxies: individual (LMC), Magellanic Clouds

1. The SAGE-LMC survey

We have performed a uniform and unbiased imaging survey of the Large Magellanic Cloud (LMC, $\sim 7° \times 7°$), using the IRAC (3.6, 4.5, 5.8 and 8 μm) and MIPS (24, 70, and 160 μm) instruments on board the Spitzer Space Telescope (*Spitzer*) in order to survey the agents of a galaxy's evolution (SAGE), through the interaction between the interstellar medium (ISM) and stars in the LMC. Three key science goals determined the coverage and depth of the survey. The detection of diffuse ISM with column densities $> 1.2 \times 10^{21}$ H cm^{-2} permits detailed studies of dust processes in the ISM. SAGE's point source sensitivity enables a complete census of newly formed stars with masses > 3 M$_\odot$ that will determine the current star formation rate in the LMC. SAGE's detection of evolved stars with mass-loss rates $> 1 \times 10^{-8}$ M$_\odot$ yr^{-1} will quantify the rate at which evolved stars inject mass into the ISM of the LMC. The observing strategy includes two epochs in 2005, separated by three months, that both mitigate instrumental artifacts and constrain source variability. The science drivers, detailed observing strategy, data processing steps and preliminary results of the SAGE survey are described by Meixner *et al.* (2006).

The SAGE data are non-proprietary and point source lists are released to the community. The most recent release included an improved epoch 1 catalog, an epoch 2 catalog, of IRAC ($\sim$4 million point sources) and MIPS 24 ($\sim$ 40,000 point sources) photometry, and MIPS 24, 70 and 160 micron images (See poster contribution by Meixner and SAGE-LMC team). Further investigations have followed on the SAGE theme such as SAGE-SMC, a Small Magellanic Cloud survey with *Spitzer* led by Gordon (see poster contribution), SAGE-Spec, a *Spitzer* IRS spectroscopic survey of the LMC led by Marwick-Kemper, and HERITAGE, a HERschel Inventory of The Agents of Galaxy Evolution in the Magellanic Clouds led by Meixner (see poster contribution). To learn more about all of these SAGE related projects, visit http://sage.stsci.edu. In order to avoid confusion amongst this expansion of SAGE projects, we have retitled ours SAGE-LMC. Several presentations at this conference have been based on the SAGE-LMC data, the SAGE-LMC team contributions are noted throughout this article. In this contribution, we provide an overview of the mass inventory of the ISM, galaxy wide star formation rate, mass-loss rate return of the asymptotic giant branch (AGB) stars as derived from the SAGE-LMC project.

2. ISM mass

A global view of the ISM dust properties derived from the SAGE-LMC data is presented by Bernard *et al.* (2008) (see also poster contribution by Meixner *et al.*). Among the results is a derivation of the total mass of the ISM (see Table 1), which appears to be a factor of two larger than currently known from gas measurements of H I (21 cm; Kim *et al.* 2003) and of CO (Fukui *et al.* 1999). The MIPS 160 μm imaging provides us with the most sensitive measures of the total ISM mass in galaxies. Because the optical depth is quite low at 160μm, its emission is sensitive to all of the dust and, by inference the gas, in a galaxy independent of the physical conditions of the gas which can cause gas tracers to be only selective measures of the ISM column density.

Bernard *et al.* (2008) created an infrared excess image (Fig. 1) that is derived as follows. The dust optical depth map at 160 μm is scaled to a total gas column density map by multiplying it by a gas-to-dust mass ratio determined from the lowest optical depth regions of the map. The ISM mass listed in Table 1 is integrated over the total area of this map. From this total gas column density map, we subtract the gas column density maps derived from the H I and CO maps. The result is the infrared excess map shown in Fig. 1. If the H I and CO maps had detected all of the ISM mass, as measured by the MIPS 160 μm map, then Fig. 1 would be blank. However, it is not blank and instead appears very similar in distribution to the H I image. This infrared excess map may be tracing the location of H$_2$ gas mass not detected by CO, or cold H I not detected by the 21 cm line emission. On the other extreme, it may be tracing (rather large) variations in the dust-to-gas mass ratio, although Bernard *et al.* (2008) tend to think the dust extinction

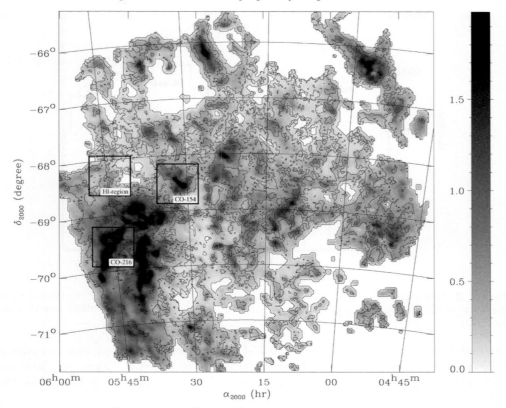

Figure 1. Map of N_H^X in units of 10^{22} H/cm^2. White regions were removed based on their high relative errors. The contours show the distribution of the H I emission of Kim *et al.* (2003). From Bernard *et al.* (2008).

maps do not support this interpretation. Additional ISM results based on the SAGE-LMC data are presented in these proceedings by Marble *et al.* (variations in Aromatic features in the ISM).

3. Star formation rate

An overview of the Star Formation as seen with the SAGE-LMC survey is presented by Whitney *et al.* (2008). The discovery of ∼1000 new candidate young stellar objects (YSOs) and our preliminary characterization of them, described in this proceedings by Indebetouw, demarks a major step forward in our understanding of star formation processes in the LMC. However, the list is just a first attempt and collects mainly the dustiest, and most massive sources. One of the properties determined from the model fits to the YSOs is the mass. Figure 2 shows the mass distribution of the published YSO candidates which reside over the whole LMC. Whitney *et al.* (2008) fitted a Kroupa (2001) initial mass function to the part of the mass distribution in which they had the most confidence, the more massive star end. They derive a star formation rate of 0.1 M$_\odot$ yr^{-1} (see Table 1). This bottoms up approach to star formation rate provides a comparable value to that measured from the Hα and far-infrared continuum emission. However, this measurement is almost certainly a lower limit, because not all of the YSOs have been found. Indeed, the poster contribution by Smith *et al.* in these proceedings shows the new

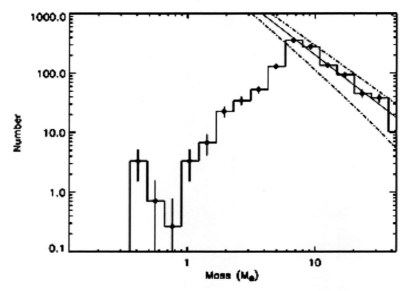

Figure 2. Histograms of stellar mass for the well-fit high-probability YSO candidates from Whitney *et al.* (2008). The dashed line in the left panel is the (Kroupa 2001) initial mass function.

mosaicked photometry catalogs from SAGE-LMC, and reveals 5 times more additional YSO candidates using the same color cut criteria as Whitney *et al.* (2008), and 30 times more additional YSO candidates if we include the fainter magnitudes in the cut for the Henize 206 and N 11 star formation regions.

4. Mass-loss rate return from AGB stars

The identification of the evolved star populations for the whole SAGE-LMC catalogs has been presented by Blum *et al.* (2006) who suggests there may be two types of oxygen-rich (O-rich) AGB stars. Srinivasan *et al.* (2008, also poster contribution in these proceeding), confirms these two types of O-rich AGB stars and also investigates the properties of carbon-rich (C-rich) and extreme AGB stars, in particular, the excess infrared emission in the IRAC and MIPS 24 μm bands. The infrared excess measured in these stars is an excess above the expected photospheric emission in these bands, as known from scaled model atmosphere profiles. The excess is due to the dusty, molecular rich mass loss experienced by AGB stars. The larger the excess, the larger the mass-loss rate the AGB star is experiencing. Srinivasan *et al.* found a positive, strong correlation between the measured excess at 8 μm of ~ 40 evolved stars for which mass-loss rates were derived from radiative transfer model fits of ISO data by van Loon *et al.* (1999). They made fits to these relations for the three classes of AGB stars under consideration: O-rich, C-rich and Extreme. Applying this relation to the entire sample of AGB stars, they estimate a mass-loss rate for each candidate AGB star in the sample. Figure 3 shows the cumulative mass-loss rate for the entire populations of AGB stars. The extreme AGB stars are the major contributors to the mass-loss rate, at $\sim 7 \times 10^{-3}$ M$_\odot$ yr^{-1}. The C–rich AGB stars and O–rich AGB stars contribute $\sim 1 \times 10^{-3}$ and 0.7×10^{-3} M$_\odot$ yr^{-1}, respectively. The total mass-loss rate is $> 8.7 \times 10^{-3}$ M$_\odot$ yr^{-1} (see Table 1). We set this as a lower limit, because the AGB stars are still candidates and need further analysis, particularly at the most luminous end. Additional evolved star results based on

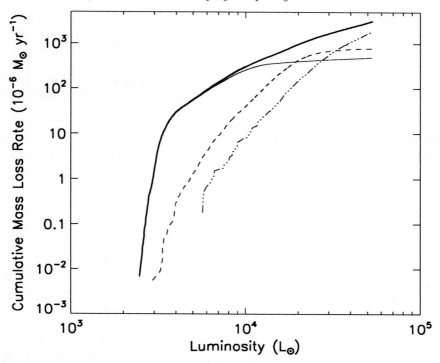

Figure 3. A preliminary cumulative mass-loss rate as a function of luminosity for the O–rich (thin solid line), C–rich (dashed line) and Extreme (dot-dashed line) AGB stars in our lists. The thick solid line is the total AGB mass-loss rate as a function of luminosity. From Srinivasan *et al.* (2008).

Table 1. SAGE-LMC Inventory of the Baryonic Mass Life Cycle

Item	Value	Reference
Stellar Mass	3×10^9 $M_\odot$	van der Marel *et al.* (2002)
ISM Mass	10^9 $M_\odot$	Bernard *et al.* (2008)
Star Formation Rate	> 0.1 $M_\odot$ yr^{-1}	Whitney *et al.* (2008)
AGB Mass Loss Return	$> 8.7 \times 10^{-3}$ $M_\odot$ yr^{-1}	Srinivasan *et al.* (2008)

the SAGE-LMC data are presented in these proceedings by Vijh *et al.* (infrared variables discovered comparing epoch 1 and 2 data), Meixner *et al.* (comparison of SAGE and MACHO data on AGB stars), Srinivasan *et al.* (near-IR spectroscopic followup), Ueta *et al.* (post-AGB stars), and Cohen *et al.* (planetary nebulae).

5. Inventory of Baryonic Mass Flow in the LMC

The mass inventory of these three components, ISM, star formation and AGB mass loss, is presented in Table 1. Our new measurement of the ISM mass makes it about one third that of the stellar mass. The formation of new stars appears to be consuming the ISM at a faster rate than the AGB stars are replenishing it. However, both of these rates are preliminary and most likely lower limits to the actual rates. Moreover, they represent a current snap shot in the lifetime of the LMC and we would need to incorporate the the star formation and evolution history to fully interpret. As we work to improve this inventory of baryonic matter, we will add the contributions of planetary nebulae, red supergiants, luminous blue variables, supernova (remnants), and infall and outflow caused by tidal interactions with nearby gas and the SMC. The SAGE-LMC and SAGE-SMC projects will be able to provide a means to calibrate such measurements from infrared observations and act as a proving ground for procedures in making such inventories in nearby galaxies. Such calibrations will be important for missions like *JWST*, which will be able to extend such studies to nearby galaxies.

Acknowledgement

The SAGE Project is supported by NASA/Spitzer grant 1275598 and NASA NAG5-12595.

References

Bernard, J.-P., Reach, W. T., Paradis, D., *et al.* 2008, *AJ*, 136, 919

Blum, R. D., Mould, J. R., Olsen, K. A., *et al.* 2006, *AJ*, 132, 2034

Fukui, Y., Mizuno, N., Yamaguchi, R., *et al.* 1999, *PASJ*, 51, 745

Kim, S., Staveley-Smith, L., Dopita, M. A., Sault, R. J., Freeman, K. C., Lee, Y., & Chu, Y.-H. 2003, *ApJS*, 148, 473

Kroupa, P. 2001, *MNRAS*, 322, 231

Meixner, M., Gordon, K., Indebetouw, R., *et al.* 2006, *AJ*, 132, 2268

Srinivasan, S., Meixner, M., Leitherer, C., *et al.* 2008, *AJ*, submitted

van der Marel, R. P., Alves, D. R., Hardy, E., & Suntzeff, N. B. 2002, *AJ*, 124, 2639

van Loon, J. Th., Groenewegen, M. A. T., de Koter, A., Trams, N. R., Waters, L. B. F. M., Zijlstra, A. A., Whitelock, P. A., & Loup, C. 1999, *A&A*, 351, 559

Whitney, B. A., Sewilo, M., Indebetouw, R., *et al.* 2008, *AJ*, 136, 18

The Magellanic System: Stars, Gas, and Galaxies
Proceedings IAU Symposium No. 256, 2008
Jacco Th. van Loon & Joana M. Oliveira, eds.

© 2009 International Astronomical Union
doi:10.1017/S1743921308028184

AKARI IRC survey of the Large Magellanic Cloud: A new feature in the infrared color—magnitude diagram

Yoshifusa Ita[1], Takashi Onaka[2], Daisuke Kato[2], and the *AKARI* LMC survey team

[1]National Astronomical Observatory of Japan, 2-21-1 Osawa, Mitaka, Tokyo, 181-8588, Japan
email: yoshifusa.ita@nao.ac.jp

[2]Department of Astronomy, Graduate School of Science, The University of Tokyo, Bunkyo-ku, Tokyo 113-0033, Japan

Abstract. We observed an area of 10 deg^2 of the Large Magellanic Cloud using the Infrared Camera on board *AKARI*. The observations were carried out using five imaging filters (3, 7, 11, 15, and 24 μm) and a dispersion prism (2 − 5 μm, $\lambda/\Delta\lambda \sim 20$) equipped in the IRC. The 11 and 15 μm data, which are unique to *AKARI* IRC, allow us to construct color-magnitude diagrams that are useful to identify stars with circumstellar dust. We found a new sequence in the color-magnitude diagram, which is attributed to red giants with luminosity fainter than that of the tip of the first red giant branch. We suggest that this sequence is likely to be related to the broad emission feature of aluminium oxide at 11.5 μm.

Keywords. stars: AGB and post-AGB, circumstellar matter, stars: mass loss, galaxies: individual (LMC), Magellanic Clouds

1. Introduction and observations

The Japan Aerospace Exploration Agency launched an Infrared Satellite, *ASTRO-F* (Murakami *et al.* 2007) at 21:28 UTC on February 21st, 2006 from the Uchinoura Space Center. Once in orbit *ASTRO-F* was renamed "AKARI". *AKARI* has a 68.5 cm telescope and two scientific instruments, namely the InfraRed Camera (IRC; Onaka *et al.* 2007) and the Far-Infrared Surveyor (FIS; Kawada *et al.* 2007). Both instruments have low- to moderate-resolution spectroscopic capability. The IRC has nine imaging bands and six dispersion elements covering from 2.5 to 26 μm wavelength range (Unfortunately, one of the dispersion elements that was expected to cover 11 to 19 μm range became opaque during the ground test operation, and defunct in orbit). The FIS observes in 4 far-infrared bands between 50 and 180 μm. One of the primary goals of the *AKARI* mission is to carry out an All-Sky survey with FIS and IRC at six bands from 9 to 180 μm (Ishihara *et al.* 2006; Kawada *et al.* 2007). As a result, the entire LMC has been mapped in 9, 18, 65, 90, 140, and 160 μm wavebands.

In addition to the All-Sky survey in mid- and far-infrared, *AKARI* carried out two large-area legacy surveys (LS) in pointed observation mode. The LMC survey project (PI. T.Onaka) is one of the two LS programs. The other is the North Ecliptic Pole survey project (PI. H.Matsuhara; Matsuhara *et al.* 2007). These survey areas are located at high ecliptic latitudes, where the visibility is high for *AKARI*'s sun-synchronous polar orbit. Therefore a batch of observing time can be allocated for pointed observations in these areas after allocating scan paths for the All-Sky survey. Using the opportunities we use the IRC to make imaging and spectroscopic mapping observations of the main part of the LMC. To cover a wide range of the spectral energy distribution of a celestial body, we obtained not only imaging data at 3 (N3), 7 (S7), 11 (S11), 15 (L15), and 24 μm (L24), but also low resolution ($\lambda/\Delta\lambda \sim 20$) $2.5 - 5$ μm (NP) spectral data at three dithered sky positions in a pointing opportunity. Details of the observations, data reduction procedures, and some initial results are described in Ita *et al.* (2008). Some first results on the spectroscopic data are presented in Shimonishi *et al.* (2008), and Shimonishi *et al.* in this volume.

2. Preliminary catalog

We constructed a preliminary photometric catalog of bright point sources for the LMC from the imaging data. More than 5.9×10^5 near-infrared and 6.4×10^4 mid-infrared point sources are detected. The 10σ detection limits of our survey are about 16.5, 14.0, 12.3, 10.8, and 9.2 in Vega-magnitude at 3, 7, 11, 15, and 24 μm, respectively. These detection limits are comparable to those of the *Spitzer* SAGE survey (Meixner *et al.* 2006). The IRC's imaging filters cover the wavelength range continuously from 2.5 to 26 μm. In particular, it has 11 μm (S11) and 15 μm (L15) imaging bands, which fill the gap between IRAC (Fazio *et al.* 2004) and MIPS (Rieke *et al.* 2004) on board *Spitzer*. The wavelength range that IRC's S11 and L15 bands cover contains interesting spectral features such as the silicate 10 and 18 μm bands. In addition, we took $2 - 5$ μm low resolution spectra ($\lambda/\Delta\lambda \sim 20$ at around 3 μm) for all bright stars that are detected. These characteristics make our survey unique and complementary to the SAGE survey. Refer to Onaka *et al.* (2007) for instrumental details and imaging performance of the IRC. General information on the IRC spectroscopic mode, particularly on the slit-less spectroscopy, is given in Ohyama *et al.* (2007). The first *AKARI*/IRC LMC point source catalog is planned to be released to the public in 2009.

3. Infrared color-magnitude diagrams

The preliminary catalog was cross-identified with the *IRSF*/SIRIUS Magellanic Clouds Point Source Catalog (Kato *et al.* 2007), which contains JHK$_s$ photometry of over 1.4×10^7 sources in the central 40 deg^2 area of the LMC. Compared to the contemporary DENIS (Cioni *et al.* 2000a) and 2MASS (Skrutskie *et al.* 2006) catalogs, the *IRSF* catalog is more than two magnitudes deeper at K$_s$-band and about four times finer in spatial resolution.

In figure 1, we show infrared color-magnitude diagrams using *AKARI* IRC and SIRIUS bands. Units of vertical axes are the apparent magnitudes. It can be scaled to the absolute magnitude by subtracting the distance modulus. The corresponding wavebands for the vertical axes are indicated at the top of each panel. The employed colors for the horizontal axes are indicated at the bottom of each panel. All of the color-magnitude planes are binned by 0.1×0.1 mag^2 to calculate the number of sources in each bin, and the fiducial color is given according to the number density levels in a logarithmic scale (see the wedge at the bottom).

The (K$_s$ – S11) vs. K$_s$ panel indicates which red giants show circumstellar dust emission. It is clear that excess in S11 ($K_s - S11 > 0.5$ mag) is seen not only among the sources brighter than the tip of the first red giant branch (TRGB, $K_s \sim 12.5$ mag, Cioni *et al.* 2000b), but also among the sources below the TRGB. Since the brighter sources exceed the K$_s$ luminosity of the TRGB, they should be an intermediate-age population and/or metal-rich old population AGB stars. On the other hand, the interpretation of the fainter sources is difficult. They can be metal-poor and old AGB stars that do not

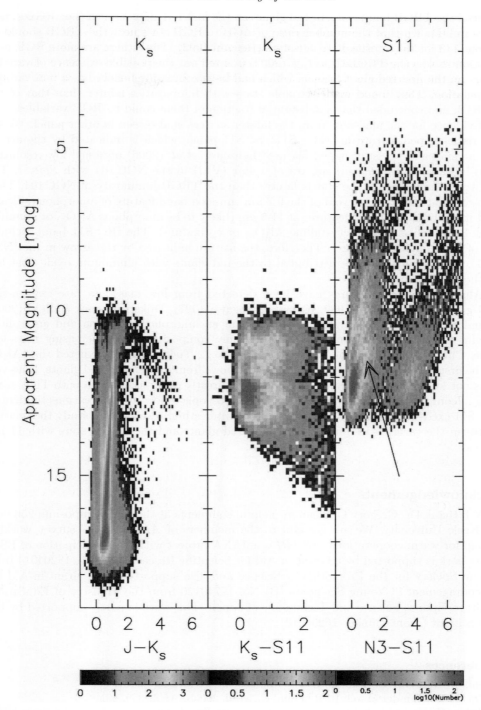

Figure 1. Color magnitude diagrams of *AKARI* LMC sources using several combinations of IRC and *IRSF*/SIRIUS bands. The vertical axis is in apparent magnitude at the corresponding wavebands, which are indicated at the top of each panel. The arrows indicate the newly found feature, which is attributed to the red giants that have luminosities below the tip of the first red giant branch.

exceed the TRGB luminosity, or red giants on the first red giant branch (i.e., RGB stars). It is well known that the number ratio of AGB to RGB stars near the TRGB should be about 1/3 for the intermediate age stars (Renzini 1992). Thus, there are more RGB populations below the TRGB. Ita *et al.* (2002) pointed out the possible existence of variable stars on the first red giant branch, which had been commonly believed as a non-variable population. They found many variable stars with luminosities fainter than that of the TRGB, and concluded that a substantial fraction of them could be RGB variables.

Evidence for the mass-loss from the fainter sources is also seen in other panel. We see a bristle-shaped feature in (N3 – S11) vs. S11 panel, which is indicated by the arrow. What makes the bristle-shaped feature? Lebzelter *et al.* (2006) obtained low-resolution mid-infrared (7.6–21.7 μm) spectra of a star (V13) in the NGC 104 with *Spitzer*. The K_s band luminosity of the star is fainter than the TRGB luminosity of NGC 104. They showed that the star is devoid of the 9.7 μm emission band feature of amorphous silicate, but it has broad emission features at 11.5 μm (likely to be amorphous Al_2O_3, or alumina) and 13 μm (likely to be crystalline Al_2O_3, or corundum). The IRC S11 band includes all of these emission features. Therefore, the feature indicated by the arrow in the (N3 – S11) vs. S11 panel may be attributed to the red giants with aluminium oxide dust but without the silicate feature.

Aluminium oxide features have been detected from low mass-loss rate oxygen-rich red giants (Onaka *et al.* 1989, Kozasa & Sogawa 1997). Dijkstra, Speck & Reid (2005) found that the dust mineralogy changes from an amorphous alumina and amorphous olivine mixture into an amorphous silicate-only composition with increasing mass-loss rate. Circumstellar dust condensation models (e.g., Tielens 1990) predicted that Al_2O_3 is the first solid to condense in the gaseous outflows from oxygen-rich red giants. However, it is not known yet whether Al_2O_3 condenses directly or by adsorption onto TiO_2 seeds (e.g., Jeong, Winters & Sedlmayr 1999). Spectroscopic follow-up observations to identify the S11 excess with the Al_2O_3 band would be interesting. Also, we will study the relation between the variabilities found in the faint red giants and the TRGB stars with 11 μm excess.

Acknowledgements

Y.I. thank Dr. Gregory C. Sloan for helpful comments at the IAU Symposium 256 held at Keele University. We are grateful to the members of *AKARI* LMC survey working group for warm cooperations. *AKARI* is a JAXA project with the participation of ESA. This work is supported by a Grant-in-Aid for Scientific Research (A) No. 18204014 from Japan Society for the Promotion of Science and also supported by a Grant-in-Aid for Encouragement of Young Scientists (B) No. 17740120 from the Ministry of Education, Culture, Sports, Science and Technology of Japan. This work is partly supported by the JSPS grant (grant number 16204013).

References

Cioni, M.-R., Loup, C., Habing, H. J., *et al.* 2000a, *A&AS*, 144, 235
Cioni, M.-R., van der Marel, R. P., Loup, C., *et al.* 2000b, *A&A*, 359, 601
Dijkstra, C., Speck, A. K., & Reid, R. B. 2005, *ApJ*, 633, L133
Fazio, G. G., Hora, J. L., Allen, L. E., *et al.* 2004, *ApJS*, 154, 10
Ishihara, D., Wada, T., Onaka, T., *et al.* 2006, *PASP*, 118, 324
Ita, Y., Tanabé, T., Matsunaga, N., *et al.* 2002, *MNRAS*, 337, 31
Ita, Y., Onaka, T., Kato, D., *et al.* 2008, *PASJ*, 60, S435

Jeong, K. S., Winters, J. M. & Sedlmayr E. 1999, in A. Lèbre, C. Waelkens & T. Le Bertre (eds.), *Asymptotic Giant Branch Stars*, Astronomical Society of the Pacific, San Francisco, California, p. 233

Kato, D., Nagashima, C., Nagayama, T., *et al.* 2007, *PASJ*, 59, 615

Kawada, M., Baba, H., Barthel, P. D., *et al.* 2007, *PASJ*, 59, S389

Kozasa, T. & Sogawa, H. 1997, *Ap&SS*, 251, 165

Lebzelter, T., Posch, T., Hinkle, K., *et al.* 2006, *ApJ*, 653, L145

Matsuhara, H., Wada, T., Pearson, C.P., *et al.* 2007, *PASJ*, 59, S543

Meixner, M., Gordon, K. D., Indebetouw, R., *et al.* 2006, *AJ*, 132, 2288

Murakami, H., Baba, H., Barthel, P., *et al.* 2007, *PASJ*, 59, S369

Ohyama, Y., Onaka, T., Matsuhara, H., *et al.* 2007, *PASJ*, 59, S411

Onaka, T., De Jong, T., & Willems, F. J. 1989, *A&A*, 218, 1690

Onaka, T., Matsuhara, H., Wada, T., *et al.* 2007, *PASJ*, 59, S401

Renzini, A., 1992, in B. Barbuy, & A. Renzini (eds.), *The Stellar Populations of Galaxies* (Dordrecht, the Netherlands: Kluwer), p. 325

Rieke, G. H., Young, E. T., Engelbracht, C. W., *et al.* 2004, *ApJS*, 154, 25

Skrutskie, M. F., Cutri, R. M., Stiening, R., *et al.* 2006, *AJ*, 131, 1163

Shimonishi, T., Onaka, T., Kato, D., *et al.* 2008, *ApJ*, 686, L99

Tielens, A. G. G. M. 1990, in M.O. Mennessier, & A. Omont (eds.), *From Miras to Planetary Nebulae: Which Path for Stellar Evolution?* (Gif-sur-Yvette, France: Frontières), p. 186

The Magellanic System: Stars, Gas, and Galaxies
Proceedings IAU Symposium No. 256, 2008
Jacco Th. van Loon & Joana M. Oliveira, eds.

© 2009 International Astronomical Union
doi:10.1017/S1743921308028196

Survey of the Magellanic Clouds at 4.8 and 8.64 GHz

John Dickel[1], Robert A. Gruendl[2], Vincent McIntyre[3], Shaun Amy[3] and Douglas Milne[3]

[1] Physics and Astronomy Bldng., MSC07-4220, U. New Mexico, Albuquerque NM 87131, USA
email: johnd@phys.unm.edu

[2] UI Astronomy Bldng, 1002 West Green St, Urbana IL 61801, USA
email: gruendl@astro.illinois.edu

[3] ATNF-CSIRO Box 76. Epping NSW 1710, Australia
email: Vincent.McIntyre@atnf.csiro.au, Shaun.Amy@atnf.csiro.au,
D_K_Milne@hotmail.com

Abstract. Detailed 4.8 and 8.64 GHz radio images of the entire Large and Small Magellanic Clouds with half-power beamwidths of $35''$ at 4.8 GHz and $22''$ at 8.64 GHz have been obtained using the Australia Telescope Compact Array. Full polarimetric observations were made. Several thousand mosaic positions were used to cover an area of $6°$ on a side for the LMC and $4.5°$ for the SMC. These images have sufficient spatial resolution (~ 8 and 5 pc, respectively) and sensitivity (3σ of 1.5 mJy beam^{-1}) to identify most of the individual supernova remnants and H II regions and also, in combination with available data from the Parkes 64-m telescope, the structure of the smooth emission in these galaxies. We have recently revised the early data analysis (Dickel *et al.* 2005) by increasing the CLEAN cutoff limit to recover more intermediate-spacing data and thus present more accurate brightnesses for extended sources. In addition, limited data using the sixth antenna at $4.5 - 6$ km baselines are available to distinguish bright point sources ($< 3''$ and $2''$, respectively) and to help estimate sizes of individual sources smaller than the resolution of the full survey. The resulting database will be valuable for statistical studies and comparisons with X-ray, optical, and infrared surveys of the LMC with similar resolution.

Keywords. instrumentation: interferometers, surveys, ISM: general, Magellanic Clouds

1. Introduction

Radio continuum observations of the Magellanic Clouds can be used to measure both thermal emission from H II regions, polarized synchrotron emission from supernova remnants (SNRs), and the distributed interstellar medium. The common distance of all the sources offers an important evaluation of their statistical properties. The thermal emission has a nearly flat radio spectrum whereas the synchrotron emission, from relativistic electrons accelerated in shocks and the magnetic fields of SNRs, has a power-law spectrum with brighter emission at lower frequencies to help distinguish the different objects.

Three important sources of stellar energy in the ISM are UV radiation, fast stellar winds, and supernovae. Thus the radio observations supply a vital component in studies of these energy sources by tracing the contributions from both the photoionized and shock-ionized material and (through polarimetry) showing the orientation of magnetic field lines, even deep within highly obscured regions like molecular clouds. In combination with optical line emission (Smith *et al.* 2005), X-ray images (Snowden 1999), and infrared images (Meixner 2008) from the new generation surveys at those wavelengths, high-resolution radio continuum imaging will help to construct a complete physical picture of stellar feedback in the Magellanic Clouds.

The survey can also probe background objects in both total intensity and polarimetry, to determine their properties and any significant effects caused by passage of their radiation through the Magellanic Clouds.

2. Equipment and Observations

The observations were made with the Australia Telescope Compact Array (*ATCA*) using two configurations of the five movable 22-m dishes to give 19 independent baselines with spacings from 30 out to 367 meters. Short spacings were recovered by adding data from the Parkes 64-m dish (Haynes *et al.* 1991). The normalization of the data sets from the two telescopes was done by using the flux density calibrations of the primary source PKS 1934–638 at each wavelength. The slight differences in frequency and a re-evaluation of the scale between the observations with the two instruments were less than 10% and were accounted for. Combination of all the data gave half-power beamwidths of 35″ at 4.8 GHz and 22″ at 8.64 GHz. Data were also available from a sixth antenna about 4.5 km from the other five to further evaluate the sizes of unresolved sources using an incomplete $2 - 3″$ interference pattern.

The final images are mosaics of 6336 overlapping pointings for the LMC and 3564 for the SMC. Each pointing was observed for $20 - 24$ seconds at each of eight approximately uniformly spaced hour angles for each of the two telescope configurations resulting in an integration time of > 320 sec per point and reasonable coverage of orientations and telescope spacings for the aperture synthesis.

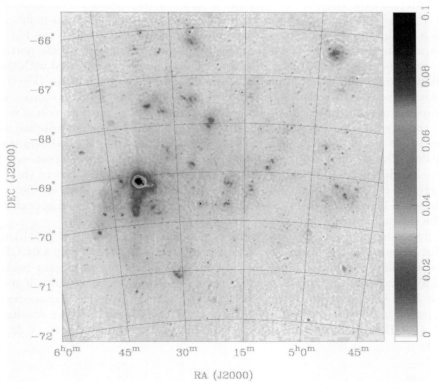

Figure 1. Total intensity image of the Large Magellanic Cloud at 4.8 GHz. The units of surface brightness in this and all subsequent images are Jy beam^{-1}.

As part of the image restoration process, they have been CLEANed (Högbom 1973) down to a cutoff level of $2\,\mathrm{mJy\,beam^{-1}}$ which is about 5 times the theoretical noise level. Experimentation with deeper clean levels resulted in the loss of some of the faint smooth emission from the shorter spacings of the *ATCA* and thus loss of structure on the extended sources.

The instrumental polarization was evaluated by observing the brightest spot in the H II region 30 Doradus. There was some off-axis polarization as well as some toward the peak but the instrumental effects everywhere were less than 0.14% at 4.8 GHz and 1.25% at 8.64 GHz. Because almost all of the sources in the images had surface brightnesses fainter than about $100\,\mathrm{mJy\,beam^{-1}}$, the polarization measurements were generally limited by the signal-to-noise ratio rather than instrumental effects. Particularly at 8.64 GHz, few sources have bright enough polarization for reliable measurement. The only polarimetric data available from Parkes are for the LMC at 4.8 GHz.

We note two features of the 8.64 GHz images. The first is that the smaller beamwidth at that frequency reduces the surface brightness per beam and thus the background noise is relatively higher. Second, atmospheric variations affect the higher frequency more so that the phase can drift more during an observation and smear a small diameter source over a larger area.

3. The Images

3.1. *LMC*

The full image of the LMC (Fig. 1) at 4.8 GHz is dominated by the 30-Dor complex (Bode 1801) and the ridge to the south of it which is very prominent in molecular line emission as first detected by Cohen *et al.* (1988). Although there are a number of H II regions and supernova remnants located along the region of the bar in the LMC, there are just as many scattered throughout the entire galaxy, particularly to the north. Thus, recent stellar birth and death does not seem to be exclusively associated with the bar.

To illustrate the full detail available in the images, we show blow-ups (Fig. 2) of the area around the N 11 (Henize 1956) complex. N 11 clearly has several compact features superimposed on a more extended fainter background. Such a phenomenon is very common to the large H II regions in the Magellanic Clouds. This smooth distributed emission is easier to see at 4.8 GHz where the low surface brightness is integrated over the larger beamwidth. The residual grating rings are also more obvious at that frequency for the same reason. In addition to N 11, the three faint blobs in the northeast of the 4.8 GHz image are the faint H II regions N 12, 13, and 14. The polarized source on the west side was labelled N 11L by Henize but is a well known supernova remnant and appears distinct from the thermal gas.

The integrated flux densities of all the components of N 11 (except N 11L) and the extended emission around them are 5.40 Jy at 4.8 GHz and 5.23 Jy at 8.64 GHz. The major uncertainty is in determining the outline of the source on the varying background of the LMC. Experimentation with various choices indicated an uncertainty of about 5 % or ~ 0.3 Jy at both frequencies. The spectral index of N 11 from these measurements is -0.04, indicating its thermal nature. N 11L, on the other hand, has flux densities of 51 mJy at 4.8 GHz and 42 mJy at 8.64 GHz producing a spectral index is -0.32, clearly non-thermal.

3.2. *SMC*

The images of the SMC (Fig. 3) are certainly dominated by the H II region N 66 at about $00^\mathrm{h}59^\mathrm{m}15^\mathrm{s}$ and $-72°10'$ just north of the apparent and dynamical center of the SMC.

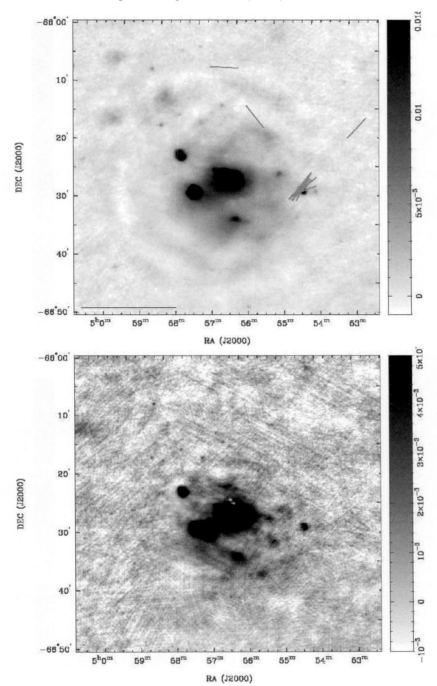

Figure 2. Total intensity image and polarization vectors of the large H II region complex N 11 in the northwest of the LMC at 4.8 GHz (top) and total intensity at 8.64 GHz (bottom). The line at the lower left of the top panel represents a polarized intensity of $5\,\mathrm{mJy\,beam}^{-1}$. The significantly polarized source is the supernova remnant N 11L.

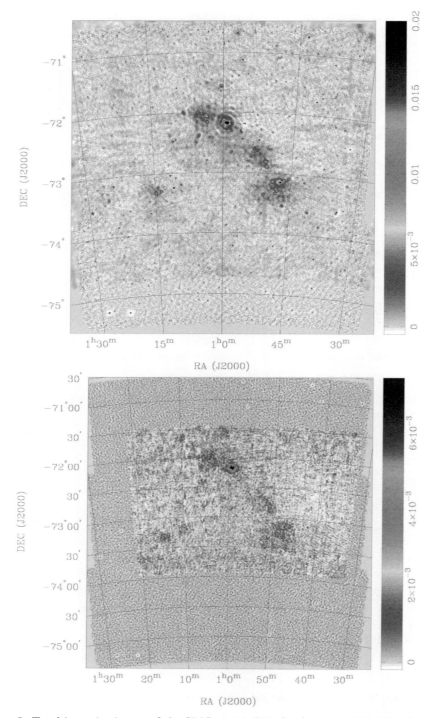

Figure 3. Total intensity image of the SMC at 4.8 GHz (top) and at 8.64 GHz (bottom).

A number of other bright sources lie just east of there, including the very bright spot at $01^{h}04^{m}02^{s}$ and $-72°01'55''$ which is the well known and brightest SNR in the SMC, E 0102.2−7219 (Seward & Mitchell 1981).

There is an indication of the bar off to the southwest ending near the bright feature N 19, a very confused area of H II regions and SNRs that still needs sorting out (Dickel *et al.* 2001) That area is near the major H I concentration in the galaxy (Stanimirović *et al.* 2004).

As with the LMC, most of the known SNRs and H II regions can be readily found in the expanded images. From the abrupt changes in the smoothness of the background, the reader can also see that the extent of the images from Parkes do not cover the full area of the *ATCA* ones. In particular, the 8.64-GHz image from Parkes is significantly smaller but still does cover most of the SMC sources. A few point sources appear around the edges where the data from Parkes are not available.

4. Data Availability

The calibrated UV data and the images, both total-intensity and polarimetric, are available in FITS format from the NCSA Astronomical Image Digital Library. The results can be downloaded from http://adil.ncsa.uiuc.edu/document/08.JD.01. The images include those from *ATCA* alone and merged ones from *ATCA* plus Parkes. We also provide filtered images so the 4.8- and 8.64-GHz images have the same *uv* range, though not identical coverage, of spacings in wavelengths to allow more careful spectral comparisons. The edited and calibrated *uv* data are also available in *MIRIAD* form from ADIL.

5. Acknowledgements

The Australia Telescope is funded by the Commonwealth of Australia as a National Facility managed by CSIRO. We have benefited from help and valuable discussions with many colleagues. They include Bob Sault, Lister Staveley-Smith, Uli Klein, Hélène Dickel, John Dickey, Robin Wark, Ray Plante, You-Hua Chu, Rosa Williams, Richard Rand, Gregory Taylor, Miroslav Filipović, and Brian Fields. Barry Parsons, Margaret House and Vicki Drazenovic helped to make the many visits to the Australia Telescope very enjoyable. Annie Hughes stimulated us to rethink the CLEANing procedures to produce more reliable images.

References

Bode, J. 1801, *General Description and Information of the Stars* (Berlin: privately published)

Cohen, R. S., Dame, T. M., Garay, G., Montani, J., Rubio, M., & Thaddeus, P. 1988, *ApJ*, 331, L95

Dickel, J. R., Williams, R. M., Carter, L. M., Milne, D. K., Petre, R., & Amy, S. 2001, *AJ*, 122, 849

Dickel, J. R., McIntyre, V. J., Gruendl, R. A., & Milne, D. K. 2005, *AJ*, 129, 790

Haynes, R. F., Klein, U., Wayte, S. R., *et al.* 1991, *A&A*, 252, 475.

Henize, K. 1956, *ApJS*, 2, 315

Högbom, J. A. 1973, *A&AS* 15, 417

Meixner, M. 2008, in Chary, R.-R., Tepliz, H., & Sheth, K. (eds.), *The Second Annual Spitzer Science Center Conference: Infrared Diagnostics of Galaxy Evolution*, ASPSC, 381, 115

Seward, F. & Mitchell, M. 1981, *ApJ*, 243, 736

Smith, R. C., Points, S., Chu, Y.-H., Winkler, P. F., Aguilera, C., & Leiton, R. 2005, *BAAS*, 37, 1403

Snowden, S. 1999, in Chu, Y.-H., Suntzeff, N., Hesser, J., & Bohlender, D. (eds), *New Views of the Magellanic Clouds, IAU Symposium 190*, ASPSC, 190, 32

Stanimirović, S., Staveley-Smith, L., & Jones, P. 2004, *ApJ*, 604, 176

The Magellanic System: Stars, Gas, and Galaxies
Proceedings IAU Symposium No. 256, 2008
Jacco Th. van Loon & Joana M. Oliveira, eds.

The X-ray stellar population of the LMC

Yaël Nazé†

Institut d'Astrophysique et de Géophysique, Université de Liège,
Allée du 6 Août 17, Bat B5C, B4000-Liège, Belgium
email: naze@astro.ulg.ac.be

Abstract. In the study of stars, the high energy domain occupies a place of choice, since it is the only one able to directly probe the most violent phenomena: indeed, young pre-main sequence objects, hot massive stars, or X-ray binaries are best revealed in X-rays. However, previously available X-ray observatories often provided only crude information on individual objects in the Magellanic Clouds. The advent of the highly efficient X-ray facilities *XMM-Newton* and *Chandra* has now dramatically increased the sensitivity and the spatial resolution available to X-ray astronomers, thus enabling a fairly easy determination of the properties of individual sources in the LMC.

Keywords. stars: early-type, galaxies: individual (LMC), X-rays: stars, X-rays: binaries

1. X-ray surveys of the LMC

In the range of astronomical tools, X-rays are of particular interest. Indeed, the high-energy domain unveils the most energetic phenomena taking place in our Universe. Such processes are usually difficult to perceive at other wavelengths, though they provide important constraints for astrophysics.

In this context, the Magellanic Clouds (MCs) are targets of choice. Their advantages are multiple: the known and small distance, together with the small angular size, small inclination, different metallicities, and recent star formation episodes are here as crucial as at other wavelengths. In addition, the low obscuration towards the MCs renders soft X-ray observations much easier while the numerous available data taken at other wavelengths ensure a correct, global analysis of the LMC X-ray sources.

X-rays associated with the LMC were first detected 40 years ago by a rocket experiment (Mark *et al.* 1969): the source appeared extended and soft, with a total luminosity estimated to 4×10^{38} erg s^{-1}. It did not take long to distinguish a few individual sources in this X-ray emission, nicknamed LMC X-1 to 6, thanks to the joint effort of the satellites *Uhuru, Copenicus, OSO-7,* and *Ariel V* (Leong *et al.* 1971; Rapley & Tuohy 1974; Markert & Clark 1975; Griffiths & Seward 1977). In the following decade, the *Einstein* observatory increased the number of known X-ray sources in the direction of the LMC to about a hundred (Long *et al.* 1981; Wang *et al.* 1991). Finally, a sensitive survey undertaken by *ROSAT* provided another ten-fold increase in the total source number (Haberl & Pietsch 1999a; Sasaki *et al.* 2000).

However, only half of the detected X-ray sources truly belonged to the LMC and an even smaller fraction appeared to be associated with LMC stars. For example, of the 758 *ROSAT*-PSPC sources, only 144 were identified at first (Haberl & Pietsch 1999a): 15 as background AGNs or galaxies behind the LMC, 57 as foreground Galactic stars, 46 as SNRs and SNR candidates, 17 as X-ray binaries (XRBs) and candidates, 9 as Supersoft sources (SSSs) and candidates. Using the observed X-ray properties (especially

† Postdoctoral Researcher FRS-FNRS

the hardness ratios), Haberl & Pietsch (1999a) further proposed additional identifications (3 AGNs, 27 foreground stars, 9 SNRs and 3 SSSs) which yields a fraction of 20% of X-ray sources associated with LMC stars. Similar results were obtained with the *ROSAT*-HRI (Sasaki *et al.* 2000): 397 detections among which 138 in common with the PSPC and 115 identified (10 AGNs, 52 foreground stars, 33 SNRs, 12 XRBs, 5 SSSs, and 3 hard sources which could either be AGNs or XRBs). If one considers the variable X-ray sources, the contamination by non-LMC objects is smaller. For the PSPC survey, the proposed identification of the 27 variable X-ray sources is 12 XRBs, 5 SSSs, 9 foreground stars and 1 Seyfert galaxy (Haberl & Pietsch 1999b); for the HRI survey, 26 variable sources were detected among which 8 XRBs, 4 SSSs, 6 foreground stars, 2 AGNs and 1 nova (Sasaki *et al.* 2000): the fraction of LMC stellar objects among variable X-ray sources is thus 60%.

The current facilities possess much higher sensitivities and spatial/spectral resolution but unfortunately they also have a smaller field-of-view. Its non-zero extension penalizes the LMC, especially in comparison with the SMC ($\sim$59 sq. deg. vs. $\sim$18). This explains why, up to now, no full survey of the LMC has been performed with *XMM-Newton* or *Chandra*. Nevertheless, smaller fields have been observed and it should be underlined that, though its coverage is patchy, the 2XMM catalog currently lists 5421 entries in the area of the *ROSAT* surveys! In addition, Haberl *et al.* (2003) reported the analysis of one deep *XMM-Newton* observation of a northern region of the LMC (on the rim of the supergiant shell LMC4). While *ROSAT* had detected 34 sources in this field, *XMM-Newton* data reveal 150 objects (detection limit 6×10^{32} erg s^{-1}). In a selection of 20 bright or peculiar sources, the majority (10) are AGNs, but there are also 3 foreground stars, 2 SNRs, 4 HMXBs, and 1 SSS†. In addition, Shtykovskiy & Gilfanov (2005) analyzed 23 *XMM-Newton* archival observations covering 3.8 sq. degrees of the LMC. With a detection limit of 3×10^{33} erg s^{-1}, they detected 460 sources in the 2–8 keV band, in vast majority AGNs, and focused on 9 good XRBs candidates and 19 possible XRBs (see below for more details). Finally, a sensitive survey of the LMC in the hard X-ray domain (15 keV– 10 MeV) has been performed with *Integral* (Götz *et al.* 2006). Only a few sources have been detected: the X-ray binaries LMC X-1, LMC X-4, and PSR B0540–69, as well as two hard sources which might correspond to LMC binaries. These encouraging first results in both the soft and hard X-ray domains enlight the detection potential of the sensitive observatories of the current generation.

2. Supersoft X-ray sources

As their names indicate, SSSs display very soft spectra: they can be fitted with thermal models with kT of only 20–80 eV. These often bright (10^{36-38} erg s^{-1}) objects became a category of their own after the discovery of several such sources in the LMC — indeed, their identification is most difficult in the Galactic plane, where high absorption prevents their detection. Including recent *XMM-Newton* detections, 18 SSSs are now identified in the LMC and several other examples are also known in >10 galaxies (Kahabka *et al.* 2008).

It is generally believed that the X-ray emission of SSSs corresponds to steady nuclear burning on the surface of a white dwarf accreting matter from its companion. SSSs would therefore trace rather old stellar populations, and they are accordingly found along the rim of the LMC optical bar (Haberl & Pietsch 1999a,b). Other possible origins for SSSs have been proposed and the LMC SSSs confirmed this fact: of the 18 known SSSs, 5 (+1?)

---

† This source was not confirmed by Kahabka *et al.* (2008).

reside in close binaries (as expected from the above picture), but 5 others correspond respectively to a post-nova object, a WD+Be binary, a symbiotic star, a planetary nebula, a transition (WD-central star of PN) object — 7 remain unidentified transients (Kahabka et al. 2008).

The X-ray spectrum of SSSs often presents "photospheric" characteristics, i.e. it consists of numerous absorption lines from highly ionized metals (generally Si, S, Ar, Ca, Fe) superimposed on a blackbody emission. One important contribution from *XMM-Newton* and *Chandra* is the availability of high-resolution spectroscopy. Combining high-resolution spectra and (preferentially NLTE) atmosphere modelling, it is possible to derive precise stellar parameters, notably the WD mass. Such an analysis was undertaken for CAL 83 by Lanz et al. (2005), who found $M_{WD} = 1.3 \pm 0.3$ M$_\odot$ and $T = 0.55 \pm 0.03$ MK. Many absorption lines were still blended, preventing any detailed chemical analysis. Looking closer at the lines, there seems to be no evidence for emission components, large line shifts or asymmetries (which would indicate fast outflows), or large line widths (which would be associated with a fast rotation). A similar study of the SSS associated with nova LMC 1995 led to $M_{WD} = 0.91$ M$_\odot$ and $T = 0.40$–0.47 MK (Orio et al. 2003) and showed that the abundance of carbon is not enhanced, suggesting the X-ray emitting matter to be accreted material from the companion.

Some SSSs present however very different spectral characteristics. At high resolution, the X-ray spectrum of CAL 87 appears composed exclusively of numerous *emission* lines (mostly from O, N, and Fe). These lines which come from recombination and resonant scattering are clearly redshifted, with a double-peaked structure observable in the brightest features (Greiner et al. 2004; Orio et al. 2004). In addition, the broad X-ray eclipse suggests the X-ray source to be extended, about 1.5 R$_\odot$ in size (to be compared to the orbital separation of 3.5 R$_\odot$). The X-rays thus originate in a fast (2000 km s^{-1}) and non-spherical outflow, e.g., an accretion disk corona.

The brightest SSS is RX J0513.9–6951. Its spectrum is a mixture of absorption and emission lines superimposed on a continuum, indicating the superposition of emissions from a hot atmosphere and from an optically-thin corona (McGowan et al. 2005; Burwitz et al. 2007). The absorption lines exhibit several blueshifted velocity components, a typical signature of an outflow, which vary with time (the deepest absorptions were observed when the X-ray flux was lowest). High abundances are observed for N, S, and Ar; they may imply that the outflowing material was affected by nova nucleosynthesis (McGowan et al. 2005). Two temperature components may in fact be present and would be linked to a fast rotation of the WD (cool equatorial regions + hot polar caps, Burwitz et al. 2007).

SSSs are also variable X-ray sources: they exhibit a variety of flux changes occuring on a variety of timescales. For example, CAL 83 displays recurrent X-ray low states and short-term variations of smaller amplitude (Lanz et al. 2005). The presence of 38-min oscillations at some epochs was also claimed for this object by Schmidtke & Cowley (2006) who attributed them to non-radial pulsations. Another SSS, RX J0513.9–6951, displays every 100–200 d optically faint/X-ray bright episodes of duration 20–40 d. The exact recurrence timescale varies, which is probably related to variations in the accretion rate by a factor of 5 (Burwitz et al. 2008). The high-resolution X-ray data were acquired during one of the low optical state but the last *XMM-Newton* observation samples the beginning of the recovering to 'normal' optical/UV intensity (McGowan et al. 2005; Burwitz et al. 2007). As the recovering progresses, the X-ray luminosity decreases, as well as the temperature of the blackbody: it thus seems that the peak emission slowly shifts towards longer wavelengths (McGowan et al. 2005). At the same time, the radius of the blackbody emitter becomes larger. This is consistent with the global picture of the WD contraction model: the accretion rate is high during the low optical state; when it drops, the WD contracts and the emission becomes more energetic; the enhanced X-ray emission then influences the WD environment and provokes a new increase in the accretion rate, thereby inflating again the WD (McGowan *et al.* 2005). Another interpretation implies changes in the accretion disk's size and the wind outflow, but observations seem to contradict the predictions from that particular model (Burwitz *et al.* 2007).

3. X-ray binaries

3.1. *HMXBs*

In high-mass X-ray binaries, the primary is a compact object (neutron star, NS, or black hole, BH) while the secondary can either be a Be star or a supergiant star. In the first case, the binary is generally eccentric and accretion onto the compact companion occurs only when the latter crosses the dense equatorial regions of the Be star, producing recurrent X-ray outbursts. Most HMXBs of the LMC are of this type. In the second case, when the secondary is a supergiant, the accretion occurs either through wind capture or Roche-lobe overflow. In the LMC, there are one or two candidates for wind accretion and two or three candidates for RLOF (see, e.g., Negueruela & Coe 2002).

As they contain massive secondaries, HMXBs reveal young stellar populations. However, they are not instantaneous tracers of star formation, since one needs to wait until the compact object forms: while the number of HMXBs certainly decreases for populations older than 20 Myr, no HMXB is expected when the stellar population is younger than 3 Myr. This explains why 30 Doradus, one of the most active star-forming regions in the LMC, contains few HMXBs while the ~10 Myr supergiant shell LMC4 harbors many of them (Haberl & Pietsch 1999a,b; Shtykovskiy & Gilfanov 2005).

While only 10 HMXBs were known in the MCs before 1995, their number now exceeds 100. In the LMC, 36 cases are listed by Liu *et al.* (2005). Shtykovskiy & Gilfanov (2005) studied the cumulative distribution of X-ray luminosities (the X-ray luminosity function) of HMXBs and candidates found in 23 *XMM-Newton* fields. They reported a possible flattening of the luminosity function at high luminosities. If confirmed, this would indicate a deficit of low-luminosity sources, probably linked to the 'propeller effect' (i.e. inhibition of accretion, and therefore decrease of the X-ray luminosity, by fast-rotating NS with low accretion rates).

Among the HMXBs of the LMC, two could contain black holes: LMC X-1 and LMC X-3 ($M_{\rm BH} \sim 4$ M$_\odot$, see, e.g., Yao *et al.* 2005). Such systems display two spectral states: a high/soft state and a low/hard state. In the former case, the X-ray luminosity is high while the X-ray spectrum is soft and the X-ray emission is believed to be mostly thermal emission from the accretion disk. In the latter case, the situation is opposite and the X-ray emission is attributed to comptonization of the soft photons from the accretion disk by hot electrons in the surrounding corona. In most cases, the spectrum is thus fitted by the combination of a multi-temperature blackbody model (where the temperature decreases while the radius increases) and a power-law. This yields rather good results, especially when the photon statistics is poor, but is clearly oversimplistic (presence of an intense gravitational field in the inner regions, interrelations between the comptonized emission and the seed blackbody spectrum,...) and can be improved in many ways (Yao *et al.* 2005, and references therein).

LMC X-1 comprises an O8 III–V star in a 4.2-d orbit around a BH (Negueruela & Coe 2002). Up to now, it has only been observed in the high/soft state but variations of its X-ray emission were detected, even for the hardest X-rays (detection in the 20–40 keV range in 2003 but not in 2004, Götz *et al.* 2006). Only a disk blackbody was required to fit its spectrum and no discrete features were found in the high-resolution X-ray spectrum, though one would have expected the Ne K edge to be detected (Cui *et al.* 2002).

LMC X-3 harbours a B star of uncertain type (B2.5 V, Negueruela & Coe 2002, or B5 IV, Wu *et al.* 2001) in a 1.7-d orbit around a BH. This system is often — but not always — seen in the high/soft state and a combination of a multicolour blackbody and a power law is needed to describe its X-ray spectrum (Cui *et al.* 2002). Peaks in the power-law flux are related to drops in the blackbody flux: accreting matter thus appears diverted from the accretion disk to feed the surrounding hot corona (Smith *et al.* 2007). The transfer of material occurs most probably through Roche lobe overflow since there is no evidence for the presence of a wind (no emission lines, no absorption edges, and low absorbing column for neutral/ionized matter compatible with Galactic values, Wu *et al.* 2001; Page *et al.* 2003). Turning to the X-ray lightcurve of LMC X-3, there appear to be no significant variations on very short timescales; however, modulations with the orbital period and with twice this period were detected (Boyd *et al.* 2001). The former, which is particularly obvious when the X-ray flux is lower, could be attributed to the presence of a hot spot (e.g., where the gas stream impacts the accretion disk). On the other hand, the abrupt flux decrease seen each $2 \times P_{\mathrm{orb}}$ may be caused by a large perturbation of the disk, such as a global density wave periodically obscuring our view of the inner regions (Boyd *et al.* 2001).

The NS involved in the other HMXBs is sometimes a pulsar. In the last decade, detailed timing studies were made possible, especially thanks to the *RXTE* satellite. In PSR B0540–69 (50 ms), the pulse profile, found to be similar in optical and in different X-ray bands, appears composed of two gaussians whose ratio does not vary with the energy considered (de Plaa *et al.* 2003). Over the 8 years of *RXTE* observations, only one change in the pulsar period ('glitch') was observed (Livingstone *et al.* 2005). Compared to the Crab, this pulsar has a much lower 'glitching' activity though otherwise both objects share similar properties (age, magnetic field strength, rotation, ...). This probably indicates that another physical parameter, overlooked up to now, plays here a crucial role. In contrast, the much older PSR J0537–6910 (16.1 ms, in N 157B) displays a large 'glitching' activity: 23 glitches were observed in the 7 years of *RXTE* data (Middleditch *et al.* 2006). The timing analysis yielded interesting results: (1) the amplitude $\Delta\nu$ of a glitch is proportional to the interval to the next glitch; (2) the longer the time before a glitch, the larger the change $|\Delta\dot{\nu}|$ but there is a maximum value for this variation; (3) the gain in $|\dot{\nu}|$ across one glitch is not completely given back before the next glitch; (4) microglitches often precede large glitches. The overall activity of PSR J0537–6910 is much higher than that of the Crab; on the other hand, its glitches are smaller but more frequent than for Vela, to which it is often compared. The analysis of PSR J0537–6910 indicates that glitch models relying on sudden onsets are not compatible with the observed glitches and that observations of PSR J0537–6910 appear in agreement with the picture of cracks at the NS surface combined to unstable vortices in the neutron superfluid (Middleditch *et al.* 2006).

The HMXB LMC X-4 also harbours a pulsar (period 13.5 s) in a 1.4-d orbit around an O8 III star (Negueruela & Coe 2002). The X-ray flux is modulated with a 30-d timescale (Lang *et al.* 1981, see also Naik & Paul 2003 and Götz *et al.* 2006), which likely reflects the precession period of the tilted accretion disk as in Her X-1. While the orbital period appears very stable, the disk precessing period may vary non-uniformly (Tsygankov & Lutovinov 2005). Pulses and eclipses are also seen in the hard X-ray range by *Integral* (Götz *et al.* 2006). The eclipses suggest a size for the hard X-ray emitting region of 0.38 $R_\odot$, i.e. larger than the size of a NS and more typical of that of a hot corona.

LMC X-4 is also varying at these high energies by an order of magnitude (Götz *et al.* 2006; Tsygankov & Lutovinov 2005). Finally, LMC X-4 experiences X-ray flares and at these times, mHz quasi-periodic oscillations have been detected (Moon & Eikenberry 2001): strong, burstlike-features with a timescale of 700–1500 s and weak oscillation with periods of 50–500 s. The former could be explained by beating frequencies between the pulsar frequency and the orbital frequencies of big clumps on the verge of being accreted, while the latter is more compatible with Keplerian periods of clumps outside the corotation radius (Moon & Eikenberry 2001).

3.2. *LMXBs*

Low-mass X-ray binaries are systems composed of a compact object (NS, BH) and a faint, low-mass (generally < 1 M$_\odot$) companion. The X-ray emission is a consequence of the mass transfer via Roche-lobe overflow towards the compact object. Only one such object (or maybe two: LMC X-2 and possibly RX J0532.7–6926, Liu *et al.* 2007)† is known in the LMC, implying that the proportion of LMXBs with respect to HMXBs is much smaller in the LMC than in the Galaxy. This can be explained by the different star formation histories of the two galaxies since LMXBs are associated with an older stellar population than HMXBs (Liu *et al.* 2005).

LMC X-2 is a binary of period 8 h (Cornelisse *et al.* 2007), similar in many respects to Sco X-1. A monitoring has indicated that, during the bright X-ray states, the optical lightcurve lags $\lesssim 20$ s behind the X-ray lightcurve (McGowan *et al.* 2003). There thus seems to be some light reprocessing in the system, but the location where it takes place is unclear since the lower limit for the delay is larger than the light travel time across the accretion disk and smaller than the light travel time to the secondary (McGowan *et al.* 2003). The X-ray spectrum is well fitted by the combination of a disk multi-temperature blackbody and a hot blackbody (kT of 1.5 keV, probably associated with regions at or close to the surface of the NS, Lavagetto *et al.* 2008). The inner parts of the disk are not seen: as the source's luminosity is close to the Eddington limit, some material can be ejected, which would obscure our view towards those inner regions. The Fe Kα line stays undetected while the O VIII Lyα is present: this is most probably a metallicity effect, which needs to be further investigated (Lavagetto *et al.* 2008).

4. Massive stars and clusters

Massive stars, of spectral types O and early B, are soft and moderate X-ray emitters. In our Galaxy, their overall luminosity scales with their bolometric luminosity ($L_{\rm X} \sim 10^{-7} L_{\rm bol}$) and their spectra reveal emission lines from an optically-thin hot plasma with $kT = 0.3$–0.7 keV. Such objects are the progenitors of SNe, GRBs, NSs and BHs, and are often responsible for the presence of diffuse X-ray emissions (SNRs and wind-blown bubbles, see Chu, these proceedings). One of their main characteristics is the presence of strong stellar winds, driven through resonant scattering of their intense UV radiation by metals. The mass-loss rate and wind velocities of massive stars are typically 10^{-6} M$_\odot$ yr^{-1} and 2000 km s^{-1}, respectively. The X-ray emission is generally believed to arise in these winds, through collisions of structures travelling at different velocities (for a review, see Güdel & Nazé, in preparation).

---

† The status of RX J0532.7–6926 is still debated. On the one hand, Haberl & Pietsch (1999b) strongly advocate in favor of a LMXB nature on the basis of the shape of the X-ray lightcurve. On the other hand, Haberl & Pietsch (1999a) and Sasaki *et al.* (2000) only categorize it as a candidate LMXB ("LMXB?") while Kahabka (2002) mentions the source as a possible LMXB or AGN. At the present time, no counterpart has been detected for this source.

Since winds are heavily dependent on the metallicity of their host galaxy, LMC observations of massive stars are crucially important. A first test can be performed on Wolf-Rayet stars, the evolved descendants of O-type stars which display the strongest and densest winds. WRs mostly come in two flavours, WN if their spectrum is enriched in nitrogen and WC in the case of a carbon enrichment. In the Galaxy, no single WC star has ever been observed in the X-ray domain, most probably because of the high absorption of their stellar winds; the situation for WN is less clear and slight differences in wind structures and composition could play an important role (for a review, see Güdel & Nazé, in preparation). For the LMC, Guerrero & Chu (2008a,b) and Guerrero *et al.* (2008) have analyzed all *ROSAT*, *Chandra*, and *XMM-Newton* observations available, which cover more than 90% of the known WRs in the LMC. Of the 125 observed objects, only 32 were detected, mostly binaries: the detection rate is 50% for binaries but only 10% for supposedly single objects. There are similarities with the Galactic case (non-detection of single WC stars, binaries preferentially detected) but there are also clear differences. Notably, the X-ray luminosity and L_X/L_{bol} ratio are larger for LMC objects, which could be explained by a lower opacity of the winds.

Peculiar phenomena can enhance and harden the stellar X-ray emission: (1) in single objects, intense magnetic fields channel the wind streams towards the equatorial regions where they collide, producing a very hot plasma; (2) in binaries composed of two massive objects, the wind of one star can interact with that of the other, again leading to the formation of a hot plasma (for a review, see Güdel & Nazé, in preparation). Bright, hard X-ray sources associated with massive stars have therefore often been attributed to one or the other phenomenon, depending on the authors involved (see, e.g., for colliding wind binaries in the LMC: Portegies Zwart *et al.* 2002). However, one must be cautious about such conclusions. First of all, a spectral monitoring in the IR, optical, or UV domain is needed to ascertain the binary nature of the object. This is however no definite proof of colliding-wind binaries, since magnetic objects (single or in binaries) are also overluminous, and it should be noted that not all massive binaries are overluminous even if both components possess significant stellar winds. Second, a monitoring is requested in the X-ray domain. Indeed, phase-locked variations are the signature of peculiar phenomena and help reject the simple line-of-sight coincidence. The X-ray emission of a magnetic oblique rotator is modulated by the (usually short) rotation period of the star, while the X-ray emission from wind-wind interactions changes with the orbital period because of varying absorptions crossing the line-of-sight or the varying distance between the two stars (hence a changing strength of the wind-wind collision). Up to now, variability as just described could be established only in one massive system of the MCs, HD 5980 in the SMC where an *XMM-Newton* monitoring was performed to ascertain the colliding-wind nature of the emission (Nazé *et al.* 2007).

Massive stars generally reside in clusters. In the LMC, only two of them were studied in the X-ray domain: 30 Doradus and N 11. A 20-ks *Chandra* observation of 30 Doradus revealed 180 sources with $L_X > 10^{33}$ erg s^{-1}, 109 of them being within 30″ of R 136 (Townsley *et al.* 2006a,b). Half of the X-ray sources possess counterparts at other wavelengths, generally massive stars: some bright, hard sources are considered as potential colliding-wind binaries. Some non-detection should also be underlined: no star from the embedded new stellar generation has been detected and not all early-type objects (e.g., O3) are detected. A longer exposure (100 ks) has now been obtained and is still under analysis. In N 11, the coarse spatial resolution of *XMM-Newton* data failed to provide clear detections of individual stars, though hints in this direction were found in the clusters LH 10 and LH 13 (Nazé *et al.* 2004). A 283-ks *Chandra* observation found 165 point sources in the central area of N 11 (clusters LH 9/LH 10) with $L_X > 10^{32}$ erg s^{-1} (Chu,

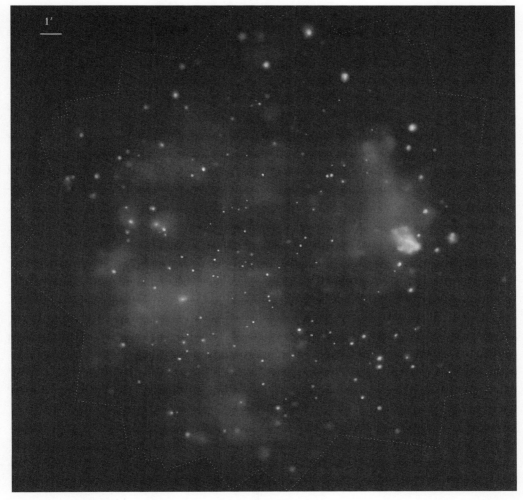

Figure 1. *Chandra* observation of N 11. Note the large number of point sources, scattered all over the field-of-view, without any correlation with the positions of the clusters (LH 10 is in the middle-left of the image and LH 9 just below), suggesting a large contamination from background/foreground objects (from Chu, Wang, Nazé, *et al.*, in preparation).

Wang, Nazé, *et al.*, in preparation). Fifteen of these are associated with massive objects (10 with O/WR, 2 with B stars, 3 with compact groups of massive stars), yielding an overall detection rate of 16% (indeed, the brightest and/or earliest objects display the highest detection fraction). Known binaries constitute only 20% of the detected objects: comparing with the clusters's binary fraction of about 36%, this suggests that massive binaries are *not* preferentially detected, contrary to what happens in the Galaxy. Moreover, if the 15 detected objects can be considered as truly typical, the $L_X - L_{bol}$ relation would be 0.4 dex higher in the LMC than in the Galaxy, again contrary to expectations. However, it remains to be confirmed that no peculiar object (magnetic star, colliding wind binary) contaminates the sample. This is indeed a plausible hypothesis since stars of apparently similar spectral types display very different X-ray fluxes (as for 30 Doradus).

5. Perspectives

Many stellar objects emit X-rays. The brightest ones, involving compact objects (XRBs and SSSs, $L_X \sim 10^{36-38}$ erg s^{-1}), have been detected in the LMC more than 3 decades ago; the current instruments have now provided the first detailed timing sequences and high-resolution spectra, which often led to changes or refinements of the initial models. For these sources, it is now necessary to reach even higher spectral resolutions and sensitivity to get more precise observational constraints. Moderate X-ray sources such as massive stars (10^{31-34} erg s^{-1}) now enter the picture. At least the brightest examples have been detected, notably in 30 Doradus and N 11. Beyond enlarging the number of X-ray detections, the future lies in getting high-resolution spectroscopy of LMC massive stars: since their X-ray emission is linked to their stellar winds, which crucially depend on metallicity, high-resolution data are necessary to test our X-ray generation models. Indeed, in the Galaxy, such high-resolution observations have already initiated a shift in thought — but it is essential to check the new theories in a different metallicity environment like that of the LMC. Finally, the future generation of X-ray telescopes should be able to detect even fainter X-ray sources such as low-mass/coronal sources (10^{26-33} erg s^{-1}) and young pre-main sequence objects (flaring T Tauri stars can reach 10^{31-32} erg s^{-1})†.

Once all this is accomplished, a full picture of the LMC at high energies will be available. Of course, acquiring such data is not a simple question of 'filling the catalogues'. It must always be kept in mind that, with its lower metallicity, the LMC provides a crucial test of theoretical models — see, e.g., the case of massive stars. The astronomical community should thus promote the advent of a new generation of X-ray facilities possessing three concommitant characteristics: high sensitivity (to detect the faintest X-ray sources in nearby galaxies), high spatial resolution (to disentangle blended stellar objects in nearby galaxies), and high spectral resolution.

Acknowledgements

YN acknowledges financial support from the Fonds de la Recherche Scientifique (FRS-FNRS Belgium), the University of Liège (through the 'patrimoine-ULg' grants), the organizers of the Symposium, and the PRODEX XMM and Integral contracts.

References

Boyd, P. T., Smale, A. P., & Dolan, J. F. 2001, *ApJ*, 555, 822

Burwitz, V., Reinsch, K., Greiner, J., Rauch, T., Suleimanov, V., Walter, F. W., Mennickent, R. E., & Predehl, P. 2007, *Adv. Sp. Res.*, 40, 1294

Burwitz, V., Reinsch, K., Greiner, J., Meyer-Hofmeister, E., Meyer, F., Walter, F. M., & Mennickent, R. E. 2008, *A&A*, 481, 193

Cornelisse, R., Steeghs, D., Casares, J., Charles, P. A., Shih, I. C., Hynes, R. I., & O'Brien, K. 2007, *MNRAS*, 381, 194

Cui, W., Feng, Y. X., Zhang, S. N., Bautz, M. W., Garmire, G. P., & Schulz, N. S. 2002, *ApJ*, 576, 357

de Plaa, J., Kuiper, L., & Hermsen, W. 2003, *A&A*, 400, 1013

Götz, D., Mereghetti, S., Merlini, D., Sidoli, L., & Belloni, T. 2006, *A&A*, 448, 873

Greiner, J., Iyudin, A., Jimenez-Garate, M., Burwitz, V., Schwarz, R., DiStefano, R., & Schulz, N. 2004, *Rev. Mexicana AyA*, 20, 18

† Note that cataclysmic variables (CVs) can also be rather bright X-ray sources, but they are difficult to identify in the Magellanic Clouds due to the faintness of their optical counterparts. Currently, there is no clear identification of a LMC X-ray source as a CV and this is why CVs were not considered in this review.

Griffiths, R. E. & Seward, F. D. 1977, *MNRAS*, 180, 75P

Guerrero, M. A. & Chu, Y.-H. 2008a, *ApJS*, 177, 216

Guerrero, M. A. & Chu, Y.-H. 2008b, *ApJS*, 177, 238

Guerrero, M. A., Carter, J. A., Chu, Y.-H., Foellmi, C., Moffat, A. F. J., Oskinova, L., & Schnurr, O. 2008, in *X-ray Universe 2008*, available on-line on http://xmm.esac.esa.int/external/xmm_science/workshops/2008symposium/guerrero_martin.pdf

Haberl, F. & Pietsch, W. 1999a, *A&AS*, 139, 277

Haberl, F. & Pietsch, W. 1999b, *A&A*, 344, 521

Haberl, F., Dennerl, K., & Pietsch, W. 2003, *A&A*, 406, 471

Kahabka, P. 2002, *A&A*, 388, 100

Kahabka, P., Haberl, F., Pakull, M., Millar, W. C., White, G. L., Filipović, M. D., & Payne, J. L. 2008, *A&A*, 482, 237

Lang, F. L., Levine, A. M., & Bautz, M., *et al.* 1981, *ApJ*, 246, L21

Lanz, T., Telis, G. A., Audard, M., Paerels, F., Rasmussen, A. P., & Hubeny, I. 2005, *ApJ*, 619, 517

Lavagetto, G., Iaria, R., D'Aı̀, A., di Salvo, T., & Robba, N. R. 2008, *A&A*, 478, 181

Leong, C., Kellogg, E., Gursky, H., Tananbaum, H., & Giacconi, R. 1971, *ApJ*, 170, L67

Liu, Q. Z., van Paradijs, J., & van den Heuvel, E. P. J. 2005, *A&A*, 442, 1135

Liu, Q. Z., van Paradijs, J., & van den Heuvel, E. P. J. 2007, *A&A*, 469, 807

Livingstone, M. A., Kaspi, V. M., & Gavriil, F. P. 2005, *ApJ*, 633, 1095

Long, K. S., Helfand, D. J., & Grabelsky, D. A. 1981, *ApJ*, 248, 925

Mark, H., Price, R., Rodrigues, R., Seward, F. D., & Swift, C. D. 1969, *ApJ*, 155, L143

Markert, T. H. & Clark, G. W. 1975, *ApJ*, 196, L55

McGowan, K. E., Charles, P. A., O'Donoghue, D., & Smale, A. P. 2003, *MNRAS*, 345, 1039

McGowan, K. E., Charles, P. A., Blustin, A. J., Livio, M., O'Donoghue, D., & Heathcote, B. 2005, *MNRAS*, 364, 462

Middleditch, J., Marshall, F. E., Wang, Q. D., Gotthelf, E. V., & Zhang, W. 2006, *ApJ*, 652, 1531

Moon, D.-S. & Eikenberry, S. S. 2001, *ApJ*, 549, L225

Naik, S. & Paul, B. 2003, *A&A*, 401, 265

Nazé, Y., Antokhin, I. I., Rauw, G., Chu, Y.-H., Gosset, E., & Vreux, J.-M. 2004, *A&A*, 418, 841

Nazé, Y., Corcoran, M. F., Koenigsberger, G., & Moffat, A. F. J. 2007, *ApJ*, 658, L25

Negueruela, I. & Coe, M. J. 2002, *A&A*, 385, 517

Orio, M., Hartmann, W., Still, M., & Greiner, J. 2003, *ApJ*, 594, 435

Orio, M., Ebisawa, K., Heise, J., & Hartmann, J. 2004, *Rev. Mexicana AyA*, 20, 210

Page, M. J., Soria, R., Wu, K., Mason, K. O., Cordova, F. A., & Priedhorsky, W. C. 2003, *MNRAS*, 345, 639

Portegies Zwart, S. F., Pooley, D., & Lewin, W. H. G. 2002, *ApJ*, 574, 762

Rapley, C. G. & Tuohy, I. R. 1974, *ApJ*, 191, L113

Sasaki, M., Haberl, F., & Pietsch, W. 2000, *A&AS*, 143, 391

Schmidtke, P. C. & Cowley, A. P. 2006, *AJ*, 131, 600

Shtykovskiy, P. & Gilfanov, M. 2005, *A&A*, 431, 597

Smith, D. M., Dawson, D. M., & Swank, J. H. 2007, *ApJ*, 669, 1138

Townsley, L. K., Broos, P. S., Feigelson, E. D., Brandl, B. R., Chu, Y.-H., Garmire, G. P., & Pavlov, G. G. 2006a, *AJ*, 131, 2140

Townsley, L. K., Broos, P. S., Feigelson, E. D., Garmire, G. P., & Getman, K. V. 2006b, *AJ*, 131, 2164

Tsygankov, S. S. & Lutovinov, A. A. 2005, *Astron. Lett.*, 31, 380

Wang, Q., Hamilton, T., Helfand, D. J., & Wu, X. 1991, *ApJ*, 374, 475

Wu, K., Soria, R., Page, M. J., Sakelliou, I., Kahn, S. M., & de Vries, C. P. 2001, *A&A*, 365, L267

Yao, Y., Wang, Q. D., & Nan Zhang, S. 2005, *MNRAS*, 362, 229

The Magellanic System: Stars, Gas, and Galaxies
Proceedings IAU Symposium No. 256, 2008
Jacco Th. van Loon & Joana M. Oliveira, eds.

The OGLE-III catalog of variable stars: First results

Igor Soszyński[1]

[1] Warsaw University Observatory, Al. Ujazdowskie 4, 00-478 Warszawa, Poland
email: soszynsk@astrouw.edu.pl

Abstract. The third phase of the Optical Gravitational Lensing Experiment (OGLE-III) has been conducted since 2001 and regularly monitors the brightness of about 200 million stars. The OGLE-III fields cover both Magellanic Clouds and a large area in the Galactic bulge and disk. Here we describe the first parts of the OGLE-III Catalog of Variable Stars which is being prepared on the basis of these data. We present the principles of the catalog and methods used to select variable stars. We expect that the whole catalog will contain at least one million variable stars of all types. The catalog includes the list of variable sources along with their basic parameters, high precision multi-epoch I and V-band photometry and accurate astrometry. All objects are classified and cross-identified with previously published catalogs. We also carry out a preliminary statistical analysis of these huge samples of variable stars.

Keywords. catalogs, Cepheids, stars: variables: other, Magellanic Clouds

1. Introduction

The Optical Gravitational Lensing Experiment is a long-term sky survey being conducted since 1992. One of the main scientific goals of the project is the searching for gravitational microlensing events (Paczyński 1986), but a huge amount of high quality photometric data collected by the OGLE survey are ideal material for many other astrophysical purposes. One of the most important results produced by the OGLE project were large catalogs of variable stars in the Magellanic Clouds and in the Galactic bulge.

The first phase of the OGLE project (OGLE-I) was conducted between 1992 and 1995 on the 1-m Swope telescope at Las Campanas Observatory in Chile. About 2 million stars were regularly monitored. This part of the survey resulted in catalogs of variable stars toward the Galactic bulge (Udalski et al. 1994), globular clusters ω Cen (Kałużny et al. 1996) and 47 Tuc (Kałużny *et al.* 1998), dwarf galaxies Sculptor (Kałużny *et al.* 1995) and Sagittarius (Mateo *et al.* 1995). In total, several thousands variables were identified on the basis of the OGLE-I observations.

In January 1997 the OGLE survey entered its second phase (OGLE-II). The upgrade included a new 1.3-m Warsaw Telescope at Las Campanas Observatory, dedicated to the project. The telescope was equipped with a "first generation" 2048×2048 CCD camera working in drift-scan mode. Comparing to the previous stage the observing capabilities of the project were increased by a factor of 30. The Magellanic Clouds were added to the list of regularly observed fields.

The outcome of variable stars increased significantly in comparison with the OGLE-I project. Thousands of Cepheids, RR Lyr stars, eclipsing binaries and long-period variables were detected in the Magellanic Clouds and Galactic bulge. Table 1 list the most important catalogs of variable stars published on the basis of the OGLE-II data. These samples often contained the largest sets of particular variable stars detected so far in any environment.

Table 1. The OGLE-II catalogs of variable stars.

Type of variable stars	Environment	Number of stars	Papers
Classical Cepheids	LMC	1335 + 81	Udalski *et al.* (1999b) Soszyński *et al.* (2000)
	SMC	2049 + 95	Udalski *et al.* (1999c) Udalski *et al.* (1999a)
Type II Cepheids	Bulge, LMC	54, 14	Kubiak & Udalski (2003)
RR Lyrae	LMC	7612	Soszyński *et al.* (2003)
	SMC	571	Soszyński *et al.* (2002)
	Bulge	2700	Mizerski (2003)
Eclipsing Binaries	LMC	2580	Wyrzykowski *et al.* (2003)
	SMC	1350	Wyrzykowski *et al.* (2004)
Miras and SRV	LMC	3221	Soszyński *et al.* (2005)
δ Scuti	Bulge	193	Pigulski *et al.* (2006)
β Cephei, SPB	LMC, SMC	98 + 90	Kołaczkowski *et al.* (2006)
all	LMC, SMC	68 000	Żebruń *et al.* (2001)
all	Bulge	200 000	Woźniak *et al.* (2002)

In 2000 the OGLE-II photometry was reprocessed using a newly developed Difference Image Analysis (DIA, Alard & Lupton 1998; Woźniak 2000). The new method of the data reduction significantly increased the quality of the photometry and opened new possibilities for the detection of variable stars. The direct result of this analysis was the preparation of the huge catalogs of variable sources in the Magellanic Clouds (Żebruń *et al.* 2001) and Galactic bulge (Woźniak *et al.* 2002).

The next upgrade of the OGLE project came in 2001. The Warsaw Telescope was equipped with a new mosaic camera consisting of eight 2048×4096 chips which increased our sky coverage by an order of magnitude. The OGLE-III project announced many important discoveries including extra-solar planets detected using two methods: transits and microlensing events. Variable stars were also analyzed using OGLE-III data, however usually as a supplement of the OGLE-II photometry (the exception are planetary transits and microlensing events which obviously also are variable stars). In this paper we describe the new sub-project being conducted with the OGLE-III data — the OGLE-III Catalog of Variable Stars (OIII-CVS).

2. OGLE-III data

OGLE-III regularly monitors the brightness of about 200 million stars in 170 square degrees of the sky. To date more than 215 000 frames have been collected which occupy about 25 TB of disk space. On average, several hundred photometric measurements per star have been secured, most of them in the standard I band. About 10% of observations were obtained with the V filter. All these data will be used for the selection and analysis of variable stars in the OGLE-III fields. Taking into account the size of the OGLE-II catalogs we expect that the number of variables detected in the OGLE-III fields will reach one million objects.

To search for variable sources we use the OGLE-III photometric data finally reduced by Udalski *et al.* (2008a). Comparing to the previous, provisional reductions, the new

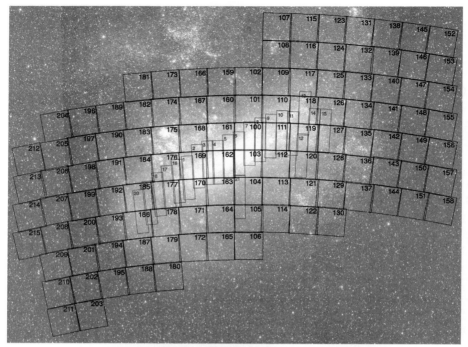

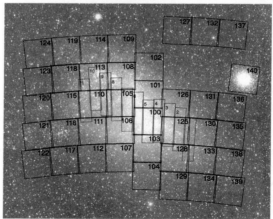

Figure 1. Contours of the OGLE-II (grey) and OGLE-III (black) fields in the LMC (upper panel) and in the SMC (lower panel) overplotted on the ASAS pictures.

data set is more accurate and well calibrated. The typical uncertainty of the photometric calibration is less than 0.02 mag. Moreover, even the gaps between the chips of the CCD mosaic are covered by our reductions, because, due to imperfections of the telescope pointing, the regions between the chips are also observed from time to time.

The OGLE-III fields cover about 40 square degrees of the densest regions in the LMC and about 14 square degrees in the SMC, including globular cluster 47 Tuc. Fig. 1 shows the contours of the OGLE-III fields plotted over pictures taken by the ASAS survey (Pojmański 1997). Very recently the OGLE-III photometric maps of the LMC were published by Udalski *et al.* (2008b). The maps contain precisely measured mean *I* and *V*

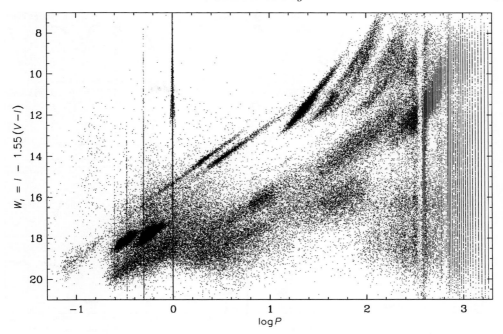

Figure 2. Period–luminosity diagram for the stars in the LMC with periodic signal-to-noise ratio larger than 7.

magnitudes of about 35 million stars. The photometric maps of the SMC will be released very soon.

3. The principles of the catalog

The principles of the OIII-CVS are as follows. We plan to detect all variables sources in the OGLE-III fields. The catalog will be divided into parts consisting of different types of variable stars in different environments (LMC, SMC and Galactic bulge). Each piece of the catalog will be successively published in the electronic form only in the OGLE Internet Archive. The catalog data set will include VI multi-epoch photometry, basic parameters of the stars (coordinates, periods, mean magnitudes, amplitudes, parameters of the Fourier light curves decompositions), and $60'' \times 60''$ finding charts.

All objects will be cross-identified with previously published samples of variable stars. Special attention will be paid to the completeness of the catalog. All previously known objects of the given type not present in our sample, will be carefully studied to find the reason of the absence. Finally, the catalog will not be a closed structure. For example, the photometry will be supplemented by new observations until the end of the OGLE-III project. Similarly, every new variable star of the given type found after the publication of the catalog will be added to the released list of objects.

All parts of the catalog will be accompanied by papers describing the methods of selection and analysis of the variable stars, showing objects of particular interest and preliminary statistical analysis.

4. Selection of variable stars

Selection and analysis of such a huge sample of variable stars obviously demands unprecedented efforts. Different types of variable stars will be searched using different

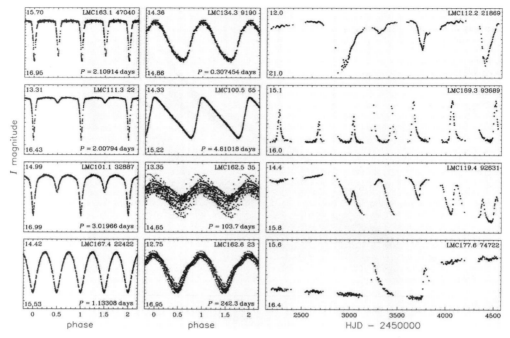

Figure 3. Exemplary light curves of variable stars in the LMC: eclipsing binary systems (left column), pulsating variables (middle column) and irregular variables (right column).

methods. To search for periodic variables, including very low amplitude and very faint stars, we conducted massive period search using supercomputers at the Interdisciplinary Centre for Mathematical and Computational Modelling (ICM). To date we derived periods for all 35 million stars in the LMC.

The results of these computations are shown in Fig. 2. We plotted here the extinction insensitive Wesenheit index $W_I = I - 1.55(V - I)$ versus logarithm of the period (in days) for the stars in the LMC with the periodic signal-to-noise ratios larger than 7. About 140 000 of 35 million stars in the LMC are presented on this raw diagram. Apart from the stars with artificial periods equal to about 1, 1/2, 1/3 or 1/4 days, one can easily recognize a series of certain period–luminosity (PL) relations for pulsating variables and binary systems. We draw the reader's attention to the sequences of classical Cepheids which are extended toward shorter periods and transform to High Amplitude δ Sct stars. Below this formation one can notice tens of thousands of RR Lyr stars. In the upper right corner of the diagram a large number of long-period variables follow a series of PL sequences. Finally, close binary systems — eclipsing and ellipsoidal variables — also delineate wide PL relations below Cepheids and long-period variables.

Fig. 3 shows the exemplary light curves of variable stars in the LMC. We present here periodic and non-periodic stars of various types.

5. First results

At present, the first part of the OIII-CVS has already been released (Soszyński *et al.* 2008b). It presents 3361 classical Cepheids in the LMC. The second part — type II Cepheids and anomalous Cepheids in the LMC — is in preparation.

Among large samples of variable stars very rare or even previously unknown types of objects can be found. For example studying classical Cepheids in the LMC we discovered two double-mode Cepheids of a new type — with the first and the third overtones simultaneously excited (Soszyński *et al.* 2008a). Besides, we identified three new triple-mode Cepheids, new classical Cepheids in eclipsing binary systems, including a system of two pulsators revealing eclipses, many Blazhko Cepheids and single-mode second-overtone Cepheids (Soszyński et al. 2008b).

6. Summary

The OGLE-III photometric database is a gold mine of information about stellar variability. Our ambitious plan to catalog all variable sources in the OGLE-III fields will be continued over the next several years. We expect that the final catalog will contain the largest number of variable stars detected so far by any large sky survey. First results show that among the large number of variable stars very uncommon or even new type of objects are present. Huge samples open new possibilities for statistical analysis of variable stars.

Acknowledgements

This work has been supported by Foundation for Polish Science through the Homing (Powroty) Program and by MNiSW grant NN203293533. The massive period searching was performed at the Interdisciplinary Centre for Mathematical and Computational Modelling (ICM).

References

Alard, C., & Lupton, R.H. 1998, *ApJ*, 503, 325
Kałużny, J., Kubiak, M., Szymański, M., *et al.* 1995, *A&AS*, 112, 407
Kałużny, J., Kubiak, M., Szymański, M., *et al.* 1996, *A&AS*, 120, 139
Kałużny, J., Kubiak, M., Szymański, M., *et al.* 1998, *A&AS*, 128, 19
Kołaczkowski, Z., Pigulski, A., Soszyński, I., *et al.* 2006, *MemSAI*, 77, 336
Kubiak, M., & Udalski, A. 2003, *AcA*, 53, 117
Mateo, M., Udalski, A., Szymański, M., *et al.* 1995, *AJ*, 109, 588
Mizerski, T. 2003, *AcA*, 53, 307
Paczyński, B. 1986, *ApJ*, 304, 1
Pigulski, A., Kołaczkowski, Z., Ramza, T., & Narwid, A. 2006, *MemSAI*, 77, 223
Pojmański, G. 1997, *AcA*, 47, 467
Soszyński, I., Udalski, A., Szymański, M., *et al.* 2000, *AcA*, 50, 451
Soszyński, I., Udalski, A., Szymański, M., *et al.* 2002, *AcA*, 52, 369
Soszyński, I., Udalski, A., Szymański, M., *et al.* 2003, *AcA*, 53, 93
Soszyński, I., Udalski, A., Kubiak, M., *et al.* 2005, *AcA*, 55, 331
Soszyński, I., Poleski, R., Udalski, A., *et al.* 2008a, *AcA*, 58, 153
Soszyński, I., Poleski, R., Udalski, A., *et al.* 2008b, *AcA*, 58, 163
Udalski, A., Kubiak, M., Szymański, M., *et al.* 1994, *AcA*, 44, 317
Udalski, A., Soszyński, I., Szymański, M., *et al.* 1999a, *AcA*, 49, 1
Udalski, A., Soszyński, I., Szymański, M., *et al.* 1999b, *AcA*, 49, 223
Udalski, A., Soszysński, I., Szymański, M., *et al.* 1999c, *AcA*, 49, 437
Udalski, A., Szymański, M.K., Soszyński, I., & Poleski, R. 2008a, *AcA*, 58, 69
Udalski, A., Soszyński, I., Szymański, M.K., *et al.* 2008b, *AcA*, 58, 89
Woźniak, P.R. 2000, *AcA*, 50, 421
Woźniak, P.R., Udalski, A., Szymański, M., *et al.* 2002, *AcA*, 52, 129
Wyrzykowski, Ł., Udalski, A., Kubiak, M., *et al.* 2003, *AcA*, 53, 1
Wyrzykowski, Ł., Udalski, A., Kubiak, M., *et al.* 2004, *AcA*, 54, 1
Żebruń, K., Soszyński, I., Woźniak, P.R., *et al.* 2001, *AcA*, 51, 317

The Magellanic System: Stars, Gas, and Galaxies
Proceedings IAU Symposium No. 256, 2008
Jacco Th. van Loon & Joana M. Oliveira, eds.

© 2009 International Astronomical Union
doi:10.1017/S1743921308028226

Significant new planetary nebula discoveries as powerful probes of the LMC

Warren A. Reid[1] and Quentin A. Parker[1,2]

[1] Dept of Physics, Macquarie University North Ryde, Sydney, NSW 2190, Australia
email: `warren@ics.mq.edu.au`

[2] Anglo-Australian Observatory, PO Box 296 Epping, NSW 1710, Australia
email: `qap@ics.mq.edu.au`

Abstract. Our discovery and analysis of 452 new planetary nebulae (PNe) in the Large Magellanic Cloud (LMC) has tripled the number of known LMC PNe, providing a powerful new resource for probing the kinematics of the LMC as well as contributing fresh insight into the PN luminosity function (PNLF) which we now extend to over 10 magnitudes in [O III] and Hα. These discoveries have resulted from a new, deep ($R \equiv 22$), high resolution Hα map of the central 25 deg^2 of the LMC, achieved by a process of multi-exposure median co-addition of a dozen 2-hour exposures. The resulting map is at least 1 magnitude deeper than the best wide-field narrow-band LMC images currently available and has proven a major resource for the discovery of emission objects of all kinds. As a result, the near complete sample of the PN population in the central 25 deg^2 of the LMC has permitted truly meaningful quantitative determinations of the PNLF, distribution, abundances and kinematics. We briefly describe the importance of these PN discoveries, the additional spectroscopic confirmation of >2,000 compact emission sources, flux calibration, the newly derived electron temperatures and electron densities.

Keywords. surveys, planetary nebulae: general, galaxies: individual (LMC), Magellanic Clouds

1. Introduction

Planetary nebulae (PNe) are a short-lived phase in the late evolution of low mass stars and are important astrophysical tools. They provide key data on the physics of stellar evolution, mass loss (Iben 1995), nucleosynthesis processes, abundance gradients and ISM chemical enrichment. They are powerful tracers of star-forming history (e.g., Maciel & Costa 2003). Accurate velocities and nebular parameters, such as excitation class and electron temperature, can be derived from their strong emission lines which are detectable at large distances. Most physical PNe properties, including ionized and total nebular masses and the brightness and evolutionary states of their central stars, depend on accurate distances (Ciardullo & Jacoby 1999). This is difficult in our own Galaxy due to inherent problems with variable extinction and lack of central star homogeneity (Terzian 1997). The well determined 50.6 kpc LMC distance (e.g., Keller & Wood 2006), modest 35 degree inclination angle and disk thickness (only $\sim$ 500 pc, van der Marel & Cioni 2001), mean that all LMC PNe are effectively co-located. Since dimming of their light by intervening gas and dust is low and uniform (e.g., Kaler & Jacoby 1990), we can better estimate absolute nebula luminosity and size.

2. An unprecedented catalogue of LMC PNe

Reid & Parker (2006b) (hereafter RPb) have constructed the most complete, least biased census of a PNe population ever compiled for a single galaxy based on discoveries from their deep *AAO/UKST* Hα multi-exposure stack of the LMC's central 25 deg^2

(Reid & Parker 2006a). This stack comprises a series of 12 repeated narrow-band 'A'-grade Hα and 6 matching broad-band 'SR' (Short Red) exposures of the central LMC field, taken over a three year period. From these exposures, deep, homogeneous, narrow-band Hα and matching broad-band 'SR' maps of the entire central 25 $\deg^2$ square of the LMC were constructed. Using these maps, over 2,000 emission objects in the central area of the LMC were identified. A major spectral confirmation program was undertaken in November and December 2004 mainly comprising 5 nights using 2dF on the Anglo-Australian Telescope but supported with 7 nights using the 1.9m at the South African Astronomical Observatory, 3 nights using the FLAMES multi-object spectrograph on the ESO Very Large Telescope, 7 nights using the 2.3m Australian National University telescope at the Siding Spring Observatory and 3 half nights using 6dF on the *UKST*.

Individual exposure times for the 18 2dF fields observed were 1200s using the 300B grating with a central wavelength of 5852 Å and wavelength range 3600–8000 Å at a dispersion of 4.30 Å/pixel. These low-resolution observations (9.0 Å FWHM) were used as the primary means of object classification. All fields were re-observed using the 1200R high resolution grating with a central wavelength of 6793 Å. These observations covered a range 6200–7300 Å with a dispersion of 1.10 Å/pixel and resolution of 2.2 Å FWHM which cleanly separated the [S II] 6716 and 6731 lines used for electron density determination. The high resolution spectra were also used for determination of accurate velocities (see RPb). In all we had 7,521 high and low resolution object spectra for LMC targets.

A combination of spectroscopy and image analysis confirmed 452 new PN candidates along with the 161 previously known PNe in the survey area. A large fraction of new LMC PNe are ⩾ 3× fainter than those previously known, effectively tripling numbers accrued from all surveys over the last 80 years. These additional objects have already led to significant advances in our understanding of the kinematical sub-structure of the central LMC including rotation, inclination and transverse velocity as well as the distribution of the old stellar population (RPb). They will assist us in refining the PNLF and physical characteristics such as temperatures, densities, nebulae masses and abundances (Reid 2008). It is our intention to publish a detailed analysis within the next 12 months.

3. Objects discovered

Table 1 provides a summary of our object classification following spectral analysis. It includes 2 previously known PNe now re-classified as H II regions, a further 4 demoted to possible PNe and 1 to likely (RPb). With the exception of previously known PNe, the numbers for other previously known objects represent those included in the catalogue and observed. They do not necessarily represent the full number for that object type that may exist within the central 25 $\deg^2$ of the LMC. Figure 1 shows an example of 6 new LMC PNe we have classified as 'true', due to their strong [O III]/Hβ ratios, lack of relatively strong continuum and clean separation from neighbouring stars. Careful re-examination of object images and spectra with the addition of IRAC false colour images from SAGE (*Spitzer*) and new high resolution radio mosaic images from Parks/ATCA have assisted us in the re-classification of several possible and likely PNe. We have re-classified RP 1495, RP 1716, RP 872, RP 1113, RP 641, RP 105, RP 1933 as compact H II regions. The high resolution *Spitzer* images have also revealed 2 objects in the position of RP 1534. We classify the larger one to the NW as a bright H II region with a central star. The PN is clearly separated (∼2 arcsec) to the SE. In the same way, SMP 48 also clearly

Table 1. Emission object classification results from spectral observations covering the central 25 deg^2, area of the main LMC bar.

Object	Previously Known	Newly Confirmed
PNe "True"	162(-2)	285
PNe "Likely"	1	53
PNe "Possible"	4	114
Emission-line stars	55	622
Late-type stars	10	247
H II regions	85	70
S/N too low for ID		32
Variable stars	61	28
Emission objects of unknown nature	12	25
SNR	9	18
Wolf-Rayet stars	14	8

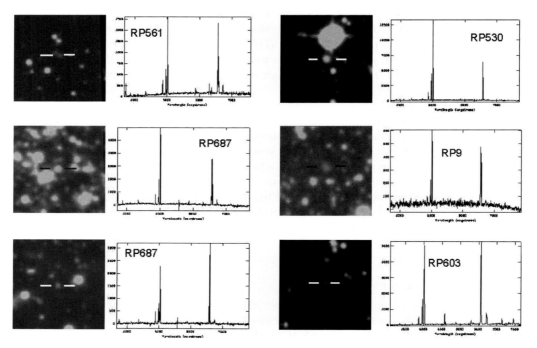

Figure 1. Six examples of PNe discovered in the RP survey of the LMC showing strong [O III]/Hβ ratios.

separates into 2 equal size objects. The object to the east is an H II region while the PN is to the west. This demonstrates the power of multiwavelength data in the identification process (e.g., Parker *et al.* 2006).

4. Flux calibration

Flux calibration of the new LMC PN spectra was a prerequisite for determining temperatures, densities, nebula masses and chemical abundances. It also facilitated conversion to magnitudes for luminosity studies. To achieve this, a flux calibration technique had to be found. This has previously proved quite difficult to do with fibre-based spectra.

As each of the 18 2dF field-plate exposures has its own relative line strengths and creates its own individual trend across the spectrum, experiments were undertaken in order to find the best method of flux calibration using the observed line intensities. The best results were obtained by individually calibrating each spectral line on each field plate to raw MCPN fluxes gained from *HST* exposures (Shaw, private communication). The known PNe included on each field plate were used as flux calibrators for each individual field. The MCPN fluxes for known PNe observed on each field plate exposure were graphed against the individual 2dF line intensities. In each case, a line of best fit was derived and the underlying linear equation extracted. This equation became the calibrator for each emission line on each individual 2dF field plate exposure. The equation was applied to each spectrum with a detectable line intensity in that field.

Fluxes for LMC PNe from other catalogues (Jacoby *et al.* 1990; Leisy *et al.* private communication; Meatheringham *et al.* 1988) were also included in order to build up the number of calibrators per field. Where only a de-reddened flux value was published, a relative reddening was applied in order to make a better correlation. These fluxes were used for double checking since each independent survey revealed many fluxes which agreed and equally many which varied considerably. A spread in published line fluxes for the same PN is common but helps us to establish the uncertainties. Since the MCPN catalogue contained the largest number of PNe and the raw fluxes were 'as measured' (not de-reddened) these were given preference where irregularities became obvious. The MCPN set also includes some ground-based fluxes from ELCAT available at http://stsdas.stsci.edu/elcat/ where the spectra are uncorrected for extinction. Following flux calibration, a number of nebula diagnostic values such as temperatures, densities and ionised masses were derived. Sections 5 and 6 give a brief introduction to some of our results.

5. Determination of nebula electron temperature

The measurement of electron temperatures (T_e) from ions in PNe is important not only as a diagnostic for understanding the photoionisation of the nebulae but is required for subsequent density and abundance determinations. In PNe, the most powerful mechanism for the loss of energy by free electrons is the excitation of forbidden lines. A T_e based on these lines is an independent parameter describing conditions in the nebula. T_e is not a strong function of the distance to the central ionising star (Stasińska 2002). It is, however, higher for high stellar temperatures ($T_\star$) and low metallicity values, Z. The [O III] 4363, 5007 Å ratio is sensitive to metallicity and effective $T_\star$. At higher metallicity, the nebulae are cooled efficiently through the [O III] 5007 line, producing cooler nebulae (Osterbrock & Ferland 2006). For O^{++}, the transition from level $^1S_0$ to the $^1D_2$ level emits the forbidden line [O III] 4363. The transition from $^1D_2$ downward to the levels $^3P_{1,2}$ creates the so called N_1 and N_2 'nebulium' ([O III] 4959 Å and 5007 Å) lines. An estimate of electron temperature can then be made by comparing the number of ions in the $^1S_0$ state with the number in the $^1D_2$ (i.e. the [O III] 4363 increase in strength over [O III] ($N_1 + N_2$) (Osterbrock 1989). However, in many PNe, this is not straightforward to measure as the 4363 Å line can range between a hundredth and a thousandth the strength of the 5007 Å line. The method applied was the temperature equation provided by Osterbrock & Ferland (2006), p. 109.

Although the [N II] 5755 Å line can also be very faint, it was clearly measurable for 159 PNe. The [N II] (6548 Å + 6583 Å)/5755 Å calculation provided us with an alternative temperature estimate (preferable to default values) for density determinations in PNe

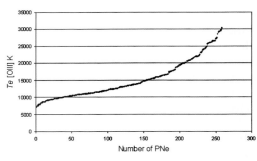
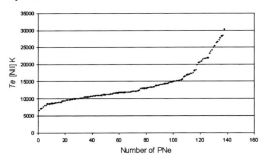

Figure 2. The derived electron temperatures for all 256 LMC PNe with measurable lines of [O III] 4363 Å, 4959 Å and 5007 Å. Temperatures range from 7,728 K to 30,107 K with 80% of PNe under 20,000 K.

Figure 3. The derived electron temperatures for all 137 LMC PNe with measurable lines of [N II] 5755 Å, 6548 Å and 6583 Å. Temperatures range from 6,646 K to 30,099 K with 85% of PNe under 20,000 K.

where [O III] 4363 was unavailable to provide an [O III] temperature. The T_e using [N II] was primarily calculated for use in the low zone abundance determinations following in the form of Osterbrock & Ferland (2006), p. 193.

Figures 2 and 3 show the resulting T_e for [O III] and [N II] respectively. High T_e between 20,000 and 30,000 K are due to both the high $T_\star$ of the PN and high electron densities in PNe, compared to H II regions as well as low Z in the LMC. Some of the large dispersion is also due to physical differences in the nebulae and may also be related to nebula evolution (e.g., Zhang et al. 2004). 108 PNe or 43% of the [O III] T_e sample are above 15,000 K (generally the highest T_e for H II regions (Osterbrock & Ferland 2006)). This compares to only 37 PNe or 28% yielding high [N II] T_e. The lower percentage of high T_e in [N II] may be due to this line arising in the outer, cooler zones of the nebulae.

6. Determination of Nebula Electron Density

Nebula electron density (n_e) is one of the key parameters needed to confidently derive chemical abundances of nebulae and to calculate the total ionised mass used in certain distance/radius relation formulae. The [S II] doublet method, using the singly ionized sulphur lines at $\lambda 6731$ and $\lambda 6717$, as given by Osterbrock & Ferland (2006), was employed to calculate n_e. This doublet is emitted at the transition from level $^2D_{3/2,5/2}$ to $^4S_{3/2}$. The dependence of n_e and T_e on the ratio of $\lambda 6717$ / $\lambda 6731$ is given by:

$$\frac{I\lambda 6717}{I\lambda 6731} = 1.5 \frac{1 + 0.35x}{1 + 0.96x} , \quad \text{where} \quad x = 10^{-2} \frac{n_e}{T_e^{1/2}}. \tag{6.1}$$

The [S II] lines were measured from the PN spectra obtained with the 2dF 1200R high resolution grating. The 1200R measurements were preferred for density estimates because the lines are always cleanly separated and no de-blending was required. Also, the [S II] lines are close in wavelength so the issue of flux calibration effects does not arise. The resulting histogram is shown in Figure 4. For comparison, the measured line intensities were also given as input to the IRAF STDAS *temden* task (Shaw & Dufour 1995), which calculated the densities using the electron temperatures previously derived. The resulting densities were effectively equivalent to those derived using equation 6.1. Further analysis

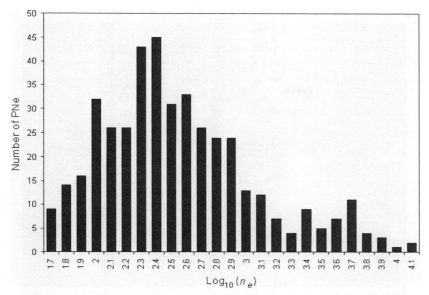

Figure 4. Histogram of electron densities for 487 LMC PNe. This is the largest and most comprehensive set of derived densities ever assembled for a PN sample. The range in the histogram is from 1.7 to 4.1 log density with a peak at 2.4 embraced by a gradual rise and fall from 1.7 to 3.3 log density. Of special interest is the second rise in the number of PNe between log 3.4 and 4.1. These PNe are amongst the brightest in Hα.

of these PNe, including nebula masses, excitation classes and a new luminosity function will be available shortly.

References

Ciardullo R. & Jacoby, G. H. 1999, *ApJ*, 515, 191

Iben I . Jr. 1995, *Phys.Reports*, 250, 2

Jacoby G. H., Walker A. R., & Ciardullo R. 1990, *ApJ*, 365, 471

Kaler J. B. & Jacoby G. H. 1990, *ApJ*, 362, 491

Keller S. C. & Wood, P. R. 2006, *ApJ*, 642, 834

Maciel W. J. & Costa R. D. 2003, in S. Kwok, M. Dopita, & R. Sutherland (eds.), *Planetary Nebulae: Their Evolution and Role in the Universe*, IAU Conf.Proc. 209 (ASP), p. 551

Meatheringham S. J., Dopita M. A., & Morgan D. H. 1988, *ApJ*, 329, 166

Osterbrock, D. E. 1989, *Astrophysics of Gaseous Nebulae and Active Galactic Nuclei* (Mill Valley: University Science Books), p. 118

Osterbrock, D. E. & Ferland, G. J. 2006, *Astrophysics of Gaseous Nebulae and Active Galactic Nuclei* (Sausalito, CA: University Science Books)

Parker Q. A., Acker A., Frew D. J., & Reid, W. A. 2006, in M.J. Barlow & R.H. Méndez (eds.), *Planetary Nebulae in our Galaxy and Beyond*, IAU Conf.Proc. 234 (Cambridge: CUP), p. 1

Reid, W. A. & Parker Q. A. 2006a, *MNRAS*, 365, 401

Reid, W. A. & Parker Q. A. 2006b, *MNRAS*, 373, 521

Reid W. A. 2008, *PhD Thesis*, Macquarie University, Sydney

Shaw, R. A. & Dufour, R. J. 1995, *PASP*, 107, 896

Stasińska G. 2002, *Rev. Mexicana AyA* 12, 62

Terzian, Y. 1997, in H. J. Habing & H. J. G. L. M. Lamers (eds.), *Planetary Nebulae*, IAU Conf.Proc. 180 (Dordrecht: Kluwer), p. 29

van der Marel, R. & Cioni, M. 2001, *AJ*, 122, 1807

Zhang Y., Liu X.-W., Wesson R., Storey P. J., Liu Y., & Danziger I. J. 2004, *MNRAS*, 351, 935

Nye Evans trying to initiate a Mexican wave, of which the others are blissfullly unaware.

Enjoying a relaxed Summer's evening outside Keele Hall.

Session II

The structure and dynamics of the
Magellanic System

The Magellanic System: Stars, Gas, and Galaxies
Proceedings IAU Symposium No. 256, 2008
Jacco Th. van Loon & Joana M. Oliveira, eds.

The spatial evolution of stellar structures in the LMC/SMC

Nate Bastian[1], Mark Gieles[2], Barbara Ercolano[1] and Robert Gutermuth[3]

[1] Institute of Astronomy, University of Cambridge, Madingley Road, Cambridge, CB3 0HA, UK
email: bastian@ast.cam.ac.uk; be@ast.cam.ac.uk

[2] European Southern Observatory, Casilla 19001, Santiago 19, Chile
email: mgieles@eso.org

[3] Harvard-Smithsonian Center for Astrophysics, 60 Garden Street, Cambridge, MA 02138, USA
email: rgutermuth@cfa.harvard.edu

Abstract. We present an analysis of the spatial distribution of various stellar populations within the Large and Small Magellanic Clouds. We use optically selected stellar samples with mean ages between ~ 9 and ~ 1000 Myr, and existing stellar cluster catalogues to investigate how stellar structures form and evolve within the LMC/SMC. We use two statistical techniques to study the evolution of structure within these galaxies, the Q-parameter and the two-point correlation function (TPCF). In both galaxies we find the stars are born with a high degree of substructure (i.e. are highly fractal) and that the stellar distribution approaches that of the "background" population on timescales similar to the crossing times of the galaxy (~ 80 Myr & ~ 150 Myr for the SMC/LMC respectively). By comparing our observations to simple models of structural evolution we find that "popping star clusters" do not significantly influence structural evolution in these galaxies. Instead we argue that general galactic dynamics are the main drivers, and that substructure will be erased in approximately the crossing time, regardless of spatial scale, from small clusters to whole galaxies. This can explain why many young Galactic clusters have high degrees of substructure, while others are smooth and centrally concentrated. We conclude with a general discussion on cluster "infant mortality", in an attempt to clarify the time/spatial scales involved.

Keywords. stars: kinematics, Magellanic Clouds, galaxies: structure

1. Introduction

Most, if not all, stars are thought to be born in a *clustered* or fractal distribution, which is usually interpreted as being due to the imprint of the gas hierarchy from which stars form (e.g., Elmegreen & Efremov 1996; Elmegreen *et al.* 2006; Bastian *et al.* 2007). Older stellar distributions, however, appear to be much more smoothly distributed, begging the questions; 1) what is the main driver of this evolution? and 2) what is the timescale for the natal structure to be erased?

It has been noted that many nearby star forming clusters also have hierarchical structure seemingly dictated by the structure of the dense gas of natal molecular clouds (e.g., Lada & Lada 2003). Using statistical techniques, Gutermuth *et al.* (2005) studied three clusters of varying degrees of embeddedness and demonstrated that the least embedded cluster was also the least dense and the least substructured of the three. That result hinted at the idea that the youngest clusters are substructured, but that dynamical interactions and ejection of the structured gas contributes to the evolution and eventual erasure of that substructure in approximately the cluster formation timescale of a few

Myr (Palla & Stahler 2000). By having an accurate model of the spatial evolution of stellar structures we can approach a number of fundamental questions about star-formation, including what is the percentage of stars born in "clusters" and whether this depends on environmental conditions. Using automated algorithms on infrared Spitzer surveys of star-forming sites within the Galaxy (e.g., Allen *et al.* 2007), such constraints are now becoming possible.

In this contribution we present results of two recent studies on the evolution of structure in the LMC (Bastian *et al.* 2009) and SMC (Gieles *et al.* 2008).

2. Datasets and techniques

Our main dataset is the Magellanic Cloud Photometric survey (Zaritsky *et al.* 2004), consisting of several million stars with UBVI photometry in the LMC/SMC. By using colour and magnitude cuts in the colour-magnitude diagrams (CMDs) of the LMC/SMC stars, we can select populations of different mean ages. An example CMD (where the grey scale is the logarithm of the density of stars at that colour/magnitude) for the LMC is shown in top panel of Fig. 1, where the bottom panels show the spatial positions of stars in a sample of age boxes. The mean ages of these boxes are derived using Monte Carlo sampling, assuming a Salpeter IMF, of theoretical stellar isochrones (Girardi *et al.* 2002). Additionally, for comparison, we use the ages and positions of stellar clusters from the Hunter *et al.* (2003) catalogue.

In order to study the evolution of structure, we employ two statistical methods. The first is the Q-parameter (Cartwright & Whitworth 2004) which is based on minimum spanning trees (MST)†, while the second is the two-point correlation function.

The Q-parameter uses the normalised mean MST branch length, $\overline{m}$, and the normalised distance between all sources within a region, $\overline{s}$. The ratio between these to quantities, $\overline{m}/\overline{s} = Q$, is able to distinguish between a power-law (centrally concentrated) profile and a profile with sub-structure (i.e. a fractal distribution). Additionally, this parameter can quantify the index of the power-law or the degree of sub-substructure, which, if one assumes is due to a fractal nature, its fractal dimension can also be estimated. Assuming a three dimensional structure, if Q is less than 0.79 then the region is fractal, larger than 0.79 refers to a power-law structure, and a value of 0.79 implies a random distribution (a 3D fractal of dimension 3 is a random distribution). However, if the distribution is two dimensional, as is approximately true if one is looking at a disk-like galaxy face on (like the LMC), then a random distribution has a Q value of 0.72.

The two-point correlation function (TPCF) determines the distance between all possible pairs of stars, shown as a histogram, which is then normalized to that of a reference distribution, i.e. N_{links} / $N_{\mathrm{reference}}$ where $N_{\mathrm{reference}}$ is taken as a smooth, centrally concentrated power-law distribution as observed in old stellar populations for each galaxy. This is similar to that done by Gomez *et al.* (1993) who used the TPCF to study the distribution of pre-main sequence stars in Taurus. We then measure the slope and zero-point of the resulting distributions.

The results from these two methods are given in Fig. 2 for the LMC and Fig. 3 for the SMC. For the LMC we see that both the Q-parameter method and both measurements from the TPCF method, show that substructure is erased (i.e. where the distributions become flat) on a timescale of ~ 175 Myr (see Bastian *et al.* 2009 for details). For

---

† An MST is formed by connecting all points (spatial positions in this case) in order to form a unified network, such that the sum of all of the connections, known as "edges" or "branches", is minimized, and no closed loops are formed.

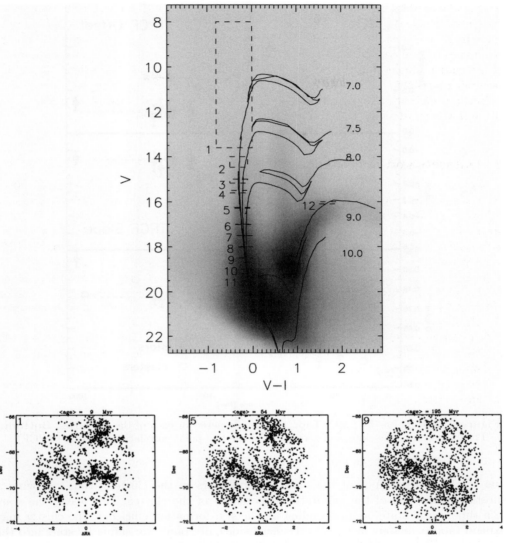

Figure 1. Top: A V–I vs. V CMD of stars in the LMC. The selected boxes are shown, along with theoretical stellar evolutionary isochrones. **Bottom:** The spatial distribution of stars in boxes 1, 5, & 9 in the LMC. The mean age of the stellar distributions is given at the top of each panel.

the SMC, substructure appears to be removed in ~ 80 Myr (see Gieles *et al.* 2008 for details).

3. Implications

Two scenarios have been put forward to explain the rapid evolution of substructure in galaxies. The first scenario, "popping star clusters" (Kroupa 2002) is based on the idea of infant mortality, or early dissolution of star clusters. When a cluster forms, a large fraction of the gas of the natal cloud or core remains after star-formation has terminated (i.e. there is a non-100% star-formation efficiency). This gas may be removed explosively

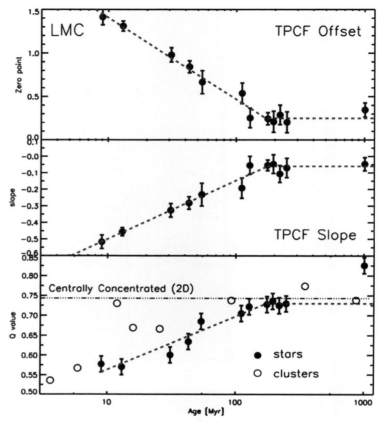

Figure 2. Results for the LMC. **Top:** The Q-parameter for each of the age boxes. **Bottom:** The evolution of the slope and zero-point of the two-point correlation function (TPCF).

due to the energy input of the most massive stars in the clusters, thus removing a substantial fraction of the gravitational potential of the cluster. This leaves the stars in a super-virial state, and depending on the star-formation efficiency, the cluster may expand (approximately at its velocity dispersion), or "pop", spreading its stars into the field. The vast majority of clusters in the LMC/SMC are low mass systems, and hence are expected to have small velocity dispersions, on the order of < 1 km/s. With these small velocities, it would take ~ 1 Gyr for a star in either galaxy to move one galactic radius. Additionally, in this scenario we would expect to see differences in the evolution of the stars vs. that of the clusters, as the clusters are not affected by this, only by the general dynamics of the galaxy. However, for both the LMC/SMC we see that the Q-parameter of the clusters and stars have a very similar evolution (see Bastian *et al.* 2009 for detailed simulations of this effect).

Hence we are left to conclude that general galactic dynamics is the main driver of the removal of substructure within galaxies. This does not imply that infant mortality does not exist within these galaxies, only that it does not have a significant impact on the evolution of substructure within the LMC/SMC. If general dynamics are the main driver, then we expect that it should be independent of spatial scale, i.e. always removed on approximately the crossing time. This can be readily seen in many young clusters in the Galaxy as well as in the LMC/SMC, where the central regions of clusters are

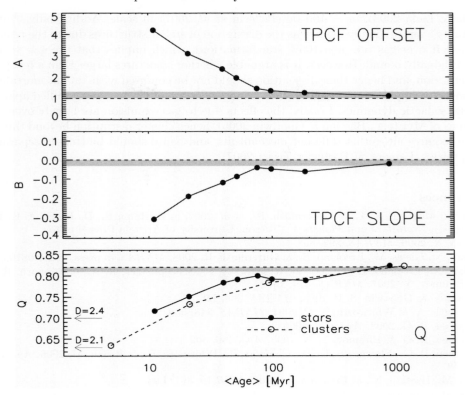

Figure 3. Results for the SMC. **Top and Middle:** The evolution of the slope and zero-point of the two-point correlation function (TPCF). **Bottom:** The Q-parameter for each of the age boxes.

smooth and show little or no substructure (i.e. where the crossing time is short) whereas in the outer regions (where the crossing time is significantly longer) substructure remains for longer periods. With this hypothesis in hand, it is relatively easy to calculate where substructure should be present on any scale, if the age, size and velocity dispersion (or crossing time) is known.

3.1. *Clustered star formation vs. star formation in clusters*

In Bastian *et al.* (2007) we showed that most, if not all, OB stars in M33 are formed in a clustered fashion, i.e. part of a hierarchical or fractal distribution down to the resolution limit of the sample. However, this is not to say that all stars form in dense (e.g. $R_{\mathrm{eff}} \sim 3$ pc) clusters, in fact no characteristic size of the star-forming regions was found. Only a small fraction of star-formation takes place in clusters which will survive long enough (i.e. > 3 Myr) and be dense enough, to be selected in optical samples (Lada & Lada 2003; Gieles & Bastian 2008; Bastian 2008), on the order of a few percent. This follows naturally from a fractal gas distribution and a threshold density for star formation along with a threshold efficiency for bound cluster formation (Elmegreen 2008).

Since young clusters do not necessarily look like older, evolved, and relaxed clusters (i.e. young embedded clusters are often extremely hierarchical) it is difficult to define which stars belong to a young "cluster", as the "clusters" themselves are somewhat arbitrarily defined. Hence the term "infant mortality" itself is fairly ill-defined, as one must pre-select the physical scale of interest, e.g., structures which are ~ 1 pc

(Lada & Lada 2003), or ~ 100 pc (Pellerin *et al.* 2007) in scale. Additionally, "infant mortality" was originally defined as the disruption of stellar structures due to the removal of the left over gas (i.e. non-100% star-formation), which implies that the gas + stars were originally bound. However, it is arguable whether structures larger than a few pc fit this criterion, and hence their dissolution should not be confused with "infant mortality", but instead simply represents an *initially* unbound system which is being pulled apart by general galactic dynamics. Finally, the effects due to gas expulsion are largely over by a few tens of Myr (e.g., Bastian & Goodwin 2006), hence cluster disruption beyond this age range requires altogether different mechanisms, and hence should be treated separately (e.g., Gieles, Lamers, & Portegies Zwart 2007).

References

Allen, L., Megeath, S. T., & Gutermuth, R., *et al.* 2007, in B. Reipurth, D. Jewitt, & K. Keil (eds.), *Protostars and Planets V* (Tucson: University of Arizona Press), p 361

Bastian, N. 2009, *MNRAS*, 392, 868

Bastian, N., Gieles, M., Ercolano, B., & Gutermuth, R. 2008, *MNRAS*, in press, arXiv:0810.3190

Bastian, N., Ercolano, B., Gieles, M., Rosolowsky, E., Scheepmaker, R. A., Gutermuth, R., & Efremov, Y. 2007, *MNRAS*, 379, 1302

Bastian, N. & Goodwin, S. P. 2006, *MNRAS*, 369, L9

Cartwright, A. & Whitworth, A. P. 2004, *MNRAS*, 348, 589

Elmegreen, B. G. 2008, *ApJ*, 672, 1006

Elmegreen, B. G. & Efremov, Y. N. 1996, *ApJ*, 466, 802

Elmegreen, B. G., Elmegreen, D. M., Chandar, R., Whitmore, B., & Regan, M. 2006, *ApJ*, 644, 879

Gieles, M., Bastian, N., & Ercolano, E. 2008, *MNRAS*, 391, L93

Gieles, M. & Bastian, N. 2008, *A&A*, 482, 165

Gieles, M., Lamers, H. J. G. L. M., & Portegies Zwart, S. F. 2007, *ApJ*, 668, 268

Girardi, L., Bertelli, G., Bressan, A., Chiosi, C., Groenewegen, M. A. T., Marigo, P., Salasnich, B., & Weiss, A. 2002, *A&A*, 391, 195

Gomez, M., Hartmann, L., Kenyon, S. J., & Hewett, R. 1993, *AJ*, 105, 1927

Gutermuth, R. A., Megeath, S. T., Pipher, J. L., Williams, J. P., Allen, L. E., Myers, P. C., & Raines, S. N. 2005, *ApJ*, 632, 397

Hunter, D. A., Elmegreen, B. G., Dupuy, T. J., & Mortonson, M. 2003, *AJ*, 126, 1836

Kroupa, P. 2002, *MNRAS*, 330, 707

Lada, C. J. & Lada, E. A. 2003, *ARAA*, 41, 57

Palla, F. & Stahler, S. W. 2000, *ApJ*, 540, 255

Pellerin, A., Meyer, M., Harris, J., & Calzetti, D. 2007, *ApJ*, 658, L87

Zaritsky, D., Harris, J., Thompson, I. B., & Grebel, E. K. 2004, *AJ*, 128, 1606

The Magellanic System: Stars, Gas, and Galaxies
Proceedings IAU Symposium No. 256, 2008
Jacco Th. van Loon & Joana M. Oliveira, eds.

© 2009 International Astronomical Union
doi:10.1017/S1743921308028251

Discovery of an extended, halo-like stellar population around the Large Magellanic Cloud

Steven R. Majewski[1], David L. Nidever[1], Ricardo R. Muñoz[1], Richard J. Patterson[1], William E. Kunkel[2] and Jeffrey L. Carlin[1]

[1] Department of Astronomy, University of Virginia, Charlottesville, VA, 22904-4325, USA
email: `srm4n, dnidever, rrm8f, rjp0i, jc4qn@virginia.edu`

[2] Las Campanas Observatory, Carnegie Institution of Washington, Casilla 601, La Serena, Chile
email: `kunkel@lco.cl`

Abstract. We describe an ongoing, large-scale, photometric and spectroscopic survey of the Large Magellanic Cloud (LMC) periphery. This survey uses Washington $M, T_2 + DDO51$ photometry to identify distant LMC red giant branch (RGB) star candidates; multi-object spectroscopy is used to confirm the stellar surface gravities of these RGB stars and their association with the LMC (e.g., through radial velocities). The survey now encompasses hundreds of fields ranging from the LMC center with full azimuthal coverage around the LMC and out to 23° from the LMC center. We have confirmed the existence of RGB stars with (the unusual) Magellanic velocities out to the radial limit of this survey coverage. From data in a subsample of these fields, we show that this extended population of stars makes up a diffuse structure enveloping the LMC with a two-dimensional distribution resembling a classical halo with a shallow de Vaucouleurs profile and a broad metallicity spread around a typical mean value of [Fe/H] ~ -1.0.

Keywords. surveys, galaxies: evolution, galaxies: halos, galaxies: individual (LMC), Magellanic Clouds, galaxies: structure

1. Introduction and motivation

Recent photometric and spectroscopic surveys have identified extended, metal-poor, Milky Way-like stellar halos around two other Local Group galaxies: M 31 (Ostheimer 2002; Guthathakurta *et al.* 2005; Chapman *et al.* 2006) and M 33 (McConnachie *et al.* 2006; Sarajedini *et al.* 2006). That M 33 is a "dwarf spiral" begs the question of how late in morphological type and low in mass can galaxies have stellar halos. Might Magellanic-sized galaxies also have stellar halos? Certainly cold dark matter (CDM) simulations predict that hierarchical structure formation occurs on *all scales* (e.g., Diemand *et al.* 2007). If the Magellanic Clouds, which probably formed outside the Milky Way (MW) and are only now making their first pass near the Sun (Kallivayalil *et al.* 2006; Besla *et al.* 2007; D'Onghia 2008; see also contributions by these authors in this proceedings), were found to have their own stellar halos, this would provide an opportunity to explore in exquisite detail (nearly as easily as for the MW) two more "independent" galaxy halos and evaluate their structure in the context of prevailing (i.e. CDM) galaxy formation models. We report here the discovery of a large, diffuse structure around the Large Magellanic Cloud (LMC) that is likely that system's stellar halo.

The dominant stellar population of the LMC is its obvious, elongated, exponential stellar disk that extends to $R \sim 9°$ with a radial scale length of $\sim 1.6°$ (van der Marel 2001; Harris 2007) and which has a typical metallicity spanning $-1 \lesssim$ [Fe/H] $\lesssim -0.4$ (Harris & Zaritsky 2004). And, while a population of old, metal-poor RR Lyrae stars with a large velocity dispersion ($\sigma_{\rm v} = 53$ km s^{-1}) has also been found in the LMC, this population apparently follows the exponential density profile of the LMC disk (Minniti *et al.* 2003; Borissova *et al.* 2004, 2006; Alves 2004).

The first hints that there may be a diffuse, dynamically warm and *extended* population around the Magellanic Clouds (MCs) came with the discovery of K giant stars having the approximate distance and mean velocity of the LMC (and SMC) in a probe of several dozen pencil-beam fields encircling the Clouds at distances of $20 - 25°$ from a point midway between the Clouds, as described in Majewski *et al.* (1999) and Majewski (2004). More recently, a significant number of such stars were seen as a colder ($\sigma_{\rm v} = 9.8$ km s^{-1}) moving group serendipitously found in the foreground of fields centered on the Carina dwarf spheroidal galaxy at a remarkably large $R \sim 22°$ from the LMC center (Muñoz *et al.* 2006). The properties of this group of stars — including a mean heliocentric velocity of 332 km s^{-1} and a color-magnitude distribution exactly matching the red clump of the LMC — as well as our ability to track the existence of similar K giants all of the way from Carina back to the LMC (Muñoz *et al.* 2006) prove the existence of MC stars out to these extreme angular separations. However, because of the limited azimuthal distribution of these discoveries about the LMC, it has not been possible to ascertain definitively whether these stars are a bound LMC halo, tidal debris from the LMC (e.g., Weinberg 2000) or SMC (e.g., Kunkel *et al.* 1997), or some other MC-associated population.

2. Systematic survey of the Magellanic periphery

The discovery of these widely separated stars apparently associated with the MCs and the uncertainty of their nature motivated us to undertake a larger and more systematic photometric and spectroscopic survey of the LMC periphery. (This larger survey is part of D. Nidever's Ph.D. thesis.) As in our previous probes around the LMC and in the direction of Carina, we used the *CTIO*-4 m+MOSAIC and *Swope*-1 m CCD cameras to obtain photometry of numerous outlying fields around the LMC in the Washington M and T_2 filters as well as the *DDO*51 filter — a combination of photometric bands that can be used photometrically to discriminate distant giant stars from nearby dwarf stars of similar apparent brightness and color (Majewski *et al.* 2000); the new photometric coverage extends to $R \sim 23°$ and has full azimuthal coverage around the LMC (Fig. 1). Stars are selected to be LMC giant star candidates not only by having color-magnitude combinations appropriate to the LMC red giant branch and red clump, but also ($M - T_2$, $M - DDO$51) combinations appropriate for giant stars (see Majewski *et al.* 2000).

Candidate LMC giant stars identified in this way are then followed up with spectroscopic observations to derive radial velocities and metallicities using a variety of southern instruments. In this contribution, we discuss primarily our first spectroscopic observations from the *CTIO*-4 m+HYDRA multi-object spectrometer in 27 fields (Fig.1).

3. The LMC density and metallicity profiles

Using our photometric and spectroscopic dataset for the 27 fields having Hydra data shown in Fig. 1, plus the MC giants found in the Carina survey of Muñoz *et al.* (2006), we have derived a density profile of the LMC over a radial range of $7-23°$ from the LMC center (Fig. 2a); these particular fields represent an azimuthal coverage of nearly 180°. Stars used for this radial profile not only pass the photometric criteria for LMC giant stars mentioned above, but also are limited to stars with radial velocities appropriate to the LMC (following a modified van der Marel (2002) LMC velocity model). Because the spectroscopic data used to check velocity membership do not completely sample the photometrically-selected LMC giant star candidates (e.g., because of limits to the Hydra

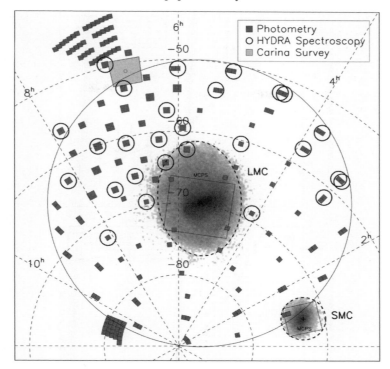

Figure 1. Map of our new survey of the LMC periphery in celestial coordinates. The background greyscale image shows the density of LMC and SMC red giant branch stars selected from 2MASS. Photometrically targeted fields from the *Blanco* telescope+MOSAIC and from the *Swope*-1 m imager are represented by dark shaded boxes lying mostly beyond the obvious 2MASS distribution. The circled boxes indicate fields with *Blanco*+HYDRA multi-object spectroscopy (27 fields). The large, lightly shaded box indicates the area covered by the Carina survey of Muñoz *et al.* (2006), where the 332 km s^{-1} group was found. The large circle has a radius of 20° from the LMC center. The extent of the Magellanic Clouds Photometric Survey (MCPS; Zaritsky *et al.* 2002, 2004) is indicated by the large boxes inside the LMC and SMC.

field size and constraints on fiber placement), we correct the giant star densities in each field for both the sampling and success fractions in each field; consequently, the derived densities shown in Fig. 2a are mostly, but not completely spectroscopically derived.

The observed density profile in Fig. 2a is inconsistent with a single density law over the radial range explored. More specifically, the inner density profile is matched by the known exponential LMC disk profile (e.g., radial scale length of 1.6°; van der Marel 2001) out to $R \sim 9°$ (the large scatter at small radii is due to the elongation and inclination of the LMC disk). However, beyond this radius, the profile flattens and begins to resemble the halos seen in other galaxies. This new, extended component of the LMC is well-fitted by either a de Vaucouleurs profile with a core radius of 2.4° or an exponential with a radial scale length of 4.1°. In the fields beyond $R = 9°$, the spectroscopically-derived radial profile (i.e. Fig. 2a) shows no obvious global asymmetry with position angle — consistent with a nominal halo but generally not for tidal debris; however, this profile is derived from the limited azimuthal coverage of the Hydra-investigated fields (see Fig. 1).

To test this azimuthal symmetry further, in Fig. 2b we plot densities for all photometrically-selected LMC giant star candidates in fields with *CTIO*-4 m+MOSAIC photometry to $R \sim 13°$ regardless of spectroscopic coverage (the densities beyond $R \sim 13°$ are so low that spectroscopic data are required to calculate them reliably). A nominal background

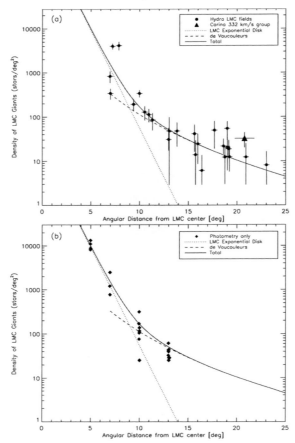

Figure 2. (a): Spectroscopically-determined density profile of LMC giants from the Hydra fields in Fig. 1. Densities in each field are incompleteness-corrected by applying RV-membership fractions in the spectroscopic sample to remaining photometric candidates without spectroscopy. Inner fields match the nominal LMC exponential disk profile with $1.6°$ scale length previously detected to $R \sim 9°$ (dotted line). The newly found, extended LMC population beyond $R \sim 9°$ can be fitted by a de Vaucouleurs profile with $2.4°$ core radius or an exponential of scale length $4.1°$ (dashed line). The initial discovery of this population in the foreground of the Carina dSph is indicated by a triangle. **(b):** LMC density profile from photometrically-selected, "LMC-like" giant stars for all survey fields with $CTIO$-$4\,\mathrm{m}$+MOSAIC data to $R \sim 13°$ with an appropriate background subtraction (see text). The $R \sim 13°$ densities clump very tightly around the de Vaucouleurs profile for all azimuthal angles (showing the radial symmetry of the density law) and well above extrapolation of the LMC disk exponential. The lines are the same as in (a).

density, calculated by using the LMC giant star selection in the $(M - T_2, M - DDO51)$ diagram but applied to other magnitudes in the color-magnitude diagram (a method we have successfully used in other studies), has been subtracted for each field. The photometrically-derived density profile (Fig. 2b) has nearly full azimuthal coverage and shows not only that the LMC giant candidate densities at $R \sim 13°$ are well above the density level predicted by extrapolation of the exponential disk, but fall along the combined two-component profile derived in Fig. 2a with very little scatter. The narrow dispersion of densities in Fig. 2b and lack of any position angle-dependencies in the spectroscopic density profile in Fig. 2a indicate that the newly found extended component of the LMC is azimuthally symmetric — as expected for a classical stellar halo.

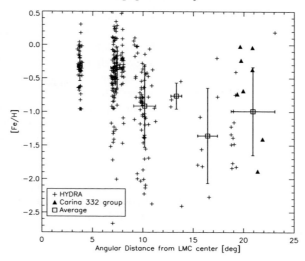

Figure 3. Spectroscopic metallicity profile of our LMC stars. Triangles represent the 332 km s^{-1} moving group found in the Carina survey. Open squares show averages in radial bins, while bars around these squares represent the dispersions in quantities in each bin.

Our spectra cover the wavelength range $4500-7300$ Å. We have calculated spectroscopic metallicities for our stars using Lick indices. Metallicity results presented here come from measurement of six iron lines in the above wavelength region calibrated against a large grid of spectroscopic standard stars (Schiavon 2007). The metallicity profile ([Fe/H] vs. radius) shown in Fig. 3 derives from those stars with the most reliable [Fe/H] measurements (with typical errors estimated to be ~ 0.4 dex), and confirms that there is a transition from dominance of the LMC disk population at [Fe/H] ~ -0.4 to a different population of lower metallicity at large radius. Stars from this more metal-poor, outer population start to appear in fields at $R \sim 7°$. Thus there is an overall radial gradient seen in the LMC profile, but at all radii there is a large spread in metallicity.

4. Origin of the new population

What is the origin of this new LMC population? An LMC halo could be created by an "outside-in" process similar to the one hypothesized for formation of the MW halo by the accretion and merger of smaller systems. Cosmologically-driven N-body simulations of structure formation (e.g., by Diemand *et al.* 2007) indicate that the satellites of MW-like galaxies (e.g., the LMC) should have their own substructure.

Clues to the origin of the outer population may lie in the metallicity profile in Fig. 3. While the sample of spectroscopic metallicities at large radii is still small, it suggests a mean metallicity for the outer LMC population of [Fe/H] ~ -1, but with stars spanning an enormous range, from [Fe/H] < -2 to about solar. This mean metallicity is higher than the metal-poor halos of either the MW or M 31 (Kalirai *et al.* 2006); if stellar halos form from the accretion of smaller systems, and such systems generally follow a mass–metallicity relationship, one might expect the mass spectrum of accreted subhalos for the LMC to promote the creation of a halo of overall lower metallicity. On the other hand, the impression from Fig. 3 is that the metallicities of outer LMC stars are not well-mixed. While admittedly it might be premature to conclude that this is a result of other than statistical fluctuations, an inhomogeneous metallicity distribution is what one expects from hierarchical halo formation, and were the LMC halo dominated by the contribution of a few of its largest subhalos, this could skew the metallicity distribution and mean metallicity over the small number of large radius fields we have sampled.

Alternatively, the newly found outer LMC population might have originated via an "inside-out" scenario — e.g., one in which the LMC disk is kinematically "puffed up" by dynamical interactions with the MW and stars at the edge of the disk are liberated into the LMC surroundings (Weinberg 2000). However, this scenario would require the LMC disk to contain rather metal-poor stars to account for those seen at large radii. Moreover, if the Magellanic Clouds are on their first passage by the MW (see references in §1), the inside-out scenario, which depends on multiple orbits of the LMC around the MW, is much less likely. Finally, the Weinberg (2000) models show a density distribution for the liberated disk stars with significant azimuthal asymmetry.

In future work we will analyze a much larger spectroscopic sample than that discussed here and derive the detailed velocity field and chemical abundance distributions of stars in this newly found MC population. Such data will help clarify the origin of these newly identified, widely separated Magellanic stars.

References

Alves, D. R. 2004, *ApJ*, 601, L151

Besla, G., Kallivayalil, N., Hernquist, L., Robertson, B., Cox, T. J., van der Marel, R. P., & Alcock, C. 2007, *ApJ*, 668, 949

Borissova, J., Minniti, D., Rejkuba, M., Alves, D., Cook, K. H., & Freeman, K. C. 2004, *A&A*, 423, 97

Borissova, J., Minniti, D., Rejkuba, M., & Alves, D. 2006, *A&A*, 460, 459

Chapman, S. C., Ibata, R., Lewis, G. F., Ferguson, A. M. N., Irwin, M., McConnachie, A., & Tanvir, N. 2006, *ApJ*, 653, 255

Diemand, J., Kuhlen, M., & Madau, P. 2007, *ApJ*, 657, 262

D'Onghia, E. 2008, *ApJ* submitted, arXiv:0802.0302

Guhathakurta, P., Ostheimer, J. C., & Gilbert, K. M., *et al.* 2005, arXiv:astro-ph/0502366

Harris, J. 2007, *ApJ*, 658, 345

Harris, J. & Zaritsky, D. 2004, *AJ*, 127, 1531

Kalirai, J. S., Gilbert, C. M., & Guhathakurta, P., *et al.* 2006, *ApJ*, 648, 389

Kallivayalil, N., van der Marel, R. P., Alcock, C., Axelrod, T., Cook, K. H., Drake, A. J., & Geha, M. 2006, *ApJ*, 638, 772

Kunkel, W. E., Demers, S., Irwin, M. J., & Albert, L. 1997, *ApJ*, 488, L129

Majewski, S. R., Ostheimer, J. C., Kunkel, W. E., Johnston, K. V., Patterson, R. J., & Palma, C. 1999, in Y.-H. Chu, N. Suntzeff, J. Hesser, & D. Bohlender (eds.), *New Views of the Magellanic Clouds*, IAU Symposium 190, p. 508

Majewski, S. R., Ostheimer, J. C., Kunkel, W. E., & Patterson, R. J. 2000, *AJ*, 120, 2550

Majewski, S. R. 2004, *PASA*, 21, 197

McConnachie, A. W., Chapman, S. C., Ibata, R. A., Ferguson, A. M. N., Irwin, M. J., Lewis, G. F., Tanvir, N. R., & Martin, N. 2006, *ApJL*, 647, L25

Minniti, D., Borissova, J., Rejkuba, M., Alves, D. R., Cook, K. H., & Freeman, K. C. 2003, *Science*, 301, 1508

Muñoz R. R., Majewski, S. R., & Zaggia, S., *et al.* 2006, *ApJ*, 649, 201

Ostheimer, J. C. 2002, *Ph. D. dissertation*, University of Virginia

Sarajedini, A., Barker, M. K., Geisler, D., Harding, P., & Schommer, R. 2006, *AJ*, 132, 1361

Schiavon, R. P. 2007, *ApJS*, 171, 146

van der Marel, R. P. 2001, *AJ*, 122, 1827

van der Marel, R. P., Alves, D. R., Hardy, E., & Suntzeff, N. B. 2002, *AJ*, 124, 2639

Weinberg, D. 2000, *ApJ*, 532, 922

Zartisky, D., Harris, J., Thompson, I. B., Grebel, E. K.. & Massey, P. 2002, *AJ*, 123, 855

Zartisky, D., Harris, J., Thompson, I. B.. & Grebel, E. K. 2004, *AJ*, 128, 1606

The Magellanic System: Stars, Gas, and Galaxies
Proceedings IAU Symposium No. 256, 2008
Jacco Th. van Loon & Joana M. Oliveira, eds.

© 2009 International Astronomical Union
doi:10.1017/S1743921308028263

New distance and depth estimates from observations of eclipsing binaries in the SMC

Pierre L. North[1], Romain Gauderon[1] and Frédéric Royer[2]

[1]École Polytechnique Fédérale de Lausanne (EPFL), Observatoire de Sauverny,
CH-1290 Versoix, Switzerland
email: **pierre.north@epfl.ch**
[2]GEPI, Observatoire de Paris – Section de Meudon,
5, place Jules Jansen, F-92195 Meudon Cedex, France
email: **Frdric.Royer@obspm.fr**

Abstract. A sample of 33 eclipsing binaries observed in a field of the SMC with FLAMES@*VLT* is presented. The radial velocity curves obtained, together with existing OGLE light curves, allowed the determination of all stellar and orbital parameters of these binary systems. The mean distance modulus of the observed part of the SMC is 19.05 mag, based on the 26 most reliable systems. Assuming an average error of 0.1 mag on the distance modulus to an individual system, and a gaussian distribution of the distance moduli, we obtain a 2-σ depth of 0.36 mag or 10.6 kpc. Some results on the kinematics of the binary stars and of the H II gas are also given.

Keywords. binaries: eclipsing, stars: distances, stars: evolution, stars: fundamental parameters, stars: kinematics, galaxies: distances and redshifts, galaxies: kinematics and dynamics, galaxies: individual (SMC), Magellanic Clouds, distance scale

1. Introduction

The last decade has seen a renewal of interest for eclipsing binary stars, thanks to the release of a huge number of light curves as a byproduct of automated microlensing surveys (EROS, MACHO, OGLE, etc.) with 1-m class telescopes . The reader can refer to the reviews from Clausen (2004), Guinan (2004) and Guinan *et al.* (2007). The interest of eclipsing binary systems resides essentially in the potential they offer to determine with excellent accuracy the masses and the radii of the stellar components. To that end, both photometric and spectroscopic (radial velocity) data are needed. If the metallicity is known, and if the surface brightness of each component is well determined (through spectroscopic or photometric estimate of effective temperature), tests of evolutionary models of single stars can be made, provided the components are sufficiently far apart ("detached" systems). Such tests have been discussed by, e.g., Andersen (1991).

Conversely, one can consider the internal structure models as reliable enough, and use them to determine both the metallicity Z and helium content Y, which give access to the relative enrichment $\Delta Y/\Delta Z$. That original approach was proposed by Ribas *et al.* (2000a); their sample included essentially Galactic binary systems, with only one belonging to the LMC. Of course, the slope $\Delta Y/\Delta Z$ would be much better constrained by adding a large number of SMC systems.

Another reason to focus on eclipsing binaries in the Magellanic Clouds is that we have a nearly complete sample of such objects to a given limiting magnitude, making them representative of a whole galaxy. Therefore, statistics of the orbital elements of detached systems can potentially yield clues about the formation mechanisms of such systems, and the study of semi-detached and contact ones may constrain scenarios of binary evolution.

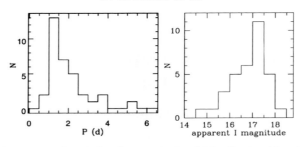

Figure 1. Left: Histogram of periods of our sample of 33 eclipsing binaries in 0.5 day bins. Right: Histogram of apparent *I* magnitudes at quadrature.

Here, we will briefly discuss the "twin hypothesis" (Pinsonneault & Stanek 2006) which suggests an excess of systems with a mass ration close to 1.

Until a purely geometrical distance determination is feasible, Paczyński (2001) considered that detached EBs are the most promising distance indicators to the Magellanic Clouds. Four B-type EB systems belonging to the Large Magellanic Cloud (LMC) were accurately characterized in a series of papers by Guinan *et al.* (1998), Ribas *et al.* (2000b, 2002) and Fitzpatrick *et al.* (2002, 2003). More recently, from high resolution, high *S/N* spectra obtained with UVES at the ESO *VLT*, the analysis of eight more LMC systems was presented by González *et al.* (2005). Harries *et al.* (2003, hereinafter HHH03) and Hilditch *et al.* (2005, hereinafter HHH05) have given the fundamental parameters of a total of 50 EB systems of spectral types O and B. The spectroscopic data were obtained with the 2dF multi-object spectrograph on the 3.9-m Anglo-Australian Telescope. This was the first use of multi-object spectroscopy in the field of extragalactic EBs. Let us also mention that the distances of an EB in M 31 (Ribas *et al.* 2005) and another in M 33 (Bonanos *et al.* 2006) were measured recently. Although the controversy about the distance to the Magellanic Clouds seems to be solved in favour of a mid position between the "short" and the "long" scales, distance data and line of sight depth remain vital for comparison with theoretical models concerning the three-dimensional structure and the kinematics of the SMC (Stanimirović *et al.* 2004).

Our contribution provides both qualitative and quantitative improvement over previous studies. Thanks to the *VLT* GIRAFFE facility, spectra were obtained with a resolution three times that in Hilditch's study. Another strong point is the treatment of nebular emission. The SMC is known to be rich in H II regions (Fitzpatrick 1985; Torres & Carranza 1987). Thus, strong emission frequently appears superposed to photospheric Balmer lines, which complicates the analysis.

2. Observations

The targets, astrometry included, were selected from the first OGLE photometric catalog. The GIRAFFE field of view (FoV) constrained us to choose systems inside a 25′ diameter circle. Other constraints were $I \leqslant 18$ mag, at least 15 well-behaved detached light curves and finally seven bump cepheids in the FoV (for another program). The observations were done in November 2003 during eight consecutive nights, and the field was observed twice a night, which makes 16 observations in all. The exposure time was 43 minutes for all but one exposure which was limited to 12 minutes because of a technical problem. Fig. 1 shows the histogram of the orbital periods of the 33 systems.

For all but two binaries, the light curves come from the new version of the OGLE-II catalog of eclipsing binaries detected in the SMC (Wyrzykowski *et al.* 2004). This catalog is based on the Difference Image Analysis (DIA) catalog of variable stars in the SMC.

The data were collected from 1997 to 2000. Two systems were selected from the first version of the catalog (standard PSF photometry) but for an unknow reason they do not appear any more in the new version.

The DIA photometry is based on I-band observations (between 202 and 312 points per curve). B and V light curves were also used in spite of a much poorer sampling (22–28 points/curve and 28–46 points/curve in B and V respectively). The objects studied in this paper have an average I magnitude and scatter (calculated from the best-fitting synthetic light curves) in the range 15.083 ± 0.009 to 18.159 ± 0.047 mag.

The spectrograph was used in the low resolution (LR2) Medusa mode: resolving power $R = 6400$, bandwidth $\Delta\lambda = 603$ Å centered on 4272 Å. The most prominent absorption lines in the blue part of early-B stars spectra are: Hϵ, He I $\lambda4026$, Hδ, He I $\lambda4144$, Hγ, He I $\lambda4388$, and He I $\lambda4471$. For late-O stars, He II $\lambda4200$ and He II $\lambda4542$ gain in importance.

Beside the spectra of the objects, 21 sky spectra were obtained for each exposure in the SMC.

3. Data reduction and analysis

The basic reduction and calibration steps including velocity correction to the heliocentric reference frame for the spectra were performed with the GIRAFFE Base Line Data Reduction Software (BLDRS). We subtracted a continuum component of the sky, obtained from an average of the 21 sky spectra measured over the whole FoV. For a given epoch the sky level varies slightly across the field, but we neglected that second order correction. Normalization to the continuum, cosmic-rays removal and Gaussian smoothing ($FWHM = 3.3$ pix) were performed with standard NOAO/PyRAF tasks. Spectral disentangling was performed with the KOREL code (Hadrava 1995, 2004), which also gives the radial velocities and orbital elements. The simultaneous analysis of light curves and RV curves was made with the 2003 version of the Wilson-Devinney (WD) Binary Star Observables Program (Wilson & Devinney 1971; Wilson 1979, 1990) via the PHOEBE interface (Prša & Zwitter 2005).

Radial velocities. Simon & Sturm (1994) were the first to propose a method allowing the simultaneous recovery of the individual spectra of the components and of the radial velocities. Another method aimed at the same results, but using Fourier transforms to save computing time, was proposed almost simultaneously by Hadrava (1995). The advantages of these methods are that they need no hypothesis about the nature of the components of the binary system, except that their individual spectra remain constant with time. Contrary to the correlation techniques, no template is needed. In addition to getting at once the radial velocities and orbital elements, one gets the individual spectra of the components ("disentangling"), with a signal-to-noise ratio which significantly exceeds that of the observed composite spectra. Other details about these techniques and their applications can be found in, e.g., Hensberge *et al.* (2000), Pavlovski & Hensberge (2005) and Hensberge & Pavlovski (2000). The radial velocities were determined from the lines of He I ($\lambda4471$, $\lambda4388$, $\lambda4144$, $\lambda4026$) only. We preferred to avoid the H Balmer lines (as did Fitzpatrick *et al.* 2002) because of moderate to strong nebular emission polluting most systems. Four 80 Å spectral ranges centered on the four He I lines were extracted from each spectrum. For each system, KOREL was run with a grid of values (K_P, q). The solution with the minimum sum of squared residuals as defined by Hadrava (2004) was retained as the best solution. For eccentric systems, a second run was performed letting K_P, q, T_0 and ω free to converge (e is fixed by photometry). It is important to notice that the four spectral regions were analyzed simultaneously, i.e. in a single run of KOREL. Each region was weighted according to the S/N of each He I line (weight $\propto (S/N)^2$). Beside the simultaneous retrieving of RV curves, orbital parameters and disentangled

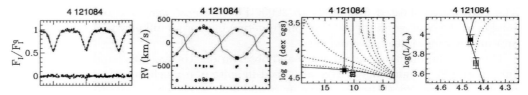

Figure 2. Example of results for the binary system 4_121084. The two left panels show the light and velocity curves as a function of the orbital phase, the lower part of each panel shows the difference $O - C$. The two right panels show the surface gravity versus mass and the luminosity versus effective temperature on a logarithmic scale (HR diagram). The isochrones shown on the 3rd panel correspond to ages $\log t = 0$ (ZAMS, continuous line), 2, 5, 10, 20, 30, 40, 50 and 100 Myr (dotted lines).

spectra, the KOREL code is able to disentangle spectra for a given orbital solution (K_P, q, T_0 and ω fixed). A final run of KOREL with this mode was then used to disentangle the regions around the Balmer and He II 4200 and 4542 lines. Indeed, He II lines and a number of Si III–IV lines are very useful to constrain the temperature of hot components.

   Light curve analysis. For each system, a preliminary photometric solution was found by the application of the method of multiple subsets (MMS) (Wilson & Biermann 1976). That allowed to provide fairly precise values of e and ω that were introduced in the KOREL analysis. Then, all three light curves and both RV curves provided by KOREL were analyzed simultaneously using the WD code. The I light curve is the most constraining one thanks to the large number of points, while the B and V light curves provide accurate out-of-eclipse B and V magnitudes. The mass ratio q was fixed to the value found by KOREL. The semi-major orbital axis a, treated as a free parameter, allows to scale the masses and radii. In a first run, the temperature of the primary was arbitrarily fixed to 26 000 K. Second-order parameters as albedos and gravity darkening exponents were fixed to 1.0. Metallicities [M/H] were set at -0.5. The limb-darkening coefficients were automatically interpolated after each fit from the Van Hamme tables (Van Hamme 1993).

   A fine tuning run was performed with the primary temperature found after analyzing the observed spectra. The standard uncertainties on the whole set of parameters were estimated in a final iteration by letting them free to converge.

4. Results and discussion

   The four panels of Fig. 2 illustrate the results we obtain for a typical binary. Among our 33 systems, 23 are detached, 9 are semi-detached and 1 is a contact one. The uncertainties range from 2 to 7% for the masses, 2 to 20% for the radii and 1 to 8% for the effective temperatures. There is an excellent overall agreement between the empirical masses and those obtained from interpolation of theoretical evolutionary tracks with $Z = 0.004$ in the HR diagram. The few discrepant points can be ascribed to incomplete lightcurves or to unrecognized third light.

   From the 21 detached systems of the HHH03/05 sample, Pinsonneault & Stanek (2006) suggest that the fraction of massive detached systems with a mass ratio close to unity is far larger than what would be expected from a classic Salpeter-like ($p(q) \propto q^{-2.35}$) or a flat ($p(q) = const$) q distribution. While the median mass ratio is high, 0.87, there are two systems with $q \sim 0.55$ only, which suggest that the high median value does not result from an observational bias. On the other hand, the q distribution of our sample of detached systems does not extend below 0.7 (see Fig. 3), which on the contrary suggests that lower q values would correspond to secondary companions too faint to be seen. Our observed q distribution is quite compatible with a flat parent distribution. For semi-detached and

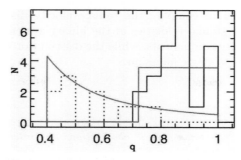

Figure 3. Distribution of the 33 observed mass ratios (continuous: detached, dashed: semi-detached/contact) with 0.05 bins. The best flat distribution (for detached) and a Salpeter-like decreasing power law (for semi-detached/contact) are over-plotted. Both distributions are truncated at a cut-off value of $q = 0.72$ and $q \sim 0.4$, respectively. Note that a flat distribution would do as well for semi-detached/contact systems.

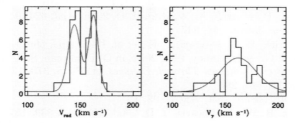

Figure 4. Distribution of the radial velocities of the ionized gas (left) and of the systemic velocities of the binary systems (right).

contact systems, it is compatible with a flat distribution extending from 0.4 to 0.7, or with a decreasing one, e.g., a Salpeter-like one, though without a theoretical justification. Therefore, the twin hypothesis is clearly not substantiated by our results, but that does not imply that the latter are compelling enough to refute it. Lucy (2006) distinguishes between the "weak" twin hypothesis (excess of binaries with $q > 0.80$) and the "strong" hypothesis (excess of binaries with $q > 0.95$). He shows that the strong hypothesis is verified on the basis of a sample of 109 Galactic binary systems, and that errors larger than ± 0.01 on q can smear out the signal. Therefore, the HHH03/05 sample, being much smaller and including low-precision q values, is far from sufficient to confirm or deny the twin hypothesis. Likewise, even though our sample includes undoubtedly better q values, it remains far from sufficient either.

The distance moduli are reliable for 26 systems in the I band, and for 25 systems in the V band. The average distance modulus is a few hundredths of a magnitude smaller for the V than for the I band. We adopt

$$DM = 19.05 \pm 0.04 \text{ mag} \quad (64.5 \pm 1.2 \text{ kpc})$$

This value is slightly higher than that adopted by HHH05 (18.912 ± 0.035 mag). The latter authors have systems scattered over the whole SMC, which should be more representative of the mean distance. On the other hand, our field is only 0.45° away from the optical centre (to the SW), so it would be difficult to reconcile the two values by a mere geometrical effect linked with the orientation of the SMC.

Assuming an average error of ~ 0.1 mag on individual distance moduli and a gaussian intrinsic dispersion of the true moduli, a quadratic difference yields a 2-σ depth of 0.36 mag or 10.6 kpc.

Finally, it is interesting to compare the radial velocity distribution of the nebular emission lines with the systemic velocities of the binary systems. Fig. 4 shows that the gas has two narrow velocity components, while the distribution of the systemic velocities is wider and can be fit by a single gaussian.

References

Andersen, J. 1991, *A&AR*, 3, 91

Andersen, J., Clausen, J. V., & Nordtsröm, B. 1980, in M. J. Plavec, D. M. Popper, & R. K. Ulrich (eds.), *Close binary stars: observations and interpretation*, IAU Conf.Proc. 88, (Dordrecht: Reidel), p. 81

Bonanos, A. Z., Stanek, K. Z., Kudritzki, R. P., et al. 2006, *ApJ*, 652, 313

Clausen, J. V. 2004, *New Astron.*, 48, 679

De Mink, S. E., Pols, O. R., & Hilditch, R. W. 2007, *A&A*, 467, 1181

Fitzpatrick, E. L. 1985, *ApJS*, 59, 77

Fitzpatrick, E. L., Ribas, I., Guinan, E. F., et al. 2002, *ApJ*, 564, 260

Fitzpatrick, E. L., Ribas, I., Guinan, E. F., Maloney, F. P., & Claret, A. 2003, *ApJ*, 587, 685

González, J. F., Ostrov, P., Morrell, N., & Minniti, D. 2005, *ApJ*, 624, 946

Guinan, E. F. 2004, *New Astron.*, 48, 647

Guinan, E. F., Fitzpatrick, E. L., DeWarf, L. E., *et al.* 1998, *ApJ*, 509, L21

Guinan, E. F., Engle, S. G., & Devinney, E. J. 2007, in O. Demircan, S.O. Selam, & B. Albayrak (eds.), *Solar and Stellar Physics Through Eclipses*, ASP-CS, 370, 125

Hadrava, P. 1995, *A&AS*, 114, 393

Hadrava, P. 2004, *Publ. Astron. Inst. ASCR*, 92, 15

Harries, T. J., Hilditch, R. W., & Howarth, I. D. 2003, *MNRAS*, 339, 157

Hensberge, H. H. & Pavlovski, K. 2007, in W. I. Hartkopf, E. F. Guinan, & P. Harmanec (eds.), *Binary stars as critical tools & tests in contemporary astrophysics*, IAU Conf.Proc. 240, (Cambridge: CUP), p. 136

Hensberge, H., Pavlovski, K., & Verschueren, W. 2000, *A&A*, 358, 553

Hilditch, R. W., Howarth, I. D., & Harries, T. J. 2005, *MNRAS*, 357, 304 (HHH05)

Lucy, L. B. 2006, *A&A*, 457, 629

Paczyński, B. 2001, *AcA*, 51, 81

Pavlovski, K. & Hensberge, H. H. 2005, *A&A*, 439, 309

Pinsonneault, M. H. & Stanek, K. Z. 2006, *ApJ*, 639, L67

Prša, A., 2006, PHOEBE Scientific Reference, Univ. of Ljubljana, available in electronic form at http://phoebe.fiz.uni-lj.si/

Prša, A. & Zwitter T. 2005, *ApJ*, 628, 426

Ribas, I., Guinan, E. F., Fitzpatrick, E. L., et al. 2000b, *ApJ*, 528, 692

Ribas, I., Jordi, C., Torra, J., & Giménez, A. 2000a, *MNRAS*, 313, 99

Ribas, I., Fitzpatrick, E. L., Maloney, F. P., Guinan, E. F., & Udalski, A. 2002, *ApJ*, 574, 771

Ribas, I., Jordi, C., Vilardell, F., Fitzpatrick, E. L., Hilditch, R. W., & Guinan, E. F. 2005, *ApJ*, 635, 37

Stanimirović, S., Staveley-Smith, L., & Jones, P. A. 2004, *ApJ*, 604, 176

Simon, K. P. & Sturm, E. 1994, *A&A*, 281, 286

Torres, G. & Carranza, G. J. 1987, *MNRAS*, 226, 513

Van Hamme, W., 1993, *AJ*, 106, 2096

Van Rensbergen, W., De Loore, C., & Vanbeveren, D. 2005, in *Interacting Binaries: Accretion, Evolution, and Outcomes*, AIP-CP, 797, 301

Van Rensbergen, W., De Loore, C., & Jansen, K. 2006, *A&A*, 446, 1071

Wilson, R. E. & Devinney, E. J. 1971, *ApJ*, 166, 605

Wilson R. E. 1979, *ApJ*, 234, 1054

Wilson R. E. 1990, *ApJ*, 356, 613

Wilson, R. E. & Biermannn, P. 1976, *A&A*, 48, 349

Wyrzykowski, L., Udalski, A., Kubiak, M., et al. 2004, *AcA*, 54, 1

The Magellanic System: Stars, Gas, and Galaxies
Proceedings IAU Symposium No. 256, 2008
Jacco Th. van Loon & Joana M. Oliveira, eds.

Line of sight depth of the Large and Small Magellanic Clouds

Annapurni Subramaniam and Smitha Subramaniam

Indian Institute of Astrophysics, Sarjapur Road, Koramangala II Block,
Bangalore -560034, India
email: purni@iiap.res.in, smitha@iiap.res.in

Abstract. We used the red clump stars from the Optical Gravitational Lensing Experiment (OGLE II) survey and the Magellanic Cloud Photometric Survey (MCPS), to estimate the line-of-sight depth. The observed dispersion in the magnitude and colour distribution of red clump stars is used to estimate the line-of-sight depth, after correcting for the contribution due to other effects. This dispersion due to depth, has a range from minimum dispersion that can be estimated, to 0.46 mag (a depth of 500 pc to 10.44 kpc), in the LMC. In the case of the SMC, the dispersion ranges from minimum dispersion to 0.35 magnitude (a depth of 665 pc to 9.53 kpc). The thickness profile of the LMC bar indicates that it is flared. The average depth in the bar region is 4.0 ± 1.4 kpc. The halo of the LMC (using RR Lyrae stars) is found to have larger depth compared to the disk/bar, which supports the presence of an inner halo for the LMC. The large depth estimated for the LMC bar and the disk suggests that the LMC might have had minor mergers. In the case of the SMC, the bar depth (4.90 ± 1.23 kpc) and the disk depth (4.23 ± 1.48 kpc) are found to be within the standard deviations. We find evidence for an increase in depth near the optical center (up to 9 kpc). On the other hand, the estimated depth for the halo (RR Lyrae stars) and disk (RC stars) for the bar region of the SMC is found to be similar. Thus, increased depth and enhanced stellar as well as H I density near the optical center suggests that the SMC may have a bulge.

Keywords. stars: horizontal-branch, galaxies: bulges, galaxies: halos, Magellanic Clouds, galaxies: stellar content, galaxies: structure

1. Introduction

The Magellanic Clouds were believed to have interactions with our Galaxy as well as between each other (Westerlund 1997). The N-body simulations by Weinberg (2000) predicted that the LMC's evolution is significantly affected by its interactions with the Milky Way and the tidal forces will thicken and warp the LMC disk. Alves & Nelson (2000) studied the carbon star kinematics and found that the scale height, h, increases from 0.3 to 1.6 kpc over the range of radial distance, R, from 0.5 to 5.6 kpc and hence concluded that the LMC disk is flared. Using an expanded sample of carbon stars van der Marel *et al.* (2002) also found that the thickness of LMC disk increases with the radius. There has not been any direct estimate of the thickness or the line-of-sight depth of the bar and disk of the L&SMC so far.

Mathewson, Ford & Visvanathan (1986) found that SMC Cepheids extend from 43 to 75 kpc with most Cepheids found in the neighbourhood of 59 kpc. Later, the line-of-sight depth of the SMC was estimated (Welch *et al.* 1987) by investigating the line-of-sight distribution and period–luminosity relation of Cepheids and found the line-of-sight depth of the SMC to be around 3.3 kpc. Hatzidimitriou & Hawkins (1989), estimated the line-of-sight depth in the outer regions of the SMC to be around 10–20 kpc. Measurement of the thickness in the central regions of the Magellanic Clouds, especially the LMC, is of strong interest to understand the contribution of LMC's self lensing to the observed microlensing events from this Galaxy.

Red Clump (RC) stars are core helium burning stars which are metal rich and slightly more massive counter parts of horizontal branch stars. In this paper, we use the dispersions in the colour and magnitude distribution of RC stars for the depth estimation. The dispersion in colour is due to a combination of observational error, internal reddening (reddening within the disk of the LMC/SMC) and population effects. The dispersion in magnitude is due to internal disk extinction, depth of the distribution, population effects and photometric errors associated with the observations. By deconvolving other effects from the dispersion of magnitude, we can estimate the depth of the disk. The advantage of choosing RC stars as proxy is that there are large numbers of these stars available to determine the dispersions in their distributions with good statistics.

2. Data

Data for the LMC are taken from the OGLE II catalogue (Udalski *et al.* 2000). The average photometric error of red clump stars in I and V bands are around 0.05 magnitude. Photometric data with an error less than 0.15 mag are considered for the analysis. The 26 strips of LMC are divided into 1664 regions. (V−I) vs. I CMDs are plotted for each region and red clump stars ($\geqslant$ 1000 stars) were identified. For all the regions, red clump stars are well within a box in the CMD with boundaries 0.65–1.35 mag in (V−I) colour and 17.5–19.5 mag in I magnitude. The OGLE data suffer from incompleteness due to crowding effects and it is corrected using the data given in Udalski *et al.* (2000).

The data for the SMC are taken from two surveys (OGLE II: Udalski *et al.* 1998, and MCPS: Zaritsky *et al.* 2002). The OGLE and MCPS photometric data with an error less than 0.15 mag are considered for the analysis. The observed regions of the SMC OGLE and MCPS data are divided into 176 and 876 sub regions respectively. In the MCPS data the regions away from the bar are less dense compared to the bar region and out of 876 regions only 755 regions which have a reasonable number of stars ($\sim$ 1000 stars) are considered for the analysis. For all the regions, red clump stars are identified in the same region in the CMD with boundaries 0.65–1.35 mag in (V−I) colour and 17.5–19.5 mag in I magnitude. OGLE data are corrected for incompleteness.

3. Analysis

A spread in magnitude and colour of red clump stars is observed in the CMDs of both the LMC and SMC. The number distribution profile against colour and magnitude roughly resembles a Gaussian. The width of the Gaussian in the distribution of colour and magnitude is obtained using a non-linear least-squares method to fit the profile.

The RC stars in the disk of the LMC/SMC constitute a heterogeneous population spanning a range in mass, age and metallicity. The density of stars in various locations will also vary with star formation rate as a function of time. Girardi & Salaris (2001) simulated the RC stars in the LMC using the star formation rate results from Holtzman *et al.* (1999) and the age–metallicity relation from Pagel & Tautvaisiene (1998). They also simulated the RC stars in the SMC using the star formation results and age–metallicity relation from Pagel & Tautvaisiene (1998). The intrinsic dispersions obtained from the above model are used in our estimation to account for the population effects. The estimated intrinsic dispersions in magnitude and colour distributions according to the theoretical model for the LMC are 0.1 and 0.025 mag respectively. In the case of the SMC, the values are 0.076 and 0.03 mag respectively.

The average photometric errors of I and V band magnitudes are calculated for each region and the errors in I and (V−I) colour are estimated. These are subtracted from the observed width of the magnitude and colour distribution respectively. After correcting for population effects and the observational error in colour, the remaining spread in colour distribution is taken as due to the internal reddening, E(V−I) . This is converted into extinction in I band using the relation $A(I) = 0.934E(V − I)$, where E(V−I) is the internal reddening value estimated for each location. The above relation is derived from the relations $E(V − I) = 1.6E(B − V)$ and $A(I) = 0.482A(V)$ (Rieke & Lebofsky 1985).

The following relations are used for estimating the resultant dispersion due to depth:

$$\sigma^2_{mag} = \sigma^2_{depth} + \sigma^2_{internal\ extinction} + \sigma^2_{intrinsic} + \sigma^2_{error}, \tag{3.1}$$

$$\sigma^2_{col} = \sigma^2_{internal\ reddening} + \sigma^2_{intrinsic} + \sigma^2_{error}. \tag{3.2}$$

4. Results: LMC

A two dimensional plot of the depth for the 1528 regions in the LMC is shown in figure 1. The optical center of the LMC is taken to be RA = $5^h19^m38^s$, Dec = $−69°27'5''.2$ (J2000, de Vaucouleurs & Freeman 1973). The plot shows a range of dispersion values from 0.033 to 0.46 mag (a depth of 700 pc to 10.5 kpc; average: 3.95±1.42 kpc) for the LMC central bar region. For the N-W disk region, the dispersion estimated ranges from minimal dispersion that can be estimated (limited by errors), to 0.33 mag (a depth of 500 pc to 7.7 kpc; average: 3.56±1.04 kpc). Regions in the bar between RA 80–84 degrees show a reduced depth (0.5–4 kpc, as indicated by yellow and black points). The regions to the east and west of the above region are found to have a larger depth (2.0–8.0 kpc, black, red and green points). Thus, the depth of the bar at its ends is larger than that near its center. The N-W region has a depth similar to the central region of the bar. In general, a thicker and heated up bar could be considered as a signature of minor mergers. Thus, the LMC is likely to have experienced minor mergers in its history.

Subramaniam (2006) studied the distribution of RR Lyrae in the bar region of the LMC. She found that the RR Lyrae stars in the bar region have a disk like distribution, but halo like location. The RR Lyrae stars are in the same evolutionary state as the RC stars, except that the RR Lyrae stars belong to an older and metal poor population and hence a proxy for the halo as it belongs to the Population II stars. We used the total depth estimated for RR Lyrae stars and compared it with the RC depth which can be considered as proxies for halo and disk respectively. It was seen that the depth estimated from RR Lyrae stars ranged between 4.0–8.0 kpc, suggesting that the RR Lyrae stars span a larger depth than the RC stars. Thus, at least in the central region of the LMC, the halo, as delineated by the RR Lyrae stars has a much larger depth than the disk, as delineated by the RC stars. This supports the idea of an inner halo for the LMC.

5. Results: SMC

Colour-coded two-dimensional plots of thickness for the two data sets are shown in figure 2 (OGLE data in the lower panel and MCPS data in the upper panel). The optical center of the SMC is taken to be RA = $00^h52^m12.5^s$, Dec = $−72°49'43''$ (J2000, de Vaucouleurs & Freeman 1973). The prominent feature in both plots is the presence of

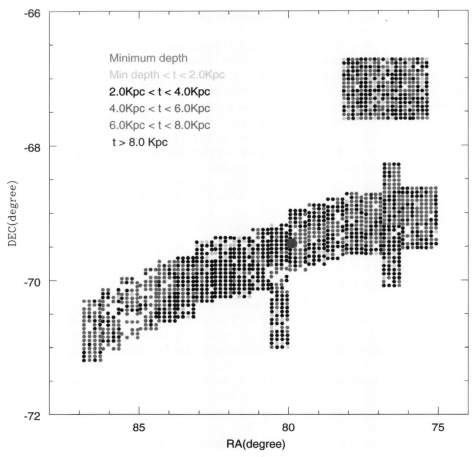

Figure 1. The line-of-sight depth in the bar region of the LMC. The colour codes used to denote the depth are explained. The optical center is shown as a purple dot.

blue and green points indicating increased depth, for regions located near the SMC optical center ($\sim$9 kpc). The net dispersions range from 0.10 to 0.35 mag (corresponding to a depth of 2.8 kpc to 9.6 kpc) in the OGLE data set and from minimum dispersion to 0.34 mag (corresponding to a depth of 665 pc to 9.47 kpc) in the MCPS data set. The average value of the SMC thickness estimated using the OGLE data set in the central bar region is 4.9±1.2 kpc and the average thickness estimated using the MCPS data set, which covers a larger area than the OGLE data, is 4.42±1.46 kpc. The average depth obtained for the bar region alone is 4.97±1.28 kpc, which is very similar to the value obtained from the OGLE data. The depth estimated for the disk alone is 4.23±1.47 kpc. Thus the disk and the bar of the SMC do not show any significant difference in the depth. Evidence of increased depth in the outer regions of the disk is also indicated, this might suggest a large depth for the outer SMC. The enhanced central depth is shown in the averaged width over a limited RA and Dec in figure 3, using the MCPS data. The increased depth near the center is clearly indicated. Thus, the depth near the center is about 9.6 kpc, which is twice the average depth of the bar region (4.9 kpc). The depth profile, especially the one along RA, is very similar to the luminosity profile of bulges and hence suggests the presence of a bulge in the SMC.

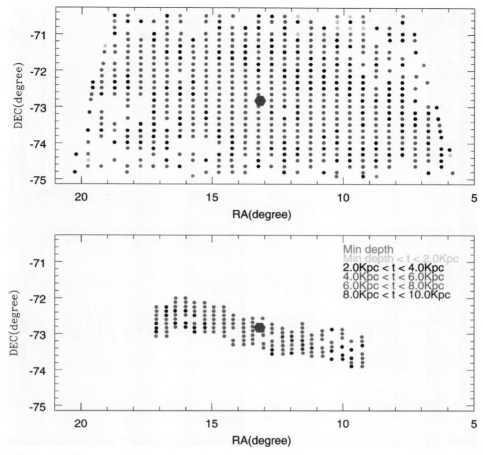

Figure 2. The line-of-sight depth in the bar and disk region of the SMC. The upper panel shows the depth derived from the MCPS data and the lower panel shows that derived from the OGLE data. The optical center is shown as a purple dot.

A comparison can be made between the halo and disk/bar of the SMC. The RR Lyrae stars from the OGLE II data were analysed similar to the procedure adopted by Subramaniam (2006) and the dispersion due to depth alone was estimated. In contrast to what is seen in the case of the LMC, both populations show a very similar dispersion in the SMC. The comparison not only suggests that the RR Lyrae stars and the RC stars occupy similar depth, but also indicates that they show a similar depth profile across the bar. The increased depth near the optical center is also closely matched. This suggests that the RR Lyrae stars and RC stars are born in the same location and occupy similar volume in the galaxy. This co-existence and the similar depth of RR Lyrae stars and the RC stars in the central region of the SMC can be easily explained, if it is the bulge. We also find that the old and young stellar density as well as the H I density show peaks near the region with bulge like depth. This supports the idea that the central region of the SMC could be its bulge. The elongation and the rather non-spherical appearance of the bulge could be due to tidal effects or minor mergers (Bekki & Chiba 2008).

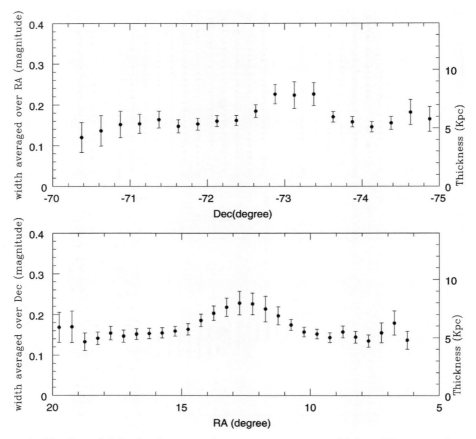

Figure 3. The line-of-sight depth averaged over a narrow range of RA and Dec near the center using the MCPS data. The errors correspond to the standard deviation over the area averaged.

References

Alves, D. R. & Nelson, C. A. 2000, *ApJ*, 542, 789
Bekki, K. & Chiba, M. 2008, *ApJ*, 679, L89
de Vaucouleurs, G. & Freeman, K. C. 1973, *Vistas Astron.*, 14, 163
Girardi, L. & Salaris, M. 2001, *MNRAS*, 323, 109
Hatzidimitriou, D. & Hawkins, M. R. S. 1989, *MNRAS*, 241, 667
Holtzman, J. A., Gallagher, J. S., III, Cole, A. A., et al. 1999, *AJ*, 118, 2262
Mathewson, D. S., Ford, V. L., & Visvanathan, N. 1986, *ApJ*, 301, 664
Pagel, B. E. J. & Tautvaisiene, G. 1998, *MNRAS*, 299, 535
Rieke, G. H. & Lebofsky, M. J. 1985, *ApJ*, 288, 618
Subramaniam, A. 2006, *A&A*, 449, 101
Udalski, A., Szymański, M., Kubiak, M., et al. 1998, *AcA*, 48,147 (SMC OGLE II data)
Udalski, A., Szymański, M., Kubiak, M., et al. 2000, *AcA*, 50,307 (LMC OGLE II data)
van der Marel, R. P., Alves, D. R., Hardy, E., & Suntzeff, N. B. 2002, *AJ*, 124, 2639
Weinberg, M. D. 2000, *ApJ*, 532, 922
Welch, D. L., McLaren, R. A., Madore, B. F., & McAlarey, C. W. 1987, *ApJ*, 321, 162
Westurlund, B. E. 1997, *The Magellanic Clouds* (Cambridge: CUP)
Zaritsky, D., Harris, J., Thompson, I. B., Grebel, E. K., & Massey, P. 2002, *AJ*, 123, 855

The Magellanic System: Stars, Gas, and Galaxies
Proceedings IAU Symposium No. 256, 2008
Jacco Th. van Loon & Joana M. Oliveira, eds.

The star clusters of the Magellanic System

Basílio X. Santiago[1]

[1]Instituto de Física, Universidade Federal do Rio Grande do Sul
Caixa Postal 15051, Porto Alegre, RS, Brazil
email: santiago@if.ufrgs.br

Abstract. More than 50 years have elapsed since the first studies of star clusters in the Magellanic Clouds. The wealth of data accumulated since then has not only revealed a large cluster system, but also a diversified one, filling loci in the age, mass and chemical abundance parameter space which are complementary to Galactic clusters. Catalogs and photometric samples currently available cover most of the cluster mass range. The expectations of relatively long cluster disruption timescales in the Clouds have been confirmed, allowing reliable assessments of the cluster initial mass function and of the cluster formation rate in the Clouds. Due to their proximity to the Galaxy, Magellanic clusters are also well resolved into stars. Analysis of colour—magnitude diagrams (CMDs) of clusters with different ages, masses and metallicities are useful tools to test dynamical effects such as mass loss due to stellar evolution, two-body relaxation, stellar evaporation, cluster interactions and tidal effects. The existence of massive and young Magellanic clusters has provided insight into the physics of cluster formation. The magnitudes and colours of different stellar types are confronted with stellar evolutionary tracks, thus constraining processes such as convective overshooting, stellar mass-loss, rotation and pre main-sequence evolution. Finally, the Magellanic cluster system may contribute with nearby and well studied counterparts of recently proposed types of extragalactic clusters, such as Faint Fuzzies and Diffuse Star Clusters.

Keywords. Magellanic Clouds, galaxies: star clusters, galaxies: stellar content

1. The census of star clusters in the Magellanic Clouds

The first papers on star clusters in the Magellanic Clouds (MCs) date from over 50 years ago. They were concerned with detecting these objects and measuring their basic properties such as positions, apparent sizes and position angles. Figure 1 shows the evolution in the census of star clusters in the Clouds from these early papers to the present. The numbers shown were taken from references in the literature spanning the entire period (Kron 1956; Lindsay 1958; Shapley & Lindsay 1963; Lyngå & Westerlund 1963; Hodge & Sexton 1966; Hodge 1986, 1988; Bica *et al.* 1999; Bica & Dutra 2000; Bica *et al.* 2008). A steady increase in the number of known clusters can be seen, especially in the case of the Large Magellanic Cloud (LMC). For the Small Magellanic Cloud (SMC), the rise has been slower. This may, at least in part, reflect the fact that some of the earlier catalogs did not separate star clusters from stellar associations.

The latest catalog has just recently been published by Bica *et al.* (2008). It includes star clusters, associations and other extended objects. It contains a total of 9305 entries, 3740 of which have been classified as star clusters. The on-sky distribution of star clusters in Bica *et al.* (2008) is shown in Figure 2. We see that the main structural components of the Magellanic System are traced by these objects, including the LMC bar and outer ring, the SMC wing and the bridge.

One important issue is whether the current census of stars clusters is complete in the MCs. A bias against faint and/or compact star clusters should exist to some extent, as

B. X. Santiago

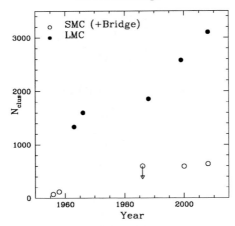

Figure 1. Census of known star clusters as a function of time. LMC: solid circles; SMC and Bridge: open circles.

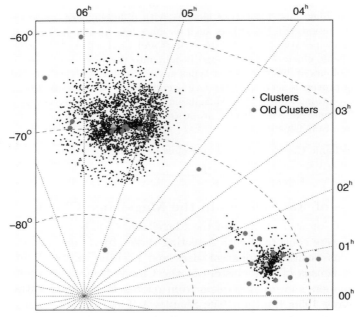

Figure 2. On-sky distribution of the 3740 star clusters in the Bica *et al.* (2008) catalogue, on a grid of equatorial coordinates. The larger dots are the clusters with estimated ages $\tau > 4$ Gyr.

these objects are harder to detect or to separate from single bright stars. The apparent diameter function of star clusters from Bica *et al.* (2008) is shown in Figure 3, separately for the LMC, SMC and Bridge clusters. The clusters in the SMC and Bridge have been scaled to the LMC distance. The diameter function shows a power law behaviour at large diameters, reaching a peak at $D_{\mathrm{app}} \simeq 0.6'$ and then dropping sharply at the small diameter end. The distribution of tidal radii of Galactic globular clusters is also shown in the figure and has a similar shape. As these latter should make up an essentially complete sample, the authors conclude that this down turn is a real feature, rather than the result of a selection bias.

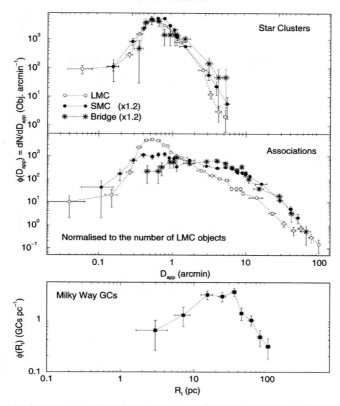

Figure 3. Top panel: apparent diameter function of star clusters of Bica *et al.* (2008). The LMC, SMC and Bridge clusters are shown in separate, as indicated. Middle panel: The diameter function of the 3326 stellar associations, again separately showing the LMC, SMC and Bridge systems. Lower panel: The distribution of tidal radii of Galactic globular clusters.

2. Basic properties of the MC star cluster system

Since the LMC and SMC are at distances of $\simeq 50$ kpc and 60 kpc, respectively, their system of star clusters can be studied in much more detail than those of more distant hosts. For instance, the photometric sample of LMC clusters by Hunter *et al.* (2003) is complete down to $M_V \simeq -3.5$. This is 4 magnitudes fainter than the absolute magnitude at the peak of the globular cluster luminosity function (GCLF) observed in other galaxies, especially in luminous early-type ones. Therefore, the LMC and SMC cluster systems are probed towards much lower masses than elsewhere.

Basic properties of hundreds of individual star clusters in both Clouds, such as ages, metallicities and masses, have been obtained from integrated photometry, integrated spectroscopy or colour-magnitude diagrams (CMDs) resulting from high-resolution imaging.

The age—metallicity relation (AMR) of rich LMC clusters compiled by Kerber *et al.* (2007) is shown in Figure 4. It allows us to discuss the main properties of the LMC cluster system. Rich clusters span a wide range in ages, $7 \leqslant \log \tau(\text{yr}) \leqslant 10$. Most clusters younger than $\simeq 3$ Gyr have metallicities between half and one third of the solar value. Only about 15 rich clusters are older than 10 Gyr and have [Fe/H] $\simeq -1.5$ or less, thus having properties similar to those of globular clusters (GCs) found in the Galactic halo. A noticeable feature is the so-called *age gap*, as only one cluster, ESO 121 SC03, is found

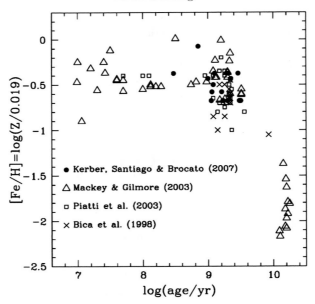

Figure 4. The age—metallicity relation for LMC clusters, taken from Kerber *et al.* (2007).
The different symbols show clusters from different samples, as indicated.

in the range 3 Gyr $\leqslant \tau \leqslant$ 10 Gyr. This cluster has recently been studied by Xin *et al.*
(2008), using Hubble Space Telescope Advanced Camera for Surveys (*HST*/ACS), and
its age has been confirmed to fall in this interval.

The SMC cluster system is also rich and diversified in terms of ages and abundances.
A recent AMR for the SMC can be found in Piatti *et al.* (2007). Despite the often
large error bars in determined cluster ages, the SMC seems to have several examples of
intermediate-age clusters, with no clear sign of an age gap. Another difference relative to
the LMC is the absence of a significant population of rich, old and metal poor clusters
similar to the Galactic GCs. Only NGC 121 has GC properties (Glatt *et al.* 2008).

3. Age and mass distributions: constraining the cluster formation rate and initial mass function

Large photometric surveys of both clouds are currently available, yielding magnitudes
and colour for millions of stars and hundreds of star clusters (Massey 2002; Zaritsky *et al.*
1997; Hunter *et al.* 2003; Rafelski & Zaritsky 2005).

The integrated magnitudes and colours of star clusters can be compared to those of
single stellar populations, as predicted by different models (Leitherer *et al.* 1999; Kurth
et al. 1999). As a result, age and mass estimates can be inferred for each cluster. The
distribution of clusters with these quantities bears a lot of information on the cluster
initial mass function (ICMF), the cluster formation history (CFR) and the timescale for
cluster disruption.

As an example, Boutloukos & Lamers (2003) have modelled the age distribution of star
clusters. The model assumes that the ICMF is a power-law with cluster mass, $\propto m^{-2}$,
and that the CFR has been uniform. Assuming also that the cluster disruption timescale,
$t_{\rm d}$, is a power-law as a function of cluster mass, the model predicts that the number of
star clusters per unit age should be described by a double power-law as a function of age,
with a shallow slope at small ages and a steeper one for the older clusters. The slower

drop at small ages results from fading: clusters become less luminous as they age, so that their number decreases in any magnitude limited sample. At larger ages, dynamical evolution leads to the disruption of less massive clusters, leading to a faster drop in the observed age distribution. This behaviour was in fact observed by Boutloukos & Lamers (2003) in several galaxies with sizable cluster samples with age estimates, including the SMC. From the age distribution of the SMC clusters, they infer a disruption timescale

$$t_{\rm d} \simeq 8\left(\frac{M}{10^4\,{\rm M_\odot}}\right)^{0.62}\,{\rm Gyr}$$

Hunter *et al.* (2003) have obtained integrated magnitudes and colours for 939 SMC and LMC clusters using the images from the Massey (2002) survey. They have modelled the age and mass distributions and found evidence for cluster fading as well as an increase in maximum cluster mass for larger ages resulting from statistical sampling, from which they were able to constrain the ICMF slope.

De Grijs & Anders (2006) re-analyzed the Hunter *et al.* (2003) data using a tool to compare clusters colours and ages to SSPs and applied the Boutloukos & Lamers (2003) model in order to infer a $t_{\rm d} = 8$ Gyr for 10^4 M$_\odot$ LMC clusters, in agreement with the results for the SMC.

In brief, these studies quantitatively confirm that star clusters tend to live much longer in the Clouds than in the Galaxy as a result of slower disruption processes. We point out that, qualitatively, the differences in age distribution between the Galaxy and the Clouds is long known (Hodge 1988). These studies based on integrated photometry also constrained the ICMF slope and confirmed a roughly constant CFR in the MCs, apart from the age gap in the LMC.

4. Magellanic star clusters in high resolution: structure and dynamics

The *HST* has allowed individual star clusters to be resolved into stars and studied in detail (Brocato *et al.* 2001; Santiago *et al.* 2001). Detailed and self-consistent modelling of the colour-magnitude diagrams provides a strong tool for determining the physical properties of rich star clusters. Self-consistency should be understood in this context as the capacity to extract the relevant parameters, such as age, metallicity, distance, extinction and mass functions at different positions, all from the same data-set, without pre-fixing any of them. This may often be the most reliable approach considering the spatial depth of the two Clouds, the granularity of the dust distribution and the range in metallicities in these galaxies. Kerber *et al.* (2002) and Kerber & Santiago (2005) presented a statistical method, based on a code from D. Valls-Gabaud, to model observed CMDs by comparing them to synthetic ones. The authors applied this tool to several LMC star clusters observed with *HST*'s Wide Field and Planetary Camera 2 (*HST*/WFPC2). Similar techniques are also successfully applied to CMDs of field stars in the Clouds and other dwarf satellites of the Galaxy, in order to reconstruct the star formation history (SFH) in them (Hernandez *et al.* 2000; Javiel *et al.* 2005; Noël et. 2007).

The MCs are an excellent laboratory to observe and constrain dynamical effects on stellar systems, in particular star clusters. Dynamical and structural evolution can be observed and modelled from analysis of high resolution images in LMC and SMC clusters.

First, clusters respond to the evolution of their stars. Strong winds from massive stars and supernovae (SN) explosions lead to mass loss and expansion in young clusters, as observed by Bastian & Goodwin (2006). Any gas not converted into stars before the first SN bursts will be swept out from the cluster. If the efficiency of star formation is low, mass loss can be severe, the remaining cluster becoming unbound and dispersing into the

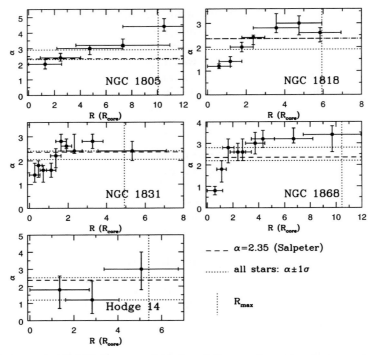

Figure 5. Variation in PDMF slope as a function of distance from the cluster centre for a sample of rich LMC clusters imaged with *HST*/WFPC2 and studied by Kerber & Santiago (2006) and de Grijs *et al.* (2002). Each panel shows the slope α vs. R relation for a different cluster, as indicated. The distances are in units of the core radius. The dashed line corresponds to $\alpha = 2.35$ and the horizontal dotted lines indicate the 1σ range around the global PDMF slope.

field at its early stages of evolution. This process is usually referred to as *infant mortality*. In fact, the infant mortality rate of Magellanic clusters, especially in the SMC, has been subject to a lot of recent controversy. Based on the decline in the number of clusters as a function of age, Chandar *et al.* (2006) have suggested that up to 90% of the young star clusters formed in the SMC do not survive the early stages. Kruijssen & Lamers (2008) also needed a similarly large infant mortality rate to accommodate the observed age distributions of SMC clusters and field stars under the same general SFH. On the other hand, Gieles *et al.* (2007) and de Grijs & Goodwin (2008) interpret the declining number of SMC clusters at young ages as being dominated by the effect of cluster fading in brightness in a magnitude-limited sample. More on infant mortality is discussed by R. de Grijs in these proceedings.

Second, two-body interactions among stars occur inside clusters, as the relaxation timescale is shorter than a Hubble time and may in fact be of a few Myr in compact and poor clusters. As a result of these interactions, more massive stars will tend to donate part of their orbital energy to lower mass ones; the former will tend to sink towards the gravitational potential well, whereas the latter will achieve less bound orbits, causing mass segregation inside the cluster. A fraction of the low mass stars will acquire escape velocity and leave the cluster (stellar evaporation).

Evidence for mass segregation and stellar evaporation is often observed in rich star clusters for which deep and high-resolution photometry is available, including young ones. Figure 5 shows that the present day mass function (PDMF) slopes of several LMC

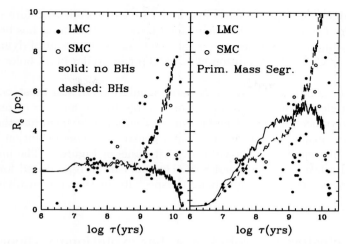

Figure 6. Left panel: Core radius (R_c, in parsecs) as a function of age ($\log \tau(\mathrm{yr})$) for LMC and SMC star clusters. The data are from Mackey *et al.* (2003a,b). The lines show the expected evolution based on N-body simulations. Solid line: no heating from stellar degenerates; dashed line: heating included. Right panel: the data are the same as in the previous panel. The simulation results now include primordial mass segregation, again with and without heating by stellar black holes.

clusters modelled by Kerber & Santiago (2006) increase steadily as a function of distance from the cluster centre. The slopes α result from power-law fits to the differential number of stars as a function of initial stellar mass, $dN/dm \propto m^{-\alpha}$, in different radial annuli around the cluster centre. At the largest distance bins, the trend flattens out or is even reversed in some cases, due to the loss of evaporating low mass stars from the cluster. Notice also that the global PDMF slopes are consistent with the Salpeter value ($\alpha = 2.35$).

For young enough clusters, the PDMF should reflect more closely the initial mass function (IMF). In particular, as the MCs are metal poorer than the Galaxy, and have a large system of young and rich clusters, mass function analysis in these clusters may help constrain variations in the IMF with environment. Kumar *et al.* (2008) have analyzed the mass function of 9 young star clusters in the LMC and found power-law slopes (in the mass range $2 < m/\mathrm{M}_\odot < 12$) consistent with one another for all but one of them. The slopes are also close to the Salpeter value, supporting the idea of a universal IMF. Schmalzi *et al.* (2008) have also found a nearly Salpeter value for the IMF slope in the SMC star forming regions NGC 602, in the mass range $1 \leqslant m/\mathrm{M}_\odot \leqslant 45$. On the other hand, at lower masses ($m/\mathrm{M}_\odot < 1$), the existence of a universal IMF is still in debate (see contribution by D. Gouliermis in these proceedings).

A clear evidence of structural evolution of star clusters is given by the relation between core size (R_c) and age. The core radius is determined by fitting an equilibrium model, such as the Elson *et al.* (1987) model, to the cluster surface brightness profile. The relation between R_c and age, using the data on rich LMC and SMC clusters from Mackey *et al.* (2003a,b), is shown in Figure 6. The mean core radius clearly increases with age, as does the spread around the mean. This shows not only that clusters change their structure with time but also that they do not follow a single path of core evolution. In fact, a large fraction of the clusters older than 10 Gyr shown in the Figure have very small cores, possibly having undergone core collapse.

Several mechanisms have been proposed to account for core evolution. Elson *et al.* (1989) explored the possibility that variations in the IMF from one cluster to another could lead to different rates of mass loss and core expansion, therefore explaining the spread in the R_c–age relation. The mean trend towards larger cores has been tentatively explained as a sampling effect (Hunter *et al.* 2003), or as the result of dynamical heating of the core by binary stars (Wilkinson *et al.* 2003) or by stellar black holes (Mackey *et al.* 2008).

Mackey *et al.* (2008) used realistic N-body simulations of star clusters, in which the number of particles in the simulation is comparable to the number of real stars. They were able to reproduce the mean trend in core size with age by including the effect of core heating by black holes that sink to the cluster centre. The core evolution in simulations with and without the effect is shown in the left panel of Figure 6. The upper boundary of the observed relation in LMC and SMC clusters can be reproduced for clusters with primordial mass segregation (right panel), especially in the case involving heating by black holes.

5. MC star clusters as probes of stellar evolutionary theory

Compared to the Galaxy, the Magellanic Clouds are gas rich and metal poor, thus providing a different environment where stars form and evolve. As such, they are a complementary and very useful laboratory to test stellar evolutionary theories. Observed cluster CMDs are particularly useful as they can be approximated as single stellar populations and compared to model isochrones. We here quote some recent examples focussed on very different stellar evolutionary phases.

Mucciarelli *et al.* (2007a,b) analyzed *HST*/ACS CMDs of two rich intemediate-age star clusters in LMC, namely NGC 1978 and NGC 1783. They compared several model expectations, with convective regions that exceed the classical one by values in the range $0 \leqslant \Lambda \leqslant 25\%$, to their high-resolution CMDs. As a result, they favour a mild or large amount of convective overshooting ($\Lambda = 0.10 - 0.25$) in the intermediate mass stars at the main sequence turn-off, sub-giant (SGB) and red-giant (RGB) phases of these clusters. They also detected the elusive, but predicted, bump along the RGB sequence. Kerber & Santiago, in these proceedings, have also tested different stellar evolution models, with and without overshooting, using a sample of 15 intermediate age clusters in the LMC.

Marigo & Girardi (2007) have developed synthetic models to describe the late evolutionary stage called thermally pulsating asymptotic giant branch (TP-AGB), where significant variability, mass loss and dredge-up events take place. Their models were calibrated using the observed distribution of both luminosities and lifetimes of carbon rich and oxygen rich (C and M, respectively) AGB stars in the two Magellanic Clouds.

Hennekemper *et al.* (2008), again using *HST*/ACS, found a large number of pre-main sequence stars (PMS) in the young massive star forming region N66/NGC 346. Comparison of the observed PMS distribution in their CMD with model predictions by Siess *et al.* (2000) indicate the existence of two recent episodes of star formation. PMS stars, with $\tau \simeq 4$ Myr, have also been recently detected by Carlson *et al.* (2007) in the star forming region NGC 602, located in the SMC wing.

6. Spectroscopy of MC clusters: a bit of kinematics and abundances

Radial velocities and detailed abundance analysis of MC clusters require use of spectroscopy. Recent detailed abundance studies of MC clusters can be found in Trundle *et al.* (2007), Hunter *et al.* (2007) and Mucciarelli *et al.* (2008). The first two analyzed over 100 B stars in several star clusters in the Galaxy and in the MCs, and with a large age span. They derived atmospheric parameters and photospheric abundances of

C, N, O, Mg, Si and Fe and investigated the effect of evolutionary processes such as rotation, mass loss and binarity on the abundance of nitrogen.

Mucciarelli *et al.* (2008) studied spectra of 27 red giant stars located in 4 rich and intermediate-age LMC clusters. Their analysis yielded abundance ratios for about 20 atomic species, including α, iron group and neutron capture elements.

Kinematic studies of the MC cluster systems date from the 1980s. Freeman *et al.* (1983) analyzed a sample of 59 LMC clusters with individual radial velocities accurate to 10–20 km s^{-1}. They concluded that the LMC cluster system is consistent with a flattened disk rotating at $\simeq 40$ km s^{-1}, although the disk geometry and systemic velocity was found to be different for young and old clusters. Schommer *et al.* (1992) used a larger cluster sample with more accurate velocities and concluded that all LMC clusters in their sample have a single disk kinematics. This result has been recently confirmed at $\simeq 2$ km s^{-1} precision by Grocholski *et al.* (2006), who obtained metallicities and velocities from Ca II triplet lines for over 200 stars in 28 populous LMC clusters observed with the Very Large Telescope (*VLT*). The authors also conclude that the LMC has no metallicity gradient.

For more on spectroscopic studies of star clusters in both Clouds, we refer to the contributions by A. Ahumada, G. Bosch, A. Grocholski and A. Mucciarelli to this symposium.

7. Comparison with other star cluster systems

Knowledge on extragalactic clusters has increased immensely in the last two decades. Among the most important breakthroughs are the universal (or nearly so) GCLF, the existence of bimodality in the distribution of globular cluster colours and, more recently, the ability to measure cluster sizes. Correlations involving cluster luminosities, colours, sizes, location and host properties provide insight into the process of galaxy and cluster formation and evolution.

An interesting and relatively unexplored piece of information is the size distribution of star clusters. The distributions of half-light radii ($R_{\rm eff}$) of massive clusters in the Magellanic Clouds and in the Galaxy are shown in the left panel of Figure 7. The figure reproduces Figure 2 from Mackey *et al.* (2008). $R_{\rm c}$ values were transformed into $R_{\rm eff}$ using the relation quoted by Larsen (2001). For the Clouds, only clusters older than $\tau > 7$ Gyr are shown, in order to make them more comparable to their Galactic counterparts, which are all globular clusters. For the Galaxy, clusters located closer than $R_{\rm g} = 15$ kpc to the centre were eliminated from the figure, as these suffer stronger tidal effects, not occuring in the Clouds, and thus tend to be smaller ($R_{\rm eff} \simeq 3$ pc). The two distributions are very similar. The peak typical of Galactic globular clusters in the inner halo is still the dominant one. Two other peaks, at $R_{\rm eff} \simeq 7$ pc and at $R_{\rm eff} \simeq 15$ pc are also seen.

The right panel shows the distribution of star clusters with measured sizes in NGC 1380, which is a lenticular galaxy in the Fornax galaxy cluster. The data are from Chies-Santos *et al.* (2007). There is no cut in distance from the host centre, which likely explains the higher fraction of clusters at $R_{\rm eff} \simeq 3$ pc. Interestingly, secondary peaks, similar to but at smaller radii than those in the Galaxy and in the Clouds, are also present.

The existence of clusters more extended than typical globulars may be evidence of distinct populations of star clusters, possibly with different formation and evolution histories. In well studied luminous early-type galaxies, which are the ones that harbour the richest cluster systems, extended clusters may also be distinguished from the more common globulars in terms of other properties. Larsen & Brodie (2000) and Brodie & Larsen (2002), for instance, found extended clusters with $7 \leqslant R_{\rm eff} \leqslant 15$ pc in two nearby lenticular galaxies. These were named Faint Fuzzies, as they also tend to be underluminous

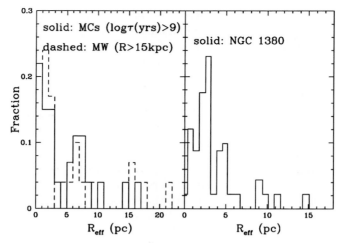

Figure 7. Left panel: Distribution of half-light radii of clusters in the Magellanic Clouds (solid) and in the Galaxy (dashed). Only clusters with $\log \tau(\text{yr}) > 9$ in the Mackey *et al.* (2003a,b) samples are included. The Galactic sample comes from Harris. (1996) (and its 2003 web update) and includes only clusters farther than 15 kpc from the Galactic centre, in order to avoid clusters more strongly affected by tidal effects. Right panel: distribution of R_{eff} for clusters in NGC 1380, which is a luminous S0 galaxy in the Fornax cluster. The data are from Chies-Santos *et al.* (2007).

compared to the dominant cluster population. Their metallicities were estimated spectroscopically, yielding [Fe/H] ~ -0.6. Finally, Burkert *et al.* (2005) found evidence that Faint Fuzzies in NGC 1023 are kinematically connected in a ring-like structure.

Diffuse star clusters (DSCs) may constitute another recently found class. These were found by Peng *et al.* (2006) in the ACS Virgo Cluster survey (Côté *et al.* 2004). Several of luminous early-type galaxies, most of them S0s, have a large population of extended clusters of much lower surface brightness ($\mu_g > 20$ mag arcsec^{-2}) than typical globulars. The DSCs are also redder than typical globular clusters, often having the same colour as the difuse light from the host.

A useful tool used by Peng *et al.* (2006) to separate different types of clusters is the luminosity-size diagram. The $M_V - R_{\text{eff}}$ plane is shown in Figure 8. Massive clusters in 3 markedly different environments for cluster formation and evolution, the Galaxy, the Magellanic Clouds and NGC 1380, are shown in the Figure. The data on M_V are from the same references as the sizes. The dominant population in the Galaxy and in NGC 1380 is made up of clusters with $R_{\text{eff}} = 1$–4 pc and $-9 \leqslant M_V \leqslant -6$. Only a handful of such globular-like clusters are found in the Clouds. The Clouds also contribute little to the FFs box, shown in the lower right of each panel. In fact, most of the clusters in the LMC and SMC tend to follow the low-surface brightness line used by Peng *et al.* (2006) to characterize the DSCs. This means that the LMC and SMC have a large number of massive clusters that can be considered as *structural counterparts* to DSCs, although they do not necessarily share the other properties associated to these objects.

Acknowledgements

I thank the SOC for the invitation to give this review talk. C. Bonatto, L. Kerber and D. Mackey kindly provided figures and data included in this review. Useful discussions preceding my conference presentation were carried out with colleagues in my home institution, especially O. Kepler, M. Pastoriza, E. Bica and C. Bonatto.

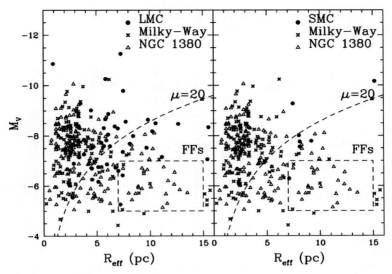

Figure 8. Left panel: M_V vs. R_{eff} relation for star clusters in the LMC (solid circles), the Galaxy (stars) and in the luminous S0 in Fornax galaxy clusters, NGC 1380 (triangles). Right panel: the same as in the previous panel, but now the SMC clusters are compared to those in the Galaxy and in NGC 1380. In both panels, the dashed box in the lower right is the region occupied by faint fuzzies. The dashed line corresponds to a surface brightness of $\mu_V = 20$ mag arcsec^{-2}, where DSCs preferentially lie.

References

Bastian, N. & Goodwin, S. 2006, *MNRAS*, 369, L9

Bica, E., Geisler, D., & Dottori, H. 1998, *AJ*, 116, 723

Bica, E., Schmitt, H., Dutra, C., & Oliveira, H. 1999, *AJ*, 117, 238

Bica, E. & Dutra, C. 2000, *AJ*, 119, 1214

Bica, E., Bonatto, C., Dutra, C., & Santos, J. 2008, *MNRAS*, 389, 678

Boutloukos, S. & Lamers, H. 2003, *MNRAS*, 338, 717

Brocato, E., Di Carlo, E., & Menna, G. 2001, *A&A*, 374, 523

Brodie, J. & Larsen, S. 2002, *AJ*, 124, 1410

Burkert, A., Brodie, J., & Larsen, S. 2005, *ApJ*, 628, 231

Carlson, L., Sabbi, E., Sirianni, M., *et al.* 2007, *ApJ*, 665, L109

Chandar, R., Fall, S. M., & Whitmore, B. 2006, *ApJ*, 650, L111

Chies-Santos, A., Santiago, B., & Pastoriza, M. 2007, *A&A*, 467, 1003

Côté, P., Blakeslee, J., Ferrarese, L., *et al.* 2004, *ApJS*, 153, 223

de Grijs, R., Gilmore, G., Mackey, A., Wilkinson, M., Beaulieu, S., Johnson, R., & Santiago, B. 2002, *MNRAS*, 337, 597

de Grijs, R. & Anders, P. 2006, *MNRAS*, 366, 295

de Grijs, R. & Goodwin, S. 2008, *MNRAS*, 383, 1000

Elson, R., Fall, M., & Freeman, K. 1987, *ApJ*, 323, 54

Elson, R., Freeman, K., & Lauer, T. 1989, *ApJ*, 347, L69

Freeman, K., Illingworth, G., & Oemler, A. 1983, *ApJ*, 272, 488

Gieles, M., Lamers, H., & Portegies Zwart, S. 2007, *ApJ*, 668, 268

Glatt, K., Gallagher, J., Grebel, E., *et al.* 2008, *AJ*, 135, 1106

Grocholski, A., Cole, A., Sarajedini, A., Geisler, D., & Smith, V. 2006, *AJ*, 132, 1630

Harris, W. 1996, *AJ*, 112, 1487

Hennekemper, E., Gouliermis, D., Henning, T., Brandner, W., & Dolphin, A. 2008, *ApJ*, 672, 914

Hernandez, X., Valls-Gabaud, D., & Gilmore, G. 2000, *MNRAS*, 317, 831

Hodge, P. & Sexton, J. 1966, *AJ*, 71, 363

Hodge, P. 1986, *PASP*, 98, 1113

Hodge, P. 1988, *PASP*, 100, 576

Hunter, D., Elmegreen, B., Dupuy, T., & Mortonson, M. 2003, *AJ*, 126, 1836

Hunter, I., Dufton, P., Smartt, S., *et al.* 2007, *A&A*, 466, 277

Javiel, S., Santiago, B., & Kerber, L. 2005 *A&A*, 431, 73

Kerber, L., Santiago, B., Castro, R., & Valls-Gabaud, D. 2002, *A&A*, 390, 121

Kerber, L. & Santiago, B. 2005, *A&A*, 435, 77

Kerber, L. & Santiago, B. 2006, *A&A*, 452, 155

Kerber, L., Santiago, B., & Brocato, E. 2007, *A&A*, 462, 139

Kron, G. 1956, *PASP*, 68, 125

Kruijssen, J. & Lamers, H. 2008, in J. G. Funes & E. M. Corsini (eds.), *Formation and Evolution of Galaxy Disks, ASP-CS*, 396, p. 149

Kumar, B., Sagar, R., & Melnick, J. 2008, *MNRAS*, 386, 1380

Kurth, O., Fritze-v, A., & Fricke, K. 1999, *A&AS*, 139, 19

Larsen, S. & Brodie, J. 2000, *AJ*, 120, 2938

Larsen, S. 2001, *AJ*, 122, 1782

Leitherer, C., Schaerer, D., Goldader, J., *et al.* 1999, *ApJS*, 123, 3

Lindsay, E. 1958, *MNRAS*, 118, 172

Lyngå, G. & Westerlund, B. 1963, *MNRAS*, 127, 31

Mackey, A. & Gilmore, G. 2003a, *MNRAS*, 338, 85

Mackey, A. & Gilmore, G. 2003b, *MNRAS*, 338, 120

Mackey, A., Wilkinson, M., Davies, M., & Gilmore, G. 2008, *MNRAS*, 386, 65

Marigo, P. & Girardi, L. 2007, *A&A*, 469, 239

Massey, P. 2002, *ApJS*, 141, 81

Mucciarelli, A., Ferraro, F., Origlia, L., & Fusi Pecci, F. 2007a, *AJ*, 133, 2053

Mucciarelli, A., Origlia, L., & Ferraro, F. 2007b, *AJ*, 134, 1813

Mucciarelli, A., Carretta, E., Origlia, L., & Ferraro, F. 2008, *AJ*, 136, 375

Noël, N., Aparício, A., Gallart, C., Hidalgo, S., Costa, E., Méndez, R. 2007, in A. Vazdekis, & R.F. Peletier (eds.), *Stellar Populations as Building Blocks of Galaxies*, IAU Conf.Proc. 241 (Cambridge: CUP), p. 373

Peng, E., Coté, P., Jordán, A., *et al.* 2006, *ApJ*, 639, 838.

Piatti, A., Bica, E., Geisler, D., & Clariá, J. 2003, *MNRAS*, 344, 965

Piatti, A., Sarajedini, A., Geisler, D., Gallart, C., & Wischnjewsky, M. 2007, *MNRAS*, 382, 1203

Rafelski, M. & Zaritsky, D. 2005, *AJ*, 129, 2701

Santiago, B., Beaulieu, S., Johnson, R., & Gilmore, G. 2001, *A&A*, 369, 74

Schmalzi, M., Gouliermis, D., Dolphin, A., & Henning, T. 2008, *ApJ*, 681, 290

Schommer, R., Olszewski, E., Suntzeff, N., & Harris, H. 1992, *AJ*, 103, 447

Siess, L., Dufour, E., & Forestini, M. 2000, *A&A*, 358, 593

Shapley, H. & Lindsay, E. 1963, *IrAJ*, 6, 74

Trundle, C., Dufton, P., Hunter, I., Evans, C., Lennon, D., Smartt, S., & Ryans, R. 2007, *A&A*, 471, 625

Wilkinson, M., Hurley, J., Mackey, A., Gilmore, G., & Tout, C. 2003, *MNRAS*, 343, 1025

Xin, Y., Deng, L., de Grijs, R., Mackey, A., & Han, Z. 2008, *MNRAS*, 384, 410

Zaritsky, D., Harris, J., & Thompson, I. 1997, *AJ*, 114, 1002

The Magellanic System: Stars, Gas, and Galaxies
Proceedings IAU Symposium No. 256, 2008
Jacco Th. van Loon & Joana M. Oliveira, eds.

© 2009 International Astronomical Union
doi:10.1017/S1743921308028299

Kinematical structure of the Magellanic System

Roeland P. van der Marel[1], Nitya Kallivayalil[2] and Gurtina Besla[3]

[1]Space Telescope Science Institute, 3700 San Martin Drive, Baltimore, MD 21218, USA
[2]MIT, Kavli Inst. for Astrophysics & Space Research, 70 Vassar Street,
Cambridge, MA 02139, USA
[3]Harvard-Smithsonian Center for Astrophysics, 60 Garden Street, Cambridge, MA 02138, USA

Abstract. We review our understanding of the kinematics of the LMC and the SMC, and their orbit around the Milky Way. The line-of-sight velocity fields of both the LMC and SMC have been mapped with high accuracy using thousands of discrete traces, as well as H I gas. The LMC is a rotating disk for which the viewing angles have been well established using various methods. The disk is elliptical in its disk plane. The disk thickness varies depending on the tracer population, with V/σ ranging from $\sim$ 2–10 from the oldest to the youngest population. For the SMC, the old stellar population resides in a spheroidal distribution with considerable line-of-sight depth and low V/σ. Young stars and H I gas reside in a more irregular rotating disk. Mass estimates based on the kinematics indicate that each Cloud is embedded in a dark halo. Proper motion measurements with HST show that both galaxies move significantly more rapidly around the Milky Way than previously believed. This indicates that for a canonical 10^{12} $M_\odot$ Milky Way the Clouds are only passing by us for the first time. Although a higher Milky Way mass yields a bound orbit, this orbit is still very different from what has been previously assumed in models of the Magellanic Stream. Hence, much of our understanding of the history of the Magellanic System and the formation of the Magellanic Stream may need to be revised. The accuracy of the proper motion data is insufficient to say whether or not the LMC and SMC are bound to each other, but bound orbits do exist within the proper motion error ellipse.

Keywords. stellar dynamics, celestial mechanics, astrometry, ISM: kinematics and dynamics, Galaxy: halo, galaxies: kinematics and dynamics, Magellanic Clouds, galaxies: structure, dark matter

1. Introduction

The Magellanic Clouds are two of the closest galaxies to the Milky Way, with the Large Magellanic Cloud (LMC) at a distance of $\sim$ 50 kpc and the Small Magellanic Cloud (SMC) at $\sim$ 62 kpc. Because of their proximity, they are two of the best-studied galaxies in the Universe. As such, they are a benchmark for studies on various topics, including stellar populations and the interstellar medium, microlensing by dark objects, and the cosmological distance scale. As nearby companions of the Milky Way with significant signs of mutual interaction, they have also been taken as examples of hierarchical structure formation in the Universe. For all these applications it is important to have an understanding of the kinematics of the LMC and the SMC, as well the kinematics (i.e., orbit) of their center of mass with respect to the Milky Way and with respect to each other. These topics form the subject of the present review. Other related topics, such as the more general aspects of the structure of the LMC and SMC, the nature of the LMC bar, the possible presence of fore- or background populations, and the large radii extent of the Clouds are not discussed here. The nature, origin, and models of the Magellanic

Stream are touched upon only briefly. All these topics are reviewed in other papers in this volume by, e.g., Majewski, Besla, Bekki, and others.

2. LMC kinematics

Kinematical observations for the LMC have been obtained for many tracers. The kinematics of gas in the LMC has been studied primarily using H I (e.g., Kim *et al.* 1998; Olsen & Massey 2007). Discrete LMC tracers which have been studied kinematically include star clusters (e.g., Schommer *et al.* 1992; Grocholski *et al.* 2006), planetary nebulae (Meatheringham *et al.* 1988), H II regions (Feitzinger, Schmidt-Kaler & Isserstedt 1977), red supergiants (Olsen & Massey 2007), red giant branch (RGB) stars (Zhao *et al.* 2003; Cole *et al.* 2005), carbon stars (e.g., van der Marel *et al.* 2002; Olsen & Massey 2007) and RR Lyrae stars (Minniti *et al.* 2003; Borissova *et al.* 2006). For the majority of tracers, the line-of-sight velocity dispersion is at least a factor ~ 2 smaller than their rotation velocity. This implies that on the whole the LMC is a (kinematically cold) disk system.

2.1. *General expressions*

To understand the kinematics of an LMC tracer population it is necessary to have a general model for the line-of-sight velocity field that can be fit to the data. All studies thus far have been based on the assumption that the mean streaming (i.e. the rotation) in the disk plane can be approximated to be circular. However, even with this simplifying assumption it is not straightforward to model the kinematics of the LMC. Its main body spans more than $20°$ on the sky and one therefore cannot make the usual approximation that "the sky is flat" over the area of the galaxy. Spherical trigonometry must be used, which yields the general expression (van der Marel *et al.* 2002; hereafter vdM02):

$$v_{\mathrm{los}}(\rho, \Phi) = s\, V(R')f \sin i \cos(\Phi - \Theta) + v_{\mathrm{sys}} \cos \rho$$
$$+ v_t \sin \rho \cos(\Phi - \Theta_t) + D_0(di/dt) \sin \rho \sin(\Phi - \Theta), \qquad (2.1)$$

with

$$R' = D_0 \sin \rho / f, \qquad f \equiv \frac{\cos i \cos \rho - \sin i \sin \rho \sin(\Phi - \Theta)}{[\cos^2 i \cos^2(\Phi - \Theta) + \sin^2(\Phi - \Theta)]^{1/2}}. \qquad (2.2)$$

Here, v_{los} is the observed component of the velocity along the line of sight. The quantities (ρ, Φ) identify the position on the sky with respect to the center: ρ is the angular distance and Φ is the position angle (measured from North over East). The kinematical center is at the center of mass (CM) of the galaxy. The quantities $(v_{\mathrm{sys}}, v_t, \Theta_t)$ describe the velocity of the CM in an inertial frame in which the sun is at rest: v_{sys} is the systemic velocity along the line of sight, v_t is the transverse velocity, and Θ_t is the position angle of the transverse velocity on the sky. The angles (i, Θ) describe the direction from which the plane of the galaxy is viewed: i is the inclination angle ($i = 0$ for a face-on disk), and Θ is the position angle of the line of nodes (the intersection of the galaxy plane and the sky plane). The velocity $V(R')$ is the rotation velocity at cylindrical radius R' in the disk plane. D_0 is the distance to the CM, and f is a geometrical factor. The quantity $s = \pm 1$ is the 'spin sign' that determines in which of the two possible directions the disk rotates.

The first term in equation (2.1) corresponds to the internal rotation of the LMC. The second term is the part of the line-of-sight velocity of the CM that is seen along the line of sight, and the third term is the part of the transverse velocity of the CM that is seen along the line of sight. For a galaxy that spans a small area on the sky (very small ρ), the second term is simply v_{sys} and the third term is zero. However, the LMC does not have a small angular extent and the inclusion of the third term is particularly important. It corresponds to a solid-body rotation component that at most radii exceeds in amplitude the contribution from the intrinsic rotation of the LMC disk. The fourth term in equation (2.1) describes the line-of-sight component due to changes in the inclination of the disk with time, as are expected due to precession and nutation of the LMC disk plane as it orbits the Milky Way (Weinberg 2000). This term also corresponds to a solid-body rotation component.

The general expression in equation (2.1) appears complicated, but it is possible to gain intuitive insight by considering some special cases. Along the line of nodes one has that $\sin(\Phi - \Theta) = 0$ and $\cos(\Phi - \Theta) = \pm 1$, so that

$$\hat{v}_{\mathrm{los}}(\mathrm{along}) = \pm[v_{tc}\sin\rho - V(D_0\tan\rho)\sin i\cos\rho]. \tag{2.3}$$

Here it has been defined that $\hat{v}_{\mathrm{los}} \equiv v_{\mathrm{los}} - v_{\mathrm{sys}}\cos\rho \approx v_{\mathrm{los}} - v_{\mathrm{sys}}$. The quantity $v_{tc} \equiv v_t\cos(\Theta_t - \Theta)$ is the component of the transverse velocity vector in the plane of the sky that lies along the line of nodes; similarly, $v_{ts} \equiv v_t\sin(\Theta_t - \Theta)$ is the component perpendicular to the line of nodes. Perpendicular to the line of nodes one has that $\cos(\Phi - \Theta) = 0$ and $\sin(\Phi - \Theta) = \pm 1$, and therefore

$$\hat{v}_{\mathrm{los}}(\mathrm{perpendicular}) = \pm w_{ts}\sin\rho. \tag{2.4}$$

Here it has been defined that $w_{ts} = v_{ts} + D_0(di/dt)$. This implies that perpendicular to the line of nodes $\hat{v}_{\mathrm{los}}$ is linearly proportional to $\sin\rho$. By contrast, along the line of nodes this is true only if $V(R')$ is a linear function of R'. This is not expected to be the case, because galaxies do not generally have solid-body rotation curves; disk galaxies tend to have flat rotation curves, at least outside the very center. This implies that, at least in principle, both the position angle Θ of the line of nodes and the quantity w_{ts} are uniquely determined by the observed velocity field: Θ is the angle along which the observed $\hat{v}_{\mathrm{los}}$ are best fit by a linear proportionality with $\sin\rho$, and w_{ts} is the proportionality constant.

2.2. *Carbon star kinematics*

vdM02 were the first to fit the velocity field expression in equation (2.1) in its most general form to a large sample of discrete LMC velocities. They modeled the data for 1041 carbon stars, obtained from the work of Kunkel, Irwin & Demers (1997) and Hardy, Schommer & Suntzeff (unpublished). The combined dataset samples both the inner and the outer parts of the LMC, although with a discontinuous distribution in radius and position angle. Figure 1 shows the data, with the best model fit overplotted. Overall, the model provides a good fit to the data. Olsen & Massey (2007) recently remodeled the same carbon star data (for which they obtained a similar fit as vdM02), as well as a large sample of red supergiant stars.

2.3. *Viewing angles and ellipticity*

The LMC inclination cannot be determined kinematically, but the line-of-nodes position angle can. vdM02 obtained $\Theta = 129.9° \pm 6.0°$ for carbon stars, whereas Olsen & Massey (2007) obtained $\Theta = 145.3°$ for red supergiants.

A more robust way to determine the LMC viewing angles is to use geometrical considerations, rather than kinematical ones (since this avoids the assumption that the orbits are circular). For an inclined disk, one side will be closer to us than the other. Tracers on that one side will appear brighter than similar tracers on the other side. To lowest order, the difference in magnitude between a tracer at the galaxy center and a similar tracer at a position (ρ, Φ) in the disk (as defined in Section 2.1) is

$$\mu = \left(\frac{5\pi}{180\ln 10}\right)\rho\tan i\sin(\Phi - \Theta), \tag{2.5}$$

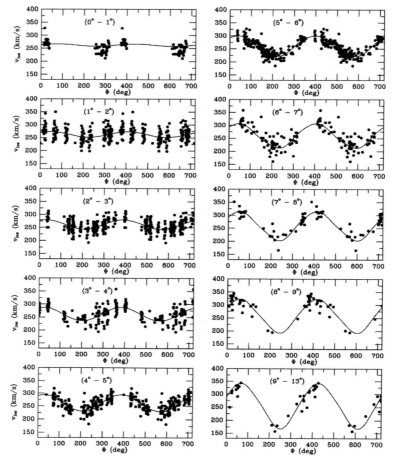

Figure 1. Carbon star line-of-sight velocity data from Kunkel *et al.* (1997) and Hardy *et al.* (unpublished), as a function of position angle Φ on the sky. The displayed range of the angle Φ is $0°$–$720°$, so each star is plotted twice. Each panel corresponds to a different range of angular distances ρ from the LMC center, as indicated. The curves show the predictions of the best-fitting circularly-rotating disk model from vdM02.

where the angular distance ρ is expressed in degrees. The constant in the equation is $(5\pi)/(180\ln 10) = 0.038$ magnitudes. Hence, when following a circle on the sky around the galaxy center one expects a sinusoidal variation in the magnitudes of tracers. The amplitude and phase of the variation yield estimates of the viewing angles (i, Θ).

Van der Marel & Cioni (2001) used a polar grid on the sky to divide the LMC area into several rings, each consisting of a number of azimuthal segments. The data from the DENIS and 2MASS surveys were used for each segment to construct near-IR color-magnitude diagrams (CMDs). For each segment both the modal magnitude of carbon stars (selected by color) and the magnitude of the RGB tip (TRGB) were determined. This revealed the expected sinusoidal variations at high significance, implying viewing angles $i = 34.7° \pm 6.2°$ and $\Theta = 122.5° \pm 8.3°$. There is an observed drift in the center of the LMC isophotes at large radii which is consistent with this result, when interpreted as a result of viewing perspective (van der Marel 2001). Also, Grocholski *et al.* (2006) found that the red clump distances to LMC star clusters are consistent with a disk-like configuration with these same viewing angles.

The aforementioned analyses are sensitive primarily to the structure of the outer parts of the LMC. Several other studies of the viewing angles have focused mostly on the region of the bar, which samples only the central few degrees. Nikolaev *et al.* (2004) analyzed a sample of more than 2000 Cepheids with lightcurves from MACHO data and obtained $i = 30.7° \pm 1.1°$ and $\Theta = 151.0° \pm 2.4°$. Persson *et al.* (2004) obtained $i = 27° \pm 6°$ and $\Theta = 127° \pm 10°$ from a much smaller sample of 92 Cepheids. Olsen & Salyk (2002) obtained $i = 35.8° \pm 2.4°$ and $\Theta = 145° \pm 4°$ from an analysis of variations in the magnitude of the red clump.

In summary, all studies agree that i is approximately in the range $30°$–$35°$, whereas Θ appears to be in the range $120°$–$150°$. The variations between results from different studies may be due to a combination of systematic errors, spatial variations in the viewing angles (warps and twists of the disk plane; van der Marel & Cioni 2001; Olsen & Salyk 2002; Subramaniam 2003; Nikolaev *et al.* 2004) combined with differences in spatial sampling between studies, contamination by possible out of plane structures, and differences between different tracer populations.

The LMC consists of an outer body that appears elliptical in projection on the sky, with a pronounced, off-center bar. The appearance in the optical wavelength regime is dominated by regions of strong star formation, and patchy dust absorption. However, when only RGB and carbon stars are selected from near-IR surveys such as 2MASS, the appearance of the LMC morphology is actually quite regular and smooth, apart from the central bar. Van der Marel (2001) found that at radii $r \gtrsim 4°$ the contour shapes converge to an approximately constant position angle $PA_{maj} = 189.3° \pm 1.4°$ and ellipticity $\epsilon = 0.199 \pm 0.008$. A disk that is intrinsically circular will appear elliptical in projection on the sky, with the major axis position angle PA_{maj} of the projected body equal to the line-of-nodes position angle Θ. The fact that for the LMC $\Theta \neq PA_{maj}$ implies that the LMC cannot be intrinsically circular. When the LMC viewing angles are used to deproject the observed morphology, this yields an in-plane ellipticity ϵ in the range ~ 0.2–0.3. This is larger than typical for disk galaxies, and is probably due to tidal interactions with either the SMC or the Milky Way.

2.4. *Transverse motion and kinematical distance*

As discussed in Section 2.1, the line-of-sight velocity field constrains the value of $w_{ts} = v_{ts} + D_0(di/dt)$. The carbon star analysis in vdM02 yields $w_{ts} = -402.9 \pm 13.0 \text{ km s}^{-1}$. With the assumptions of a known LMC distance $D_0 = 50.1 \pm 2.5$ kpc (based on the distance modulus $m - M = 18.50 \pm 0.10$ adopted by Freedman *et al.* (2001) on the basis of a review of all published work) and a constant inclination angle with time $(di/dt = 0)$ this yields an estimate of one component of the LMC transverse velocity. Some weaker constraints can also be obtained for the second component. The resulting region in LMC transverse velocity space implied by the carbon star velocity field is shown in Figure 8 of vdM02. This region is entirely consistent with the Hubble Space Telescope (*HST*) proper motion determination discussed in Section 4.1 below, and therefore provides an important consistency check on the latter. Alternatively, one can use the *HST* proper motion determination with the measured w_{ts} and the assumption that $di/dt = 0$ to obtain a kinematic distance estimate for the LMC. This yields $m - M = 18.57 \pm 0.11$, quite consistent with the Freedman *et al.* (2001) value.

Previous proper motion estimates for the LMC were lower than the current *HST* measurements. This introduced artifacts in previous analyses of the internal LMC velocity field (which ultimately depends on subtraction of the v_t term in eq. [2.1] from the observed line-of-sight velocity field). For example, the H I velocity field of the LMC presented by Kim *et al.* (1998) showed a pronounced S-shape in the zero-velocity contour. Olsen &

Massey recently showed that this S-shape straightens out when the LMC *HST* proper motion measurement is used instead. So this too provides an independent consistency check on the validity of the *HST* proper motion measurement.

2.5. *Rotation curve and mass*

The rotation curve of the LMC rises approximately linearly to $R' \approx 4$ kpc, and stays roughly flat at a value $V_{\rm rot}$ beyond that. The carbon star analysis of vdM02 with the HST proper motion measurement yields $V_{\rm rot} = 61 \, {\rm km \, s^{-1}}$. By contrast, for H I one obtains $V_{\rm rot} = 80 \, {\rm km \, s^{-1}}$ and for red supergiants $V_{\rm rot} = 107 \, {\rm km \, s^{-1}}$ (Olsen & Massey 2007). The random errors on these numbers are only a few km/s in each case, due to the large numbers of independent velocity samples. Piatek, Pryor & Olszewski (2008; hereafter P08) recently argued for an even higher $V_{\rm rot} = 120 \pm 15 \, {\rm km \, s^{-1}}$ based on rotation measurements in the plane of the sky (based on the same proper motion observations discussed in Section 4.1 below). All of these measurements are influenced by uncertainties in the LMC inclination. However, the uncertainties have different sign for the $V_{\rm rot}$ values inferred from line-of-sight velocities than and for those inferred from proper motions. The $V_{\rm rot}$ measurement of P08 becomes more consistent with the line-of-sight measurements if the inclination is lower than the canonical values quoted in Section 2.3. The $V_{\rm rot}$ estimates from line-of-sight velocities all have an additional uncertainty due to uncertainties in the LMC transverse motion. In the end, all $V_{\rm rot}$ estimates therefore have a systematic error of $\sim 10 \, {\rm km \, s^{-1}}$.

The differences between the $V_{\rm rot}$ estimates for various tracers are significant, and cannot be attributed to either random or systematic errors. The fact that the H I gas rotates faster than the carbon stars can probably be largely explained as a result of asymmetric drift, with the velocity dispersion of the carbon stars being higher than that of the H I gas (see Section 2.6 below). The difference between the rotation velocities of H I and red supergiants is more puzzling, and may point towards non-equilibrium dynamics. There are in fact clear disturbances in the kinematics of the various tracers (Olsen & Massey 2007). Moreover, the dynamical center of the H I is offset by ~ 1 kpc from the dynamical and photometric center of the stars (see, e.g., Cole *et al.* (2005) for a visual representation of the various relevant centroids of the LMC). All this complicates the inference of the underlying circular velocity of the gravitational potential.

If we use the $V_{\rm rot}$ of the H I as a proxy for the circular velocity, and use the fact that the carbon star rotation curve remains flat out to the outermost datapoint at ~ 9 kpc, then the implied LMC mass is $M_{\rm LMC}(9 \, {\rm kpc}) = (1.3 \pm 0.3) \times 10^{10} \, {\rm M_\odot}$. The mass will continue to rise linearly beyond that radius for as long as the rotation curve remains flat. By contrast, the total stellar mass of the LMC disk is $\sim 2.7 \times 10^9 \, {\rm M_\odot}$ and the mass of the neutral gas in the LMC is $\sim 0.5 \times 10^9 \, {\rm M_\odot}$ (Kim *et al.* 1998). The combined mass of the visible material in the LMC is therefore insufficient to explain the dynamically inferred mass, and the LMC must be embedded in a dark halo.

2.6. *Velocity dispersion and vertical structure*

As in the Milky Way, younger populations have a smaller velocity dispersion (and hence a smaller scale height) than older populations. Measurements, in order of increasing dispersion, include: $\sim 9 \, {\rm km \, s^{-1}}$ for red supergiants (Olsen & Massey 2007); $\sim 16 \, {\rm km \, s^{-1}}$ for H I gas (Kim *et al.* 1998); $\sim 20 \, {\rm km \, s^{-1}}$ for carbon stars (vdM02); $\sim 25 \, {\rm km \, s^{-1}}$ for RGB stars (Zhao *et al.* 2003; Cole *et al.* 2005); $\sim 30 \, {\rm km \, s^{-1}}$ for star clusters (Schommer *et al.* 1992; Grocholski *et al.* 2006); $\sim 33 \, {\rm km \, s^{-1}}$ for old long-period variables (Bessell, Freeman & Wood 1986); $\sim 40 \, {\rm km \, s^{-1}}$ for the lowest metallicity red giant branch stars with [Fe/H] < -1.15 (Cole *et al.* 2005); and $\sim 50 \, {\rm km \, s^{-1}}$ for RR Lyrae stars (Minniti *et al.* 2003; Borissova *et al.* 2006).

For the stars with the highest dispersions it has been suggested that they may form a halo distribution, and not be part of the LMC disk. On the other hand, this remains unclear, since the kinematics of these stars have typically been observed only in the central region of the LMC. Therefore, it is not known whether the rotation properties of these populations are consistent with being a separate halo component. In fact, the surface density distribution of the LMC RR Lyrae stars is well fit by an exponential with the same scale length as inferred for other tracers known to reside in the disk (Alves 2004). Either way, the vertical extent of all LMC populations is certainly significant. For example, even the (intermediate-age) carbon stars only have $V/\sigma \approx 3$. For comparison, the thin disk of the Milky Way has $V/\sigma \approx 9.8$ and its thick disk has $V/\sigma \approx 3.9$.

The velocity residuals with respect to a rotating disk model do not necessarily follow a Gaussian distribution. Although Zhao *et al.* (2003) did not find large deviations from a Gaussian for RGB stars, Graff *et al.* (2000) found that the carbon star residuals are better fit by a sum of Gaussians. More recently, Olsen & Massey (2007) showed that some fraction of both carbon stars and red supergiants have peculiar kinematics that suggest an association with tidally disturbed features previously identified in H I.

3. SMC kinematics

The SMC structure and kinematics are less well studied and understood than those of the LMC. The morphological appearance in blue optical light is patchy and irregular. Kinematical observations of H I and young stars reveal ordered rotation that indicates that these tracers may reside in a disk. However, detailed velocity field fits using equation (2.1) have not been attempted. Stanimirović, Staveley-Smith & Jones (2004) found that the H I rotation curve in the SMC rises almost linearly to $V_{\rm rot} \approx 50$ km s^{-1} at the outermost datapoint (~ 3.5 kpc), with no signs of flattening. The implied dynamical mass inside this radius is 2.4×10^9 M$_\odot$. By contrast, the total stellar mass of the SMC is $\sim 3.1 \times 10^8$ M$_\odot$ and the mass of the neutral gas is 5.6×10^8 M$_\odot$. The combined mass of the visible material in the SMC is therefore insufficient to explain the dynamically inferred mass, and the SMC must be embedded in a dark halo.†

Evans & Howarth (2008) obtained velocities for 2045 young (O, B, A) stars in the SMC, and found a velocity gradient of similar slope as seen in the H I gas. Surprisingly though, they find a position angle for the line of maximum velocity gradient that is quite different — almost orthogonal to that seen in H I. This may be an artifact of the different spatial coverage of the two studies (Evans & Howarth did not observe in the North-East region where H I velocities are largest), since it would be hard to find a physical explanation for a significant difference in kinematics between H I gas and young stars.

When the old red stars that trace most of the stellar mass are isolated using CMDs, the morphological appearance of the SMC is more spheroidal (Zaritsky *et al.* 2000, 2002; Cioni, Habing & Israel 2000; Maragoudaki *et al.* 2001). Harris & Zaritsky (2006) studied the kinematics of 2046 RGB stars and inferred a velocity dispersion $\sigma = 27.5 \pm 0.5$km s^{-1}. This is similar to the dispersion of the young stars observed by Evans & Howarth (2008), but unlike the young stars, the older RGB stars do not show much rotation. Their low $V_{\rm rot}/\sigma$ is consistent with what is typical for dE and dSph galaxies. Hence, the SMC may be more akin to those galaxy types than to other more irregular systems.

Studies of the distances of individual tracers in the SMC have shown it to be much more vertically extended than would be expected for a disk galaxy. Crowl *et al.* (2001)

---

† These values are based on the analysis in Stanimirović *et al.* (2004), although those authors do not draw the same conclusion.

mapped the distances of star clusters using red clump magnitudes. They argued that the SMC has axial ratios of 1:2:4, and is viewed almost pole on. While different authors have found a range of other axial ratios using different types of tracers, most authors agree that the SMC has a considerable line-of-sight depth.

4. Orbital history of the Magellanic Clouds

To understand the history of the Magellanic System and the origin of the Magellanic Stream it is important to know the orbit of the Magellanic Clouds around the Milky Way. This requires for each Cloud knowledge of all three of the velocity components of the center of mass. Line-of-sight velocities can be accurately determined from the Doppler velocities of tracers. However, determination of the velocity in the plane of the sky through proper motions is much more difficult. This was the primary obstacle for a long time, but recent breakthroughs have now yielded considerable progress.

4.1. *Proper motions*

Previous attempts at measuring the proper motions (PMs) of the LMC and SMC from the ground or with Hipparcos were reviewed in vdM02. However, this earlier work has now been largely superseded by the studies performed with the ACS/HRC on *HST* by Kallivayalil *et al.* (2006a,b; hereafter K06a,b). Two epochs of data were obtained with a ~ 2 year time baseline for 21 fields in the LMC and 5 fields in the SMC, all centered on background quasars identified from the MACHO database (Geha *et al.* 2003). PMs were obtained for each field by measuring the average shift of the stars with respect to the background quasar. Upon correction for the orientation and rotation of the LMC disk, each field yields an independent estimate of the center-of-mass PM. The average for the different fields yields the final estimate, while the RMS among the results from the N different fields yields the PM error $\text{RMS}/\sqrt{N}$.

K06a,b obtained for the PMs in the West and East directions that

$$\mu_W = -2.03 \pm 0.08 \,\text{mas}\,\text{yr}^{-1}, \qquad \mu_N = 0.44 \pm 0.05 \,\text{mas}\,\text{yr}^{-1} \qquad (LMC)$$
$$\mu_W = -1.16 \pm 0.18 \,\text{mas}\,\text{yr}^{-1}, \qquad \mu_N = -1.17 \pm 0.18 \,\text{mas}\,\text{yr}^{-1} \qquad (SMC). \quad (4.1)$$

The same data were reanalyzed more recently by P08. Using an independent analysis with different software and point spread function models they obtained that

$$\mu_W = -1.96 \pm 0.04 \,\text{mas}\,\text{yr}^{-1}, \qquad \mu_N = 0.44 \pm 0.04 \,\text{mas}\,\text{yr}^{-1} \qquad (LMC)$$
$$\mu_W = -0.75 \pm 0.06 \,\text{mas}\,\text{yr}^{-1}, \qquad \mu_N = -1.25 \pm 0.06 \,\text{mas}\,\text{yr}^{-1} \qquad (SMC). \quad (4.2)$$

P08 made magnitude-dependent corrections for small charge transfer inefficiency effects. By contrast, K06a,b assumed that these effects (always along the detector y axis) average to zero over all fields because of the random telescope orientations used for different fields. The fact the the results from these two studies are generally in good agreement for the LMC confirms the validity of this assumption. However, the explicit correction applied by P08 does yield better agreement between different fields, and therefore smaller errorbars. For the SMC, the results for the individual fields are in good agreement between the studies. However, the studies used different methods for weighted averaging of the fields, with K06a,b being more conservative and allowing for potential unknown systematic effects. This produces larger errorbars than in P08 and a significant difference in μ_W.

Transformation of the PMs to a space velocity in $\text{km}\,\text{s}^{-1}$ requires knowledge of the distance D_0. For the LMC, $D_0 \approx 50.1\,\text{kpc}$, so that $1\,\text{mas}\,\text{yr}^{-1}$ corresponds to $238\,\text{km}\,\text{s}^{-1}$. For the SMC, $D_0 \approx 61.6\,\text{kpc}$, so that $1\,\text{mas}\,\text{yr}^{-1}$ corresponds to $293\,\text{km}\,\text{s}^{-1}$. After transformation of the PM to $\text{km}\,\text{s}^{-1}$, it can be combined with the observed center-of-mass line-of-sight velocity to obtain the full three-dimensional velocity vector. For the LMC, $v_{\rm sys} = 262.2 \pm 3.4 \, {\rm km \, s^{-1}}$ (vdM02); and for the SMC, $v_{\rm sys} = 146 \pm 0.6 \, {\rm km \, s^{-1}}$ (Harris & Zaritsky 2006). The resulting vectors can be corrected for the solar reflex motion and transformed to the Galactocentric rest-frame as described in vdM02. For the LMC this yields that the motion has a radial component of $V_{\rm rad} = 89 \pm 4 \, {\rm km \, s^{-1}}$ pointing away from the Galactic center, and a tangential component of $V_{\rm tan} = 367 \pm 18 \, {\rm km \, s^{-1}}$.†

4.2. *Orbit around the Milky Way*

The combination of a small but positive radial velocity and a tangential velocity that exceeds the circular velocity of the Milky Way halo implies that the Clouds must be just past pericenter. The calculation of an actual orbit requires detailed knowledge of the gravitational potential of the Milky Way dark halo. Past work had generally assumed that the dark halo can be approximated by a spherical logarithmic potential. Estimates of the transverse velocities of the Clouds based on models of the Magellanic Stream had suggested that for the LMC $V_{\rm tan} = 287 \, {\rm km \, s^{-1}}$ (e.g., Gardiner, Sawa & Fujimoto 1994; Gardiner & Noguchi 1996). This then yielded an orbit with an apocenter to pericenter ratio of $\sim 2.6 : 1$ and an orbital period of $\sim 1.6 \, {\rm Gyr}$. This orbit was adopted by most subsequent modeling studies of the Magellanic System. However, the new *HST* PM measurements significantly revise this view. The observed $V_{\rm tan}$ is $80 \pm 16 \, {\rm km \, s^{-1}}$ larger than the Gardiner & Noguchi (1996) value, and therefore inconsistent with it at $\sim 5\sigma$. The observed value implies a much larger apocenter distance (in excess of 200 kpc) at which the assumption of a logarithmic potential is not a good assumption.

Motivated by these considerations, Besla *et al.* (2007) performed a new study of the Magellanic Cloud orbits using an improved Milky Way model, combined with the K06a,b *HST* PMs. The Milky Way model was chosen similar to that proposed in Klypin, Zhao & Somerville (2002). It consists of disk, bulge, hot gaseous halo, and dark halo components. The dark halo has a ΛCDM-motivated NFW potential with adiabatic contraction. In the fixed Milky Way potential, the orbits of the LMC and SMC were integrated backwards in time, starting from the current observed positions and velocities. The extent of the galaxies was taken into account in the calculation of their mutual gravitational interaction, and a parameterized prescription was used to account for dynamical friction. The gravitational influence of M 31 can be taken into account in this formalism as well, but this make little difference to the results (Kallivayalil 2007; Shattow & Loeb 2009).

The most favored Milky Way model presented by Klypin *et al.* (2002) has a total mass $M = 10^{12} \, {\rm M_\odot}$. In this model, the escape velocity at 50 kpc is $\sim 380 \, {\rm km \, s^{-1}}$. This is very similar to the observed $V_{\rm tan}$ of the LMC, and as a result, the inferred orbit is approximately parabolic, with no previous pericenter passage. In other words, the Magellanic Clouds are passing by the Milky Way now for the first time. To obtain an orbit that is significantly bound, μ_W would have to be larger by $\sim +0.3 \, {\rm mas \, yr^{-1}}$ (4σ with the K06a errorbar, or 7σ with the P08 errorbar). Alternatively, it is possible that the Milky Way is more massive (Smith *et al.* 2007; Shattow & Loeb 2008). A mass of $M = 2 \times 10^{12} \, {\rm M_\odot}$ is more or less the largest mass consistent with the available observational constraints (see also the discussion in van der Marel & Guhathakurta 2008). This would produce a bound orbit. However, with either the larger Milky Way mass or with the larger μ_W, the orbit would still be quite different than has been previously assumed in models of the Magellanic System. There would be only 1 previous pericenter passage, the apocenter distance would be 400 kpc or more, and the period would be

---

† These velocities are based on the K06a PM values. However, use of the P08 PM values would yield similar velocities that would not alter the arguments in the remainder of the text.

6−7 Gyr.‡ Therefore, the new PM results drastically alter our view of the history of the Magellanic System.

The view that the Magellanic Clouds may be passing by the Milky Way for the first time may seem revolutionary at first. However, there are arguments to consider this reasonable. Van den Bergh (2006) pointed out that the LMC and SMC are unusual in that they are the only satellites in the Local Group that are both gas rich and located close to their parent galaxy. He suggested based on this that the Magellanic Clouds are interlopers that were originally formed in the outer reaches of the Local Group. Moreover, cosmological simulations show that: (a) accretion of LMC-sized subhalos by Milky-Way sized halos is common since $z \sim 1$; and (b) finding long-term satellites with small pericenter distances around Milky-Way sized halos is rare (Kazantzidis, Zentner & Bullock 2008). Therefore, a scenario in which the Magellanic Clouds are passing by the Milky Way now for the first time seems more likely from a purely cosmological perspective than a scenario in which they have been satellites for many orbital periods of 1–2 Gyr each.

4.3. *LMC–SMC orbit*

Although the Magellanic Clouds may not be bound to the Milky Way, it would be much more unlikely for the Magellanic Clouds not to be bound to each other. The likelihood of two satellite galaxies running into each other by chance is quite low. Also, various properties of the Clouds (such as their common H I envelope) suggest that they have been associated with each other for a significant time. The K06a,b PMs imply a relative velocity between the SMC and LMC of $105 \pm 42 \, \mathrm{km \, s^{-1}}$. Orbit calculations (K06b; Besla *et al.* 2007) show that the error bar on this is too large to say with any confidence whether or not they are indeed bound. However, binary orbits do exist within the 1σ error ellipse, so there seems little reason to depart from this null hypothesis. Indeed, there are allowed orbits that have close passages between the Clouds at ~ 0.3 Gyr and ~ 1.5 Gyr in the past (Besla *et al.*, these proceedings). These are the time scales that have been previously associated with the formation of the Magellanic Bridge and Stream, respectively.

4.4. *Magellanic Stream*

The Magellanic Stream is discussed in detail in other contributions in this volume. One important thing to note though in the present context is that the new insights into the orbit of the Magellanic Clouds around the Milky Way drastically affect our understanding of the Magellanic Stream. The Stream does not lie along the projected path on the sky traced by the LMC and SMC orbits, and the HI velocity along the Stream is not as steep as that along the orbits (Besla *et al.* 2007). This is inconsistent with purely tidal models of the stream (e.g., Gardiner & Noguchi 1996; Connors, Kawata & Gibson 2006). Moreover, the more limited number of passages through the Milky Way disk, and the larger radius at which this occurs, imply that the ram pressure models that have been proposed (e.g., Moore & Davis 1994; Mastropietro *et al.* 2005) probably won't work either. It is therefore essential that models of the Magellanic Stream be revisited, with an eye towards inclusion of new physics and exploration of new scenarios (see Besla *et al.*, these proceedings). The recent finding by Nidever, Majewski & Burton (2008) that one filament of the Magellanic Stream, containing more than half its gas mass, can be traced back to the 30 Doradus star forming region in the LMC is particularly interesting in this context. This indicates that an outflow may have created or contributed to the Stream, which has not been addressed in previous models.

---

‡ if this were the case, then the mass build-up of the Milky Way with time, as in, e.g., Wechsler *et al.* (2002), would also have to be taken into account for calculation of an accurate orbit.

5. Concluding remarks

The kinematics of the LMC are now fairly well understood, with velocities of thousands of individual tracers of various types having been fitted in considerable detail with (thick) disk models. Questions that remain open for further study include the reality and origin of kinematical differences between different stellar tracer populations, the differences between the gaseous and stellar kinematics, and the amount and origin of non-equilibrium features in the kinematics. The kinematics of the SMC are understood more poorly, but appear generally consistent with being a spheroidal system of old stars with an embedded irregular disk of gas and young stars.

The *HST* PM work has provided the most surprising results in recent years, with important implications for both the history of the Magellanic System and the origin of the Magellanic Stream. Of course, it is natural in discussions about this to wonder about the robustness of the observational results. It should be noted in this context that many experimental features and consistency checks are built in that support the general validity of the *HST* PM results. These include: (1) the use of random telescope orientations causes systematic errors tied to the detector frame to cancel out when averaging over all fields; (2) the final PM errors are based on the observed scatter between fields, with no assumptions about the source and nature of the underlying errors; (3) two groups used different methods to analyze the same data and obtained consistent results; (4) P08 managed to measure a PM rotation curve for the LMC that is broadly consistent with expectation, which would have been impossible if the PM errors were in reality larger than claimed; (5) the difference between the LMC and SMC PMs is more or less consistent with expectation for a binary orbit, which would not generally have been the case if the measurements suffer from unknown systematics; (6) the LMC PM is consistent with expectation based on the line-of-sight velocity field of carbon stars (see Section 2.4); and (7) the LMC PM leads to an H I velocity field with a straight zero-velocity curve, by contrast to previously assumed values (see Section 2.4).

One interesting feature in the observational PM results is that with the P08 PM values, there are no bound LMC-SMC orbits, given their different μ_W and smaller error bars for the SMC compared to the K06b results. However, the SMC PM is significantly less certain that that for the LMC, due to the smaller number of fields observed with *HST*, and the fact that most of them were observed at a similar telescope orientation (which implies that potential systematic errors that are fixed in the detector frame do not average out when the results from different fields are combined). This underscores the need for additional PM observations. A third epoch of observations for most fields has already been obtained with *HST*/WFPC2, and preliminary analysis supports the validity of the results based on the first two epochs (Kallivayalil *et al.*, these proceedings). A fourth epoch is planned with *HST*/ACS and *HST*/WFC3 in 2009. With the increased time baselines and use of multiple different instruments it will be possible to further reduce random errors and constrain possible systematic errors. In turn, this will allow new scientific problems to be addressed, such as the internal proper motion kinematics of the Clouds, and their rotational parallax distances (the distances obtained by equating the line-of-sight and proper motion rotation curves).

References

Alves, D. R. 2004, *ApJ*, 601, L151

Besla, G., Kallivayalil, N, Hernquist, L., Robertson, B., Cox, T. J., van der Marel, R. P., & Alcock, C. 2007, *ApJ*, 668, 949

Bessell, M. S., Freeman, K. C., & Wood, P. R. 1986, *ApJ*, 310, 710

Borissova, J., Minniti, D., Rejkuba, M., & Alves, D. 2006, *A&A*, 460, 459

Cioni, M. -R. L., Habing, H. J., & Israel, F. P. 2000, *A&A*, 358, L9

Cole, A. A., Tolstoy, E., Gallagher, J. S., III, & Smecker-Hane, T. A. 2005, *AJ*, 129, 1465

Connors, T. W., Kawata, D., & Gibson, B. K. 2006, *MNRAS*, 371, 108

Crowl, H. H., Sarajedini, A., Piatti, A. E., Geisler, D., Bica, E., Claria, J. J., & Santos, J. F. C., Jr. 2001, *AJ*, 122, 220

Evans, C. J. & Howarth, I. D. 2008, *MNRAS*, 386, 826

Feitzinger, J. V., Schmidt-Kaler, T., & Isserstedt, J. 1977, *A&A*, 57, 265

Freedman, W. L., Madore, B. F., Gibson, B. K., *et al.* 2001, *ApJ*, 553, 47

Gardiner, L. T. & Noguchi, M. 1996, *MNRAS*, 278, 191

Gardiner, L. T., Sawa, T., & Fujimoto, M. 1994, *MNRAS*, 266, 567

Geha, M., Alcock, C., Allsman, R. A., *et al.* 2003, *AJ*, 125, 1

Graff, D. S., Gould, A. P., Suntzeff, N. B., Schommer, R. A., & Hardy, E. 2000, *ApJ*, 540, 211

Grocholski, A. J., Cole, A. A., Sarajedini, A., Geisler, D., & Smith, V. V. 2006, *AJ*, 132, 1630

Harris, J. & Zaritsky, D. 2006, *AJ*, 131, 2514

Kallivayalil, N., van der Marel, R. P., Alcock, C., Axelrod, T., Cook, K. H., Drake, A. J., & Geha, M. 2006a, *ApJ*, 638, 772 (K06a)

Kallivayalil, N., van der Marel, R. P., & Alcock, C. 2006b, *ApJ*, 652, 1213 (K06b)

Kallivayalil, N. 2007, PhD thesis, Harvard University

Kazantzidis, S., Zentner, A. R., & Bullock, J. S. 2008, *ApJ*, in press [arXiv:0807.2863]

Kim, S., Staveley-Smith, L., Dopita, M. A., Freeman, K. C., Sault, R. J., Kesteven M. J., & McConnell, D. 1998, *ApJ*, 503, 674

Klypin, A., Zhao., H. S., & Somerville, R. S. 2002, *ApJ*, 573, 597

Kunkel, W. E., Irwin, M. J., & Demers, S. 1997, *A&AS*, 122, 463

Maragoudaki, F., Kontizas, M., Morgan, D. H., Kontizas, E., Dapergolas, A., & Livanou, E. 2001, *A&A*, 379, 864

Mastropietro, C., Moore, B., Mayer, L., Wadsley, J., & Stadel, J. 2005, *MNRAS*, 363, 509

Meatheringham, S. J., Dopita, M. A., Ford, H. C., & Webster, B. L. 1988, *ApJ*, 327, 651

Minniti, D., Borissova, J., Rejkuba, M., Alves, D. R., Cook, K. H., & Freeman, K. C. 2003, *Science*, 301, 1508

Moore, B. & Davis, M. 1994, *MNRAS*, 270, 209

Nidever, D. L., Majewski, S. R., & Burton, W. B. 2008, *ApJ*, 679, 432

Nikolaev, S., Drake, A. J., Keller, S. C., Cook, K. H., Dalal, N., Griest, K., Welch, D. L., & Kanbur, S. M. 2004, *ApJ*, 601, 260

Olsen, K. A. G. & Salyk, C. 2002, *AJ*, 124, 2045

Olsen, K. A. G. & Massey, P. 2007, *ApJ*, 656, L61O

Persson, S. E., Madore, B. F., Krzeminski, W., Freedman, W. L., Roth, M., & Murphy, D. C. 2004, *AJ*, 128, 2239

Piatek, S., Pryor, C., & Olszewski, E. W. 2008, *ApJ*, 135, 1024 (P08)

Schommer, R. A., Suntzeff, N. B., Olszewski, E. W., & Harris, H. C. 1992, *AJ*, 103, 447

Shattow, G. & Loeb, A. 2009, *MNRAS*, 392, L21

Smith, M. C., Ruchti, G. R., Helmi, A., *et al.* 2007, *MNRAS*, 379, 755

Stanimirović, S., Staveley-Smith, L., & Jones, P. 2004, *ApJ*, 604, 176

Subramaniam, A. 2003, *ApJ*, 598, L19

van den Bergh, S. 2006, *AJ*, 132, 1571

van der Marel, R. P. & Cioni, M.-R. 2001, *AJ*, 122, 1807

van der Marel, R. P. 2001, *AJ*, 122, 1827

van der Marel, R. P., Alves, D. R., Hardy, E., & Suntzeff, N. B. 2002, *AJ*, 124, 2639 (vdM02)

van der Marel, R. P. & Guhathakurta, P. 2008, *ApJ*, 678, 187

Wechsler, R. H., Bullock, J. S., Primack, J. R., Kravtsov, A. V., & Dekel, A. 2002, *ApJ*, 568, 52

Weinberg, M. D. 2000, *ApJ*, 532, 922

Zaritsky, D., Harris, J., Grebel, E. K., & Thompson, I. B. 2000, *ApJ*, 534, L53

Zaritsky, D., Harris, J., Thompson, I. B., Grebel, E. K., & Massey, P. 2002, *AJ*, 123, 855

Zhao, H., Ibata, R. A., Lewis, G. F., & Irwin, M. J. 2003, *MNRAS*, 339, 701

The Magellanic System: Stars, Gas, and Galaxies
Proceedings IAU Symposium No. 256, 2008
Jacco Th. van Loon & Joana M. Oliveira, eds.

© 2009 International Astronomical Union
doi:10.1017/S1743921308028305

New analysis of the proper motions of the Magellanic Clouds using *HST*/WFPC2

Nitya Kallivayalil[1], Roeland P. van der Marel[2], Jay Anderson[2], Gurtina Besla[3] and Charles Alcock[3]

[1] Pappalardo Fellow, MIT Kavli Inst. for Astrophysics & Space Research, 70 Vassar Street, Cambridge, MA, 02139, USA; email: nitya@mit.edu

[2] Space Telescope Science Institute, 3700 San Martin Drive, Baltimore, MD 21218, USA

[3] Harvard-Smithsonian Center for Astrophysics, 60 Garden Street, Cambridge, MA 02138, USA

Abstract. In *HST* Cycles 11 and 13 we obtained two epochs of ACS/HRC data for fields in the Magellanic Clouds centered on background quasars. We used these data to determine the proper motions of the LMC and SMC to better than 5% and 15% respectively. The results had a number of unexpected implications for the Milky Way-LMC-SMC system. The implied three-dimensional velocities were larger than previously believed and close to the escape velocity in a standard 10^{12} M$_\odot$ Milky Way dark halo, implying that the Clouds may be on their first passage. Also, the relative velocity between the LMC and SMC was larger than expected, leaving open the possibility that the Clouds may not be bound to each other. To further verify and refine our results we requested an additional epoch of data in Cycle 16 which is being executed with WFPC2/PC due to the failure of ACS. We present the results of an ongoing analysis of these WFPC2 data which indicate good consistency with the two-epoch results.

Keywords. astrometry, galaxies: interactions, galaxies: kinematics and dynamics, Magellanic Clouds

1. Introduction

The Large and Small Magellanic Clouds (LMC & SMC) at distances of ~ 50 kpc from the Sun, and ~ 25 kpc from the Galactic Plane, provide one of our best probes of the composition and properties of the Galactic dark halo, and have long been upheld as the poster-child for a strongly interacting system, both with each other and with the Milky Way (MW). It has commonly been assumed that the Clouds have made multiple pericentric passages about the MW, and indeed current formation theories for the Magellanic Stream, which may involve tidal or ram-pressure forces, require multiple pericentric passages in order to be viable stripping mechanisms (Gardiner & Noguchi 1996, hereafter GN96; Connors *et al.* 2006; Mastropietro *et al.* 2005; Yoshizawa & Noguchi 2003; Lin *et al.* 1995; Moore & Davis 1994; Heller & Rohlfs 1994; Lin & Lynden-Bell 1982; Murai & Fujimoto 1980).

However, recent high-precision proper motion measurements for the Clouds made by our group with two epochs of ACS High Resolution Camera (HRC) data in Cycles 11 and 13, where we measured the proper motion of LMC stars relative to background quasars, imply that the LMC tangential velocity is ~ 370 km s^{-1}, approximately 100 km s^{-1} higher than previously thought (Kallivayalil *et al.* 2006a, Kallivayalil *et al.* 2006b; hereafter K1 & K2). The proper motion values of GN96, which have been adopted in all theoretical models of the formation of the Stream thus far, are not consistent with the new *HST* result. In particular, for the LMC there is a 7σ difference. The values for the

SMC are in more acceptable agreement (3σ difference). The new measurements also indicate a significant relative velocity between the LMC & SMC of 105 ± 42 km s^{-1}. This has been assumed to be closer to ~ 60 km s^{-1} in theoretical models, i.e., approximately the value for the SMC to be on a circular orbit around the LMC.

These results have surprising physical implications which require a reconsideration of the formation mechanism for the Stream. These include the possibility that the LMC may only be on its first passage about the MW (Besla *et al.* 2007; hereafter B07). B07 demonstrated this by studying the past orbital paths of the LMC using our observed proper motions and errors in a ΛCDM-motivated dark halo with a NFW profile (Navarro *et al.* 1996). This gave rise to starkly different trajectories for the LMC than those produced in a simple isothermal halo potential: even in the "best case" scenario (proper motion in the west direction, μ_W, $+4\sigma$), the LMC only completes 1 orbit within 10 Gyr and reaches an apogalacticon distance of 550 kpc (see Figure 4 in B07). Subsequently, this has led to a series of papers exploring whether the LMC is indeed *bound* to the MW (e.g., Shattow & Loeb 2009; Wu *et al.* 2008). Perhaps even more provocative is the possibility that that the LMC & SMC may have only recently become a binary system (K2, although see Besla *et al.* these proceedings).

Such large motions were unexpected in light of our previous understanding of the MW-LMC-SMC system, and it is therefore crucial that they be verified and further improved through the acquisition of additional data. Because of the large distances involved, even small differences in the proper motions can give vastly different orbits for the Clouds (1 mas yr^{-1} ≈ 238 km s^{-1} at the distance of the LMC). We thus applied for and were successful in getting a third epoch of snapshot imaging for our quasar-fields in Cycle 16. These executed with WFPC2 due to the failure of ACS. In this paper we present the preliminary results of an on-going analysis of these WFPC2 data. In § 2 we present the WFPC2 data and analysis strategy with special attention to the relative size of the position errors vis-a-vis ACS. We describe the main sources of systematic error in both the ACS and WFPC2 data. In § 3 we present results based on simple cuts aimed at minimizing these systematic errors, and discuss the expected overall improvement that this additional epoch affords. Since our analysis is still in the process of being refined, we present the results only in comparative fashion, both to our ACS-only two-epoch results (K1 & K2), and to those of Piatek *et al.* 2008 (hereafter P08), who recently re-analyzed our ACS data using their own methods to obtain results that are consistent with ours. A brief summary and future prospects are presented in § 4.

2. WFPC2 data & analysis

2.1. *Description of observations*

In our first epoch program (Cycle 11, PI: Alcock), we imaged fields around 40 quasars behind the Magellanic Clouds (Geha *et al.* 2003) with the ACS High Resolution Channel (HRC) in snapshot mode (54 targets had been approved and we achieved a 74% completion rate). Each field was imaged in the V-band (F606W) with an eight-point dither pattern, to perform astrometry. We also obtained two I-band (F814W) exposures for each field, to allow the construction of color-magnitude diagrams (CMDs) and investigate any color-dependent effects. For the second epoch (Cycle 13, PI: Alcock) we proposed the same observational strategy (except without F814W). For the third epoch (Cycle 16, PI: Kallivayalil), however, due to the recent failure of ACS we proposed to use the WFPC2/PC. We requested the 40 targets observed in the first epoch in snapshot mode again. Our observational approach is similar to that in epochs 1 & 2. We used the V-band (F606W) filter on the Planetary Camera (PC) and a 5 or 6-point dither pattern.

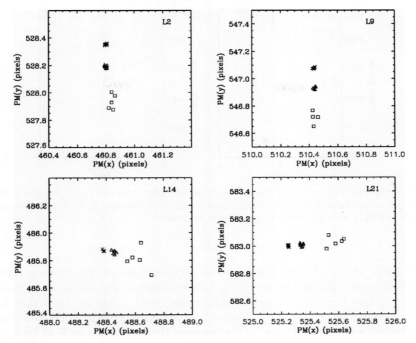

Figure 1. Positions, in a reference frame based on the 1st epoch images, of the quasar over time for 4 randomly chosen LMC fields showing ACS epoch 1 (*crosses*), ACS epoch 2 (*triangles*) and PC epoch 3 (*open squares*). The scatter per epoch gives the relative size of the rms errors in the position of the quasar relative to the star-field, given that the observation has been repeated many times per epoch.

The first concern in switching from the HRC to the PC was one of sensitivity. However our target quasars are relatively bright, ranging from $16.4 \leqslant V \leqslant 20$. The astrometric error is inversely proportional to the S/N for the quasar. Aiming at $S/N \sim 100 - 200$ with the F606W filter yielded total science exposure times ranging from 2.8 to 20 minutes, thereby making these attractive snapshot targets. New data are still being obtained: so far we have obtained 16 out of an expected ~ 20 targets, 13 in the LMC and 3 in the SMC.

2.2. Astrometric precision: Random errors

There are many stars in each field, so the random errors in our transformations will be very small. The astrometric accuracy is therefore dominated by the accuracy with which we can measure the position of the quasar in each field relative to the surrounding stars. In Figure 1 we show the position of the quasar over time, in a reference frame based on the 1st epoch images, for 4 randomly chosen LMC fields (see K1 for definitions of these fields), showing ACS epoch 1 (*crosses*), ACS epoch 2 (*triangles*) and PC epoch 3 (*open squares*). The rms error in the position of the quasar is roughly 3 times as large for WFPC2 as for ACS (~ 0.021 HRC pixels versus 0.008 HRC pixels).

2.3. Astrometric precision: Systematic errors

One of the main drivers in applying for a third epoch was to verify that there were no residual systematic effects in our measurements in addition to those that we already

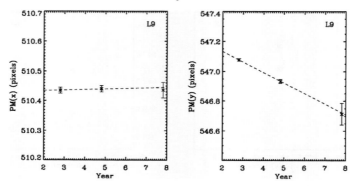

Figure 2. Geometrically-corrected positions (x on the left, y on the right) for the QSO in field L9 for the three epochs of data. The line connects the first two epoch ACS measurements. The third epoch WFPC2 measurement is consistent with this.

accounted for in our error bars (e.g., residual geometric distortion solution errors). The third epoch does give us a handle on this as follows: there is always a straight line through two measurements, but if the third measurement doesn't line up then that is a clear indication of some systematic effect. Figure 2 shows geometrically-corrected coordinates for the quasar as a function of time, in the frame of the 1st-epoch, for field L9 for all three epochs. This is essentially a measure of the reflex proper motion of the field. The line connects the two ACS epochs and is not a fit. Thus even without detailed calculations it is easy to see that the third epoch is lining-up well, which rules out the presence of any *major* residual systematic errors.

We do know that there are astrometric errors due to charge-transfer-efficiency degradation (CTE) as well as other magnitude-dependent effects, mostly along the detector y-axis for ACS and along both x and y for the PC. These trends are small: < 0.02 pixels for ACS, and < 0.03 pixels for PC but are significant when considering the accuracy we are aiming for. Our strategy of observing our N fields at random roll-angles of the telescope means that detector-frame-based errors average to zero as $N^{-1/2}$. However, they do introduce scatter. P08 showed that by accounting for these effects the rms scatter between fields can be reduced. Our current analysis also indicates that explicitly calibrating these effects can improve random errors.

3. Results

Since our analysis is still being fine-tuned, we focus in this section on demonstrating that there is generally good consistency between results from the third epoch of data and those from the first two epochs, once some simple cuts in magnitude-space to minimize systematic effects have been adopted (analogous to those in P08). Figure 3 is a field-by-field comparison for the LMC of the two-epoch proper motion results (*blue squares*) and the three-epoch results (*pink diamonds*). Proper motion in the north direction, μ_N, is shown on top and proper motion in the east direction, μ_E, is shown on the bottom. Lines show simple averages for the two-epoch case (*dot-dashed*), the three-epoch case (*dashed*), and for the P08 analysis of the first two epochs of these fields (*solid*). There is generally good agreement to within 1σ.

The proper motion values for the LMC quasar-fields with three epochs of data are shown in the (μ_W, μ_N)-plane in Figure 4. They are compared to the residual proper

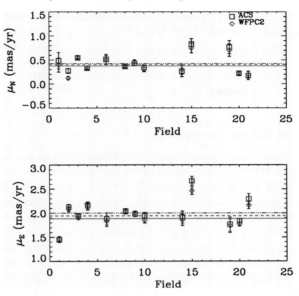

Figure 3. Field-by-field comparison for μ_N (*top*) and μ_E (*bottom*) for the LMC (see K1 for a definition of field numbers). Blue squares indicate two-epoch values, pink diamonds show three-epoch values. Lines are straight averages, dot-dashed for two-epoch, dashed for three-epoch and solid for the P08 analysis of these fields from our two-epoch data.

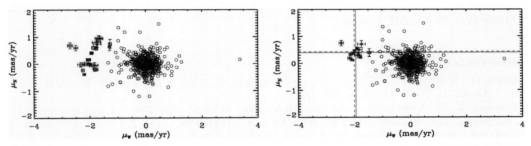

Figure 4. (*Left*) The observed PM (μ_W, μ_N) for all the quasar fields with three epochs of data (*filled circles*). Over-plotted are the corresponding values from the two-epoch analysis (*squares*). (*Right*) The estimates PM_{est}(CM) of the LMC center of mass proper motion after corrections for viewing perspective and internal rotation have been made for the three-epoch fields. The residual PMs of the LMC stars in all the fields are plotted with open circles in both panels. The reflex motions of the quasars clearly separate from the star motions. The straight lines mark the averages, dashed for the two-epoch analysis and solid for the three-epoch one.

motions of all the LMC stars that were used in the transformations. The stars are shown with open circles and the quasars with filled ones. Figure 4 (*left*) shows the observed PM values for all fields with three-epochs of data; filled circles show three-epoch analysis values while squares show the corresponding two-epoch analysis values. Figure 4 (*right*) shows the center of mass proper motion estimates, PM_{est}(CM), just for the three-epoch fields, derived from the observed values after correcting for viewing perspective and internal rotation (see K1, van der Marel *et al.* 2002). The reflex motion of the quasars clearly separates from the star-fields in both panels. The solid line marks the average of the fields in the three-epoch analysis, while the dashed lines shows the corresponding average for the two-epoch data. For clarity, error bars are not plotted on the proper motions of the individual stars. However, these motions are generally consistent with zero given the error bars. The RMS errors calculated from the spread in Figure 4 (*right*) are currently a factor of ~ 1.5 better than our two-epoch results, but we expect to improve on this as we develop better models for CTE and other magnitude-dependent effects. SMC three-epoch results show a similar consistency with the two-epoch ones but are not shown here in the interest of brevity.

4. Summary and future prospects

We have presented an ongoing analysis of the proper motions of the Magellanic Clouds using a third epoch of WFPC2 data centered on background quasars. At present the rms error in the position of the quasar is roughly 3 times as large for WFPC2 as for ACS (~ 0.021 pixels versus 0.008 pixels). However, with an improved method to deal with CTE and magnitude-related effects, and with the increase in time-baseline from 2 to 5 years, we expect final error bars for the proper motions that are smaller by a factor of $\ll 2$ from the two-epoch analysis: the two-epoch error bars in the north and east directions are $(0.05, 0.08 \text{ mas yr}^{-1})$, while we expect three-epoch error bars of $(0.04, 0.04 \text{ mas yr}^{-1})$. This will have several benefits, not the least of which is an important consistency check on our earlier results. This will allow us to better understand the orbit of the Clouds around each other. Combined with our understanding of the properties of the Magellanic Stream, this will allow us to better constrain the MW dark halo potential, as well as weigh into whether the LMC is indeed bound to the MW. Finally, we have a fourth epoch of observations scheduled in Cycle 17 with ACS and WFC3. The expected improvement in accuracy will continue to provide fundamental new insights into the unique and enigmatic Milky Way-LMC-SMC system.

References

Besla, G., Kallivayalil, N., Hernquist, L., Robertson, B., Cox, T. J., van der Marel, R. P., & Alcock, C. 2007, *ApJ*, 668, 949 (B07)
Connors, T. W., Kawata, D., & Gibson, B. K. 2006, *MNRAS*, 371, 108
Gardiner, L.T. & Noguchi, M. 1996, *MNRAS*, 278, 191 (GN96)
Geha, M., Alcock, C., Allsman, R. A., *et al.* 2003, *AJ*, 125, 1
Heller, P. & Rohlfs, K. 1994, *A&A*, 291, 743
Kallivayalil, N., van der Marel, R. P., Alcock, C., Axelrod, T., Cook, K. H., Drake, A. J., & Geha, M. 2006a, *ApJ*, 638, 772 (K1)
Kallivayalil, N., van der Marel, R. P., & Alcock, C. 2006b, *ApJ*, 652, 1213 (K2)
Lin, D. N. C., & Lynden-Bell, D. 1982, *MNRAS*, 198, 707
Lin, D. N .C., Jones, B. F., & Klemola, A. R. 1995, *ApJ*, 439, 652
Mastropietro, C., Moore, B., Mayer, L., Debattista, V. P., Piffaretti, R., & Stadel, J. 2005, *MNRAS*, 364, 607
Moore, B. & Davis, M. 1994, *MNRAS*, 270, 209
Murai, T. & Fujimoto, M. 1980, *PASJ*, 32, 581
Navarro, J. F., Frenk, C. S., & White, S. D. M. 1996, *ApJ*, 462, 563
Piatek, S., Pryor, C., & Olszewski, E. W. 2008, *AJ*, 135, 1024 (P08)
Shattow, G. & Loeb, A. 2009, *MNRAS* 392, L21
van der Marel, R. P., Alves, D. R., Hardy, E., & Suntzeff, N. B. 2002, *AJ*, 124, 2639
Wu, X., Famaey, B., Gentile, G., Perets, H., & Zhao, H. 2008, *MNRAS*, 386, 2199
Yoshizawa, A. M. & Noguchi, M. 2003, *MNRAS*, 339, 1135

The Magellanic System: Stars, Gas, and Galaxies
Proceedings IAU Symposium No. 256, 2008
Jacco Th. van Loon & Joana M. Oliveira, eds.

© 2009 International Astronomical Union
doi:10.1017/S1743921308028317

The binarity of the Clouds and the formation of the Magellanic Stream

Gurtina Besla[1], Nitya Kallivayalil[2], Lars Hernquist[1], Roeland P. van der Marel[3], T.J. Cox[1], Brant Robertson[4] and Charles Alcock[1]

[1] Harvard Smithsonian Center for Astrophysics, 60 Garden St., Cambridge, MA, 02138, USA
email: gbesla@cfa.harvard.edu

[2] Pappalardo Fellow, MIT Kavli Institute for Astrophysics and Space Research,
70 Vassar St., 37-66H, Cambridge, MA, 02139, USA

[3] Space Telescope Science Institute, 3700 San Martin Dr., Baltimore, MD, 21218, USA

[4] Spitzer Fellow; Kavli Institute for Cosmological Physics and Department of Astronomy and
Astrophysics, University of Chicago, 933 East 56th St., Chicago, IL, 60637, USA

Abstract. The *HST* proper motion (PM) measurements of the Clouds have severe implications for their interaction history with the Milky Way (MW) and with each other. The Clouds are likely on their first passage about the MW and the SMC's orbit about the LMC is better described as quasi-periodic rather than circular. Binary L/SMC orbits that satisfy observational constraints on their mutual interaction history (e.g., the formation of the Magellanic Bridge during a collision between the Clouds $\sim$ 300 Myr ago) can be located within 1σ of the mean PMs. However, these binary orbits are not co-located with the Magellanic Stream (MS) when projected on the plane of the sky and the line-of-sight velocity gradient along the LMC's orbit is significantly steeper than that along the MS. These combined results ultimately rule out a purely tidal origin for the MS: tides are ineffective without multiple pericentric passages and can neither decrease the velocity gradient nor explain the offset stream in a polar orbit configuration. Alternatively, ram pressure stripping of an extended gaseous disk may naturally explain the deviation. The offset also suggests that observations of the little-explored region between RA $21^{\rm h}$ and $23^{\rm h}$ are crucial for characterizing the full extent of the MS.

Keywords. galaxies: interactions, galaxies: kinematics and dynamics, Magellanic Clouds

1. Introduction

The recent high-precision proper motion (PM) measurements of the L/SMC determined by Kallivayalil *et al.* (2006a, 2006b — hereafter K06a and K06b; see also these proceedings) imply that the Magellanic Clouds are moving $\sim$ 100 km/s faster than previously estimated and now approach the escape velocity of the Milky Way (MW). Besla *et al.* (2007) (hereafter B07) re-examined the orbital history of the Clouds using the new PMs and a ΛCDM-motivated MW model and found that the L/SMC are either on their first passage about the MW or, if the mass of the MW is $> 2 \times 10^{12}$ M$_\odot$, that their orbital period and apogalacticon distance are a factor of three larger than previously estimated. This means that models of the Magellanic Stream (MS) need to reconcile the fact that although the efficient removal of material via tides and/or ram pressure requires multiple pericentric passages through regions of high gas density, the PMs imply that the Clouds did not pass through perigalacticon during the past $\geqslant$ 5 Gyr (this is true even if a high mass MW model is adopted). While the most dramatic consequence of the new PMs is the limit they place on the interaction timescale of the Clouds with the MW, there are a number of other equally disconcerting implications: the relative velocity between the

Clouds has increased such that only a small fraction of the orbits within the PM error space allow for stable binary L/SMC orbits (K06b; B07); the velocity gradient along the orbit is much steeper than that observed along the MS; and the past orbits are not co-located with the MS on the plane of the sky (B07). In these proceedings the listed factors are further explored and used to argue that the MS is not a tidal tail.

2. Do the Clouds form a binary system?

Doubt concerning the binarity of the Clouds is particularly troubling, as a recent chance encounter between dwarf galaxies in the MW's halo is improbable if they did not have a common origin. To address this issue, ten thousand points were randomly drawn from the SMC PM error space (K06b), each corresponding to a unique velocity vector and orbit (Fig. 1). Bound orbits are identified and color coded based on the number of times the separation between the Clouds reaches a minimum, assuming a mass ratio of 10:1 between the L/SMC (although the mass ratio is not well constrained). Orbits with only one close encounter (like for the SMC PM determined in the re-analysis of the K06b data by Piatek *et al.* 2008, hereafter P08) are not stable binary systems. The new LMC PM also implies that orbits where the SMC traces the MS on the plane of the sky (like that chosen by Gardiner & Noguchi 1996, hereafter GN96) are no longerbinary orbits. It

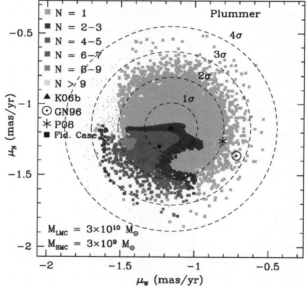

Figure 1. Ten thousand points randomly sampled from the (4σ) K06b PM error space for the SMC (where the mean value is indicated by the triangle). Each corresponds to a unique 3D velocity vector and is color coded by the number of times the separation between the Clouds reaches a minimum within a Hubble time. The circled dot indicates the GN96 PM for the SMC and the asterisk corresponds to the mean of the Piatek *et al.* (2008) (P08) re-analysis of the K06b data - neither correspond to long-lived binary states. The Clouds are modeled as Plummer potentials with masses of $M_{\rm LMC} = 3 \times 10^{10}$ M$_\odot$ and $M_{\rm SMC} = 3 \times 10^9$ M$_\odot$ and the MW is modeled as a NFW halo with a total mass of 10^{12} M$_\odot$ as described in B07. The LMC is assumed to be moving with the mean K06a PM (v = 378 km s^{-1}). The black square represents a solution for the SMC's PM that allows for close passages between the Clouds at characteristic timescales (see Fig. 2) and is our fiducial case.

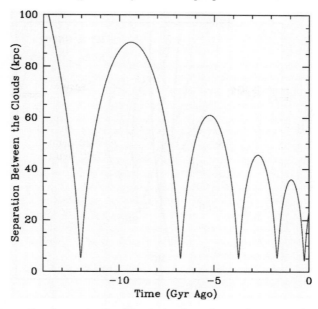

Figure 2. The separation between the Clouds is plotted as a function of time in the past for the fiducial SMC orbit indicated by the black square in Fig. 1 and assuming a mass ratio of 10:1 between the L/SMC. The separation reaches a minimum at ∼ 300 Myr and ∼ 1.5 Gyr in the past, corresponding to the formation times for the bridge and the MS.

is clear from Fig. 1 that stable binary orbits exist within 1σ of the mean K06b value — however, in all cases the SMC's orbit about the LMC is highly eccentric (Fig. 2), which differs markedly from the conventional view that the SMC is in a circular orbit about the LMC (GN96, Gardiner *et al.* 1994). It should also be noted that the likelihood of finding a binary L/SMC system that is stable against the tidal force exerted by the MW decreases if the MW's mass is increased.

We further require that the last close encounter between the Clouds occurred ∼ 300 Myr ago, corresponding to the formation timescale of the Magellanic Bridge (Harris 2007), and that a second close encounter occurs ∼ 1.5 Gyr ago, a timeframe conventionally adopted for the formation of the MS (GN96). A case that also satisfies these constraints is indicated in Fig. 1 by the black square and will be referred to as our fiducial SMC orbit. The corresponding orbital evolution of the SMC about the LMC is plotted in Fig. 2: the new PMs are not in conflict with known observational constraints on the mutual interaction history of the Clouds.

3. Consequences for the Magellanic Stream

The spatial location of the fiducial orbit on the plane of sky and the line-of-sight velocity gradient along it are compared to the observed properties of the MS. The GN96 orbits were a priori chosen to trace both the spatial location and velocity structure of the MS, but this is an assumption. Indeed, from Fig. 3, the LMC's orbit using the new PM is found to be offset from the MS (indicated by the GN96 orbits) by roughly 10°. The offset arises because the north component of the LMC's PM vector as defined by K06a, the re-analysis by P08, *and* the weighted average of all PM measurements prior to 2002 (van der Marel *et al.* 2002), is not consistent with 0 (which was the assumption made by GN96): this result is thus independent of the MW model (B07). Furthermore, the SMC

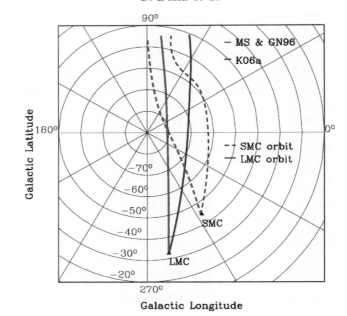

Figure 3. The past orbits of the LMC (solid lines) and SMC (dashed lines) for our fiducial PMs (blue) and the GN96 PMs (magenta) are mapped as a polar projection in galactic (l,b) coordinates. The orbits are followed backwards in time from the Clouds' current positions (filled triangles) until they extend 100° in the sky. The fiducial orbits deviate markedly from the current location of the MS, which is traced by the GN96 orbits.

must have a similar tangential motion as the LMC in order to maintain a binary state, meaning that our fiducial SMC orbit deviates even further from the MS than that of the LMC. In addition, the line-of-sight velocity gradient along the LMC's orbit is found to be significantly steeper than that of the MS, reaching velocities $\sim 200 \mathrm{~km \, s^{-1}}$ larger than that observed at the same position along the MS (Fig. 4).

The offset and steep velocity gradient are unexplainable in a tidal model. While tidal tails may deviate from their progenitor's orbit, they remain confined to the orbital plane (Choi *et al.* 2007): since the Clouds are in a polar orbit no deviation is expected in projection in a tidal model. Furthermore, material in tails is accelerated by the gravitational field of the progenitor — however, to explain the observed velocity gradient the opposite needs to occur. Coupling these factors to a first passage scenario strongly suggests that, while tides likely help shape the stream (e.g., the leading arm feature), hydrodynamic processes are the *primary* mechanism for the removal of material from the Clouds and for shaping its velocity structure (e.g., via gas drag).

The main difficulty for ram pressure stripping in a first passage scenario is the low halo gas densities at large galactocentric distances. The efficiency of stripping may be improved if material is given an additional kick by stellar feedback (e.g., Nidever *et al.* 2008; Olano 2004) or if the LMC initially possessed an extended disk of H I like those observed in isolated dIrrs — note that the latter is not a viable initial condition if the LMC were not on its first passage and the former may violate metallicity constraints on the MS which indicate the MS is metal poor (Gibson *et al.* 2000). If ram pressure stripping is efficient, the offset may occur naturally: Roediger & Brüggen (2006) have shown that material can be removed asymmetrically from gaseous disks that are inclined relative to their line of motion (the LMC's disk is inclined by 30°) and caution that tails

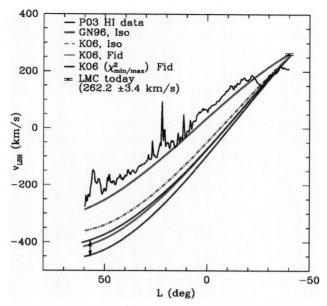

Figure 4. The line-of-sight velocities (v_{LSR}) with respect to the local standard of rest are plotted as a function of Magellanic longitude (L) (see B07, Fig. 20) along the LMC's orbit. The orbital velocity corresponding to the mean PM for the B07 fiducial (isothermal sphere) MW model is indicated by the solid red (dashed red) line. The best and worst χ^2 fits to the MS data within the PM error space are indicated by the arrows and the blue lines. The black line indicates the H I data of the MS from Putman *et al.* (2003) (P03). The velocity gradients along the new orbits in both the isothermal sphere and fiducial NFW MW models are significantly steeper than the GN96 results (magenta line), which were contrived to trace the velocity data.

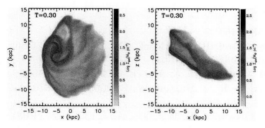

Figure 5. Ram pressure stripping studies of an LMC model with a gaseous disk component that is extended 3x the optical extent and rotating clockwise. The LMC is modeled as an NFW halo with an exponential disk and a total mass of 3×10^{10} M$_\odot$. The left panel is the face-on view and the right is edge-on: in all cases the LMC is moving to the left at 380 km s^{-1} and is inclined $30°$ relative to its direction of motion. The snapshots indicated the evolution of the gas surface density after 0.3 Gyr. The face-on projection illustrates how material is preferentially removed from the side of the disk rotating in-line with the ram pressure wind. In the edge-on projection, material from the leading edge lags behind that removed from the trailing edge.

do not always indicate the direction of motion of the galaxy. These authors considered ram pressure stripping in the context of massive galaxies in cluster environments. We are currently conducting simulations of the formation of the MS via the ram pressure stripping of the Clouds, assuming they initially entered the MW system with extended gaseous disks. Fig. 5 illustrates the proposed mechanism at work: here the LMC has been moving at 380 km s^{-1} through a box of gas at a uniform temperature of 10^6 K and density of 10^{-4} cm^{-3} for 300 Myr. Once the material is removed beyond the LMC's tidal radius, the MW's tidal force may then be able to stretch the material to its full extent — but now since the material is removed asymmetrically it will not trace the orbit in projection.

4. Conclusions

The new PMs have dramatic implications for phenomenological studies of the Clouds that assume they have undergone multiple pericentric passages about the MW and/or that the SMC is in a circular orbit about the LMC. The orbits deviate spatially from the current location of the MS on the plane of the sky and the velocity gradient along the orbit is much steeper than that observed. These results effectively rule out a purely tidal model for the MS and lend support for hydrodynamical models, such as ram pressure stripping. The offset further suggests that the Clouds have travelled across the little explored region between RA 21$^\mathrm{h}$ and 23$^\mathrm{h}$ (i.e. the region spanned by the blue lines in Fig. 3). Putman *et al.* (2003) detected diffuse H I in that region that follow similar velocity gradients as the main stream (their Fig. 7), but otherwise material in that region has been largely ignored by observers and theorists alike. The offset orbits suggest that the MS may be significantly more extended than previously believed and further observations along the region of sky they trace are warranted.

References

Besla, G., Kallivayalil, N., Hernquist, L., Robertson, B., Cox, T. J., van der Marel, R. P., & Alcock, A. 2007, *ApJ*, 668, 949 (B07)

Choi, J.-H., Weinberg, M., & Katz, N. 2007, *MNRAS*, 381, 987

Gardiner, L. T. & Noguchi, M. 1996, *MNRAS*, 278, 191, (GN96)

Gardiner, L. T., Sawa, T., & Fujimoto, M. 1994, *MNRAS*, 266, 567

Gibson, B. K., Giroux, M. L., Penton, S. V., Putman, M. E., Stocke, J. T., & Shull, J. M. 2000, *ApJ*, 120, 1830

Harris, J. 2007, *ApJ*, 658, 345

Kallivayalil, N., van der Marel, R. P., Alcock, C., Axelrod, T., Cook, K. H., Drake, A. J., & Geha, M. 2006a, *ApJ*, 638, 772 (K06a)

Kallivayalil, N., van der Marel, R. P., & Alcock, C. 2006b, *ApJ*, 652, 1213 (K06b)

Nidever, D. L., Majewski, S. R., & Burton, W. B. 2008, *ApJ*, 679, 432

Olano, C. A. 2004, *A&A*, 423, 895

Piatek, S., Pryor, C., & Olszewski, E. W. 2008, *ApJ*, 135, 1024 (P08)

Putman, M. E., Staveley-Smith, L., Freeman, K. C., Gibson, B. K., & Barnes, D. G. 2003, *ApJ*, 586, 170

Roediger, E. & Brüggen, M. 2006, *MNRAS*, 369, 567

van der Marel, R. P., Alves, D. R., Hardy, E., & Suntzeff, N.B. 2002, *AJ*, 124, 2639

The Magellanic System: Stars, Gas, and Galaxies
Proceedings IAU Symposium No. 256, 2008
Jacco Th. van Loon & Joana M. Oliveira, eds.

Models for the dynamical evolution of the Magellanic System

Kenji Bekki

School of Physics, University of New South Wales, Sydney, NSW, 2052, Australia
email: bekki@phys.unsw.edu.au

Abstract. I discuss the following five selected topics on formation and evolution of the LMC and the SMC based on fully self-consistent chemodynamical simulations of the Magellanic Clouds (MCs): (1) formation of bifurcated gaseous structures and young stars in the Magellanic bridge (MB), (2) formation of the Magellanic stream (MS) due to the tidal interaction between the LMC, the SMC, and the Galaxy within the last 2 Gyrs, (3) origin of the observed kinematical differences between H I gas and stars in the SMC, (4) formation of stellar structures dependent on their ages and metallicities in the LMC, and (5) a new common halo model explaining both the latest *HST* ACS observations on the proper motions of the LMC and the SMC and the presence of the MS in the Galactic halo. I focus exclusively on the latest developments in numerical simulations on formation and evolution of the Magellanic system.

Keywords. galaxies: evolution, galaxies: halos, galaxies: interactions, galaxies: kinematics and dynamics, Magellanic Clouds, galaxies: stellar content, dark matter

1. Introduction

Tidal interaction between the Magellanic Clouds and the Galaxy has long been considered to play vital roles in different aspects of their evolution, such as the formation of the Magellanic stream and the Magellanic Bridge from the SMC (e.g., Murai & Fujimoto 1980, Gardiner & Noguchi 1996, GN; Muller & Bekki 2007), long-term star formation histories (e.g., Yoshizawa & Noguchi 2003), chemical evolution (e.g., Hill *et al.* 2000), formation of a bar and a thick disk in the LMC (e.g., Bekki & Chiba 2005; BC), and globular cluster formation (Bekki *et al.* 2004a, b). Previous observational and theoretical studies suggested that hydrodynamical interaction between the gas disk of the LMC and the hot halo gas of the Galaxy can be responsible not only for the formation of the MS (Mastropietro *et al.* 2005) but also for the recent star formation history of the LMC (de Boer *et al.* 1998; Mastropietro, these proceedings). Although the past orbits of the MCs are key parameters that can determine the roles of the above tidal and hydrodynamical effects in the Magellanic evolution (BC), they are not so well constrained by observations.

Recent proper motion measurements of the MCs by the Advanced Camera for Surveys (ACS) on the *Hubble Space Telescope (HST)* have reported that the LMC and the SMC have significantly high Galactic tangential velocities ($367\pm18\,\mathrm{km\,s^{-1}}$ and $301\pm52\,\mathrm{km\,s^{-1}}$, respectively) and thus suggested that the MCs could be unbound from each other (Kallivayalil *et al.* 2006, K06; Piatek *et al.* 2008). These new observations have strongly suggested that previous orbital models adopted in the MS formation models (e.g., BC) are not consistent with the observed proper motions and thus that new physical ingredients would need to be included in any dynamical modes of the MCs for self-consistency with the *HST* results. Recent numerical simulations have demonstrated that the MS models consistent with the *HST* proper motion measurements can not explain self-consistently the locations of the MS and the leading arms projected onto the sky (Bekki & Chiba 2008a).

Keeping these observational and theoretical developments in mind, I discuss the latest observational results on structures, kinematics, and star formation histories of the MCs based on the results of numerical simulations of dynamical evolution of the MCs. I particularly focus on (i) stellar and gaseous distributions of the MB (Muller *et al.* 2003; Harris 2007), (ii) fine structures in the MS (e.g., Brüns *et al.* 2005), (iii) rotational and non-rotational kinematical in gas and stars of the SMC (Harris & Zaristky 2006), (iv) structural differences between different stellar populations (van der Marel 2001; Olsen & Massey 2007), such as young super-giants, AGB stars, and old PNe populations in the LMC (Reid & Parker 2006), and (v) possible presence of a common dark matter halo surrounding the MCs (Bekki 2008).

2. The Bridge

Although the origin of the MB has been suggested to be closely associated with the LMC-SMC interaction about 0.2 Gyr ago (e.g., GN; Yoshizawa & Noguchi 2003), detailed comparison between the simulated and observed properties of the MB has not been done until quite recently. Key questions as to the MB raised by previous observations include (i) how the tidal interaction model explains both the apparent kpc-scale giant H I hole in the MB and the bifurcated kinematics in a self-consistent manner (e.g., Muller *et al.* 2003), (ii) why the metallicities of young stars in the inter-Cloud and the Bridge regions are significantly lower than those ($<$ [Fe/H] ~ -0.6) of stars in the SMC (e.g., Rolleston *et al.* (1999), (iii) why the MB appears to consist almost purely of gas (e.g., Harris 2007), and (iv) how molecular clouds can be formed in a very low density gaseous region of the MB during the last LMC-SMC interaction (e.g., Mizuno *et al.* 2006). I here discuss these questions based on "the tidal interaction model" in which the MCs interact strongly with each other about 0.2 Gyr ago to form the MB from the gas disk of the SMC. The details of the SMC model are given in Bekki & Chiba (2007a) and the orbits of the models discussed here are shown in Fig. 1 for initial velocities from GN and K06. Fig. 2 shows

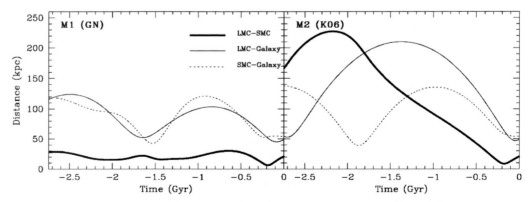

Figure 1. The time evolution of the distances between LMC-SMC (thick solid), LMC-Galaxy (thin solid), and SMC-Galaxy (dotted) for the last 2.7 Gyr in the models M1 (left, referred to as GN) consistent with orbits of MCs by GN and M2 (right, K06) consistent with those by K06. Note that the M1 model can keep the binary status of the MCs whereas the M2 one does not, which suggests that the MS would not be formed from the outer gas disk of the SMC in models with the K06 orbital types.

an example of the final particle distributions of the models in which both the MS and the MB can be formed from the SMC.

Recent numerical simulations have shown that (i) the MB is one of two tidal arms (or tails) formed from the outer gas disk of the SMC as a result of the last LMC-SMC tidal interaction about 0.2 Gyr ago and (ii) the observed apparent H I hole (or loop) can be reproduced simply as a projection of the counter-arm (Muller & Bekki 2007). Chemodynamical numerical simulations (Bekki & Chiba 2008a) showed that young stars with ages less than 0.1 Gyr formed in the MB during the last interaction can originate from the very outer part of the gas disk in the SMC so that they have very low metallicities ([Fe/H]~ -0.8). They also suggested that the metallicity distribution function of the young stars strongly depends on the initial metallicity gradient of gas in the SMC. The latest results of fully self-consistent chemodynamical simulations by Bekki & Chiba (2008a) have demonstrated that the MB inevitably contains old stars with the surface mass densities of $6 - 300 \times 10^4$ M$_\odot$ deg^{-2} depending on initial stellar distributions of the modeled SMC. The results of these numerical simulations based on the tidal interaction model are well consistent with observations on physical properties of the MB (Muller *et al.* 2003) and thus suggest that the origin of the MB can be understood in the context of the last LMC-SMC interaction without invoking energy-deposition processes such as supernova events and stellar winds.

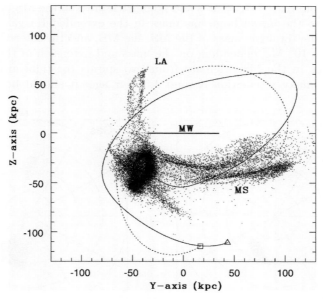

Figure 2. The final distribution of gas of the SMC projected onto the y-z plane for the new "dwarf spheroidal model" (Bekki & Chiba 2007a; Bekki & Chiba 2008a) in which the SMC has both a stellar spheroid and an extended H I gas. This is the distribution after the strong tidal interaction between the MCs and the Galaxy around 1.5 Gyr ago and that between the LMC and the SMC around 0.2 Gyr ago. Owing to the interaction, both the MS and the MB can be formed (only the MS can be clearly seen owing to a larger scale of view). The orbits of the LMC (triangle) and the SMC (square) with respect to the center of the Galaxy are shown by solid and dotted lines, respectively. For comparison, the Galaxy with the disk size of 17.5 kpc is shown by a thick solid line. The GN orbital type is adopted here in order to show a successful model in reproducing the formation of the MS, the leading arms, and the MB. The masses of the LMC and the SMC are assumed to be 2×10^{10} M$_\odot$ and 3×10^9 M$_\odot$, respectively.

3. The Stream

One of observational evidences that strongly support the tidal interaction model for the MS formation is the presence of the leading arm (e.g., Putman *et al.* 1998): this can not be simply explained by the ram pressure model in which the MS originates from the LMC. Recent observational studies have confirmed that the MS shows bifurcation in the structure and kinematics (e.g., Brüns *et al.* 2005) and it also has multiple streams (Stanimirović *et al.* 2008). These new observational results are consistent with the results of numerical simulations on the MS formation from the past LMC-SMC-Galaxy interaction about 1.5 Gyr ago (Connors *et al.* 2006). Although previous tidal interaction models are quite successful in explaining a number of fundamental properties of the MS, it has the following three problems: (i) the SMC is modeled as a rotating disk galaxy (e.g., GN), which is inconsistent with observations (e.g., Harris & Zaritsky 2006), (ii) the total gas mass in the MS and inter-Cloud region appears to be significantly smaller than the observed one (Yoshizawa & Noguchi 2003) for the model with the mass of the SMC being 3×10^9 M$_\odot$, and (iii) the simulated location of the "kink" in the leading arm around $b = 30°$ is not consisted with the observed one ($b = 0°$). In order to overcome these problems, we have constructed a new SMC model in which the SMC is represented by a "dwarf spheroidal" with an extended massive H I gas disk. These characteristics in the new SMC models are consistent with the latest observations (e.g., Harris & Zaristky 2006; Stanimirović *et al.* 2004).

Fig. 3 shows that the new model with the GN orbital type ("classical" MS model) can explain both the bifurcated structures and no or little stars in the MS in a self-consistent manner. Owing to the initial large gas mass in the extended H I gas disk around the stellar spheroid, the total gas mass of the MS, the MB, and the inter-Cloud regions can be as large as 5×10^8 M$_\odot$ depending on the mass and extension of the H I disk. These two main streams in the MS originate from two spirals (or tidal arms) formed from the LMC-SMC-Galaxy interaction around 1.5 Gyr ago: it should be noted here that

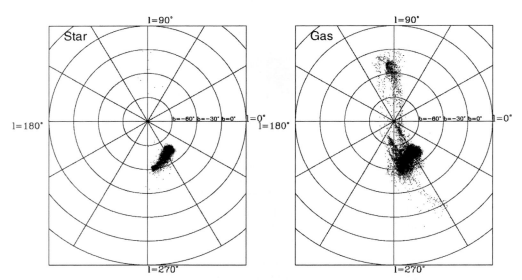

Figure 3. The projected distributions of stars (left) and gas (right) in the new dwarf spheroidal model for the SMC with the total mass of 3×10^9 M$_\odot$ and the gas mass fraction of 2 (i.e., gas mass twice as large as stellar mass). Clearly bifurcation of the MS can be seen in the vicinity of the SMC, and the leading arm also can be well reproduced. Note that the tidal interaction model does *not* predict the presence of a significant number of stars in the MS.

the leading arm is also composed of two streams, though this is not clear in the (l, b) projection. The observed $b = 0$ kink in the leading arm can not be explained simply by the new models: recent simulations by Bekki *et al.* (2008), however, have demonstrated that hydrodynamical interaction between the outer gas of the Galaxy and the leading arm can be responsible for the formation of the $b = 0$ kink. Our extensive parameter survey for different past orbits of the MCs have led us to conclude that the new models with the K06 orbital types can not explain the locations of the MS and the leading arm at all. This result implies that if the observed proper motions of the MCs are true ones *for the centers of mass* for the MCs, we would need to consider new physical ingredients in the tidal interaction model for self-consistency.

4. Kinematical differences between H I and older stars in the SMC

The latest survey of 2046 red giant stars has reported that the older stellar components of the SMC have a velocity dispersion (σ) of $\sim 27.5\,\mathrm{km\,s^{-1}}$ and a maximum possible rotation of $\sim 17\,\mathrm{km\,s^{-1}}$ (Harris & Zaritsky 2006). This result is consistent with other kinematical studies based on radial velocities of other old and intermediate-age stellar populations such as PNe and carbon stars (Russell & Dopita 1992; Hatzidimitiriou *et al.* 1993) and thus suggests that the older stellar component is a spheroid that is primarily supported by its velocity dispersion. Recent high-resolution H I observations have revealed that the SMC has a significant amount of rotation with a circular velocity (V_c) of $\sim 60\,\mathrm{km\,s^{-1}}$ (Stanimirović *et al.* 2004). These observations on stellar and gaseous kinematics in the SMC suggest that there is a remarkable difference in kinematics between older stellar populations and H I gas in the SMC.

It would be possible that the rotating gas disk of the SMC can be gradually formed via gas accretion *after the formation of the older spheroidal component* (Bekki & Chiba 2008b). As a result of this, directions of intrinsic spin axes of older stars and gas in the SMC could be significantly different to each other, like polar-ring galaxies. In this accretion scenario, there needs to be a fine-tuning of the projected spin directions of stellar and gaseous disks in the SMC for explaining the observed kinematical properties. Although this scenario is not unreasonable, it is unclear why the SMC has a stellar spheroid only for older stellar population in this scenario. Furthermore, it seems unlikely that gas can be accreted gradually onto the SMC until recently, because the tidal field of the Galaxy can strongly suppress the gas accretion onto the SMC from its outer halo.

Recent numerical studies have proposed a new scenario in which the SMC could have experienced a major merger event ("dwarf-dwarf merging") between two gas-rich dwarf irregulars (dIs) long time ago in which both the older stellar spheroid and the rotating gas disk were created (Bekki & Chiba 2008b). In this scenario, both the stellar spheroid and the rotating H I disk can be formed almost simultaneously in the last merger event, though the epoch of the merger event can not be specified. Fig. 4 describes how dwarf-dwarf merging can transform two dIs into a new dwarf with a central spheroid and an extended gas disk. Owing to strong violent relaxation in the central region of the merger, the inner stellar disks are completely destroyed and form a slightly flattened spheroidal component with a half-mass radius of 2.0 kpc. Although the gas disk of the larger dI can be temporarily disturbed strongly by the merging, it finally becomes a new extended gas disk after dissipative merging with that of the smaller dI. The gaseous component shows rotation with the maximum rotational velocity of $59\,\mathrm{km\,s^{-1}}$ and a small central velocity dispersion of $\sigma = 24\,\mathrm{km\,s^{-1}}$ (i.e., $V/\sigma \sim 2.5$) whereas the stellar one shows a smaller amount of rotation of $V \sim 20\,\mathrm{km\,s^{-1}}$ and a larger maximum velocity dispersion of $\sigma \sim 48\,\mathrm{km\,s^{-1}}$ (i.e., $V/\sigma \sim 0.4$).

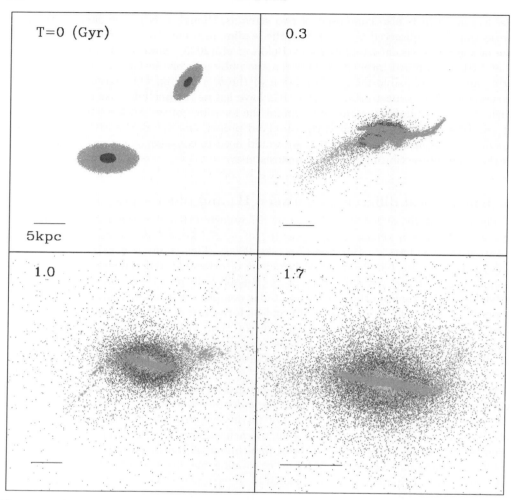

Figure 4. Mass distributions of old stars (magenta), gas (cyan), and new stars (yellow) of the unequal-mass dI-dI merger with the mass ratio of 0.5 projected onto the x-z plane at four different time steps. The time T in units of Gyr is shown in the upper left corner of each panel. Frames measure 40 kpc in the first three and 20 kpc in the last so that the extended disk can be more clearly seen at $T = 1.7$ Gyr. Only old stars can be seen for the central 1.25 kpc of two dIs in the first frame, because the stellar particles are overlaid on gaseous ones: gas particles exist in the central 1.25 kpc.

Thus the ancient merger scenario can explain well the observed kinematical properties of the SMC and thus suggests that the SMC was previously two dIs. The scenario also provides the following three implications on the origin of the MCs. Firstly, there should be a negative metallicity gradient within the central 2 kpc ($\Delta[\mathrm{Fe/H}]/\Delta R \sim -0.05$ dex kpc^{-1}) for the gaseous component (in the very early history of the SMC) in the sense that the inner part is more metal-rich. The outer part of the remnant ($R > 3$ kpc), which is composed mostly of gas ($f_\mathrm{g} > 0.6$), shows $[\mathrm{Fe/H}] < -0.95$ owing to severe suppression of star formation and the resultant much less efficient chemical enrichment. These results imply that if the MS can be formed from tidal stripping of gas from the SMC, the metallicity of the MS should be quite low. Secondly, the dI-dI merging leading to the formation of the SMC occurred long before the strong LMC-SMC interaction commenced about $3 - 4$ Gyr ago (BC): possibly this merger event might have occurred far away from the Galaxy in order to have the low relative velocity between two merging dIs. We consider that the low relative velocity would be possible, because the two dIs were either initially a pair or in a very small group (of galaxies) with a smaller circular velocity that merged the outer region of the Galaxy's halo long time ago.

Thirdly, there should be a "dip" in the age—metallicity relation (AMR) for stellar populations in the SMC: at the epoch of the last merger event, the mean metallicity of newly born stars can be significantly lower than those formed before the merging. This dip is due to star formation from metal-poor gas transfered from outer gas disks or merger progenitor dIs in the merger scenario: the dip would be possibly seen in the AMR of star clusters in the SMC. If this scenario is correct, a key question is when the SMC experienced such a dwarf-dwarf merger event. Since stellar populations formed before the merger event should have dynamically hot kinematics in this scenario, the youngest age of stellar populations that show *both* spheroidal distributions and no or little rotation can correspond to the epoch when the merging occurred. Recent observations of AGB stars in the SMC have reported that (i) the average age of the old and intermediate stellar populations is $7 - 9$ Gyr and (ii) the stars have a more regular distribution and appear to be a slightly flattened ellipsoid (Cioni *et al.* 2000): the merger event might have occurred about $7 - 9$ Gyr ago.

5. Different dynamical properties between different stellar populations in the LMC

Recent numerical simulations have shown that formation of a stellar bar, a thick stellar disk, and a dynamically hot stellar halo in the LMC is a natural result of the past tidal interaction between the LMC and the Galaxy (BC). Therefore, it is not surprising that the LMC is observed to have the above three dynamical properties: I here do not intend to discuss the origin of these (see BC for more discussion). One of the very intriguing observational results is that there appears to be a significant difference in projected distributions between stellar populations with different ages. H I gas and young stellar populations are observed to show clearly the presence of peculiar spirals arms that are the most likely to be formed from the last LMC-SMC interaction about 0.2 Gyr ago (Staveley-Smith *et al.* 2003; Olsen & Massey 2007). The intermediate-age stellar populations such as AGB/RGB/carbon stars show no spiral structures but clearly have a off-center bar (van der Marel 2001). The projected distribution of the PNe, which includes possibly very old stellar populations in the stellar halo, shows neither spiral arms nor a bar (e.g., Reid & Parker 2006), which is remarkably different from those of young and intermediate-age stars (e.g., Olsen & Massey 2007). It remains unclear why there can be a significant difference in structural properties between stellar populations with different ages in the LMC.

We consider that this structural difference has something to do with different kinematics in stellar populations with different ages. It is likely that the LMC can have an AMR similar to that in the Galaxy in the sense that older stars have higher velocity dispersions. If this is the case, it is an interesting dynamical problem how differently stellar populations having different initial velocity dispersions *before the commencement of tidal interaction* response to the tidal perturbation. We have constructed a new "two-component" disk model of the LMC in which the stellar disk have both cold (i.e., smaller velocity dispersion) and hot (larger) components. We have investigated (i) dynamical responses of the cold component with no bar and the hot one with a bar to tidal perturbation from the SMC about 0.2 Gyr ago and (ii) those of the cold and hot components

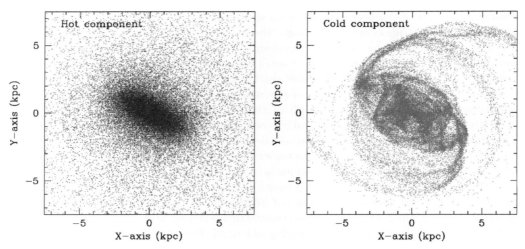

Figure 5. Final distributions of stars in the hot (left) and cold (right) component of the LMC disk interacting with the SMC (after the last ∼ 0.5 Gyr interaction). The strong LMC-SMC interaction occurs around 0.2 Gyr ago in this model. Note that tidal arms can be seen only in the cold component.

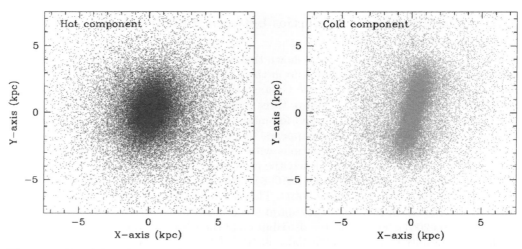

Figure 6. Final distributions of stars in the hot (left) and cold (right) component of the LMC disk interacting with the Galaxy. Note that a strong bar can be seen only in the cold component.

with neither bars nor spiral arms to the tidal perturbation of the Galaxy about 5 Gyr ago.

Fig. 5 shows final distributions of stars in the cold and hot components for the LMC after the LMC-SMC interaction about 0.2 Gyr ago. It is clear that although the cold component shows spiral arms in the central region of the disk, the hot one does not (even though it initially has a bar that can induce spiral arm formation). This is mainly because higher velocity dispersion of stars in the hot component can strongly suppress the formation of spiral arms (or tidal arms) during the LMC-SMC interaction. Fig. 6 shows final distributions of stars in the cold and hot components for the LMC after the LMC-Galaxy interaction about 5 Gyr ago. The hot component does not clearly show a stellar bar whereas the cold one clearly has a strong bar. This is mainly because higher velocity dispersion of stars in the hot component can strongly suppress the formation of a bar during the LMC-Galaxy interaction.

These results imply that the absence of spiral arms in intermediate-age stellar populations such as AGB/RGB stars can be understood, if these stars already had a higher velocity dispersion before the LMC-SMC interaction about 0.2 Gyr ago. Equally, the absence of a bar in the present PNe population can be understood, if their progenitor low-mass stars already had a high velocity dispersion before the LMC-Galaxy interaction about 5 Gyr ago. Thus our simulations have first shown that dynamical responses of stellar populations with different velocity dispersions to tidal perturbation to the SMC and the Galaxy can be different and thus that there can be significant difference in structural properties between stars with different ages. An important observational question related to the above explanations is whether older stars really have higher velocity dispersion in the LMC: it is doubtlessly worthwhile to investigate observationally how stellar kinematics can depend on ages of stars in the LMC.

6. A possible common halo of the MCs

Recent cosmological N-body simulations of the pair galaxy formation based on a Λ cold dark matter (ΛCDM) cosmology have shown that the pair formation like the MCs can occur at $z < 0.33$ corresponding to less than 3.7 Gyr ago for a canonical set of cosmological parameters (Ishiyama *et al.* 2008). Li & Helmi (2008) have investigated merging histories of subhalos in a Milky Way-like halo using high-resolution simulations based on a ΛCDM model and thereby demonstrated that about one-third of the subhalos have been accreted in groups. These results imply that the MCs can have a common dark matter halo that was either formed during the LMC-SMC binary formation or was a previously central part of the much larger Magellanic group which previously had the MCs and other satellite galaxies in the Galaxy. Since previous dynamical models of the MCs do not consider the common halo at all, it is a very interesting problem how the orbits of the MCs can change if they have a common halo.

We have investigated this problem by using the canonical backward integration scheme (Murai & Fujimoto 1980) for the derivation of the past orbits of the MCs in the models with the mass of the common halo ranging from 10^{10} M$_\odot$ to 8×10^{10} M$_\odot$ (see the detail of the models in Bekki 2008). In order to demonstrate more clearly the roles of the common halo in keeping the binary status of the MCs, we have also investigated the models without the common halo. It should be stressed here that initial velocities in these models are consistent with those in K06. Fig. 7 shows that the model with the common halo can keep their binary status (i.e., the LMC-SMC distance being less than 50 kpc) for more than 2 Gyr whereas the model without the common halo can not. These results clearly demonstrate that the common halo can play a role in keeping the binary status of the MCs through its gravitational influence. These thus imply that the MS formation models consistent with K06 can be constructed, if a common halo of the MCs is considered in the models.

If the MCs have a common halo, their evolution can be influenced by the gravitational field of the halo to some extent. One of possible effects of the common halo is that it can shorten the time scale of LMC-SMC merging owing to more efficient dynamical friction. Another effect would be that the gravitational field of the halo can dynamically heat up the disk of the LMC more strongly than the Galaxy does. These possible effects have been already investigated by our N-body simulations on the evolution of the Magellanic group consisting of the MCs and their common halo (yet without other dwarfs) around the Galaxy. The main results of the simulations are briefly summarized as follows. Firstly, it is

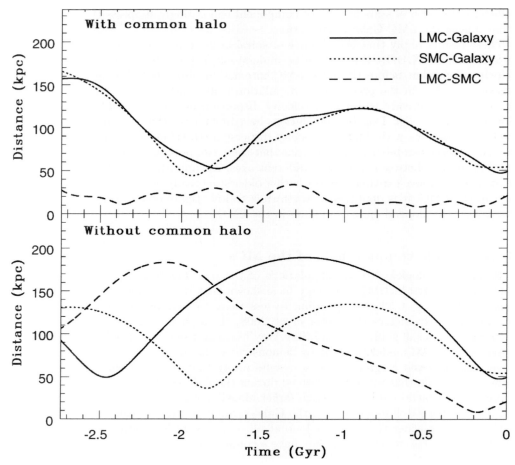

Figure 7. Time evolution of the distances between the LMC and the Galaxy (solid), the SMC and the Galaxy (dotted), and the LMC and the SMC (dashed) for the last ~ 2.7 Gyr in the model with (upper) and without (lower) the common halo of the MCs. The time $T = 0$ Gyr and $T = -2.7$ Gyr mean the present and 2.7 Gyr ago, respectively, in this figure. For these models, the masses of the LMC, the SMC, and the common halo and the scale length of the halo are $M_L = 2.0 \times 10^{10}$ M$_\odot$, $M_S = 3.0 \times 10^9$ M$_\odot$, $M_{ch} = 4.0 \times 10^{10}$ M$_\odot$, and $a_{ch} = 10$ kpc, respectively. Note that the MCs can keep their binary status for the last 2.7 Gyr only for the model with the common halo.

possible for us to construct a model with a massive common halo in which the LMC-SMC do not merge within 2 Gyr owing to tidal heating of the group by the Galaxy. Secondly, stellar tidal streams can be formed as a result of interaction between the LMC and the common halo for most models: this would be a problem if there are no observational evidences for the presence of tidal streams from the LMC. Thirdly, the LMC can show a higher degree of lopsidedness owing to the tidal field of the common halo.

7. Summary: future works

Recent chemodynamical simulations significantly improved their predictive power of the dynamical models of the MCs and thereby addressed a number of problems that had not been discussed at all until recently: bifurcated structures in the MB and the MS, chemical abundances of the MCs and the inter-Cloud regions, long-term star formation histories of the MCs , origin of thick disks and bars in the LMC, and stellar and gaseous kinematics of the SMC. Although these numerical simulations have explained a number of fundamental observations of the MCs (e.g., the MS), they have not been so successful in reproducing some of important observations. For example, the observed off-center bar in the LMC has not been well reproduced yet, though the last LMC-SMC interaction is demonstrated to play a role in forming the off-center bar (Bekki & Chiba 2007b). Also, the rotating kinematics of the old globular cluster in the LMC, which is in a striking contrast with the dynamically hot stellar halo, has not been reproduced by simulations (Bekki 2007). The models for hydrodynamical interaction between the Galactic halo gas and the MCs (Mastropietro *et al.* 2005) needs to be significantly improved, because such interaction can be potentially important for the formation of fine structures in the MS and the leading arms. Formation of star clusters, *which are fundamental building blocks of the MCs,* needs to be investigated by high-resolution *galaxy-scale* simulations which enable us to discuss the importance of the Magellanic environments in the formation of star clusters from giant molecular clouds (Hurley & Bekki 2008). It seems to me that currently the most realistic dynamical model of the Magellanic system is not so self-consistent as to explain variously different observational properties of the MCs *simultaneously*: a lot of work is ahead of us.

References

Bekki, K. 2007, *MNRAS*, 380, 1669

Bekki, K. 2008, *ApJ*, 684, L87

Bekki, K., Beasley, M. A., Forbes, D. A., & Couch, W. J. 2004a, *ApJ*, 602, 730

Bekki, K., Couch, W. J., Beasley, M. A., Forbes, D. A., Chiba, M., & Da Costa, G. 2004b, *ApJ*, 610, L93

Bekki, K. & Chiba, M. 2005, *MNRAS*, 356, 680 (BC)

Bekki, K. & Chiba, M. 2007a, *MNRAS*, 381, L11

Bekki, K. & Chiba, M. 2007b, *PASA*, 24, 21

Bekki, K. & Chiba, M. 2008a, *PASA*, submitted

Bekki, K. & Chiba, M. 2008b, *ApJ*, 679, L89

Bekki, K., Chiba, M., & McClure-Griffiths, N.M. 2008, *ApJ*, 672, L17

Brüns, C., Kerp, J., Staveley-Smith, L., *et al.* 2005, *A&A*, 432, 45

Cioni, M. -R. L., Habing, H. J., & Israel, F. P. 2000, *A&A*, 359, L9

Connors, T. W., Kawata, D., & Gibson, B. K. 2006, *MNRAS*, 371, 108

de Boer, K. S., Braun, J. M., Vallenari, A., & Mebold, U. 1998, *A&A*, 329, L49

Gardiner, L.T. & Noguchi, M., 1996, *MNRAS*, 278, 191 (GN)

Harris, J. 2007, *ApJ*, 658, 345

Harris, J. & Zaritsky, D. 2006, *AJ*, 131, 2514

Hatzidimitriou, D., Croke, B. F., Morgan, D. H., & Cannon, R. D. 1997, *A&AS*, 122, 507

Hill, V., Francois, P., Spite, M., Primas, F., & Spite, F. 2000, *A&A*, 364, L19

Hurley, J. R. & Bekki, K. 2008, *MNRAS*, 389, L61

Ishiyama, T. *et al.* 2008, in B. Koribalski & H. Jerjen (eds.), *Galaxies in the Local Volume*, Astrophysics and Space Science Proceedings), in press

Kallivayalil, N., van der Marel, R. P., & Alcock, C. 2006, *ApJ*, 652, 1213

Li, Y. & Helmi, A. 2008, *MNRAS*, 385, 1365

Mastropietro, C., Moore, B., Mayer, L., Wadsley, J., & Stadel, J. 2005, *MNRAS*, 363, 509

Mizuno, N., Muller, E., Maeda, H., *et al.* 2006, *ApJ*, 643, L107

Muller, E. & Bekki, K. 2007, *MNRAS*, 381, L11

Muller, E., Stanimirović, S., Zealey, W., & Staveley-Smith, L., 2003, *MNRAS*, 339, 105

Murai, T. & Fujimoto, M. 1980, *PASJ*, 32, 581

Olsen, K. A. G. & Massey, P. 2007, *ApJ*, 656, L61

Piatek, S., Pryor, C., & Olszewski, E. W. 2008, *AJ*, 135, 1024

Putman, M. E., Gibson, B. K., Staveley-Smith, L., *et al.* 1998, *Nature*, 394, 752

Reid, W. A. & Parker, Q.A. 2006, *MNRAS*, 365, 401

Rolleston, W. R. J., Dufton, P. L., McErlean, N. D., & Venn, K. A. 1999, *A&A*, 348, 728

Russell, S. C. & Dopita, M. A. 1992, *MNRAS*, ApJ, 384, 508

Stanimirović, S., Staveley-Smith, L., & Jones, P. A. 2004, *ApJ*, 604, 176

Stanimirović, S., Hoffman, S., Heiles, C., *et al.* 2008, *ApJ*, 680, 276

Staveley-Smith, L., Kim, S., Calabretta, M. R., Haynes, R. F., & Kesteven, M. J. 2003, *MNRAS*, 339, 87

van der Marel, R. P. 2001, *AJ*, 122, 1827

Yoshizawa, A. & Noguchi, M. 2003, *MNRAS*, 2003, 339, 1135

Keele Hall, with dusk setting in.

The Magellanic System: Stars, Gas, and Galaxies
Proceedings IAU Symposium No. 256, 2008
Jacco Th. van Loon & Joana M. Oliveira, eds.

Modeling a high velocity LMC: The formation of the Magellanic Stream

Chiara Mastropietro[1,2]

[1] LERMA, Observatoire de Paris, UPMC, CNRS, 61, A. de l'Observatoire,
75014, Paris, France,
email: `chiara.mastropietro@obspm.fr`

[2] Universitäts Sternwarte München, Scheinerstr.1, D-81679 München, Germany

Abstract. I use high resolution N-body/SPH simulations to model the new proper motion of the Large Magellanic Cloud (LMC) within the Milky Way (MW) halo and investigate the effects of gravitational and hydrodynamical forces on the formation of the Magellanic Stream (MS). Both the LMC and the MW are fully self consistent galaxy models embedded in extended cuspy ΛCDM dark matter halos. I find that ram-pressure from a low density ionized halo is sufficient to remove a large amount of gas from the LMC's disk forming a trailing Stream that extends more than 120 degrees from the Cloud. Tidal forces elongate the satellite's disk but do not affect its vertical structure. No stars become unbound showing that tidal stripping is almost effectless.

Keywords. methods: n-body simulations, Galaxy: halo, galaxies: individual (LMC), galaxies: interactions, galaxies: kinematics and dynamics, Magellanic Clouds

1. Introduction

Recent *HST* proper motion measurements of the Magellanic Clouds by Kallivayalil *et al.* (2006), Kallivayalil, van de Marel & Alcock (2006) and Piatek *et al.* (2008) indicate that they are presently moving at velocities substantially higher (almost 100 km s^{-1}) than those provided by previous observational studies (van der Marel *et al.* 2002, Kroupa & Bastian 1997). Such high velocities ($v = 378$ km s^{-1} and $v = 302$ km s^{-1} for the LMC and SMC, respectively) are close to the escape velocity of the Milky Way and consistent with the hypothesis of a first passage about the Galaxy (Besla *et al.* 2007). A single perigalactic passage has serious implications for the origin of the Magellanic Stream. It definitely rules out the tidal stripping hypothesis (Růžička, Theis & Palouš 2008) since in this scenario the loss of mass is primarily induced by tidal shocks suffered by satellites at the pericenters (Mayer *et al.* 2006) and the Stream would not have time to form before the present time. Indeed kinematical data suggest that the Clouds are now just after a perigalactic passage. On the other hand, ram-pressure scales as v^2, where v is the relative velocity between satellites and the ambient medium. The high velocities of the Clouds could therefore compensate the effect of the reduced interaction time with the hot halo of the MW and hydrodynamical forces would play a determinant role in forming the Stream.

In Mastropietro *et al.* (2005, hereafter M05) we have performed high resolution N-body/SPH simulations to study the hydrodynamical and gravitational interaction between the LMC and the MW using orbital constraints by van der Marel *et al.* (2002) and a present time satellite velocity of 250 km s^{-1}. We found that, after two perigalactic passages, the combined effect of tidal forces and ram-pressure stripping can account for the majority of the LMC's internal features and for the formation of the MS. More in detail, ram-pressure stripping of cold gas from the LMC's disk produces a Stream with morphology and kinematics similar to the observed ones, while tidal stripping has longer time-scales and is not efficient in forming stellar debris, consistently with the lack of stars observed in the Stream. Nevertheless, at each pericentric passage the LMC suffers tidal heating which perturbs the overall structure of the satellite reducing the gravitational restoring force and therefore indirectly contributing to the loss of gas.

The main objection to this model, in light of the new proper motion measurements of the LMC, is that the time spent by the LMC within the hot halo of the MW would be to short to cover the full extension of the Stream (more than 100 degrees) by ram-pressure mechanisms (Besla *et al.* 2007). Moreover, hydrodynamical forces would affect a galaxy only weakly perturbed by the gravitational interaction, and stripping would result more difficult.

In this work I present the results of N-body/SPH simulations where the interaction between the MW and the LMC is modeled according to the new proper motion measurements of Kallivayalil *et al.* (2006a).

2. Galaxy models

The initial condition of the simulations are constructed using the technique described by Hernquist (1993). Both the MW and the LMC are multi-component galaxy models with a stellar and gaseous disk embedded in a spherical dark matter halo. The density profile of the NFW halo is adiabatically contracted due to baryonic cooling. Stars and cold gas in the disks follow the same exponential surface density profile. We also explore the eventuality of an extended LMC gaseous disk. In this model the gaseous disk is characterized by an additional constant density layer which extends up to eight times the scale length of the exponential disk. The MW model comprises also a small stellar bulge and an extended low density ($n = 2 \times 10^{-5}$ cm^{-3} within 150 kpc from the Galactic center and $n = 8.5 \times 10^{-5}$ cm^{-3} at 50 kpc) hot ($T = 10^6$ K) halo in hydrostatic equilibrium inside the Galactic potential (M05). The MW model, with virial mass 10^{12} M$_\odot$ and concentration $c = 11$, is similar to model A1 of Klypin *et al.* (2002) while the structural parameters of the LMC are chosen in such a way that the resulting rotation curve resemble that of a typical bulge-less late-type disk galaxy. In detail, the satellite has virial mass 2.6×10^{10} M$_\odot$, concentration $c = 9.5$ and the same amount of mass in the stellar and gaseous disk component ($\sim 10^9$ M$_\odot$). The Toomre's stability criterion is always satisfied and the parameter Q set equal to 1.5 and 2.0 at the disk scale radius in the different LMC models.

3. Simulations

I performed adiabatic simulations using GASOLINE, a parallel tree-code with multi-stepping (Wadsley*et al.* 2004). High resolution runs have 2.46×10^6 particles, of which 3.5×10^5 are used for the disks and 5×10^5 for the hot halo of the MW. The gravitational spline softening is set equal to 0.5 kpc for the dark and gaseous halos, and to 0.1 kpc for stars and gas in the disk and bulge components.

In my best model the LMC approaches the MW on an unbound orbit (Fig. 1) with initial Galactocentric distance of 400 kpc and velocity ~ 190 km s^{-1}. After the perigalactic passage (at ~ 40 kpc) the velocity decreases faster than for a ballistic orbit as a result of dynamical friction. The escape velocity at a given LMC position is indicated by a red curve in the right panel and calculated assuming a spherical unperturbed host potential. Due to the effects of dynamical friction, at late times the satellite lies on a nearly parabolic orbit. At the present time ($t \sim 1.78$ Gyr, vertical lines in the plots) it

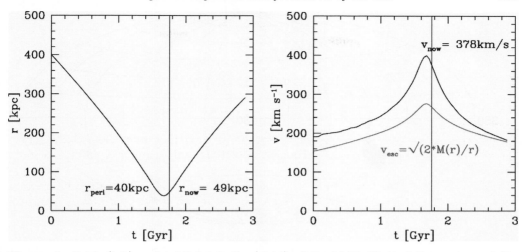

Figure 1. Orbit (left) and orbital velocity (right) of the LMC. Vertical lines represent the present time, just after the perigalacticon. Present values and values at the perigalacticon are indicated.

reaches a velocity of 378 km s^{-1} at 49 kpc from the Galactic center, in good agreement with the new proper motion measurements.

In choosing the initial inclination of the LMC I made the approximation that it does not change during the interaction due to the effects of precession or nutation of the disk plane. At the beginning of the simulation the disk moves almost face-on through the external medium, and ram-pressure affecting the whole disk perpendicularly. In proximity to the perigalactic passage the velocity vector changes rapidly and the angle between the satellite's disk and the proper motion is close to zero. At the present time the simulated disk has an inclination of about 30 degrees with respect to the orbital motion (Kallivayalil *et al.* 2006a) and is indeed moving nearly edge-on through the external hot gas, with ram-pressure compressing its eastern side.

Fig. 2 illustrates the present time distribution of stars and gas originating from the LMC's disk. The stellar disk becomes elongated while tidal debris start forming after the perigalactic passage. But all stars stay bound to the satellite. Tidal heating does not perturb significantly the vertical structure of the disk that remains thin and does not create a warp, unlike what is observed in M05. Bar instability develops at the perigalacticon only in the case of $Q = 1.5$.

Ram-pressure strips nearly 2×10^8 M$_\odot$ of gas from the LMC's disk forming a continuous Stream that lies in a thin plane perpendicular to the disk of the MW and extending up to ~ 140 degrees from the LMC (Fig. 3). The location of the Stream in the Southern Galactic hemisphere is comparable to the values of b and l provided by observations. Contrary to M05 there is no LMC's gas above the Galactic plane. In M05 the material lying in the Northern hemisphere is stripped from the satellite during the orbital period preceding the present one and the Stream forms a great polar circle around the Galaxy. The lack of gas at $b > 0$ in Fig. 3 is not due to inefficient ram-pressure at early times, but to the fact that, in order to reproduce the current location and velocity of the LMC, the satellite enters the MW halo exactly at $b = 0$.

The morphology of the Stream does not change significantly adopting an extended gaseous disk model (left panel of Fig. 3), except for the region at the head of the Stream, which appears broader (right).

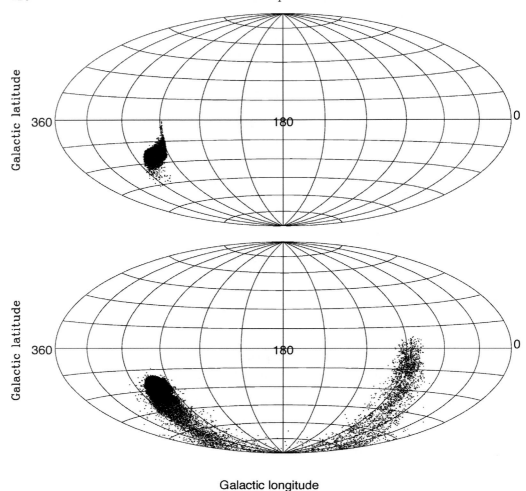

Figure 2. Present time distribution of stars (top) and gas (bottom) from the LMC disk in Galactic coordinates.

4. Conclusions

I carried out high resolution gravitational/hydrodynamical simulations of the interaction between the LMC and the MW using the orbital parameters suggested by the new HST proper motion measurements. I find that ram-pressure stripping exerted by a tenuous MW hot halo during a single perigalactic passage forms a Stream whose extension and location in the Sky are comparable to the observed ones. The stellar structure of the satellite is only marginally affected by tidal forces.

Acknowledgements

The numerical simulations were performed on the l SGI-Altix 3700 Bx2 at the University Observatory in Munich. This work was partly supported by the DFG Sonderforschungsbereich 375 "Astro-Teilchenphysik".

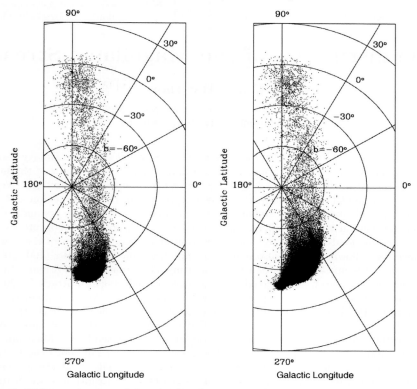

Figure 3. Polar projection of the simulated stream in Galactic coordinates. Both the pure exponential LMC model (left) and the model with extended disk (right) are shown.

References

Besla, G., Kallivayalil, N., Hernquist, L., Robertson, B., Cox, T. J., van der Marel, R. P., & Alcock, C. 2007, *ApJ*, 668, 949

Hernquist, L. 1993, *ApJS*, 86, 389

Kallivayalil, N., van der Marel, R. P., Alcock, C., Axelrod, T., Cook, K. H., Drake, A. J., & Geha, M. 2006a, *ApJ*, 638, 772

Kallivayalil, N., van der Marel, R. P., & Alcock, C. 2006b, *ApJ*, 652, 1213

Klypin, A., Zhao, H., & Somerville, R. S. 2002, *ApJ*, 573, 597

Kroupa, P. & Bastian, U. 1997, *New Astronomy*, 2, 77

Mastropietro, C., Moore, B., Mayer, L., Wadsley, J., & Stadel, J. 2005, *MNRAS*, 363, 509

Mayer, L., Mastropietro, C., Wadsley, J., Stadel, J., & Moore, B. 2006, *MNRAS*, 369, 1021

Piatek, S., Pryor, C., & Olszewski, E. W. 2008, *AJ*, 135, 1024

Růžička, A., Theis, C., & Palouš, J. 2008, *ApJ*, in press, arXiv:0810.0968

van der Marel R. P., Alves D. R., Hardy E., & Suntzeff N. B. 2002, *AJ*, 124, 2639

Wadsley, J. W., Stadel, J., & Quinn, T. 2004, *New Astronomy*, 9, 137

The Magellanic System: Stars, Gas, and Galaxies
Proceedings IAU Symposium No. 256, 2008
Jacco Th. van Loon & Joana M. Oliveira, eds.

© 2009 International Astronomical Union
doi:10.1017/S1743921308028342

The disruption of the Magellanic Stream

J. Bland-Hawthorn

School of Physics, University of Sydney, NSW 2006, Australia
email: jbh@physics.usyd.edu.au

Abstract. We present evidence that the accretion of warm gas onto the Galaxy today is at least as important as cold gas accretion. For more than a decade, the source of the bright $H\alpha$ emission (up to 750 mR†) along the Magellanic Stream has remained a mystery. We present a hydrodynamical model that explains the known properties of the $H\alpha$ emission and provides new insights on the lifetime of the Stream clouds. The upstream clouds are gradually disrupted due to their interaction with the hot halo gas. The clouds that follow plough into gas ablated from the upstream clouds, leading to shock ionisation at the leading edges of the downstream clouds. Since the following clouds also experience ablation, and weaker $H\alpha$ (100−200 mR) is quite extensive, a disruptive cascade must be operating along much of the Stream. In order to light up much of the Stream as observed, it must have a small angle of attack ($\approx 20°$) to the halo, and this may already find support in new H I observations. Another prediction is that the Balmer ratio ($H\alpha/H\beta$) will be substantially enhanced due to the slow shock; this will soon be tested by upcoming WHAM observations in Chile. We find that the clouds are evolving on timescales of 100−200 Myr, such that the Stream must be replenished by the Magellanic Clouds at a fairly constant rate ($\gtrsim 0.1$ M$_\odot$ yr^{-1}). The ablated material falls onto the Galaxy as a warm drizzle; diffuse ionized gas at 10^4 K is an important constituent of galactic accretion. The observed $H\alpha$ emission provides a new constraint on the rate of disruption of the Stream and, consequently, the infall rate of metal-poor gas onto the Galaxy. When the ionized component of the infalling gas is accounted for, the rate of gas accretion is $\gtrsim 0.4$ M$_\odot$ yr^{-1}, roughly twice the rate deduced from H I observations alone.

Keywords. hydrodynamics, instabilities, shock waves, galaxies: evolution, galaxies: interactions, Magellanic Clouds

1. Introduction

It is now well established that the observed baryons over the electromagnetic spectrum account for only a fraction of the expected baryon content in Lambda Cold Dark Matter cosmology. This is true on scales of galaxies, in particular, within the Galaxy where easily observable phases have been studied in great detail over many years. The expected baryon fraction ($\Omega_B/\Omega_{DM} \approx 0.17$) of the dark halo mass (1.4×10^{12} M$_\odot$; Smith *et al.* 2007) leads to an expected baryon mass of 2.4×10^{11} M$_\odot$ but a detailed inventory reveals only a quarter of this mass (Flynn *et al.* 2006). Moreover, the build-up of stars in the Galaxy requires an accretion rate of 1−3 M$_\odot$ yr^{-1} (Williams & McKee 1997; Binney *et al.* 2000), at least a factor of 4 larger than what can be accounted for from direct observation. The derived baryon mass may be a lower bound if the upward correction in the LMC-SMC orbit motion reflects a larger halo mass (Kallivayalil *et al.* 2006; Piatek *et al.* 2008; cf. Wilkinson & Evans 1999). Taken together, these statements suggest that most of the baryons on scales of galaxies have yet to be observed.

---

† 1 Rayleigh (R) $= 10^6/4\pi$ photons cm^{-2} s^{-1} sr^{-1}, equivalent to 5.7×10^{-18} erg cm^{-2} s^{-1} arcsec^{-2} at $H\alpha$.

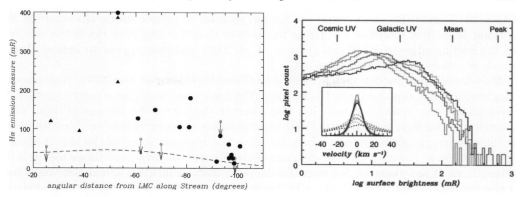

Figure 1. Left: Hα measurements and upper limits along the Stream. The filled circles are from the WHAM survey by Madsen *et al.* (2002); the filled triangles are from the TAURUS survey by Putman *et al.* (2003). The dashed line model is the Hα emission measure induced by the ionizing intensity of the Galactic disk (Bland-Hawthorn & Maloney 1999, 2002); this fails to match the Stream's Hα surface brightness by at least a factor of 3. **Right:** the evolving distribution of projected Hα emission as the shock cascade progresses. The timesteps are 70 (red), 120 (magenta), 170 (blue), 220 (green) and 270 Myr (black). The extreme emission measures increase with time and reach the observed mean values after 120 Myr; this trend in brightness arises because denser material is ablated as the cascade evolves. The mean and peak emission measures along the Stream are indicated, along with the approximate contributions from the cosmic and Galactic UV backgrounds. **Inset:** The evolving Hα line width as the shock cascade progresses; the velocity scale is with respect to the reference frame of the initial H I gas. The solid lines are flux-weighted line profiles; the dashed lines are volume-weighted profiles that reveal more extreme kinematics at the lowest densities.

So how do galaxies accrete their gas? Is the infalling gas confined by dark matter? Does the gas arrive cold, warm or hot? Does the gas rain out of the halo onto the disk or is it forced out by the strong disk-halo interaction? These issues have never been resolved, either through observation or through numerical simulation. H I observations of the nearby universe suggest that galaxy mergers and collisions are an important aspect of this process, but tidal interactions do not guarantee that the gas settles to one or other galaxy. The most spectacular interaction phenomenon is the Magellanic H I Stream that trails from the LMC-SMC system (10:1 mass ratio) in orbit about the Galaxy. Since its discovery in the 1970s, there have been repeated attempts to explain the Stream in terms of tidal and/or viscous forces (q.v. Mastropietro *et al.* 2005; Connors *et al.* 2006). Indeed, the Stream has become a benchmark against which to judge the credibility of N-body+gas codes in explaining gas processes in galaxies. A fully consistent model of the Stream continues to elude even the most sophisticated codes.

Here, we demonstrate that Hα detections along the Stream (Fig. 1) are providing new insights on the present state and evolution of the H I gas. At a distance of $D \approx 55$ kpc, the expected Hα signal excited by the cosmic and Galactic UV backgrounds are about 3 mR and 25 mR respectively (Bland-Hawthorn & Maloney 1999, 2002), significantly lower than the mean signal of 100−200 mR, and much lower than the few bright detections in the range 400−750 mR (Weiner *et al.* 2002). This signal cannot have a stellar origin since repeated attempts to detect stars along the Stream have failed.

Some of the Stream clouds exhibit compression fronts and head-tail morphologies (Brüns *et al.* 2005) and this is suggestive of confinement by a tenuous external medium. But the cloud:halo density ratio ($\eta = \rho_c/\rho_h$) necessary for confinement can be orders of magnitude *larger* than that required to achieve shock-induced Hα emission (e.g. Quilis & Moore 2001). Indeed, the best estimates of the halo density at the distance of the Stream

($\rho_h \sim 10^{-4}$ cm^{-3}; Bregman 2007) are far too tenuous to induce strong Hα emission at a cloud face. It is therefore surprising to discover that the brightest Hα detections lie at the leading edges of H$\,$I clouds (Weiner *et al.* 2002) and thus appear to indicate that shock processes are somehow involved.

We summarize a model, first presented in Bland-Hawthorn *et al.* (2007), that goes a long way towards explaining the Hα mystery. The basic premise is that a tenuous external medium not only confines clouds, but also disrupts them with the passage of time. The growth time for Kelvin-Helmholtz (KH) instabilities is given by $\tau_{KH} \approx \lambda \eta^{0.5}/v_h$ where λ is the wavelength of the growing mode, and v_h is the apparent speed of the halo medium ($v_h \approx 350$ km s^{-1}; see §2). At the distance of the Stream, the expected timescale for KH instabilities is less than for Rayleigh-Taylor (RT) instabilities. For cloud sizes of order a few kiloparsecs and $\xi \approx 10^4$, the KH timescale can be much less than an orbital time ($\tau_{MS} \approx 2\pi D/v_h \approx 1$ Gyr). Once an upstream cloud becomes disrupted, the fragments are slowed with respect to the LMC-SMC orbital speed and are subsequently ploughed into by the following clouds.

2. A new hydrodynamical model

We investigate the dynamics of the Magellanic Stream with two independent hydrodynamics codes, FYRIS and RAMSES, that solve the equations of gas dynamics with adaptive mesh refinement. The results shown here are from the FYRIS code because it includes non-equilibrium ionization, but we get comparable gas evolution from either code†.

The brightest emission is found along the leading edges of clouds MS$\,$II, III and IV with values as high as 750 mR for MS$\,$II. The Hα line emission is clearly resolved at $20-30$ km s^{-1} FWHM, and shares the same radial velocity as the H$\,$I emission within the measurement errors (Weiner *et al.* 2002; G. Madsen 2007, personal communication). This provides an important constraint on the physical processes involved in exciting the Balmer emission.

In order to explain the Hα detections along the Stream, we concentrate our efforts on the disruption of the clouds labelled MS I–IV (Brüns *et al.* 2005). The Stream is trailing the LMC-SMC system in a counter-clockwise, near-polar orbit as viewed from the Sun. The gas appears to extend from the LMC dislodged through tidal disruption although some contribution from drag must also be operating (Moore & Davis 1994). Recently, the Hubble Space Telescope has determined an orbital velocity of 378±18 km s^{-1} for the LMC. While this is higher than earlier claims, the result has been confirmed by independent researchers (Piatek *et al.* 2008). Besla *et al.* (2007) conclude that the origin of the Stream may no longer be adequately explained with existing numerical models. The Stream velocity along its orbit must be comparable to the motion of the LMC; we adopt a value of $v_{MS} \approx 350$ km s^{-1}.

Here we employ a 3D Cartesian grid with dimensions $18 \times 9 \times 9$ kpc $[(x, y, z) = (432, 216, 216)$ cells] to model a section of the Stream where x is directed along the Stream arc and the z axis points towards the observer. The grid is initially filled with two gas components. The first is a hot thin medium representing the halo corona.

Embedded in the hot halo is (initially) cold H$\,$I material with a total H$\,$I mass of 3×10^7 M$_\odot$. The cold gas has a fractal distribution and is initially confined to a cylinder with a diameter of 4 kpc and length 18 kpc; the mean volume and column densities are 0.02 cm^{-3} and 2×10^{19} cm^{-2} respectively. The 3D spatial power spectrum ($P(k) \propto k^{-5/3}$)

---

† Further details on the codes and comparative simulations are provided at http://www.aao.gov.au/astro/MS.

describes a Kolmogorov turbulent medium with a minimum wavenumber k corresponding to a spatial scale of 2.25 kpc, comparable to the size of observed clouds along the Stream.

We consider the hot corona to be an isothermal gas in hydrostatic equilibrium with the gravitational potential, $\phi(R, z)$, where R is the Galactocentric radius and z is the vertical scale height. We adopt a total potential of the form $\phi = \phi_d + \phi_h$ for the disk and halo respectively; for our calculations at the Solar Circle, we ignore the Galactic bulge. The galaxy potential is defined by

$$\phi_d(R, z) = -c_d v_{circ}^2 / \left(R^2 + \left(a_d + \sqrt{z^2 + b_d^2} \right)^2 \right)^{0.5} \tag{2.1}$$

$$\phi_h(R, z) = c_h v_{circ}^2 \ln((\psi - 1)/(\psi + 1)) \tag{2.2}$$

and $\psi = (1 + (a_h^2 + R^2 + z^2)/r_h^2)^{0.5}$. The scaling constants are $(a_d, b_d, c_d) = (6.5, 0.26, 8.9)$ kpc and $(a_h, r_h) = (12, 210)$ kpc with $c_h = 0.33$ (e.g., Miyamoto& Nagai 1975; Wolfire *et al.* 1995). The circular velocity $v_{circ} \approx 220$ km s^{-1} is now well established through wide-field stellar surveys (Smith *et al.* 2007).

We determine the vertical acceleration at the Solar Circle using $g = -\partial\phi(R_o, z)/\partial z$ with $R_o = 8$ kpc. The hydrostatic halo pressure follows from

$$\frac{\partial\phi}{\partial z} = -\frac{1}{\rho_h}\frac{\partial P}{\partial z} \tag{2.3}$$

After Ferrara & Field (1994), we adopt a solution of the form $P_h(z) = P_o \exp((\phi(R_o, z) - \phi(R_o, 0))/\sigma_h^2)$ where σ_h is the isothermal sound speed of the hot corona. To arrive at P_o, we adopt a coronal halo density of $n_{e,h} = 10^{-4}$ cm^{-3} at the Stream distance (55 kpc) in order to explain the Magellanic Stream Hα emission (Bland-Hawthorn *et al.* 2007), although this is uncertain to a factor of a few. We choose $T_h = 2 \times 10^6$ K to ensure that O VI is not seen in the diffuse corona consistent with observation (Sembach *et al.* 2003); this is consistent with a rigorously isothermal halo for the Galaxy.

A key parameter of the models is the ratio of the cloud to halo pressure, $\xi = P_c/P_h$. If the cloud is to survive the impact of the hot halo, then $\xi \gtrsim 1$. A shocked cloud is destroyed in about the time it takes for the internal cloud shock to cross the cloud, during which time the cool material mixes and ablates into the gas streaming past. Only massive clouds with dense cores can survive the powerful shocks. An approximate lifetime† for a spherical cloud of diameter d_c is

$$\tau_c = 60(d_c/2 \text{ kpc})(v_h/350 \text{ km s}^{-1})^{-1}(\eta/100)^{0.5} \text{ Myr.} \tag{2.4}$$

For η in the range of $100-1000$, this corresponds to $60-180$ Myr for individual clouds. With a view to explaining the Hα observations, we focus our simulations on the lower end of this range.

For low η, the density of the hot medium is $n_h = 2 \times 10^{-4}$ cm^{-3}. The simulations are undertaken in the frame of the cold H I clouds, so the halo gas is given an initial transverse velocity of 350 km s^{-1}. The observations reveal that the mean Hα emission has a slow trend along the Stream which requires the Stream to move through the halo at a small angle of attack ($20°$) in the plane of the sky in order to explain the more distribution emission. Independent evidence for this appears to come from a wake of low column clouds along the Stream (Westmeier & Koribalski 2008). Thus, the velocity of the hot gas as seen by the Stream is $(v_x, v_y) = (-330, -141)$ km s^{-1}. The adiabatic sound

---

† Here we correct a typo in equation (1) of Bland-Hawthorn *et al.* (2007).

speed of the halo gas is 200 km s^{-1}, such that the drift velocity is mildly supersonic (transsonic), with a Mach number of 1.75.

A unique feature of the FYRIS simulations is that they include non-equilibrium cooling through time-dependent ionisation calculations (cf. Rosen & Smith 2004). When shocks occur within the inviscid fluid, the jump shock conditions are solved across the discontinuity. This allows us to calculate the Balmer emission produced in shocks and additionally from turbulent mixing along the Stream (e.g., Slavin et al. 1993). We adopt a conservative value for the gas metallicity of [Fe/H]$= -1.0$ (cf. Gibson et al. 2000); a higher value accentuates the cooling and results in denser gas, and therefore stronger Hα emission along the Stream.

2.1. *Results*

The main results of the simulations are presented elsewhere (Bland-Hawthorn et al. 2007: see animations at http://www.aao.gov.au/astro/MS). In our model, the fractal Stream experiences a "hot wind" moving in the opposite direction. The sides of the Stream clouds are subject to gas ablation via KH instabilities due to the reduced pressure (Bernouilli's theorem). The ablated gas is slowed dramatically by the hot wind and is transported behind the cloud. As higher order modes grow, the fundamental mode associated with the cloud size will eventually fragment it. The ablated gas now plays the role of a "cool wind" that is swept up by the pursuing clouds leading to shock ionization and ablation of the downstream clouds. The newly ablated material continues the trend along the length of the Stream. The pursuing gas cloud transfers momentum to the ablated upstream gas and accelerates it; this results in Rayleigh-Taylor (RT) instabilities, especially at the stagnation point in the front of the cloud. We rapidly approach a nonlinear regime where the KH and RT instabilities become strongly entangled, and the internal motions become highly turbulent. The simulations track the progression of the shock fronts as they propagate into the cloudlets.

Bland-Hawthorn et al. (2007, Fig. 2) show the predicted conversion of neutral to ionized hydrogen due largely to cascading shocks along the Stream. The drift of the peak to higher columns is due to the shocks eroding away the outer layers, thereby progressing into increasingly dense cloud cores. The ablated gas drives a shock into the H I material with a shock speed of v_s measured in the cloud frame. At the shock interface, once ram-pressure equilibrium is reached, we find $v_s \approx v_h \eta^{-0.5}$. In order to produce significant Hα emission, $v_s \gtrsim 35$ km s^{-1} such that $\eta \lesssim 100$. In Fig. 1, we see the predicted steady rise in Hα emission along the Stream, reaching $100-200$ mR after 120 Myr, and the most extreme observed values after 170 Myr. The power-law decline to bright emission measures is a direct consequence of the shock cascade. The shock-induced ionization rate is 1.5×10^{47} phot s^{-1} kpc^{-1}. The predicted luminosity-weighted line widths of 20 km s^{-1} FWHM are consistent with the Hα kinematics. In our models, much the Hα lies at the leading edges of clouds, although there are occasional cloudlets where ionized gas dominates over the neutral column. Some of the brightest emission peaks appear to be due to limb brightening, while others arise from chance alignments.

2.2. *Discussion*

We have seen that the brightest Hα emission along the Stream can be understood in terms of shock ionization and heating in a transsonic (low Mach number) flow. For the first time, the Balmer emission (and associated emission lines) provides diagnostic information at any position along the Stream that is independent of the H I observations. Slow Balmer-dominated shocks of this kind (e.g., Chevalier & Raymond 1978) produce partially ionized media where a significant fraction of the Hα emission is due to collisional excitation. This can lead to Balmer decrements (Hα/Hβ ratio) in excess of 4, i.e. significantly enhanced over the pure recombination ratio of about 3, that will be fairly straightforward to verify in the brightest regions of the Stream.

The shock models predict a range of low-ionization emission lines (e.g., O I, S II), some of which will be detectable even though suppressed by the low gas-phase metallicity. There are likely to be EUV absorption-line diagnostics through the shock interfaces revealing more extreme kinematics, but these detections (e.g., O VI) are only possible towards fortuitous background sources (Sembach *et al.* 2001; Bregman 2007). The predicted EUV/X-ray emissivity from the post-shock regions is much too low to be detected in emission.

The characteristic timescale for large changes is roughly 100$-$200 Myr, and so the Stream needs to be replenished by the outer disk of the LMC at a fairly constant rate (e.g., Mastropietro *et al.* 2005). The timescale can be extended with larger η values (equation (2.4)), but at the expense of substantially diminished Hα surface brightness. In this respect, we consider η to be fairly well bounded by observation and theory.

What happens to the gas shedded from the dense clouds? Much of the diffuse gas will become mixed with the hot halo gas suggesting a warm accretion towards the inner Galactic halo. If most of the Stream gas enters the Galaxy via this process, the derived gas accretion rate is ~ 0.4 M$_\odot$ yr^{-1}. The higher value compared to H I (e.g., Peek *et al.* 2008) is due to the gas already shredded, not seen by radio telescopes now. In our model, the HVCs observed today are unlikely to have been dislodged from the Stream by the process described here. These may have come from an earlier stage of the LMC-SMC interaction with the outer disk of the Galaxy.

The "shock cascade" interpretation for the Stream clears up a nagging uncertainty about the Hα distance scale for high-velocity clouds. Bland-Hawthorn *et al.* (1998) first showed that distance limits to HVCs can be determined from their observed Hα strength due to ionization by the Galactic radiation field, now confirmed by clouds with reliable distance brackets from the stellar absorption line technique (Putman *et al.* 2003; Lockman *et al.* 2008; Wakker *et al.* 2007). HVCs have smaller kinetic energies compared to the Stream clouds, and their interactions with the halo gas are not expected to produce significant shock-induced or mixing layer Hα emission, thereby supporting the use of Hα as a crude distance indicator.

If we are to arrive at a satisfactory understanding of the Stream interaction with the halo, future deep Hα surveys will be essential. It is plausible that current Hα observations are still missing a substantial amount of gas, in contrast to the deepest H I observations. We can compare the particle column density inferred from H I and Hα imaging surveys. The limiting H I column density is about $N_{\rm H} \approx \langle n_{\rm H} \rangle L \approx 10^{18}$ cm^{-2} where $\langle n_{\rm H} \rangle$ is the mean atomic hydrogen density, and L is the depth through the slab. By comparison, the Hα surface brightness can be expressed as an equivalent emission measure, $E_{\rm m} \approx \langle n_{\rm e}^2 \rangle L \approx \langle n_{\rm e} \rangle N_{\rm e}$. Here $n_{\rm e}$ and $N_{\rm e}$ are the local and column electron density. The limiting value of $E_{\rm m}$ in Hα imaging is about 100 mR, and therefore $N_{\rm e} \approx 10^{18}/\langle n_{\rm e} \rangle$ cm^{-2}. Whether the ionized and neutral gas are mixed or distinct, we can hide a lot more ionized gas below the imaging threshold for a fixed L, particularly if the gas is at low density ($\langle n_{\rm e} \rangle \ll 0.1$ cm^{-3}). A small or variable volume filling factor can complicate this picture but, in general, the ionized gas still wins out because of ionization of low density H I by the cosmic UV background (Maloney 1993). In summary, even within the constraints of the cosmic microwave background (see Maloney & Bland-Hawthorn 1999), a substantial fraction of the gas can be missed if it occupies a large volume in the form of a low density plasma.

References

Besla, G., Kallivayalil, N., Hernquist, L., *et al.* 2007, *ApJ*, 668, 949

Binney, J., Dehnen, W., & Bertelli, G. 2000, *MNRAS*, 318, 658

Bland-Hawthorn, J., Veilleux, S., Cecil, G. N., Putman, M. E., Gibson, B. K., & Maloney, P. R. 1998, *MNRAS*, 299, 611

Bland-Hawthorn, J. & Maloney, P. R. 1999, *ApJ*, 510, L33

Bland-Hawthorn, J. & Maloney, P. R. 2002, in J. S. Mulchaey & J. Stocke (eds.), *Extragalactic Gas at Low Redshift*, ASP-CS, 254, 267

Bland-Hawthorn, J., Sutherland, R., Agertz, O., & Moore, B. 2007, *ApJ*, 670, L109

Bregman, J. N. 2007, *ARAA*, 45, 221

Brüns, C., Kerp, J., Staveley-Smith, L., *et al.* 2005, *A&A*, 432, 45

Chevalier, R. A. & Raymond, J. C. 1978, *ApJ*, 225, L27

Connors, T. W., Kawata, D., & Gibson, B. K. 2006, *MNRAS*, 371, 108

Ferrara, A. & Field, G. B. 1994, *ApJ*, 423, 665

Flynn, C., Holmberg, J., Portinari, L., Fuchs, B., & Jahreiß, H. 2006, *MNRAS*, 372, 1149

Gibson, B. K., Giroux, M. L., Penton, S. V., Putman, M. E., Stocke, J. T., & Shull, J. M. 2000, *AJ*, 120, 1830

Kallivayalil, N., van der Marel, R. P., & Alcock, C. 2006, *ApJ*, 652, 1213

Lockman, F. J., Benjamin, R. A., Heroux, A. J., & Langston, G. I. 2008, *ApJ*, 679, L21

Madsen, G. J., Haffner, L. M., & Reynolds, R. J. 2002, in A. R. Taylor, T. L. Landecker, & A. G. Willis (eds.), *Seeing Through the Dust. The Detection of H*I *and the Exploration of the ISM in Galaxies*, ASP-CS, 276, 96

Maloney, P. R. & Bland-Hawthorn, J. 1999, *ApJ*, 522, L81

Maloney, P. 1993, *ApJ*, 414, 41

Mastropietro, C., Moore, B., Mayer, L., Wadsley, J., & Stadel, J. 2005, *MNRAS*, 363, 509

Miyamoto, M. & Nagai, R. 1975, *PASJ*, 27, 533

Moore, B. & Davis, M. 1994, *ApJ*, 270, 209

Peek, J. E. G., Putman, M. E., & Sommer-Larsen, J. 2008, *ApJ*, 674, 227

Piatek, S., Pryor, C., & Olszewski, E. W. 2008, *AJ*, 135, 1024

Putman, M. E., Bland-Hawthorn, J., Veilleux, S., Gibson, B. K., Freeman, K. C., & Maloney, P. R. 2003, *ApJ*, 597, 948

Quilis, V. & Moore, B. 2001, *ApJ*, 555, L95

Rosen, A. & Smith, M. D. 2004, *MNRAS*, 347, 1097

Sembach, K. R., Howk, J. C., Savage, B. D., Shull, J. M., & Oegerle, W. R. 2001, *ApJ*, 561, 573

Sembach, K. R., Wakker, B. P., Savage, B. D., *et al.* 2003, *ApJS*, 146, 165

Slavin, J. D., Shull, J. M., & Begelman, M. C. 1993, *ApJ*, 407, 83

Smith, M. C., Ruchti, G. R., Helmi, A., *et al.* 2007, *MNRAS*, 379, 755

Wakker, B. P., York, D. G., Howk, J. C., *et al.* 2007, *ApJ*, 207, 670, L113

Weiner, B. J., Vogel, S. N., & Williams, T. B. 2002, in J. S. Mulchaey & J. Stocke (eds.), *Extragalactic Gas at Low Redshift*, ASP-CS, 254, 256

Westmeier, T. & Koribalski, B. S. 2008, *MNRAS*, 388, L29

Wilkinson, M. I. & Evans, N. W. 1999, *MNRAS*, 310, 645

Williams, J. P. & McKee, C. F. 1997, *ApJ*, 476, 166

Wolfire, M. G., McKee, C. F., Hollenbach, D., & Tielens, A. G. G. M. 1995, *ApJ*, 453, 673

The Magellanic System: Stars, Gas, and Galaxies
Proceedings IAU Symposium No. 256, 2008
Jacco Th. van Loon & Joana M. Oliveira, eds.

© 2009 International Astronomical Union
doi:10.1017/S1743921308028354

The many streams of the Magellanic Stream

Snežana Stanimirović[1], Samantha Hoffman[1], Carl Heiles[2], Kevin A. Douglas[3], Mary Putman[4] and Joshua E. G. Peek[2]

[1]Department of Astronomy, University of Wisconsin-Madison, 475 North Charter Street, Madison, WI 53706, USA
email: sstanimi@astro.wisc.edu

[2]Department of Astronomy, UC Berkeley, 601 Campbell Hall, Berkeley, CA 94720, USA

[3]Space Sciences Laboratory, University of California, Berkeley, CA 94720, USA

[4]University of Michigan, Department of Astronomy, 500 Church St., Ann Arbor, MI 48109, USA

Abstract. As a part of the ongoing H I survey by the consortium for Galactic studies with the *Arecibo* L-band Feed Array (GALFA-HI), we have recently imaged the tip of the MS and found several long filamentary structures. This demonstrates that the northern portion of the MS, which has been interacting with the Galactic halo for a long time, is more extended than previously thought and in the form of highly organized H I structures. The observed filaments, and especially the kinematic dichotomy of H I clouds observed for the first time, agree with predictions by the Connors, Kawata & Gibson (2006) tidal model. However, specific time-stamps in the history of the Magellanic System are required to explain these phenomena. The 20-degree long filaments are accompanied by a large population of small H I clouds. We investigate the observed properties of these clouds and explore various instabilities that affect a warm tail of gas trailing through the Galactic halo. Interestingly, if the observed H I structure is mainly due to thermal instability, then the tip of the MS is at a distance of ∼70 kpc.

Keywords. ISM: structure, Galaxy: halo, galaxies: interactions, intergalactic medium, galaxies: ISM, Magellanic Clouds

1. Introduction

The Magellanic Stream (MS) is a huge neutral hydrogen (H I) structure representing the most fascinating signature of the wild past interaction of our Galaxy with the Magellanic Clouds (MCs), and of the MCs with each other. Many past and recent H I studies have provided illuminating clues regarding the origin and evolution of the MS. One of the first H I images of the MS was obtained with the *Parkes* telescope (angular resolution of 15′) by Mathewson & Ford (1984), revealing complex diffuse structure and six discrete concentrations (labeled as MS I to VI). More recent and fully-sampled *Parkes* observations by Putman *et al.* (1998, 2003) and Brüns *et al.* (2005) showed interesting H I sub-structure in the form of two 100-degree long interwoven (helix-like) filaments. It was thought that the filaments are becoming overwhelmed by the Galactic halo around Dec ∼ 0°, ending up in a chaotic network of small filaments and clumps. A strong velocity gradient was observed from about +400 km s⁻¹ at the location of the Large Magellanic Cloud (LMC), to −400 km s⁻¹ at the tip of the MS, the farthest away from the MCs.

However, the latest observations of the MS tip, obtained with the *Arecibo* telescope, show a highly organized structure instead of a chaotic H I distribution (Stanimirović *et al.* 2008). Several filamentary structures extend to the north up to Dec ∼ 30°, and reach a heliocentric velocity of −420 km s⁻¹. The filaments have a great deal of small-scale structure, mainly in the form of discrete H I clouds. These observations are part of the ongoing Galactic H I survey (GALFA-HI; Stanimirović *et al.* 2006) and cover an area of 870 square degrees. When completed, the GALFA-HI survey will fully sample the declination range from $-1°$ to $38°$, with a $(3-\sigma)$ sensitivity of a few $\times 10^{18}$ cm^{-2} (over ~ 20 km s^{-1}).

The northern extension of the MS has been a subject of sporadic attention in the past. Mirabel (1981) suggested that nearly all high-negative velocity gas in the MS direction may be originating from the MS. Lockman *et al.* (2002) noticed several lines of sight outside of the traditional MS borders, up to $b = -30°$, while Braun & Thilker (2004) suggested that the diffuse MS extends up to Dec $40°$. Recently, Westmeier & Koribalski (2008) detected about 150 isolated H I clouds tens of degrees away (RA 0^h to 2^h) from the traditional MS, but organized in several long filaments running almost parallel to the MS.

In this paper, we focus on the northern extension of the MS by summarizing the main results from Stanimirović *et al.* (2008). We provide a brief comparison of H I observations with recent numerical simulations, and investigate physical processes that may be responsible for the small-scale H I structure of the MS. We also use the small-scale H I structure of the MS to place constraints on the distance of the MS tip.

2. H I filaments at the MS tip

With *Arecibo*'s angular resolution of 3.5', four separate filaments were revealed at the MS tip, each with distinct H I morphology and velocity gradient (Figure 1). In terms of morphology, the three streams starting from RA 23^h appear similar and are clumpy, while the stream starting at RA 23^h40^m appears to consist of mainly diffuse H I. This could be interpreted as the more diffuse stream being younger and less fragmented than the other three. The velocity gradient of the three clumpy streams subsequently decreases in steepness.

The only numerical study that addressed in detail the internal MS structure is that by Connors *et al.* (2006). While considering only gravitational interactions, Connors *et al.* were able to reproduce the observationally-inferred two long MS filaments by Putman *et al.* (1998, 2003). They also predicted the existence of the kinematic bifurcation along the MS. Within this numerical framework, several important events needed to happen in the MS history and leave strong marks on the current Magellanic System. In the close passage between the Small Magellanic Cloud (SMC) and the LMC 2.2 Gyrs ago, the H I gas was pulled out of the SMC outskirts. This resulted in the formation of a very distant stream which is now located at a distance of 170–220 kpc. The major perigalacticon about 1.5 Gyrs ago resulted in the formation of the main MS. This was followed by two close passages between the SMC and the LMC, which resulted in the spatial, and then kinematic bifurcation of the main MS filament. The two bifurcated filaments follow each other along most of the MS, however have different velocities; while one filament follows roughly the LMC's orbit, the other one is at a higher negative velocity reaching almost -500 km s^{-1}. Only < 200 Myr ago, two additional tidal tails were drawn from the Magellanic Bridge and follow spatially the main MS.

Within this framework, a plausible scenario could be that the three new clumpy streams represent a 3-way splitting of the main MS, which happened 0.5–1 Gyr ago, while the diffuse (fourth) stream represents one of the younger tidal tails, formed < 200 Myr ago. While it is highly encouraging that tidal effects alone can produce large-scale spatial and kinematic sub-structure of the MS, many details still need to be worked out. From an observational perspective, the number and full extent of the new filaments have to be

GALFA–HI RA+DEC Tile 324.00+02.3

Figure 1. An H I image of the MS at a LSR velocity of -386 km s^{-1} obtained with the *Arecibo* telescope. To enhance our sensitivity to diffuse emission, the image was smoothed to $10'$, and then median filtered. Two white areas at RA $\sim 23^{\rm h}45^{\rm m}$, and a grey strip across the whole image centered at Dec $9°$, currently have no data. Note that the GALFA-HI survey is still under way and better sensitivity images, with a full coverage, will be available in the near future.

constrained with future high resolution observations. It is not clear at the moment how are all these filaments related, whether they are localized in certain areas of the MS or are present along the whole length of the MS. From a theoretical perspective, a potential problem is the need of separate special events (encounters) between the SMC and the LMC to produce the small-scale filamentary structure of the MS. Considering that the new proper motion measurements (Kallivayalil, van der Marel & Alcock 2006) imply a different history of the SMC-LMC-MW interactions (Besla *et al.* 2007), it is not clear how many actual encounters were available in the past for shaping of the MS. Therefore, other and/or additional structuring mechanisms may be required.

3. Clumpy H I structure

Besides several long H I filaments, GALFA-HI observations also show a wealth of small H I clumps. As the clumpiness of the MS is visually striking we produced a catalog of H I clouds and measured their basic observed properties. The cloud angular size distribution peaks at about $10'$, while the H I column density peaks at about 10^{19} cm^{-2}. If at a distance of 60 kpc, then typical H I clouds have a mass of $\sim 10^3$ M$_\odot$. The cloud central velocity decreases with Galactic latitude, from -300 km s^{-1} at $b = -50°$, to -420 km s^{-1} at $b = -20°$. This is shown in Figure 2 (left), where we plot H I clouds from Stanimirović

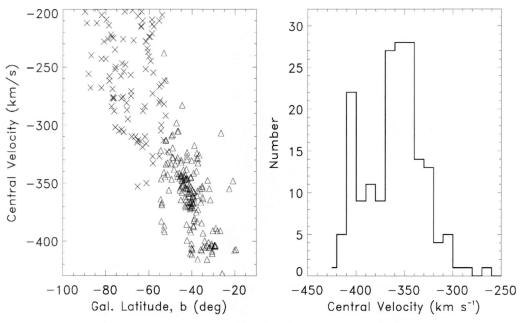

Figure 2. (left) Central velocity of MS clouds as a function of Galactic latitude. Crosses are from the Putman *et al.* (2002) catalog, based on observations with the *Parkes* telescope, while triangles are from Stanimirović *et al.* (2008), based on the observations with the *Arecibo* telescope. (right) Histogram of the cloud central velocity, from Stanimirović *et al.* (2008).

et al. (2008) and Putman *et al.* (2002). The two catalogs merge smoothly, confirming that the new cloud catalog represents an northern extension of the MS population. The same figure (right) also shows the histogram of H I central velocities. Interestingly, there are two strong peaks at −350 and −405 km s^{-1}. This velocity dichotomy may correspond to the kinematic bifurcation of the MS predicted by Connors *et al.* (2006) and never previously confirmed observationally.

Another interesting phenomenon is the multi-phase H I structure. About 15% of clouds in our sample have velocity profiles whose fitting requires two temperature components. This suggests the existence of the multi-phase medium at a significant distance from the Galactic plane. We find evidence for a warm gas, with a FWHM of about 25 km s^{-1}, and a cooler component, with a FWHM of 10–13 km s^{-1}. The cooler component is most likely in the regime of the thermally unstable WNM. Kalberla & Haud (2006) investigated velocity profiles along the MS based on the Leiden/Argentine/Bonn data (Kalberla *et al.* 2005). They found that 27% of MS profiles at positive LSR velocities, and 12% of profiles at negative LSR velocities, require two temperature components.

4. Physical processes responsible for small-scale structure

One of the crucial issues regarding the structure of the MS is to what extent interactions with the halo determine or influence the MS gas. This is especially important for the MS tip as this part of the MS has been immersed in the hot halo for a long time. In trying to understand the origin and evolution of the small-scale H I structure in the MS, we ask the question of what happens to a warm stream of gas, resulted from tidal or hydrodynamical interactions, as it moves through a hot ambient medium?

There are at least three hydrodynamical effects that play important roles. (i) Thermal instability (TI) develops due to gas cooling and results in the fragmentation of the warm stream. Assuming that the stream has properties similar to those found in the outskirts of the SMC, the TI fragmentation will occur on timescales of a few tens to a few hundreds of Myrs (for details please see Stanimirović *et al.* 2008). (ii) Kelvin-Helmholtz instability (KHI) occurs at the interface between the moving warm stream and the hot ambient medium and provides a continuous stripping mechanism. The KHI timescale depends on the properties of the halo gas as well (temperature, density), which are not well constrained observationally. However, its typical timescale is in the range of a few hundreds to a few thousands of Myrs. (iii) The small fragments made by TI and/or KHI will be subjected to the heat transfer from the much warmer ambient medium. In the classical evaporation by heat conduction (McKee & Cowie 1977), the H I clouds are undergoing evaporation but the evaporation timescale is long, > 1 Gyr. In the case of turbulent mixing layers the evaporation timescale would decrease, while an inclusion of a magnetic field would make clouds even longer lived.

To conclude, TI and KHI must have had important effects on the shaping of the small-scale H I structure over the MS lifetime (in most theoretical frameworks). As TI typically operates on shorter timescales than KHI, future high resolution observations along the MS should be able to explore the evolution of the MS gas and place constraints on the effectiveness of these instabilities. Also, while undergoing evaporation, the H I clouds can survive for a long time, and therefore it may not be surprising to observe such clumpy morphology at the MS tip.

If we assume that TI is the dominant shaping agent, then we can predict a typical size of thermal fragments. For the density and temperature conditions characteristic of the SMC outskirts, "typical" thermal fragments should be about 200 pc in size. A comparison with the peak of the cloud angular size distribution, suggests that the MS tip is at a distance of $\sim$ 70 kpc. While this simple, back-of-the-envelope calculation is only demonstrative, it is interesting that our distance estimate agrees well with the predictions from tidal models. Even more impressively, our distance estimate is in agreement with the recent estimate of 75 kpc based on a model by Jin & Lynden-Bell (2008). This model assumed that energy and angular momentum are conserved along the MS, and that the MS is trailing on a planar orbit around the Galactic center.

Additionally, an upper limit on the MS distance can be placed by considering how far away from the Galactic plane the conditions are reasonable for the existence of a multi-phase medium. Based on the cooling/heating processes in the Galactic halo, Wolfire *et al.* (1995) suggested that pressure-confined multi-phase clouds should not be found at distances larger than 25 kpc. Sternberg, McKee & Wolfire (2002) considered both pressure and dark matter confinement of halo clouds, and concluded that the multi-phase medium can survive at distances < 150 kpc. Therefore, the MS tip is likely to be closer than 150 kpc. This causes difficulties for the existence of a very distant MS filament as suggested by Connors *et al.* Also, the latest orbit calculations by Besla *et al.* (2007) imply a significant distance to the MS, $\sim$ 150 kpc, and may require refinements.

5. Conclusions

Several recent observational studies have shown that the northern portion of the MS is significantly more extended and consists of long filamentary structures. The velocity dichotomy of H I clumps at the MS tip, observed for the first time, agrees with predictions by the Connors *et al.* (2006) tidal model. However, special encounters in the Magellanic history are necessary to explain detail kinematic and spatial structure of the MS in this framework. While tides may be the dominant structuring agent on large-scales, we have demonstrated that hydrodynamical instabilities are the key for explaining the small-scale structure of the MS, and in turn could provide important constraints on the formation and evolution of the MS.

A few important questions remain for the future: how much more of the low column density MS debris remain to be discovered with future observations? what is the fate of this material? why are the head-tail and cometary H I clouds absent in high-resolution observations? do these morphological features really trace the cloud-halo interactions, or could they be possibly caused by cloud-cloud interactions? how turbulent is the MS gas, and how does this turbulence evolve along the MS?

Acknowledgements

The Arecibo Observatory is part of the National Astronomy and Ionosphere Center, operated by Cornell University under a cooperative agreement with the National Science Foundation. It is a pleasure to acknowledge other members of the GALFA-HI survey, and Tim Connors, Fabian Heitsch, and Jay Gallagher for many stimulating discussions. SS is thankful to conference organizers for partial travel support. Support by NSF grants AST-0097417, AST-0707679, and AST-0709347 is gratefully acknowledged.

References

Besla, G., Kallivayalil, N., Hernquist, L., Robertson, B., Cox, T. J., van der Marel, R. P., & Alcock, C. 2007, *ApJ*, 668, 949
Braun, R. & Thilker, D. A. 2004, *A&A*, 417, 421
Brüns, C., Kerp, J., Staveley-Smith, L., *et al.* 2005, *A&A*, 432, 45
Connors, T. W., Kawata, D., & Gibson, B. K. 2006, *MNRAS*, 371, 108
Jin, S. & Lynden-Bell, D. 2008, *MNRAS*, 383, 1686
Kalberla, P. M. W., Burton, W. B., Hartmann, D., Arnal, E. M., Bajaja, E., Morras, R., Pöppel, W. G. L. 2005, *A&A*, 440, 775
Kalberla, P. M. W. & Haud, U. 2006, *A&A*, 455, 481
Kallivayalil, N., van der Marel, R. P., & Alcock, C. 2006, *ApJ*, 652, 1213
Lockman, F. J., Murphy, E. M., Petty-Powell, S., & Urick, V. J. 2002, *ApJS*, 140, 331
Mathewson, D. S. & Ford, V. L. 1984, in S. van den Bergh, & K. S. de Boer (eds.) *Structure and evolution of the Magellanic Clouds*, IAU Symposium 108 (Dordrecht: Reidel), p. 125
McKee, C. F. & Cowie, L. L. 1977, *ApJ*, 215, 213
Mirabel, I. F. 1981, *ApJ*, 250, 528
Putman, M. E., Gibson, B. K., Staveley-Smith, L., *et al.* 1998, *Nature*, 394, 752
Putman, M. E., de Heij, V., Staveley-Smith, L., et al. 2002, *AJ*, 123, 873
Putman, M. E., Staveley-Smith, L., Freeman, K. C., Gibson, B. K., & Barnes, D. G. 2003, *ApJ*, 586, 170
Stanimirović, S., Putman, M., Heiles, C., et al. 2006, *ApJ*, 653, 1210
Stanimirović, S., Hoffman, S., Heiles, C., Douglas, K. A., Putman, M., & Peek, J. E. G. 2008, *ApJ*, 680, 276
Sternberg, A., McKee, C. F., & Wolfire, M. G. 2002, *ApJS*, 143, 419
Westmeier, T. & Koribalski, B. S. 2008, *MNRAS*, 388, 29
Wolfire, M. G., McKee, C. F., Hollenbach, D., & Tielens, A. G. G. M. 1995, *ApJ*, 453, 673

Session III

The properties of the interstellar medium

The Magellanic System: Stars, Gas, and Galaxies
Proceedings IAU Symposium No. 256, 2008
Jacco Th. van Loon & Joana M. Oliveira, eds.

© 2009 International Astronomical Union
doi:10.1017/S1743921308028378

Dust in the Magellanic Clouds

François Boulanger

Institut d'Astrophysique Spatiale (IAS), UMR 8617, CNRS & Université Paris-Sud 11,
Bâtiment 121, 91405 Orsay Cedex, France
email: francois.boulanger@ias.u-psud.fr

Abstract. The Magellanic Clouds are important templates for studying the role interstellar dust plays as actor and tracer of galaxy evolution. Due to their proximity, the Large and Small Magellanic clouds are uniquely suited to put detailed Galactic dust studies in a global context. With a metal abundance lower than that of the Sun, the Magellanic Clouds also permit to characterize interstellar matter composition and structure as a function of metallicity. The presentation of spectacular results from the *AKARI* and *Spitzer* surveys was one of the highlights of this Magellanic Clouds meeting. This paper puts these results in context. I discuss UV extinction and IR emission signatures of carbon and silicate dust. I present diverse evidence of dust processing in the ISM. I illustrate the correlation between the mm emission of dust, and gas column density using Milky Way surveys. I conclude with three main results. Dust in the SMC is not carbon poor. The composition of interstellar dust reflects its processing in interstellar space and thereby depends on local conditions and its past history. In the Magellanic Clouds, far-IR and sub-mm observations are indicating that there may be significantly more cold interstellar matter, cold H I and H_2 gas, than estimated from H I and CO observations.

Keywords. dust, extinction, ISM: evolution, Magellanic Clouds, infrared: galaxies

1. Introduction

Infrared space observatories from *IRAS* to *Spitzer* have laid the groundwork for our current understanding of the properties of interstellar dust. Prior to the opening of infrared space astronomy, dust was mainly known from extinction studies. For most observational astronomers it was considered a mere nuisance in their lives. Today, we realize that dust plays a prominent role as actor and tracer of the structure of matter and of its physical and chemical evolution, as well as of the formation of galaxies, stars and planets.

The Magellanic Clouds are important templates for understanding galaxy evolution. As nicely summarized in SAGE, the *Spitzer* surveys acronym, infrared dust observations bear on three of the key agents of galaxy evolution: interstellar matter, star formation, and the late stages of stellar evolution. Due to their proximity, the Large and Small Magellanic clouds (LMC and SMC) are uniquely suited to put detailed Galactic studies in a global context. For interstellar matter, the *Spitzer* resolution in the near and mid-IR permits to separate individual molecular clouds and star forming regions from the diffuse interstellar medium. Herschel will soon provide comparable sensitivity and resolution in the far-IR. Sub-mm ground observations are also advancing with the LABOCA bolometer array camera on the southern *APEX* sub-mm telescope. With a metal abundance about a factor three (LMC) and eight (SMC) lower than that of the Sun, the Magellanic Clouds permit to characterize interstellar matter composition and structure as a function of metallicity. The small, metal-poor and gas rich SMC is a nearby environment which has some of the characteristics of the early stages of galaxy formation.

The spectacular *Spitzer* color image of the LMC was put forward to announce the meeting. During the conference, several talks and posters detailed the diverse research

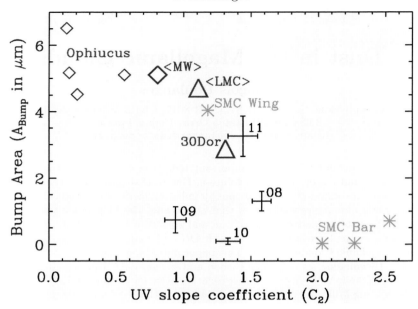

Figure 1. This figure illustrates the difference in the UV extinction curve (UV slope and bump area) between the Milky Way (red diamonds) and the SMC. The mean LMC and 30 Doradus extinction curves from Misselt *et al.* (1999) are plotted with triangles. The stars represent the data gathered by Gordon and Clayton (1998): three bump-free extinction curve towards star forming regions in the SMC Bar and one 30 Doradus-like curve towards the Wing. The numbered points with error bars are results from STIS/*Hubble Space Telescope* observations of a small group of reddened stars located towards the SMC B1 molecular cloud. The four stars are located within a 20″ area. The STIS data analysis has been performed by Jesus Appelaniz and will be presented in a forthcoming paper (Boulanger *et al.* in preparation).

which is being carried out with *AKARI* and *Spitzer* observations, in particular the SAGE LMC survey (Meixner *et al.* 2006) and the S^3MC and SAGE SMC surveys (Bolatto *et al.* 2007); Gordon *et al.*, these proceedings). For the first time, I saw dust observations in the fore-front of an extragalactic conference. With this paper, I wish to provide a broader context including Galactic studies, which complements the more specific reports on the Magellanic Clouds data analysis presented in these proceedings.

2. Carbon and silicate dust

The observational constraints on dust, its composition and size distribution, are element abundances and depletions, extinction and scattering properties and the spectral energy distribution (SED) of its emission in the infrared. Most of what we know on interstellar dust comes from Galactic observations.

Interstellar dust comprises several components, including carbonaceous and amorphous silicate grains that are frosted with icy mantles in dense clouds. The amorphous structure of interstellar grains is in contrast with crystalline silicates which have been discovered around some oxygen-rich, evolved stars and comets by the *ISO* satellite.

The smallest grains are aromatic and amorphous hydrocarbon particles containing tens to thousands of carbon atoms per particle. The smallest (sizes less than ~ 1000 atoms) carbon particles are the carriers of the mid-IR bands and are referred to as interstellar Polycyclic Aromatic Hydrocarbons (PAHs). We refer to nanometer size particles as Very Small Grains (VSGs). Both PAHs and VSGs have small heat capacities due to their small

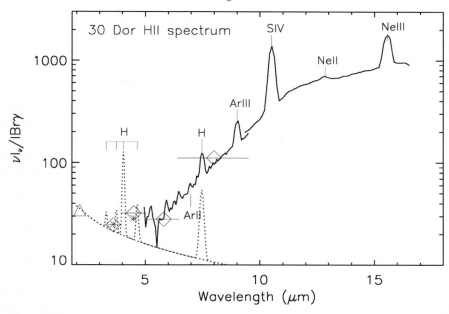

Figure 2. Mid-IR emission spectrum of H II gas near the 30 Doradus super star cluster obtained by correlating IRAC/*Spitzer* (diamonds) and *ISO* spectro-imaging observations with an image in the Brγ hydrogen line. The gas free-free plus the hydrogen lines contribution are shown as a dashed line. This spectrum illustrates PAH destruction in H II gas. The large grains are not hot enough to account for the mid-IR continuum. This continuum must come from VSGs. Note that the continuum is featureless.

dimensions and therefore undergo stochastic heating upon the absorption of photons from the ambient interstellar radiation field (ISRF). They reach peak temperatures from hundreds up to one thousand degrees Kelvin, depending upon their size, and therefore tend to emit most of their thermal energy at wavelengths shortward of 60 μm. In the Solar Neighborhood, stochastically heated grains with sizes smaller than ~ 1.5 nm account for about 10% of the dust mass and 30% of the power radiated by dust. Since the 9.6 and 18 μm silicate features are not seen in emission — outside high radiation field environments where the equilibrium temperature of dust grains is high enough for them to emit at mid-IR wavelengths — VSGs are thought to be carbon dust.

The larger interstellar particles with dimensions of the order of 100 nm, the "big grains", are in thermal equilibrium with the ISRF. In the Solar Neighborhood, big grains in the diffuse ISM penetrated by UV radiation have temperatures of typically 18 K, and dominate the emission in the far-IR to sub-mm range (the peak of their thermal emission occurs at about 150 μm). The big grains temperature rises with ISRF intensity, $G_{\rm UV}$, as $G_{\rm UV}^{1/\beta}$ wite β in the range 5 to 6.

The Magellanic Clouds are becoming a reference for extragalactic dust. In particular, the 30 Doradus and SMC Bar extinction curves are thought to apply to starburst and low metallicity galaxies in general (Calzetti *et al.* 1994). The Milky Way, LMC and SMC extinction curves have been modeled with a size distribution of carbon and silicate dust (Weingartner and Draine 2001). In these models the 220 nm bump is directly related to the fraction of the dust mass in carbon grains smaller than about 5 nm. Both small silicates and carbon grains in this size range contribute to the UV slope. The differences in the 220 nm bump and the UV slope (Fig. 1) are interpreted as indicating a change in the relative fraction of mass in small carbon and silicate dust, with more small silicate dust and less small carbon dust in the SMC bar.

Infrared observations provide complementary evidence for a reduction of the fraction of small carbon dust in low metallicity galaxies like the SMC. Several *ISO* and *Spitzer* spectroscopic and photometric observations show a drop of the PAH emission bands in galaxies of low metallicity (Engelbracht *et al.* 2005). Modeling of the infrared SEDs quantifies this drop in terms of the fraction of the dust mass in PAHs, q_{PAH} (Draine *et al.* 2007). For the SINGS survey galaxies, the median value of q_{PAH} is 3.6% for galaxies with $12 + \log_{10}(O/H)_{gas} > 8.1$ and 1% for lower metallicity galaxies.

The weakness of the PAHs emission bands in the SEDs of low metallicity galaxies has received two interpretations. (1) It traces enhanced destruction of PAHs in regions penetrated by hard ionizing radiation (Madden *et al.* 2006). (2) It reflects a general deficiency of carbon dust due to the delayed injection of carbon dust by AGB winds into the interstellar medium (Galliano *et al.* 2008). The analysis of *Spitzer* observations of evolved stars in the SMC do not support this second interpretation. The first interpretation is supported by the observed destruction of PAHs in H II gas (Contursi *et al.* 2000 and Fig. 2). PAH destruction by ionizing photons may be enhanced in low metallicity galaxies because stars have a stronger ionizing flux but, possibly, mainly because they are small galaxies with a low ISM pressure. The low pressure implies a lower ionized gas density than in large spirals and thereby a larger ionized gas mass for a given ionizing flux.

3. Interstellar dust life cycle

The composition of interstellar dust reflects the action of interstellar processes that contribute to break and re-build grains over timescales much shorter than the renewal timescale by stellar ejecta (Fig. 3). If there is a wide consensus on this conclusion among dust experts, the processes that drive dust evolution in interstellar space are still poorly understood. Understanding interstellar dust evolution is a major challenge underlying many interstellar processes and the interpretation of a wealth of *Spitzer* observations including the Magellanic Clouds survey data.

Dust is subject to processing in the ISM through gas-grain, grain-grain and photon-grain interactions. The degree and nature of the processing depends on the rate and the energy of these interactions both of which are related to the density structure and dynamics of the ISM. High energy gas-grain collisions lead to the erosion of some of the dust mass by sputtering, while low energy collisions lead to the reverse process of gas accretion onto dust. For grain-grain collisions, above some velocity threshold the dust is shattered into smaller fragments, while at lower velocities coagulation occurs. Finally, UV and X-ray photons can alter dust by inducing photon-driven physical and chemical changes including destruction.

High velocity supernova shock waves are thought to be the dominant dust destruction mechanism in the warm ISM. The timescale for dust destruction by sputtering becomes shorter than the gas cooling time for gas temperatures larger than 10^6 K. Gas shock-heated to higher temperatures loses its dust content before it has time to cool (Smith *et al.* 1996). The large variations in the elemental depletions observed in the Galaxy but also in the Magellanic Clouds (Sofia *et al.* 2006) are evidence for the efficiency of dust erosion in the diffuse ISM.

Lower velocity shocks and interstellar turbulence create relative motions between grains that may lead to grain shattering in grain-grain collisions (Falgarone & Puget 1995; Jones *et al.* 1996; Guillet *et al.* 2007). Turbulence might affect the dust evolution more

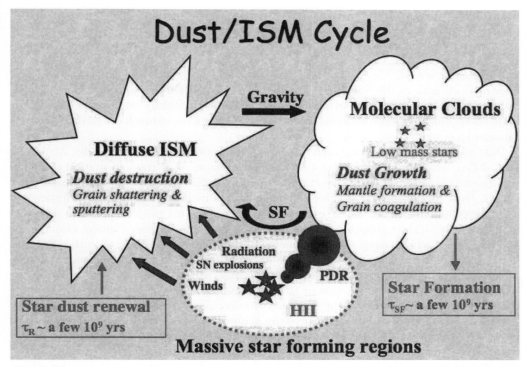

Figure 3. Schematic cartoon of the dust evolution that goes together with the cycling of interstellar matter between diffuse and dense gas phases. The gas cycle and dust evolution is driven by the radiative and mechanical impact of massive star forming regions on their environment. This cycling of matter from the diffuse ISM to molecular clouds occurs on timescales ($< 10^7$ yr) commensurate with the lifetime of massive stars. It is happening on timescales three orders of magnitude shorter than the time for dust renewal by stellar ejecta. Observations provide evidence for dust destruction in the diffuse ISM and dust growth in molecular clouds. The composition of interstellar dust reflects its processing in interstellar space and thereby depends on local conditions and its past history.

frequently and thereby more deeply on average than supernovae shock waves. It also continuously cycles dust grains through a variety of physical conditions. Grain shattering impacts the dust size distribution and might be the dominant source of the small interstellar dust particles. Absorption spectroscopy in the mid-IR demonstrates the ubiquitousness of hydrogenated amorphous carbons (a-C:H) in the diffuse interstellar medium of galaxies (Dartois *et al.* 2007). Based on comparison with laboratory samples, Dartois *et al.* (2005) conclude that the interstellar hydrocarbon material consists of aromatic structures bound together with aliphatic carbon chains. Since this material is the main reservoir of interstellar carbon dust, it is considered a likely precursor of VSGs and PAHs. Galactic observations provide spectroscopic evidence for an evolution of small carbon dust, with production of interstellar PAHs out of larger particles, at the surface of molecular clouds (Berné *et al.* 2007).

Evolution processes leave specific signatures on the dust size distribution and thereby on the dust SED which Magellanic Clouds Spitzer surveys are well suited to map. With spectral bands measuring specifically the emission features from PAHs, the mid-IR emission from VSGs and the far-IR emission from big grains, the *Spitzer Space Telescope* imaging instruments IRAC and MIPS are particularly appropriate to map the relative abundance of dust in these three size bins. Observations with the IRS spectrometer can provide spectroscopic insight on PAHs and VSGs.

Figs. 1, 2 and 4 illustrate diverse evidence of dust evolution in the Magellanic Clouds. Fig. 1 shows that the UV extinction curve varies from one line of sight to another on small angular scales. Fig. 2 illustrates the destruction of PAHs in H II regions. Fig. 4 illustrates changes in the fraction of the dust mass in PAHs from molecular clouds to the diffuse ISM. Fig. 5 suggests that dust processing can significantly impact the dust-to-gas mass ratio. More evidence was presented in the conference and much of the data analysis remains to be done.

Dust has been introduced in Galactic chemical evolution models (Dwek 1998; Zhukovska et al. 2008). These models consider dust mass return from AGB stars, formation of dust in supernovae explosions, dust destruction in the diffuse ISM and dust growth in molecular clouds. Dust processing in the interstellar medium is included with these two processes. The galaxy metallicity affects the evolution of the gas-to-dust mass ratio in two ways. (1) The timescale for dust growth scales as the inverse of the refractory element abundances while the destruction timescale is independent of metallicity. (2) Dust evolution depends on the ratio between the time dust resides in the diffuse ISM versus in molecular clouds, schematically, between interstellar space where dust is destroyed versus where it grows. This ratio depends on the fraction of the gas mass that is in molecular clouds. In large star forming galaxies, most of the gas mass is molecular. In small irregular galaxies like the SMC, possibly because the ISM pressure is lower, most of the gas mass is observed to be in the diffuse ISM. In the bottom plot in Fig. 5, we have used the formula in section 4 of Zhukovska et al. (2008) to illustrate the impact these two factors may have on the dust-to-gas mass ratio: its mass weighted value and the difference between the diffuse ISM and molecular clouds. This calculation includes a 10% fraction of the dust mass in dust grains — possibly metallic oxides (Sofia et al. 2006) — that resist dust destruction. It is essential to keep nucleation sites to rebuild dust in molecular clouds.

4. Tracing gas with dust emission

Among the various observational means of imaging the structure of interstellar matter, observations of the IR dust emission remain unique for tracing interstellar matter over a wide range of conditions and, in particular, across the key H I to H_2 chemical transition where neither H I nor CO lines are good tracers.

Low metallicity star forming dwarf galaxies are gas rich but have weak CO emission. It is commonly thought that they contain little molecular gas but is this really true? Much of the mass could reside in H_2 gas where CO is photo-dissociated. In low metallicity galaxies, contrary to the Milky Way, CO molecules can only survive in dense clumps of molecular clouds while they are photo-dissociated in the less dense "inter-clump" gas (Lequeux et al. 1994). Bot et al. (2007) and Leroy et al. (2007) address this important question using SMC ground-based sub-mm dust and MIPS 160 μm observations.

Fig. 6 illustrates the correlation between mm dust emission, H I and CO emission in the Solar Neighborhood and in the Galactic plane. The tight correlation between the dust and gas emission supports the use of the long wavelength dust emission as a tracer of gas on large scales in galaxies. Clearly, the mm dust emission is a better tracer than the far-IR emission because it depends linearly on the dust temperature. The accuracy to which the dust optical depth may be determined using far-IR data is limited by differences in dust temperatures among ISM components which are most often ignored. A significant result of the Milky-Way correlation analysis is that the mm dust emissivity per hydrogen atom increases by a factor 2 from the H I gas to molecular clouds. This change can be

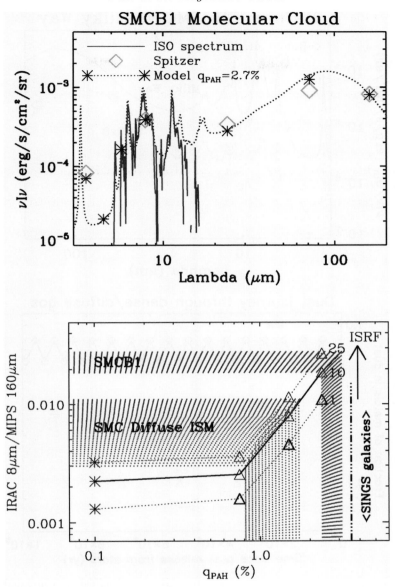

Figure 4. Top: The *ISO* mid-IR spectrum of the SMC molecular cloud SMC B1 (Rubio *et al.* 1996) from Reach *et al.* (2000). The *ISO* spectrum is complemented with IRAC and MIPS photometry obtained with the S^3MC survey (Bollato *et al.* 2007). The dashed line is a Draine & Li (2007) model for a 30 Doradus (LMC2) extinction curve with a fraction of the dust mass in the form of PAHs $q_{PAH} = 2.7\%$, and a radiation field $G_{UV} = 10$ in Solar Neighborhood units. Bottom: The IRAC $8\,\mu$m to MIPS $160\,\mu$m color ratio is plotted versus q_{PAH} with $G_{UV} = 1$, 10 and 25. The stars are the models of Draine and Li for the linear, 220 nm bump-free SMC UV extinction curve. The triangles are models that fit the 30 Doradus extinction curve. The color ratios measured for the SMC B1 cloud and the diffuse ISM by Bot *et al.* (2004) are marked with their error bar by the horizontal hatched area. The models are used to convert these colors in estimates of q_{PAH}. The median value found in Solar metallicity SINGs galaxies (Draine *et al.* 2007) is marked with a dot dashed line.

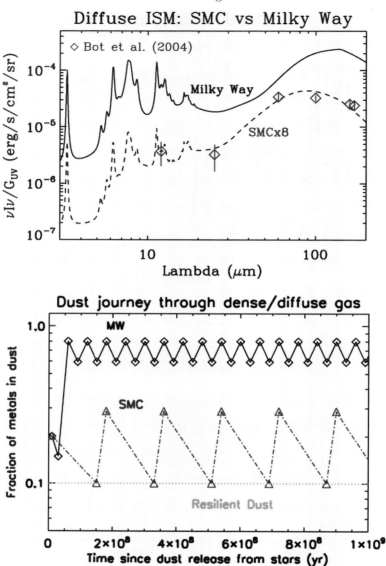

Figure 5. Top: The diffuse ISM SMC SED from Bot *et al.* (2004) is compared to that of the Solar Neighborhood represented here by the Draine and Li (2007) model fit. The model is plotted for a gas column density of 10^{21} H cm^{-2}. The dashed line is a Draine and Li model — for the 30 Doradus extinction curve and $q_{PAH} = 1.5\%$ and a radiation field $G_{UV} = 10$ in Solar Neighborhood units — which fits the mid-IR and far-IR SMC data. The SMC data points and model have been scaled down by this G_{UV} factor and multiplied by 8 to account for the difference in radiation field intensity and metallicity, respectively. The systematic shift between the two SEDs has been interpreted as indicating a drop in dust-to-gas mass ratio from the Milky Way to the SMC larger than the metallicity difference (Bot *et al.* 2004).

Bottom: Schematic model to illustrate the impact of the metallicity on the evolution of the dust-to-gas mass ratio between the diffuse ISM, where dust destruction occurs, and molecular clouds, where refractory gas atoms accrete on dust. This calculation is based on the Zhukovska *et al.* (2008) dust evolution model. Note the difference in the mean fraction of metals in dust and the range of variations introduced by ISM processing. The initial fraction is assumed to be 20%. The large number of cycles illustrates the fact that the cycling time is much smaller than the dust renewal time by stellar ejecta.

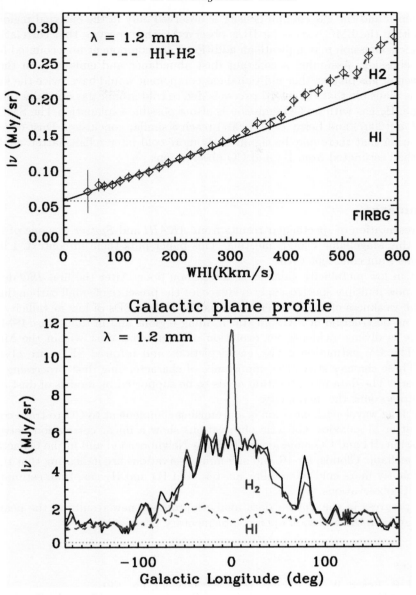

Figure 6. Correlation between 1.2 mm dust (*FIRAS/COBE* data) and gas emission (H I Leiden Dwingeloo survey and CO from Dame *et al.* 2001) in the Milky Way at high Galactic latitude ($|b| > 10°$) and along the Galactic plane. The red lines (dashed in the top figure and solid in the bottom one) is a data fit where the dust emission is expressed as a linear combination of the H I and CO emission. In the top figure the H I emission is the integrated line emission in $K\,km\,s^{-1}$. The offset is the far-IR extragalactic background (FIRBG). Note the absence of a dust counterpart to the CO intensity peak in the Galactic center. This is also seen when comparing CO and γ rays emission (Blitz *et al.* 1985). It reflects a much larger CO luminosity to H_2 mass ratio in the Galactic center than in the Molecular Ring.

partially accounted by an increase in the dust-to-gas mass ratio associated with grain growth in dense gas. It is likely that it also reflects a change in emission properties.

In their analysis of the diffuse emission in the SAGE LMC data, Bernard *et al.* (2008) report a similar study but a different result. A linear correlation between the far-IR optical depth and the gas column density is observed only in the external regions of the LMC. Within the LMC, excess far-IR is observed with respect to that correlation. This far-IR excess emission may indicate an additional gas component unaccounted for by H I and CO emission. Assuming a constant dust abundance and emissivity in the atomic and molecular gas phases, this additional gas component would have twice the known H I mass. It is plausible that the far-IR excess is due to cold atomic gas optically thick in the 21 cm line. H_2 gas with no CO emission is also a possible explanation.The SMC studies of Bot *et al.* (2007) and Leroy *et al.* (2007) reach a similar conclusion. Dust observations are indicating that there may be significantly more cold interstellar matter, cold H I and H_2 gas, than estimated from H I and CO observations.

5. Summary

The presentation of spectacular results from *AKARI* and *Spitzer* was one of the highlights of this Magellanic Clouds meeting. This paper puts them in context. I list a few conclusions I am confident with.

• Dust in low metallicity galaxies is not carbon poor. After the first *ISO* detections, there are now multiple spectroscopic evidence for the presence of small carbon dust in the SMC. The weakness of the PAH emission bands in the SEDs of low metallicity galaxies results from destruction of PAHs in star forming regions and in the diffuse ISM.

• There is diverse evidence for evolution of interstellar dust within the Magellanic Clouds. The UV extinction curve, gas depletions and infrared SEDs are observed to change. These changes stress the importance of characterizing dust processing in interstellar space. The data interpretation needs to be supported by models of dust evolution taking into account the metallicity.

• Dust long wavelength emission is a promising complement to CO to trace cold interstellar matter in galaxies. Galactic observations show a linear correlation between mm dust emission, H I and CO emission in the Solar Neighborhood and in the Galactic plane. In the Magellanic Clouds, far-IR and sub-mm observations are indicating that there may be significantly more cold interstellar matter, cold H I and H_2 gas, than estimated from H I and CO observations.

• The progress in the *Spitzer* data analysis promises new results in the near future. *Herschel* and *ALMA* will soon open complementary perspectives.

References

Bernard, J. P., Reach, W. T., Paradis, D., *et al.* 2008, *AJ*, 136, 819
Berné, O., Joblin, C., Deville, Y., *et al.* 2007, *A&A*, 469, 575
Blitz, L., Bloemen, J. B. G. M., Hermsen, W., & Bania, T. M. 1985, *A&A*, 143, 267
Bolatto, A. D., Simon, J. D., Stanimirović, S., *et al.* 2007, *ApJ*, 655, 212
Bot, C., Boulanger, F., Lagache, G., Cambrésy, L., & Egret, D. 2004, *A&A*, 423, 567
Bot, C., Boulanger, F., Rubio, M., & Rantakyro, F. 2007, *A&A*, 471, 103
Calzetti, D., Kinney, A. L., & Storchi-Bergmann, T. 1994, *ApJ*, 429, 582
Contursi, A., Lequeux, J., Cesarsky, D., *et al.* 2000, *A&A*, 362, 310
Dame, T. M., Hartmann, D., & Thaddeus, P. 2001, *ApJ*, 547, 792
Dartois, E., Munoz Caro, G. M., Deboffle, D., *et al.* 2005, *A&A* 432, 895
Dartois, E. & Muñoz-Caro, G. M. 2007, *A&A* 476, 1235
Draine, B. T. & Li, A. 2007, *ApJ*, 657, 810
Draine, B. T., Dale, D. A., Bendo, G., *et al.* 2007, *ApJ*, 663, 866
Dwek, E. 1998, *ApJ*, 501, 643

Engelbracht, C. W., Gordon, K. D., Rieke, G. H., Werner, M. W., Dale, D. A., & Latter, W. B. 2005, *ApJ*, 628, L29

Falgarone, E. & Puget, J. L. 1995, *A&A* 293, 840

Galliano, F., Dwek, E., & Chanial, P. 2008, *ApJ*, 672, 214

Gordon, K. D. & Clayton, G. C. 1998, *ApJ*, 500, 816

Guillet, V., Pineau Des Forêts, G., & Jones, A. P. 2007, *A&A*, 476, 263

Jones A. P., Tielens, A. G. G. M., & Hollenbach D. J. 1996, *ApJ*, 469, 740

Lequeux, J., Le Bourlot, J., Pineau Des Forêts, G., Roueff, E., Boulanger, F., & Rubio, M. 1994, *A&A*, 292, 371

Leroy, A., Bolatto, A., Stanimirović, S., Mizuno, N., Israel, F., & Bot, C. 2007, *ApJ*, 658, 1027

Madden, S., Galliano, F., Jones, A. P., & Sauvage, M. 2006, *A&A*, 446, 877

Meixner, M., Gordon, K. D., Indebetouw, R., *et al.* 2006, *AJ*, 132, 2268

Misselt, K. A., Clayton, G. C., & Gordon, K. D. 1999, *ApJ*, 515, 128

Reach, W. T., Boulanger, F., Contursi, A., & Lequeux, J. 2000, *A&A*, 361, 895

Rubio, M., Lequeux, J., Boulanger, F., *et al.* 1996, *A& AS*, 118, 263

Smith, R. K., Krsewina, L. G., Cox, D. P., Edgar, R. J., & Miller, W. W. I. 1996, *ApJ*, 473, 864

Sofia, U. J., Gordon, K. D., Clayton, G. C., *et al.* 2006, *ApJ*, 636, 753

Weingartner, J. C. & Draine, B. T. 2001, *ApJ*, 548, 296

Zhukovska, S., Gail, H. P., & Trieloff, M. 2008, *A&A*, 479, 453

The Magellanic System: Stars, Gas, and Galaxies
Proceedings IAU Symposium No. 256, 2008
Jacco Th. van Loon & Joana M. Oliveira, eds.

© 2009 International Astronomical Union
doi:10.1017/S174392130802838X

Tracing the cold molecular gas reservoir through dust emission in the SMC

Caroline Bot[1], Mónica Rubio[2], François Boulanger[3], Marcus Albrecht[4,5], Frank Bertoldi[4], Alberto D. Bolatto[6] and Adam K. Leroy[7]

[1] UMR7550, Observatoire Astronomique de Strasbourg, F-67000 Strasbourg, France
email: `bot@astro.u-strasbg.fr`

[2] Departamento de Astronomia, Universidad de Chile, Casilla 36-Dm Santiago, Chile

[3] Institut d'Astrophysique Spatiale, Université de Paris-Sud, F-91405, Orsay, France

[4] Argelander-Institut für Astronomie, Universität Bonn, Germany

[5] Universidad Católica del Norte, Chile

[6] University of Maryland, MD, USA

[7] Max Planck Institute for Astronomy, Heidelberg, Germany

Abstract. The amount of molecular gas is a key for understanding the future star formation in a galaxy. However, this quantity is difficult to infer as the cold H_2 is almost impossible to observe and, especially at low metallicities, CO only traces part of the clouds, keeping large envelopes of H_2 hidden from observations. In this context, millimeter dust emission tracing the cold and dense regions can be used as a tracer to unveil the total molecular gas masses. I present studies of a sample of giant molecular clouds in the Small Magellanic Cloud. These clouds have been observed in the millimeter and sub-millimeter continuum of dust emission: with SIMBA/*SEST* at 1.2 mm and the new LABOCA bolometer on *APEX* at 870 μm. Combining these with radio data for each cloud, the spectral energy distribution of dust emission are obtained and gas masses are inferred. The molecular cloud masses are found to be systematically larger than the virial masses deduced from CO emission. Therefore, the molecular gas mass in the SMC has been underestimated by CO observations, even through the dynamical masses. This result confirms what was previously observed by Bot *et al.* (2007). We discuss possible interpretations of the mass discrepancy observed: in the giant molecular clouds of the SMC, part of cloud's support against gravity could be given by a magnetic field. Alternatively, the inclusion of surface terms in the virial theorem for turbulent clouds could reproduce the observed results and the giant molecular clouds could be transient structures.

Keywords. dust, extinction, ISM: clouds, ISM: molecules, galaxies: individual (SMC), galaxies: ISM, Magellanic Clouds, sub-millimeter

1. Introduction

Star formation is fueled by the cold interstellar medium from the molecular clouds where this process takes place. The star formation rate in a galaxy therefore depends on the amount of molecular gas available. Most of the molecular gas in a galaxy lies in giant molecular clouds, but the quantity of H_2 remains difficult to estimate precisely since it is almost impossible to observe directly in cold interstellar regions. In this context, one has to rely on tracers and the CO molecule is the most widely used one from our solar neighborhood to distant galaxies.

In particular, the CO luminosity can be converted to a mass of molecular hydrogen through the X_{CO} factor defined as $N(H_2)/I_{CO}$. This factor has been well calibrated in our galaxy with different methods and a consensus is reached for $X_{CO} \sim 2 \times 10^{20}$ mol

cm^{-2} K km s^{-1}. However, variations of this factor with the environment is expected and is still debated. In particular, at low metallicities, the relationship between H_2 and CO remains unclear. Alternatively, one can use the CO data to observed line widths and therefore probe the dynamics of the clouds. Simplified forms of the virial theorem ($M = 190\,R\,\Delta\,V^2$, MacLaren *et al.* 1988) are largely used to determine molecular cloud masses. This method is applied to various scales ranging from cores, to giant molecular clouds, to galaxies as a whole. However, this assumes that clouds are virialized (bound) and the effects of external pressure or magnetic fields are ignored.

The Small Magellanic Cloud (SMC) is one of the closest and easiest to observe, low metallicity galaxy. Due to its proximity, molecular clouds in this galaxy can be resolved at various wavelengths by the observations. Numerous molecular clouds in the Magellanic Clouds were observed in CO lines (Rubio *et al.* 1993a,b, 1996; Mizuno *et al.* 2001) and were found to have weak CO emission. If the CO molecule is tracing accurately the molecular gas, then the star formation per unit molecular gas in the SMC is higher than in most galaxies. Such an effect is observed in other low metallicity dwarf galaxies (Leroy *et al.* 2006a) and could be an evolutionary effect or the fact that CO observations largely underestimate the amount of molecular gas in these conditions. Indeed, at low metallicities, CO can be photodissociated and trace only the densest parts of the clouds, while H_2 molecules have the capacity to self-shield and would be present at larger radius than the CO molecule.

To better understand the total molecular content and its relationship to CO emission, other tracers of the molecular gas at low metallicities have been looked for. Dust continuum emission can be used to trace the dense and cold interstellar medium where the gas is molecular and it can therefore be used to unveil the total mass of molecular gas in clouds. Leroy *et al.* (2006b) attempted to do so using far infrared emission observed with Spitzer data in the SMC. In dense regions, they observed large quantities of dust not associated to either H I or CO emission. However, the dust emission in the far-infrared is sensitive to the dust temperature, leading to large uncertainties in the cloud's masses. In the sub-millimeter and the millimeter range, dust emission is difficult to observe but depends only linearly on the dust temperature and is optically thin. Rubio *et al.* (2004) observed for the first time with SIMBA on the *SEST* telescope, the 1.2 mm emission from dust in a quiescent molecular cloud in the SMC. The mass of molecular gas deduced was several times higher than the mass deduced from CO observations in this cloud. Bot *et al.* (2007) extended this study with SIMBA data to a larger sample of molecular clouds in the south-west region of the SMC. In all clouds, the gas mass deduced from the dust emission is systematically larger than the virial mass deduced from CO, even for conservative values of the dust emissivities and free-free contribution to the millimeter emission. All these studies come to agreement in the fact that molecular clouds in the SMC are more massive than what can be deduced from CO emission or dynamical masses. This mass difference is understood in a scenario where CO traces only the densest parts of the molecular clouds and the clouds are partially supported by a magnetic field (i.e. the motions in the cloud do not trace the gravitational potential). However, these results need to be confirmed with observations in the sub-millimeter range in order to more certainly exclude the contamination of the fluxes by free-free emission, as well as unknown instrumental effects, etc.

2. 870 μm emission in the south west region of the SMC

We present new observations of the sample of molecular clouds in the south west region of the SMC taken with the LABOCA camera on the *APEX* telescope (Fig. 1). These observations at $870\,\mu$m complement the SIMBA data at 1.2 mm. Sub-millimeter emission is detected in the whole region and all the giant molecular clouds that were observed in CO(1–0) are detected.

For every cloud detected in CO, we compute the integrated emission at $870\,\mu$m as observed with LABOCA and compare it to the emission at 1.2 mm as observed by SIMBA and radio emission at 4.8 and 8.64 GHz from *ATCA* (Dickel *et al.*, this volume). Fig. 2 presents the spectral energy distribution obtained that way for each giant molecular cloud. By extrapolating the radio emission, we observe that the sub-millimeter and millimeter emission are dominated by dust emission and the contribution of free-free emission to the (sub-)millimeter fluxes is negligible. Furthermore, the $870\,\mu$m and 1.2 mm fluxes are consistent with a standard dust emission law, i.e. a modified blackbody with a spectral index of 2 ($I_\nu \propto \nu^2 B_\nu T_{\rm dust}$). The LABOCA fluxes in south west region of the Small Magellanic Cloud can therefore safely be used to deduce molecular gas masses given the dust emissivity at $870\,\mu$m is known.

3. Molecular cloud masses

We computed a reference emissivity at $870\,\mu$m for dust associated to molecular gas. This was done using FIRAS data at $850\,\mu$m and correlating it with LAB H I (Kalberla *et al.* 2005) and CO data (Dame *et al.* 2001) in the molecular ring of the Milky Way. The method used is similar to the one described in Bot *et al.* (2007) and we obtain: $\epsilon_{870}({\rm H}_2) = (2.32 \pm 0.03) \times 10^{-26}$ at^{-1} cm^2. Taking this value and assuming a dust to gas ratio 0.17 times solar, we deduce molecular gas masses from the LABOCA $870\,\mu$m emission for each giant molecular cloud (as defined by CO) in the south west region of the SMC.

The masses obtained from the dust sub-millimeter emission are compared to virial masses deduced from the CO line-widths for the same clouds. The comparison is shown in Fig. 3: for all the clouds detected in CO in this region of the SMC, the masses deduced

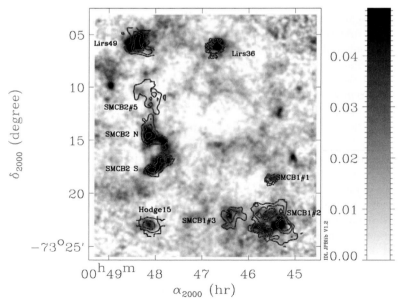

Figure 1. LABOCA image in the south west region of the SMC with CO contours overlaid. All the molecular clouds observed in CO are detected at $870\,\mu$m.

from the $870\,\mu$m dust emission are systematically larger than the one deduced from the virial theorem (black rhombs). These results are completely consistent with what is obtained from SIMBA data (green triangles). For comparison, the same method was applied to a sample of similar clouds in the Galaxy (Bot *et al.* 2007, noted BBR07) using FIRAS data at $850\,\mu$m. For the galactic clouds, the trend observed is the complete opposite: the virial masses are always larger than the mass deduced from the dust sub-millimeter emission. These results confirm what was observed previously by Bot *et al.* (2007). In the low metallicity environment of the SMC, the molecular gas mass of giant molecular clouds is much larger than what is inferred from CO data, even when computed from the dynamical masses.

Furthermore, there seems to be a trend between the mass ratio $(M_{\rm vir}/M_{\rm mm})$ with respect to the mass deduced from the millimeter (believed to be the true mass of the clouds): the larger the mass of a giant molecular cloud, the smaller the mass ratio $(M_{\rm vir}/M_{\rm mm})$.

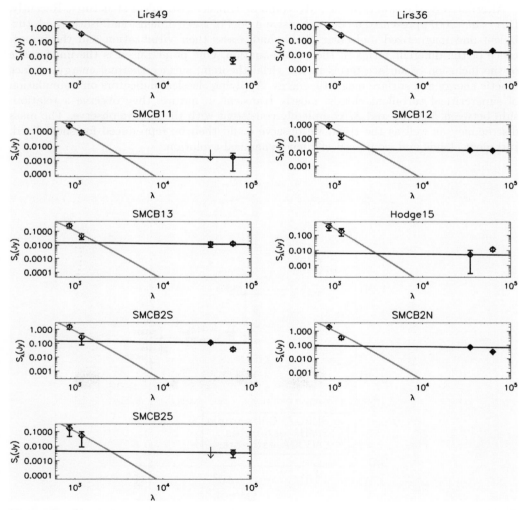

Figure 2. Spectral energy distribution from the far-infrared to the radio observed for each cloud. The fluxes observed at $870\,\mu$m and $1.2\,$mm are consistent with dust emission with an emissivity index of 2. Radio fluxes for the same regions clearly show that the sub-millimeter and millimeter emission detected is dominated by dust emission rather than free-free.

4. Discussion

The overestimation of the mass of the clouds by the virial theorem for the Galactic cloud, was observed previously (Solomon *et al.* 1987) and is interpreted as the effect of external pressure on the molecular clouds. On the other hand, the underestimation of the mass of the clouds by the virial theorem in the giant molecular clouds of the SMC was unexpected and is more difficult to explain. Bot *et al.* (2007) used the general version of the virial theorem and showed that additional support to the cloud by a magnetic field could explain this effect. Such a magnetic support has been observed in our galaxy, but on much smaller scales (e.g., Wong *et al.* 2008). This magnetic field support could be more salient in the SMC due to the fact that the giant molecular clouds are denser than their galactic counterparts. However, the continuity between the galactic pressure confined molecular clouds and the SMC magnetically supported clouds is difficult to understand in that scheme.

Alternatively, the clouds in the SMC could be transient turbulent structures. Recently, Dib *et al.* (2007) presented a comprehensive model of three dimensional, isothermal, turbulent and magnetized molecular clouds and assess their virialization and the importance of the different terms in the virial equation. One point raised is the importance of the inclusion of surface terms in the virial theorem: surface thermal energy, surface kinetic energy and surface magnetic energy. Applying classical indicators on a simulation of supercritical turbulent clouds, mostly transient in nature, they observe a relationship between M_{vir}/M and M completely consistent with the one we observe. The mass discrepancy as well as the trend we observe could then be reproduced for short lived, turbulent clouds and would reflect their dynamical evolution.

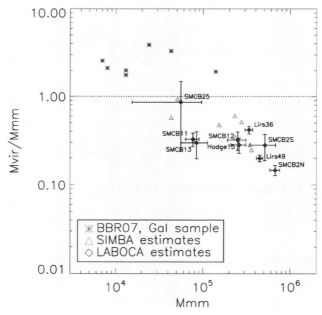

Figure 3. Comparison between the two molecular mass estimates: dust emission and virial masses. Galactic giant molecular clouds studied with the same method are shown for comparison (Bot *et al.* 2007). In the SMC, the molecular mass deduced from the $870\,\mu$m emission (black rhombs) or 1.2 mm emission (green triangles) are systematically larger than virial masses.

5. Conclusion

We presented new LABOCA data in the south west region of the Small Magellanic Cloud. All the giant molecular clouds that were observed in CO are detected. By comparing to SIMBA and radio data, we show that the sub-millimeter and millimeter emission of these clouds are dominated by normal dust emission.

For each giant molecular cloud observed both with LABOCA and CO (1–0), we compute a molecular gas mass from the sub-millimeter emission and compare it to the virial mass deduced from the CO line widths in the same regions. We confirm the results observed by Bot *et al.* (2007): in the SMC, the virial mass systematically underestimate the true mass of the molecular clouds. Applying the same method to a sample of similar clouds in our galaxy shows the reverse (the virial masses under-estimate the true mass of the clouds). We also observe a possible trend from the Galactic clouds to the SMC one: the mass ratio M_{vir}/M seems to decrease with the true molecular mass of the clouds.

The mass discrepancy between the two methods (from the dust emission or the virial theorem) can be understood as follows: in the galaxy, the clouds are pressure confined, while in the SMC, the clouds could be partially supported by magnetic fields. The difference between these two regimes could be due to the different densities between the two clouds samples. However, an alternative model where clouds are transient and the surface terms in the virial theorem are important could also reproduce the mass discrepancy and the trend observed.

This study shows clearly how sub-millimeter and millimeter observations of molecular clouds can shed new light on the molecular content and the state of the molecular clouds. Similar studies on wider samples of clouds and in various environments will have to be pursued to bring new insights on the molecular content of galaxies and the relationship with star formation.

References

Bot, C., Boulanger, F., Rubio, M., & Rantakyro, F. 2007, *A&A*, 471, 103

Dame, T. M., Hartmann, D., & Thaddeus, P. 2001, *ApJ*, 547, 792

Dib, S., Kim, J., Vasquez-Semadeni, E., Burkert, A., & Shadmehri, M. 2007 *ApJ*, 661, 262

Kalberla, P. M. W., Burton, W. B., Hartmann, D., Arnal, E. M., Bajaja, E., Morras, R., & Pöppel, W.G.L. 2005, *A&A*, 440, 775

Leroy, A., Bolatto, A., Walter, & F., Blitz, L. 2006a, *ApJ*, 643, 825

Leroy, A., Bolatto, A., Stanimirović, S., Mizuno, N., Israel, F.P., & Bot, C. 2006b, *ApJ*, 658, 1027

MacLaren, I., Richardson, K. M., & Wolfendale, A. W. 1988, *ApJ*, 333, 821

Mizuno, N., Rubio, M., Mizuno, A., Yamaguchi, R., Onishi, T., & Fukui, Y. 2001, *PASJ*, 53, L45

Rubio, M., Lequeux, J., & Boulanger, F. 1993, *A&A*, 271, 9

Rubio, M., Lequeux, J., Boulanger, F., *et al.* 1993, *A&A*, 271, 1

Rubio, M., Lequeux, J., Boulanger, F., *et al.* 1996, *A&A*, 118, 263

Rubio, M., Boulanger, F., Rantakyro, F., & Contursi, A. 2004, *A&A*, 425, L1

Solomon, P. M., Rivolo, A. R., Barrett, J., & Yahil, A. 1987 *ApJ*, 319, 730

Wong, T., Ladd, E. F., Brisbin, D., *et al.*, 2008 *MNRAS*, 386, 1069

The Magellanic System: Stars, Gas, and Galaxies
Proceedings IAU Symposium No. 256, 2008
Jacco Th. van Loon & Joana M. Oliveira, eds.

© 2009 International Astronomical Union
doi:10.1017/S1743921308028391

The state of molecular gas in the Small Magellanic Cloud

Adam K. Leroy[1], Alberto D. Bolatto[2], Erik Rosolowsky[3], Snežana Stanimirović[4], Norikazu Mizuno[5], Caroline Bot[6], Frank Israel[7], Fabian Walter[1] and Leo Blitz[8]

[1] Max Planck Institute for Astronomy, Heidelberg, Germany

[2] Department of Astronomy and Laboratory for Millimeter-wave Astronomy, University of Maryland, USA

[3] Department of Mathematics, Statistics, and Physics, University British Columbia at Okanagan, Canada

[4] Department of Astronomy, University of Wisconsin, USA

[5] Department of Astrophysics, Nagoya University, Japan

[6] Observatoire Astronomique de Strasbourg, France

[7] Sterrewacht Leiden, The Netherlands

[8] Department of Astronomy and Radio Astronomy Laboratory, U.C. Berkeley, USA

Abstract. We compare the resolved properties of giant molecular clouds (GMCs) in the Small Magellanic Cloud (SMC) and other low mass galaxies to those in more massive spirals. When measured using CO line emission, differences among the various populations of GMCs are fairly small. We contrast this result with the view afforded by dust emission in the Small Magellanic Cloud. Comparing temperature-corrected dust opacity to the distribution of H I suggests extended envelopes of CO-free H_2, implying that CO traces only the highest density H_2 in the SMC. Including this CO-free H_2, the gas depletion time, H_2-to-H I ratio, and H_2-to-stellar mass/light ratio in the SMC are all typical of those found in more massive irregular galaxies.

Keywords. ISM: clouds, galaxies: dwarf, galaxies: individual (SMC), galaxies: ISM, Magellanic Clouds

1. Introduction

In the Milky Way, most star formation occurs in giant molecular clouds (GMCs). Because H_2 does not readily emit under the conditions in these clouds, they are usually observed via line emission from tracer molecules — most commonly the lowest rotational transition of CO. From these observations we know that most of the H_2 in the Milky Way lies in gravitationally bound GMCs with masses from 10^5 to 10^6 $M_\odot$. Their luminosities, line widths, and sizes obey certain scaling relations — commonly referred to as "Larson's Laws" (Larson 1981; see reviews by, e.g., Blitz 1993 and McKee & Ostriker 2007).

An open question is how the properties of GMCs are affected by their environment. Star formation over the history of the universe has occurred under a vast range of conditions — from chemically pristine gas to violent galaxy mergers. Even "normal" disk galaxies host a wide range of radiation fields, metallicities, pressures, and dynamical states. Environment may affect the fraction of gas converted to stars (and thus stellar clustering and feedback) or the initial distribution of stellar masses. In order to do so, these conditions must first affect GMCs, the structures out of which stars form.

In particular, the relationship between metallicity (and the closely related dust abundance) and GMC structure is both intriguing and difficult to approach. There are fundamental reasons to expect a relationship: at all but the lowest metallicities, H I is converted to H_2 on the surface of dust grains and dust can help shield H_2 from dissociating radiation. Further, metallicity impacts thermal structure of the atomic ISM, so one may expect that clouds form under different conditions in low metallicity environments. Star formation obviously *does* proceed at low metallicities: even in the nearby universe, there are numerous vigorously star-forming low-metallicity galaxies. Unfortunately, it is difficult to characterize molecular gas in these galaxies because their CO emission is faint (e.g., Taylor *et al.* 1998). This is partially because low metallicity galaxies also tend to have low masses (e.g., Lee *et al.* 2006), making them less luminous at all wavelengths. However, very low-metallicity star-forming galaxies show distinctly low *normalized* CO emission — i.e., their CO emission is low compared to their star formation rate (SFR) or stellar mass. Diminished dust shielding is probably as responsible for this as the underabundance of C and O (e.g., Maloney & Black 1988). Regardless of the cause, the practical results are that it is difficult to observe molecular gas in very low metallicity systems and that the standard tracers of H_2 must be employed with caution.

As the nearest low-metallicity, actively star-forming system, the Small Magellanic Cloud (SMC) is key to understand the effect of metallicity on GMC structure. Some effects are clearly present: the SMC's normalized CO emission ($L_{CO} \sim 10^5$ K km s^{-1} pc^{-2}, Mizuno *et al.* 2001) is quite low compared to its other properties. For example, Wilke *et al.* (2004) estimate the SFR in the SMC to be ~ 0.05 M$_\odot$ yr^{-1}. For a standard Galactic CO-to-H_2 conversion factor, this implies a molecular gas depletion time (M_{H2}/SFR) of $\sim 10^7$ years. This is about two orders of magnitude lower than that observed in most spiral galaxies (e.g., Young *et al.* 1996, Kennicutt 1998). The implied ratio of H_2-to-H I is also strikingly small, $\sim$ 1-to-1000 (Stanimirović *et al.* 2004), about two orders of magnitude lower than that in more massive irregular galaxies (e.g., Young & Scoville 1991). In Fig. 1, we show that the ratio of CO emission to stellar light, which varies only weakly among massive star forming galaxies is similarly low in the SMC (blue circle).

These ratios place the SMC in the company of only a few very nearby low metallicity galaxies that have observed —- but very faint — CO emission. More distant analogs to these systems tend to be CO non-detections. The SMC is unique among these objects because of its proximity, which allows even single-dish millimeter-wave telescopes to achieve good spatial resolution. This has allowed extensive studies of of molecular gas on the scale of individual GMCs (e.g., Rubio *et al.* 1993a,b; Mizuno *et al.* 2001; Bolatto *et al.* 2003; Rubio *et al.* 2004; Bot *et al.* 2007). In these proceedings, we summarize two recent results : *1)* that the properties of resolved GMCs — as measured from CO emission — in the SMC and other dwarf galaxies are quite similar to those in the Milky Way, M 31, and M 33; and *2)* that dust emission suggests large, extended reservoirs of CO-free H_2 surrounding these Galactic-looking CO clouds.

2. GMC scaling relations in low mass galaxies

A basic test of the state of molecular gas is to resolve CO emission into individual GMCs and compare their properties to those of Milky Way GMCs. Bolatto *et al.* (2008, B08) recently attempted this test by measuring GMC properties in 11 nearby dwarf galaxies (including the SMC) and comparing these to results from the Milky Way, M 31, and M 33. The data are a mixture of new and previously published observations† obtained

† *Spirals:* Milky Way, Solomon *et al.* (1987); M 33, Rosolowsky *et al.* (2003); M 31, Rosolowsky *et al.* (2007). *Magellanic Clouds:* LIRS 36 & LIRS 49, Rubio *et al.* (1993a,b); N 159, Bolatto *et al.* (2000); N83, Bolatto *et al.* (2003). *Local Group Dwarfs:* NGC 185 and NGC 205, Young

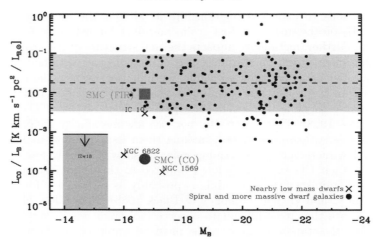

Figure 1. The ratio of CO to B-band luminosity vs. B-band absolute magnitude in nearby galaxies with detected CO emission (adapted from Leroy *et al.* 2007b). Several nearby, low-metallicity dwarfs are highlighted, including the SMC (blue circle). These systems show faint CO emission relative to their other properties (they also show very low CO/SFR and CO/H I). We also plot the normalized CO content one would expect if all of the H_2 inferred from dust emission exhibited a Galactic CO-to-H_2 conversion factor (red square). The ratio is close to that in larger galaxies, suggesting that CO is more strongly affected by environmental differences than H_2.

with interferometers (*BIMA*, *OVRO*, and *PdBI*) and single dish telescopes (*SEST*). The targets span metallicities from $12 + \log O/H = 8.02$ (the SMC) to 8.85 (i.e., slightly above solar) and distances out to ~ 4 Mpc. B08 measured GMC properties using the CPROPS algorithm (Rosolowsky & Leroy 2006). CPROPS makes a conservative decomposition of emission into individual GMCs and then uses moment methods, extrapolation to ideal sensitivity, and a simple quadratic deconvolution to derive sizes, line widths, and luminosities (corrected for resolution and sensitivity). The goal of this approach is a consistent intercomparison of observations at mixed resolution and signal-to-noise. Blitz *et al.* (2007) recently present a complementary review and analysis (also using CPROPS) that focused on complete surveys of Local Group galaxies.

B08 compared the properties of extragalactic GMCs to the scaling relations measured for Milky Way GMCs, essentially asking whether GMCs in dwarf galaxies are consistent with being drawn from the population of Milky Way GMCs. Broadly, the answer is "yes". GMCs from the Magellanic Clouds, Local Group dwarfs, and more distant dwarfs tend to lie within a factor of ~ 2 of the Milky Way GMC scaling relations. That is, *CO emission from dwarf galaxy GMCs closely resembles that from GMCs in the Milky Way and other Local Group spirals.* We show this in Fig. 2, which presents GMC line width as a function of size (left) and virial mass as a function of CO luminosity (right). Similarities in the line width-size relation may reflect a similar character of turbulence in all of the systems surveyed. The correlation between virial mass and CO luminosity is often interpreted to mean that GMCs are in approximate virial equilibrium and with CO luminosity a good tracer of cloud mass. A particularly surprising result is that low metallicity GMCs (e.g., from the SMC or IC 10) show roughly the same virial mass-to-luminosity relation seen in

et al. (2001); IC 10, Walter *et al.* (2003), Leroy *et al.* (2006). *Dwarfs Beyond the Local Group:* NGC 1569, Taylor *et al.* (1999); NGC 4214, Walter *et al.* (2001), Bolatto *et al.* (2008); NGC 4605, Bolatto *et al.* (2002); NGC 3077, Walter *et al.* (2002), Bolatto *et al.* (2008); NGC 4449, Bolatto *et al.* (2008); NGC 2976, Simon *et al.* (2003).

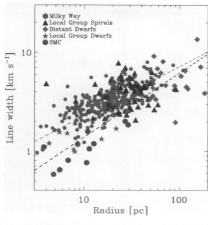 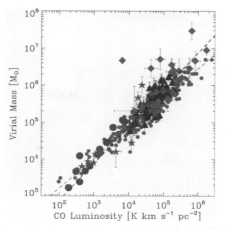

Figure 2. GMC scaling relations in the Milky Way and other galaxies (adapted from Bolatto *et al.* 2008): (*left*) line width vs. cloud radius and (*right*) virial mass ($\propto R\sigma^2$) vs. CO luminosity. Milky Way clouds are light gray circles, M 31 and M 33 are blue triangles, Local Group dwarfs are green stars, and more distant dwarfs are red diamonds. SMC clouds are purple circles.

more massive galaxies. If GMCs are virialized, this implies that the CO-to-H_2 conversion factor *for resolved, CO-bright clumps* has little dependence on metallicity.

Despite overall agreement, B08 do find some differences among the populations surveyed. For example, the left panel in Fig. 2 shows that SMC clouds have lower line widths than Galactic clouds of the same size. Possible explanations are increased magnetic support in SMC clouds (e.g., Bot *et al.* 2007) or a simple lack of virial equilibrium (implying short-lived GMCs). An alternative explanation is that the Milky Way scaling relations derived by Solomon *et al.* (1987) are in need of revision: Heyer *et al.* (2008) recently re-measured the properties of the Solomon *et al.* clouds using the Galactic Ring Survey (Jackson *et al.* 2006) and found a relationship among size, line width, and surface density that agrees well with the B08 data.

3. H_2 traced by dust in the SMC

Thus GMC properties measured from CO suggest that molecular gas in the SMC is similar to that in the Galaxy (with perhaps small differences). However, as we have already emphasized, CO is a suspect tracer of molecular gas at low metallicities. Israel (1997) used IRAS emission (dust continuum) to trace molecular gas in dwarf irregulars and found evidence for large amounts of CO-free H_2. His study was limited to relatively large scales by the available data. Subsequently, improved surveys of the SMC have been carried out in CO (Mizuno *et al.* 2001), H I (Stanimirović *et al.* 2004), and the infrared (the *Spitzer* Survey of the SMC, Bolatto *et al.* 2007). In Leroy *et al.* (2007a), we combined these data to use dust as a probe of the H_2 distribution.

Our technique was to estimate the surface density of dust, Σ_{Dust}, everywhere in the SMC using 100 and 160 μm emission (two or more bands are needed to account for varying dust temperature). We then identified likely H_2 peaks from CO and measured the dust-to-gas ratio (DGR) near these peaks, but displaced enough that H I (and not H_2) is likely to dominate the ISM. We combined these locally measured DGRs with the Σ_{Dust} near molecular peaks to estimate the total gas surface density. By subtracting the measured contribution of H I, we derived a dust-based map of H_2, i.e., $\Sigma_{\mathrm{H2}} = DGR^{-1} \times \Sigma_{\mathrm{Dust}} - \Sigma_{\mathrm{HI}}$.

This approach yields $\sim 3 \times 10^7$ M$_\odot$ of H_2, much higher than using CO alone. This is close to what one would infer from the SMC's other properties: the implied H_2 depletion

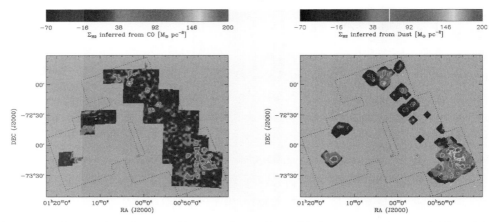

Figure 3. Mass surface density of H_2 as inferred from CO (Mizuno *et al.* 2001, left) and dust emission (right). CO has been translated to H_2 surface density using the average CO-to-H_2 conversion factor implied by the dust map (so that the two maps contain the same total mass). H_2 inferred from dust shows the same peaks as CO, but has a more extended distribution. The overall mass of H_2 implied by dust, $\sim 3 \times 10^7$ $M_\odot$, is also much larger than one would infer from CO emission alone ($\sim 5 \times 10^5$ $M_\odot$ for a Galactic CO-to-H_2 conversion factor; see Fig. 1).

time is $\sim 6 \times 10^8$ years, within a factor of $\sim 2-3$ of that found for spirals (i.e., the SMC obeys approximately the same "molecular Schmidt law" as larger galaxies). The implied H_2-to-H I ratio is $\sim 1 : 10$, consistent with larger irregulars, and the ratio of H_2 to B-band luminosity also matches that in larger galaxies (red square in Fig. 1).

This result is quite distinct from what B08 find using CO and the detailed distribution of H_2 compared to CO suggest a possible reason. Although H_2 and CO share roughly the same peaks, we find a distribution of H_2 that is more extended than that of CO (compare the left and right panels in Fig. 3). Indeed, about several H_2 peaks, we measure H_2 to be more extended than CO by a factor of ~ 1.5 in radius (a number revised from Leroy *et al.* 2007 to include the effects of H I opacity estimated by Dickey *et al.* 2000; also likely a lower limit given the relatively large physical size of the *NANTEN* beam). This difference suggests the selective photodissociation of CO in SMC clouds (Maloney & Black 1988), i.e. that in the outer parts of clouds H_2 self-shields while CO — which relies largely on dust for shielding — is destroyed by dissociating radiation.

If CO is preferentially destroyed in the outer parts of SMC clouds, then our conclusions may not be contradictory at all. The similarity in GMC properties measured from CO implies that the densest parts of these clouds resemble entire Milky Way GMCs. Similar situations are already observed in Milky Way clouds: the line width-size relation appears to extend from the scale of whole clouds down to less than a parsec (Heyer & Brunt 2004; Rosolowsky *et al.* 2008) and substructures within Milky Way GMCs can appear virialized (e.g., Rosolowsky *et al.* 2008). In this case the *intercloud* dispersion (rather than the CO line width of individual clouds) may offer an independent way to trace the full molecular mass and indeed measurements of SMC clouds by Rubio *et al.* (1993b) and Bolatto *et al.* (2003) find the ratio of CO luminosity to virial mass to be a strong function of scale. That CO-free H_2 is needed to establish rough agreement between H_2 depletion times in the SMC and larger galaxies is also not as surprising as it may first appear. In the Milky Way, little or no star formation is actually associated with the bulk of CO emission; instead, it tends to occur only towards the highest density/extinction peaks (e.g., Johnstone *et al.* 2004).

References

Blitz, L. 1993, in *Protostars and Planets III*, p. 125

Blitz, L., Fukui, Y., Kawamura, A., Leroy, A., Mizuno, N., & Rosolowsky, E. 2007, in B. Reipurth, D. Jewitt, & K. Keil (eds.), *Protostars and Planets V* (Tucson: University of Arizona Press), p. 81

Bolatto, A. D., Jackson, J. M., Israel, F. P., Zhang, X., & Kim, S. 2000, *ApJ*, 545, 234

Bolatto, A. D., Simon, J. D., Leroy, A., & Blitz, L. 2002, *ApJ*, 565, 238

Bolatto, A. D., Leroy, A., Israel, F. P., & Jackson, J. M. 2003, *ApJ*, 595, 167

Bolatto, A. D., Simon, J. D., Stanimirović, S., *et al.* 2007, *ApJ*, 655, 12

Bolatto, A. D., Leroy, A. K., Rosolowsky, E., Walter, F., & Blitz, L. 2008, *ApJ*, 686, 948

Bot, C., Boulanger, F., Rubio, M., & Rantakyro, F. 2007, *A&A*, 471, 103

Dickey, J. M., Mebold U., Stanimirović, S., & Staveley-Smith, L. 2000, *ApJ*, 536, 756

Heyer, M. H. & Brunt, C. M. 2004, *ApJ*, 615, L45

Heyer, M., Krawczyk, C., Duval, J., & Jackson, J. M. 2008, *ApJ* submitted

Israel, F. P. 1997, *A&A*, 328, 471

Jackson, J. M., Rathborne, J. M., Shah, R. Y., *et al.* 2006, *ApJS*, 163, 145

Johnstone, D., Di Francesco, J., & Kirk, H. 2004, *ApJ*, 611, L45

Kennicutt, R. C., Jr. 1998, *ApJ*, 498, 541

Larson, R. B. 1981, *MNRAS*, 194, 809

Lee, H., Skillman, E. D., Cannon, J. M., Jackson, D. C., Gehrz, R. D., Polomski, E. F., & Woodward, C. E. 2006, *ApJ*, 647, 970

Leroy, A., Bolatto, A., Walter, F., & Blitz, L. 2006, *ApJ*, 643, 825

Leroy, A., Bolatto, A., Stanimirović, S., Mizuno, N., Israel, F., & Bot, C. 2007a, *ApJ*, 658, 1027

Leroy, A., Cannon, J., Walter, F., Bolatto, A., & Weiss, A. 2007b, *ApJ*, 663, 990

Maloney, P. & Black, J. H. 1988, *ApJ*, 325, 389

McKee, C. F. & Ostriker, E. C. 2007, *ARAA*, 45, 565

Mizuno, N., Rubio, M., Mizuno, A., Yamaguchi, R., Onishi, T., & Fukui, Y. 2001, *PASJ*, 53, L45

Simon, J. D., Bolatto, A. D., Leroy, A., & Blitz, L. 2003, *ApJ*, 596, 957

Solomon, P. M., Rivolo, A. R., Barrett, J., & Yahil, A. 1987, *ApJ*, 319, 730

Stanimirović, S., Staveley-Smith, L., & Jones, P. A. 2004, *ApJ*, 604, 176

Rosolowsky, E., Engargiola, G., Plambeck, R., & Blitz, L. 2003, *ApJ*, 599, 258

Rosolowsky, E. & Leroy, A. 2006, *PASP*, 118, 590

Rosolowsky, E. 2007, *ApJ*, 654, 240

Rosolowsky, E. W., Pineda, J. E., Kauffmann, J., & Goodman, A. A. 2008, *ApJ*, 679, 1338

Rubio, M., Lequeux, J., Boulanger, F. *et al.* 1993a, *A&A*, 271, 1

Rubio, M., Lequeux, J., & Boulanger, F. 1993b, *A&A*, 271, 9

Rubio, M., Boulanger, F., Rantakyro, F., & Contursi, A. 2004, *A&A*, 425, L1

Taylor, C. L., Kobulnicky, H. A., & Skillman, E. D. 1998, *AJ*, 116, 2746

Taylor, C. L., Hüttemeister, S., Klein, U., & Greve, A. 1999, *A&A*, 349, 424

Walter, F., Taylor, C. L., Hüttemeister, S., Scoville, N., & McIntyre, V. 2001, *AJ*, 121, 727

Walter, F., Weiss, A., Martin, C., & Scoville, N. 2002, *AJ*, 123, 225

Walter, F. 2003, in *Star Formation at High Angular Resolution*, IAU Symposium 221, p. 176

Wilke, K., Klaas, U., Lemke, D., Mattila, K., Stickel, M., & Haas, M. 2004, *A&A*, 414, 69

Young, J. S. & Scoville, N. Z. 1991, *ARAA*, 29, 581

Young, J. S., Allen, L., Kenney, J. D. P., Lesser, A., & Rownd, B. 1996, *AJ*, 112, 1903

Young, L. M. 2001, *AJ*, 122, 1747

The Magellanic System: Stars, Gas, and Galaxies
Proceedings IAU Symposium No. 256, 2008
Jacco Th. van Loon & Joana M. Oliveira, eds.

© 2009 International Astronomical Union
doi:10.1017/S1743921308028408

The *Spitzer* spectroscopic survey of the Small Magellanic Cloud: polycyclic aromatic hydrocarbon emission from SMC star-forming regions

Karin M. Sandstrom[1], Alberto D. Bolatto[2], Snežana Stanimirović[3], J. D. T. Smith[4], Jacco Th. van Loon[5] and Adam K. Leroy[6]

[1]Dept. of Astronomy, University of California, Berkeley, 601 Campbell Hall,
Berkeley CA 94720 USA
email: karin@astro.berkeley.edu

[2]Dept. of Astronomy & Laboratory for Millimeter-wave Astronomy, University of Maryland,
College Park, College Park, MD 20742 USA

[3]Dept. of Astronomy, University of Wisconsin, Madison, Madison, WI 53706 USA

[4]Ritter Astrophysical Research Center, University of Toledo Toledo, OH 43603 USA

[5]Astrophysics Group, Lennard-Jones Laboratories, Keele University,
Staffordshire ST5 5BG, UK

[6]Max-Planck-Institut für Astronomie, Königstuhl 17, D-69117 Heidelberg, Germany

Abstract. Because of its proximity, the Small Magellanic Cloud provides a unique opportunity to map the polycyclic aromatic hydrocarbon (PAH) emission from photo-dissociation regions (PDRs) in a low-metallicity $(12 + \log(\text{O/H}) \sim 8)$ galaxy at high spatial resolution in order to learn about their abundance and physical state. We present mid-IR spectral mapping observations of star-forming regions in the Small Magellanic Cloud obtained as part of the *Spitzer* Spectroscopic Survey of the SMC (S^4MC) project. These observations allow us to map the distribution of PAH emission in these regions and the measure the variation of PAH band strengths with local physical conditions. In these proceedings we discuss preliminary results on the physical state of the PAHs, in particular their ionization fraction.

Keywords. dust, extinction, HII regions, ISM: molecules, galaxies: individual (SMC), Magellanic Clouds, infrared: ISM

1. Introduction

Dust grains span a range of sizes from the ~ 0.1 μm "classical" grains which provide most of the optical extinction to the few Å-sized polycyclic aromatic hydrocarbons (PAHs), which straddle the line between very small dust grains and large molecules. PAHs are ubiquitous in the Milky Way, as shown by the widely-observed mid-IR emission bands (for a recent review see Tielens 2008). PAHs have significant roles in ISM heating, particularly in photodissociation regions (PDRs) where PAHs can be the dominant source of photoelectrons; grain surface chemistry; and charge balance, particularly in low-UV flux regions where they can efficiently mediate charge exchange reactions (Bakes & Tielens 1998). The physical state of PAHs is intimately tied to their roles in the ISM. For example, the charge state of PAHs in a PDR determines the efficiency of the photoelectric effect in the far-UV dominated regions of the PDR. The physical state of PAHs also has a direct impact on their destruction by UV radiation fields. PAHs that are small, highly ionized, and/or dehydrogenated are more susceptible to photo-dissociation (Allain *et al.* 1996).

Observations with *Spitzer* and *ISO* have convincingly demonstrated that there is a deficiency in PAH emission from low-metallicity galaxies (Madden *et al.* 2006; Engelbracht *et al.* 2005, 2008; Draine *et al.* 2007). This deficit has been attributed to a variety of causes: delayed injection of PAHs into the ISM because of the long lifetimes of their evolved star progenitors (Galliano *et al.* 2008a), destruction of PAHs by supernova shock waves (O'Halloran *et al.* 2006) and/or destruction by harder and more intense UV radiation fields (Madden *et al.* 2006). In the SMC, the deficiency of PAH emission has been observed in the 8-to-24 μm ratio in the Southwest Bar region by Bolatto *et al.* (2007). The SMC is a particularly interesting object for studying PAH emission from low-metallicity galaxies because its metallicity of $12 + \log(O/H) \sim 8$ puts it right at the observed transition where more metal-rich galaxies show normal PAH emission and more metal-poor galaxies have mid-IR colors and spectra consistent with litte to no PAH emission.

If PAHs are being destroyed by more intense UV radiation fields in low-metallicity galaxies, we may expect to see some signature of this process in their ionization state or size distribution. These properties can be diagnosed by examining the ratios of the mid-IR emission bands (Allamandola *et al.* 1999). Ionized PAHs exhibit higher ratios of C–C/C–H band strengths compared to neutral PAHs. For instance, the ratio of the 6.2 μm and 7.7 μm C–C bands versus the 11.3 μm C–H band traces the ionization of the PAHs, (see recent work by Galliano *et al.* 2008b). In the following, we present observations of the mid-IR emission bands from PAHs in star-formation regions in the SMC obtained as part of the Spitzer Spectroscopic Survey of the SMC (S^{4}MC). We discuss preliminary results on the ionization state of PAHs and the variations of the mid-IR band ratios.

2. Observations & data reduction

S^{4}MC covered six of the major star-forming regions of the SMC with spectral mapping observations using the low-resolution orders (SL and LL) of the InfraRed Spectrograph (IRS) on *Spitzer*. The maps were fully sampled spatially and had integration times per slit of 14 and 30 seconds for SL and LL orders, respectively. The maps cover typically ~ 70 arcmin^2 in LL and ~ 14 arcmin^2 in SL. The SL map of the SMC B1 region covers ~ 6 arcmin^2 with 60 second integrations per slit in order to get higher signal-to-noise in this unique region. A dedicated "off" position was used to remove the Galactic and zodiacal light foregrounds from the observations. A more detailed discussion of the S^{4}MC observations and data reduction can be found in Sandstrom *et al.* (2008).

The spectral maps were calibrated and assembled in IDL using the Cubism package (Smith *et al.* 2007). In the following we discuss only the data in the SL orders, which range from 5–14 μm. Each wavelength slice of the cubes was convolved to match the lowest resolution point spread function of the SL observations (i.e. that of the $\lambda = 14.74$ μm) using convolution kernels derived from the predictions of sTinyTim †. More information on the convolution kernels can be found in Sandstrom *et al.* (2008). The cubes were aligned to match the SL1 pixel grid. The PAH feature strengths were measured in IDL using the program Pahfit (Smith *et al.* 2007). Figure 1 shows the spectrum of the peak of PAH emission in SMC B1 with the best-fit results of Pahfit overplotted.

† http://ssc.spitzer.caltech.edu/archanaly/contributed/stinytim/index.html

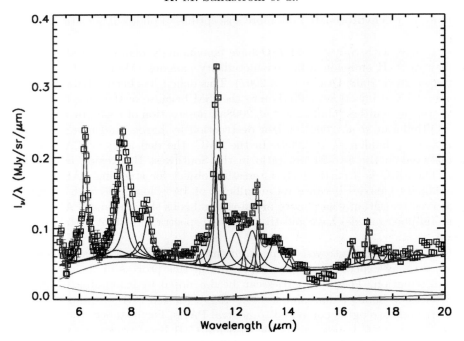

Figure 1. The PAH emission spectrum of the brightest position in SMC B1 overlayed with the results of Pahfit. Pahfit decomposes the spectrum into contributions from starlight continuum, dust continuum, emission lines and PAH band features.

3. Results

3.1. *SMC B1 molecular cloud*

SMC B1#1 is a quiescent molecular cloud in the Southwest Bar of the SMC. There are no H II regions or massive stars in its vicinity, so the dust is heated by a combination of the overall SMC radiation field and possibly a few nearby stars. The first detection of PAH emission from the SMC was from this region by Reach *et al.* (2000) using *ISO* observations. They noted the unusual strength of the 11.3 μm feature relative to the 7.7 μm feature. We also observe a low ratio of the 7.7/11.3 bands in this region. Figure 2 shows a map of the integrated intensity of the 11.3 μm band with a few selected regions highlighted. Figure 3 shows the average spectrum in the highlighted regions.

The 7.7/11.3 band ratio in SMC B1 has an average value of ~ 2 and does not show any significant spatial variations. Comparing these ratios to those in Galliano *et al.* (2008b) we find that the PAH emission from SMC B1 is consistent with a low ionization level.

3.2. *N 66 star-forming region*

N 66 is the most luminous H II region in the SMC containing 33 O stars (Massey *et al.* 1989). PAH emission from N 66 has been studied previously by Contursi *et al.* (2000). The higher angular resolution of *Spitzer* allows us to observe PAH emission tracing out the filamentary structures of the H II region. Figure 4 shows the integrated intensity of the 11.3 μm feature in this region with boxes showing the regions over which we have extracted spectra. Figure 5 shows a few of those spectra.

The 7.7/11.3 band ratio shows a wider range of variation in N 66 than in SMC B1. This is consistent with some regions having a higher PAH ionization fraction than achieved in SMC B1. However, we also observe band ratios as low as those observed in SMC B1. This

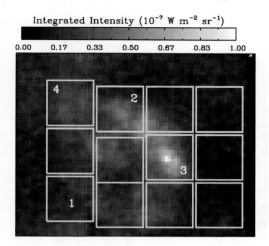

Integrated Intensity (10^{-7} W m^{-2} sr^{-1})

| 0.00 | 0.17 | 0.33 | 0.50 | 0.67 | 0.83 | 1.00 |

Figure 2. This figure shows the integrated intensity of the 11.3 μm PAH feature in the SMC B1 region. In order to increase the signal-to-noise we have extracted spectra from the cube in 30″ boxes. The spectra from the numbered boxes are plotted in Figure 3.

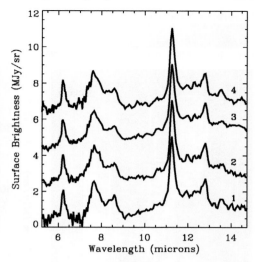

Figure 3. Spectra from the numbered boxes in Figure 2. The spectra are normalized to the 11.3 μm band and then shifted by 2 MJy sr^{-1} for clarity. The band strengths in SMC B1 show little variation across the region.

is particularly interesting because previous theoretical studies have had difficulty reproducing the band ratios in SMC B1 without adjusting the intrinsic band strengths (Li & Draine 2002). Excess hydrogenation and/or excitation by reddened starlight (rather than the average SMC radiation field) have been proposed as explanations for the band ratios in SMC B1. Seeing similar 7.7/11.3 ratios in both N 66 and SMC B1, two very different regions, suggests that different intrinsic band strengths may be the best explanation.

3.3. *Band ratio variations*

In addition to variations in the 7.7/11.3 band ratios, we observe variations in the ratios of different C–C bands as well, something that has not been observed in previous studies of more metal-rich galaxies (Galliano *et al.* 2008b). Figure 6 shows the variations in the 7.7/11.3 and 7.7/6.2 ratio, the latter of which is a ratio between two C–C emission bands. In our observations these band ratios appear to be positively correlated in N 66 and N 83, with N 66 offset towards larger 7.7/6.2 ratios. It is interesting that the band ratios from these regions seem to fall in distinct groups on the plot, suggesting different physical states for the PAHs, possibly related to different processing by the UV radiation field in each region.

4. Summary & conclusions

We detect emission from PAHs in all of the regions mapped by the S^{4}MC project. Although it has been shown that PAH emission is deficient in the SMC relative to more metal-rich galaxies Bolatto *et al.* (2007), PAHs are not absent from the SMC. In this proceedings we present a preliminary investigation of the band ratio variations. We find 7.7/11.3 ratios that are consistent with a range of ionization states for the PAHs across the SMC. In the quiescent molecular cloud SMC B1 the 7.7/11.3 ratio indicates a lower ionization level. Our observations of the band ratios in SMC B1 agree with the values determined by Reach *et al.* (2000). In addition, however, we see similarly low 7.7/11.3

Integrated Intensity (10^{-7} W m^{-2} sr^{-1})

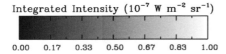

0.00 0.17 0.33 0.50 0.67 0.83 1.00

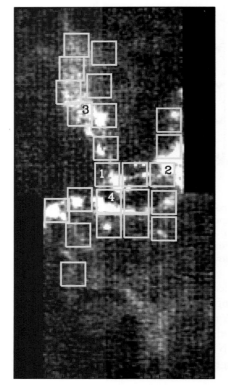

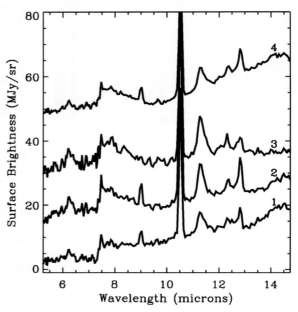

Figure 4. Integrated intensity of the 11.3 μm band in N 66.

Figure 5. Spectra of selected regions from N 66, normalized to the 11.3 μm band strength and then shifted by 15 MJy sr^{-1}. The strong emission line at 10.5 μm is from [S IV].

ratios even in N 66, suggesting that the intrinsic band strengths may be different in the SMC and the Milky Way, rather than SMC B1 having excess PAH hydrogenation or some other unique excitiation conditions. We also observe variations in the 7.7/6.2 ratio which are positively correlated with the 7.7/11.3 ratio. These observations will be explored further in an upcoming paper (Sandstrom *et al.* 2009, in preparation).

References

Allain, T., Leach, S., & Sedlmayr, E. 1996, *A&A*, 305, 616

Allamandola, L. J. and Hudgins, D. M., & Sandford, S. A. 1999, *ApJ*, 511, L115

Bakes, E. L.O. & Tielens, A. G. G. M. 1998, *ApJ*, 499, 258

Bolatto, A. D., Simon, J. D., Stanimirović, S., *et al.* 2007, *ApJ*, 655, 212

Contursi, A., Lequeux, J., Cesarsky, D., *et al.* 2000, *A&A*, 362, 310

Draine, B. T., Dale, D. A., Bendo, G., *et al.* 2007, *ApJ*, 663, 866

Engelbracht, C. W., Gordon, K. D., Rieke, G. H., Werner, M. W., Dale, D. A., & Latter, W. B. 2005, *ApJ*, 628, L29

Engelbracht, C. W., Rieke, G. H., Gordon, K. D., Smith, J.-D. T., Werner, M. W., Moustakas, J., Willmer, C. N. A., & Vanzi, L. 2008, *ApJ*, 678, 804

Galliano, F., Dwek, E., & Chanial, P. 2008a, *ApJ*, 672, 214

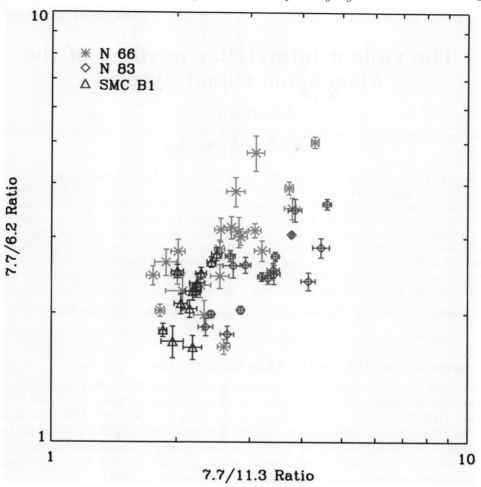

Figure 6. A plot of the 7.7/6.2 vs. 7.7/11.3 ratio in the N 66, N 83 and SMC B1 regions (the points are extracted from the boxes shown on previous plots). The ratios are positively correlated in N 66 and N 83.

Galliano, F., Madden, S. C., Tielens, A. G. G. M., Peeters, E., & Jones, A. P. 2008b, *ApJ*, 679, 310

Li, A. & Draine, B. T. 2002, *ApJ*, 576, 762

Madden, S. C., Galliano, F., Jones, A. P., & Sauvage, M. 2006, *A&A*, 446, 877

Massey, P., Parker, J. W., & Garmany, C. D. 1989, *AJ*, 98, 1305

O'Halloran, B., Satyapal, S., & Dudik, R. P. 2006, *ApJ*, 641, 795

Reach, W. T., Boulanger, F., Contursi, A., & Lequeux, J. 2000, *A&A*, 361, 895

Sandstrom, K. M., Bolatto, A. D., Stanimirović, S., van Loon, J.Th, & Smith, J. D. T. 2008, *ApJ*, submitted [ArXiv:0810.2803]

Smith, J. D. T., Armus, L., & Dale, D. A. 2007, *PASP*, 119, 1133

Tielens, A. G. G. M. 2008, *ARAA*, 46, 289

The Magellanic System: Stars, Gas, and Galaxies
Proceedings IAU Symposium No. 256, 2008
Jacco Th. van Loon & Joana M. Oliveira, eds.

The violent interstellar medium of the Magellanic Cloud System

You-Hua Chu[1]

[1] Astronomy Department, University of Illinois, 1002 W. Green Street, Urbana, IL 61801, USA
email: yhchu@illinois.edu

Abstract. The interstellar gas of the Magellanic System is subject to the harassment of tidal interactions on galaxy-wide scales and stellar energy feedback on sub-galactic scales. H I surveys of the Magellanic System have produced spectacular images of the tidally displaced interstellar gas in the Magellanic Bridge and Streams. Multi-wavelength observations of the interstellar gas in the Magellanic Clouds have revealed gas components in physical conditions ranging from cold molecular cloud to hot ionized coronal gas. While stellar energy feedback is responsible for heating and dispersing interstellar gas, it can also compress ambient cloud to form stars. I will use *Chandra, XMM-Newton, FUSE, HST, Spitzer, ATCA*, and other ground-based observations to illustrate the interplay among massive stars, interstellar medium, and star formation.

Keywords. ISM: bubbles, ISM: evolution, ISM: general, galaxies: ISM, Magellanic Clouds

1. Introduction: ISM in the Magellanic System

The tidal interactions among the Magellanic Clouds and the Galaxy have wildly disrupted the interstellar medium (ISM) in this system. As show in the H I *Parkes* All-Sky Survey (HIPASS), H I is distributed not only within the LMC and SMC, but also in the Magellanic Bridge between them and the leading and trailing Streams outside them (Putman *et al.* 1998). Interstellar absorption observations of abundances in the Magellanic Streams rule out a primordial origin of the gas (Gibson *et al.* 2000), and suggest that the gas has been pulled out of the SMC (Sembach *et al.* 2001).

Since the last IAU Symposium on the Magellanic Clouds in 1998, the *Australia Telescope Compact Array (ATCA)* surveys of H I in the LMC and SMC have been combined with Parkes single-dish observations to recover large-scale diffuse emission. Figure 1 shows the new H I column density maps of the SMC (Stanimirović *et al.* 2004) and the LMC (Kim *et al.* 2003; Staveley-Smith *et al.* 2003). The effects of tidal interactions are clearly evidenced in the position-velocity plots of the SMC (Fig. 2-left), where double velocity components are seen everywhere. The position-velocity plots of the LMC show high-velocity clouds (HVCs); while some HVCs result from tidal interactions, others are associated with supergiant shells, such as LMC 3 and LMC 8, where stellar energy feedback may be responsible for the HVCs (see Fig. 2-right; Staveley-Smith *et al.* 2003).

In the bodies of the Magellanic Clouds, over scales of a few to ∼1000 pc, the structure of the ISM is mainly determined by stellar energy feedback via fast stellar winds, supernova explosions, and UV radiation. The energetic interactions have produced multiple phase components of the ISM, which require observations at multiple wavelengths. For example, cold molecular gas requires mm-wavelength CO observations, interstellar dust requires far-IR observations, H I gas requires 21-cm line observations, warm (10^4 K) ionized gas is best observed with optical recombination and forbidden lines, hot ($10^6 - 10^7$ K) ionized gas requires X-ray observations, and the interface between hot ionized gas and cooler gas is best studied in far-UV.

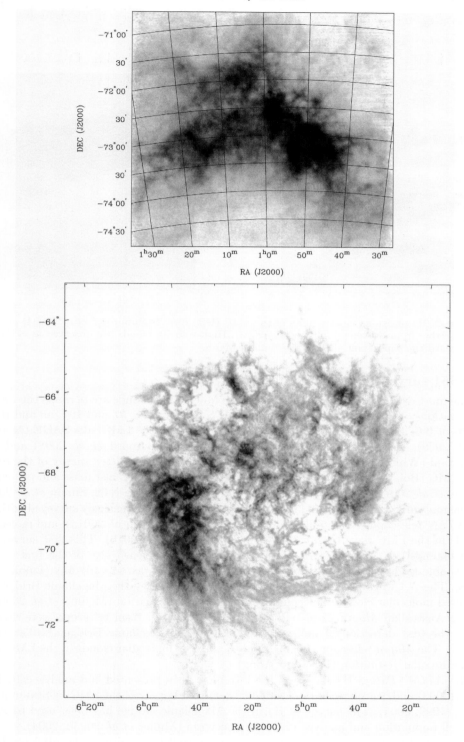

Figure 1. H I column density maps of the SMC (top) and the LMC (bottom) from Stanimirović *et al.* (2004) and Kim *et al.* (2003), respectively.

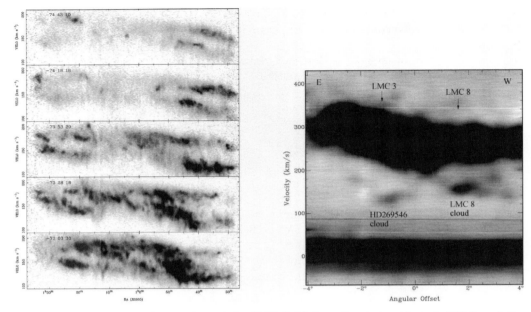

Figure 2. H I position-velocity plots of the SMC (left) from Stanimirović *et al.* (2004) and the LMC (right) from Staveley-Smith *et al.* (2003). High-velocity clouds are observed at ∼150 km s^{-1} toward the supergiant shells LMC 3 and LMC 8.

2. ISM surveys in the last decade

The most exciting surveys of the ISM in the Magellanic Clouds are perhaps the *Spitzer* observations of the dust component in the MIPS bands at 24, 70, and 160 μm and IRAC bands at 3.6, 4.5, 5.8, and 8.0 μm. The *Spitzer* survey of the LMC, aka SAGE (Meixner *et al.* 2006), has been used to study dust properties (Bernard *et al.* 2008) and star formation (Whitney *et al.* 2008; Gruendl & Chu 2008). The *Spitzer* survey of the SMC, aka S^{3}MC (Bolatto *et al.* 2007), has also been used to study the dust mass and properties (Leroy *et al.* 2007) and star formation (in the giant H II region N 66, Simon *et al.* 2007).

The molecular clouds in the Magellanic Clouds have been extensively surveyed with the *NANTEN* telescope (LMC: Fukui *et al.* 2001; SMC: Mizuno *et al.* 2001a), and molecular clouds in the LMC have been cataloged by Mizuno *et al.* (2001b). This first survey has been extended to *NANTEN II*, which reaches twice the sensitivity of the first survey and is able to detect molecular clouds with masses as low as ∼2×10^4 $M_\odot$ (Fukui *et al.* 2008). The *NANTEN* telescope has also been used to survey the Magellanic Bridge and detected molecular clouds that could be sites for star formation (Mizuno *et al.* 2006).

The Australian *Mopra* 22-m telescope and ESO *SEST* 15-m telescope were used to make the first detection of molecular clouds in the Magellanic Bridge (Muller *et al.* 2003a). The *Mopra* telescope is being used to map the molecular clouds in the LMC with higher angular resolution (PI: Tony Wong).

The *ATCA+Parkes* H I survey of the LMC has been presented and analyzed by Kim *et al.* (2003) and Staveley-Smith *et al.* (2003), and the survey of the SMC by Stanimirović *et al.* (1999, 2004). The survey of H I in the Magellanic Bridge has been used to study its shell population and analyze its power spectrum (Muller *et al.* 2003b, 2004).

The Magellanic Clouds Emission-Line Survey (MCELS; Smith & the MCELS Team 1999) of the warm (10^4 K) ionized gas in the Magellanic Clouds has been completed. The spectacular color composites of the Hα, [O III], and [S II] images are available at http://www.noao.edu/image_gallery/html/im0993.html (SMC, made by F. Winkler *et al.*), and http://www.noao.edu/image_gallery/html/im0994.html (LMC, made by C. Smith, S. Points, *et al.*).

ROSAT surveys of large-scale diffuse X-ray emission from hot gas in the LMC have been reported by Points *et al.* (2001) and Sasaki *et al.* (2002). The *Chandra* X-ray Observatory and the *XMM-Newton* X-ray Observatory have higher sensitivity and angular resolution, but their small field-of-views make it difficult to carry out large-scale surveys. Observations have been made only for regions of interest, such as 30 Dor (Dennerl *et al.* 2001; Townsley *et al.* 2006).

The *Far Ultraviolet Spectroscopic Explorer* (*FUSE*) has provided a great opportunity to use interstellar O VI $\lambda\lambda$ 1031.9, 1037.6 Å lines to probe the 10^5 K gas at interfaces between 10^6 K gas and cooler gas. An atlas of early *FUSE* spectra of Magellanic Clouds targets was presented by Danforth *et al.* (2002). The *FUSE* Magellanic Clouds Legacy Project (http://archive.stsci.edu/prepds/fuse_mc/) has made spectra available for 186 sightlines in the LMC and 100 sightlines in the SMC (Blair *et al.* 2007). *FUSE* observations have detected shocked gas in supernova remnants (Blair *et al.* 2006), interfaces in superbubbles (Sankrit & Dixon 2007), and hot gas halos (Howk *et al.* 2002) of the Magellanic Clouds, as well as ionized gas in the Magellanic Bridge (Lehner *et al.* 2008) and the Magellanic System outside the Clouds themselves (Dixon & Sankrit 2008).

3. Stellar energy feedback and star formation

Massive stars inject energy into the ISM and dynamically change the distribution and physical conditions of the ambient ISM. The interactions frequently produce interstellar shell structures with sizes that depend on the star formation history and distribution of massive stars. Single massive stars can blow bubbles with sizes up to a few $\times 10$ pc, and the bubble ages are observed to be 10^3-10^5 yr. When a single massive star explodes as a supernova, the ejecta will interact with the ambient medium and form a supernova remnant (SNR). Multiple massive stars in an OB association can jointly blow a superbubble with diameter up to $\sim 10^2$ pc; superbubble ages are commonly observed to be of order 10^6 yr. Supernovae that have exploded in the central cavity of a superbubble will not produce strong optical forbidden line emission or nonthermal radio emission that are characteristic of isolated SNRs; only X-ray emission from the shock-heated, 10^6-10^7 K, gas can be detected. When multiple generations of massive star formation have occurred, a supergiant shell (SGS) with size approaching 10^3 pc can be produced; the dynamic ages of such SGSs are usually a few $\times 10^7$ yr.

The interstellar shell structures are prevalent in the LMC, as shown in the MCELS Hα image in Figure 3 (top). A close-up of the SGS LMC 4 region is shown in the bottom panels: the Hα image to the left adequately illustrates the relative sizes of SGS, superbubble, bubble, and SNR, while the H I column density map to the right shows a neutral shell exterior to the ionized SGS LMC 4. The rim of the SGS LMC 4 is dotted with OB associations and H II regions with ages of a few Myr, suggesting that the star formation may have been triggered by the expansion of the SGS. Interestingly, the pattern that stellar energy feedback leads to star formation is observed at many different scales.

3.1. *SNRs and star formation*

Star formation has been observed to take place in the vicinity of SNRs. It is unlikely that SNR shocks have triggered the star formation, because the identifying characteristics of SNRs are produced by strong shocks and strong shocks disrupt molecular clouds rather than compressing them to form stars. In H II regions around OB associations, SNRs may

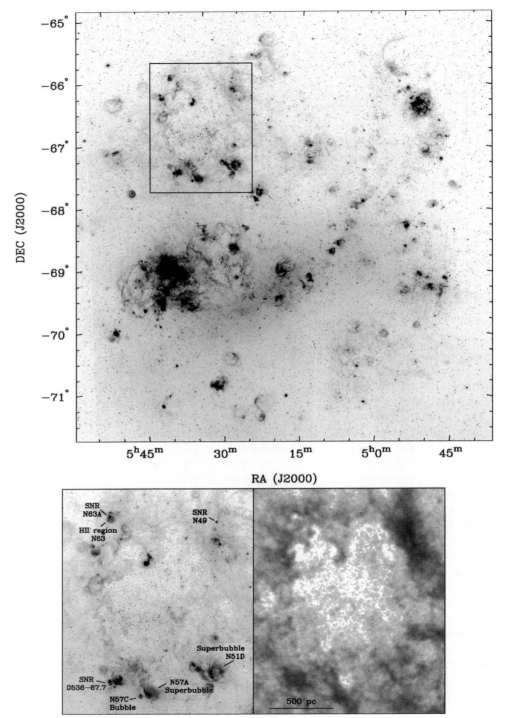

Figure 3. MCELS Hα image of the LMC (top). A closeup of the SGS LMC 4 region is shown in the bottom left panel with examples of bubbles and SNRs marked. The H I column density map in the bottom right panel shows a neutral shell exterior to the ionized shell of SGS LMC 4.

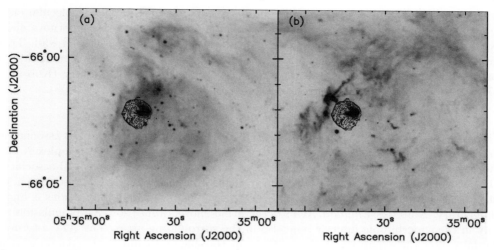

Figure 4. MCELS Hα (left) and *Spitzer* IRAC 8.0 μm (right) images of N 63. Overplotted are X-ray contours, extracted from a *Chandra* ACIS observation, to show the location of the SNR N 63A. Young stellar objects are detected along the northeast boundary of the H II region. The SNR is obviously not responsible for triggering the formation of these young stars.

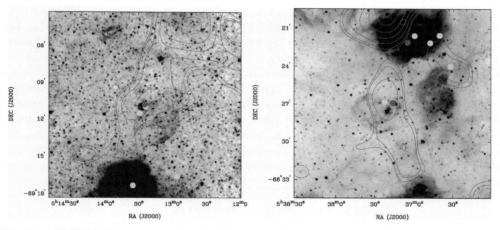

Figure 5. MCELS Hα images of SNRs 0513−69.2 (left) and DEM L256 (right). Overplotted are *NANTEN* CO contours. Young stellar objects are marked by filled circles. Young stellar objects are observed along the rim of the SNR projected within a molecular cloud.

exist near young stars and the supernova ejecta may even enrich the star- and planet-forming material, but the SNR shocks are not responsible for the onset of star formation. It is more likely that the gentle expansion of an H II region compresses ambient clouds and triggers star formation. As shown in Figure 4, young stellar objects (YSOs) are present along the rim of the H II region N 63, but not the SNR N 63A (Caulet *et al.* 2008). The YSOs are just superposed near the SNR N 63A by chance.

Some isolated SNRs show YSOs along their rims, for example, SNRs 0513−69.2 and DEM L256 in Figure 5. These two SNRs not only have YSOs, but also appear to be interacting with molecular clouds. The large sizes, ~ 50 pc in diameter, suggest that they are old ($> 10^5$ yr) SNRs, but their X-ray emission suggests that they are still young ($10^3 - 10^4$ yr). Only a supernova explosion within the cavity of a pre-existing shell can explain both the large size and X-ray emission. It is most likely that the supernova progenitor blew a bubble, the expanding bubble compressed the ambient molecular cloud to form stars, the supernova exploded in the bubble interior, and the supernova ejecta expands quickly through in a low-density medium to reach the dense shell wall. Thus, the supernova progenitors, rather than the SNRs, were responsible for triggering the star formation. Similar phenomenon has been observed in SNRs in the Galaxy (Koo *et al.* 2008) and near the giant H II region N 66 in the SMC (Gouliermis *et al.* 2008).

3.2. *Energy feedback and star formation in superbubbles*

The dynamic impact of fast stellar winds and supernova explosions from OB associations can be well illustrated by the multi-wavelength observations of the H II complex N 11 in the LMC (see Figure 6). The Hα image shows a superbubble around the OB association LH 9 at the center. Two bright H II regions, each containing a young OB association, exist along the superbubble shell rim. The ionized gas in the surrounding regions is highly filamentary. A 300 ks *Chandra* ACIS observation shows diffuse soft X-ray emission not only in the superbubble interior, but also in the surrounding areas. The *ATCA+Parkes* H I data show the expansion of the central superbubble and additional expanding structures to the north. Clearly, the ISM in N 11 has been highly disturbed by the energies injected by massive stars. There are still molecular clouds in N 11, and star formation is still raging on: numerous YSOs are observed in regions where the superbubble shell expands into molecular clouds (8 μm point sources along the superbubble shell rim).

The hot gas content of superbubbles in the LMC has been surveyed with *ROSAT* observations (Dunne *et al.* 2001). A detailed analysis of the superbubble N 51D has been carried out by Cooper *et al.* (2004); they find that the total kinetic and thermal energy retained in the interstellar gas is only about 1/3 of the stellar mechanical energy injected. A remarkable feature of the X-ray emission from N 51D is that its X-ray spectrum requires a nonthermal component to explain the emission in $1-3$ keV. Nonthermal X-ray emission has also been detected in the LMC superbubbles 30 Dor C (Bamba *et al.* 2004; Smith & Wang 2004) and N 11 (Maddox *et al.* 2008). N 51D does not have detectable molecular clouds around its superbubble shell rim; however, there is still on-going star formation. Some YSOs are projected within the superbubble; high-resolution *Hubble Space Telescope* images show that these YSOs are formed in dust globules, possibly triggered by photo-implosion (Chu *et al.* 2005).

3.3. *Energy feedback and star formation in supergiant shells*

The hot gas content of SGSs in the LMC has been studied with *ROSAT* observations (Points *et al.* 2001). The physical structure of the SGS LMC 2 has been analyzed in great detail (Points *et al.* 1999, 2000). Using the improved *ATCA+Parkes* H I data of the LMC, Book *et al.* (2008) re-examined the physical structure of all nine optically identified SGSs in the LMC. As shown in Figure 7, the H I structure of LMC 2 is quite complex exhibiting expansion over scales that extend beyond the optical boundaries of the SGS. *XMM-Newton* observations of LMC 2 detect diffuse X-ray emission from the SGS interior. There is a good correlation between low X-ray surface brightness and high H I column density, which can be used to determine their relative locations along the line of sight.

Only SGSs LMC 1, 4, 5, and 6 show simple expansion, making it possible to assess their relationship with the on-going star formation (Book *et al.* 2009). Comparing the YSO and molecular cloud contents of these simple SGSs, it is concluded that triggered star formation is prevalent. Figure 8 shows the SGSs LMC 4 and 5 and their YSOs. There is on-going star formation along the rims of these SGSs and particularly in the interaction zone between these two SGSs.

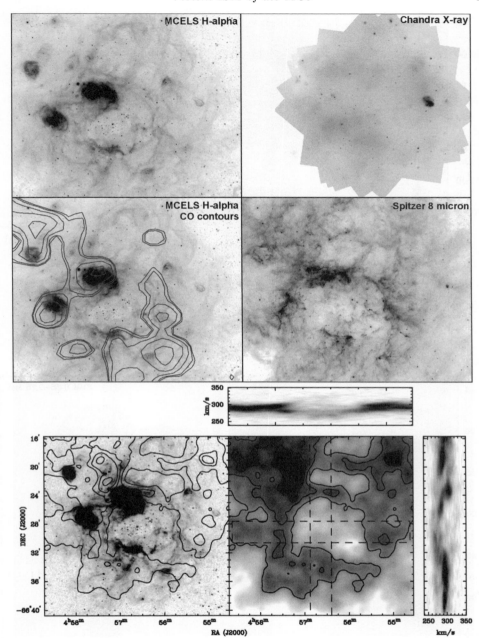

Figure 6. The N 11 H II complex. The top four panels are MCELS Hα image, *Chandra* ACIS image in soft X-rays, Hα image superposed with *NANTEN* CO contours, and *Spitzer* IRAC 8.0 μm image. The bright X-ray source on the west side of N 11 is the known SNR N 11L. The bottom left panel is an Hα image superposed with H I contours, and the bottom right panel is an H I column density map extracted from *ATCA+Parkes* survey (Kim *et al.* 2003). Two position-velocity plots are presented on the sides of the H I map; the dashed lines in the map mark the locations of the position-velocity plots.

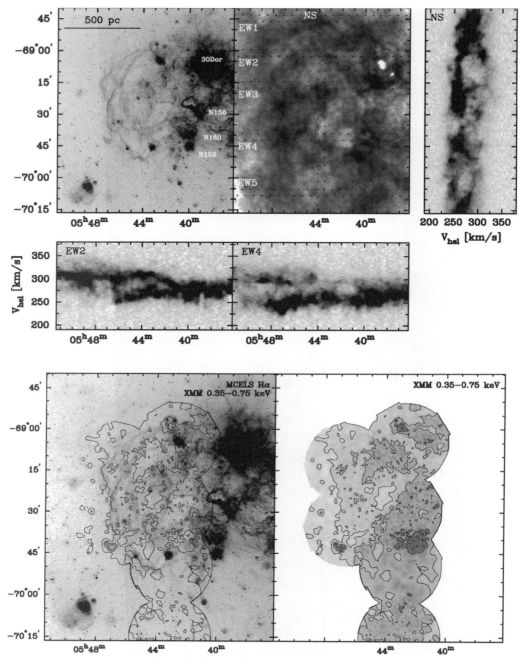

Figure 7. Supergiant shell LMC 2. MCELS Hα image and *ATCA+Parkes* H I column density map. The locations of the position-velocity plots are marked on the H I map. The lower panels are MCELS Hα image and *XMM-Newton* X-ray (0.35−0.75 keV) image of LMC 2; the X-ray contours are overplotted on both images for comparison.

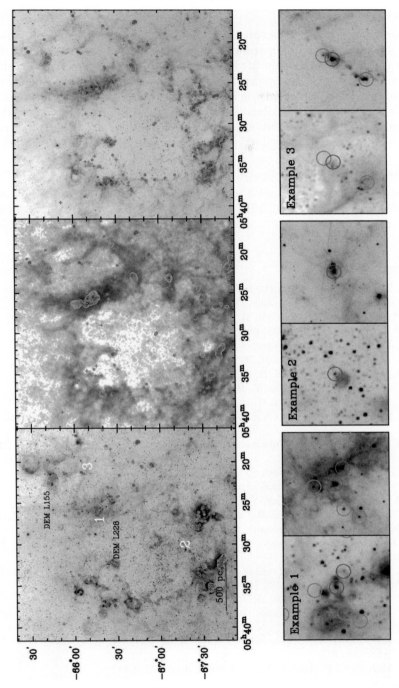

Figure 8. The supergiant shells LMC 4 and 5. Top row (left to right) — MCELS Hα image, *ATCA+Parkes* H i column density map, and *Spitzer* 8 μm image, superposed by *NANTEN* CO contours and marked with locations of YSOs. The probable YSOs are marked by circles and the less certain ones in crosses. The bottom row shows three examples of YSOs with pairs of Hα (left) and 8 μm (right) images. The locations of these examples are marked in the Hα image in the top row. This figure is from Book *et al.* (2009).

4. Future work

The interstellar space is permeated with hot coronal gas, a product of violent shocks from fast stellar winds and supernova explosions. The hot gas component is perhaps the least well studied component of the ISM. While there are X-ray images showing the existence and distribution of hot gas, the spectral information is always of low resolution, which limits our ability to extract accurate temperature and abundance information of the hot gas. Sensitive, high-resolution X-ray spectra of the hot ISM are needed so that the thermal and chemical evolution of the hot gas can be studied with accuracy. It is particularly important to study how supernovae enrich their ambient medium.

Acknowledgements

This research has been supported by NASA grants SAO GO7-8091A, NNG04GF34G, JPL 1264494, and JPL 1290956. YHC thanks R. Gruendl for making many of the figures in this paper.

References

Bamba, A., Ueno, M., Nakajima, H., & Koyama, K. 2004, *ApJ*, 602, 257
Bernard, J.-P., Reach, W. T., Paradis, D., *et al.* 2008, *AJ*, 136, 919
Blair, W. P., Ghavamian, P., Sankrit, R., & Danforth, C. W. 2006, *ApJS*, 165, 480
Blair, W. P., *et al.* 2007, *BAAS*, 38, 898
Bolatto, A. D., Simon, J. D., Stanimirović, S., *et al.* 2007, *ApJ*, 655, 212
Book, L. G., Chu, Y.-H., & Gruendl, R. A. 2008, *ApJS*, 175, 165
Book, L. G., Chu, Y.-H., Gruendl, R. A., & Fukui, Y. 2009, *AJ*, in press
Caulet, A., Gruendl, R. A., & Chu, Y.-H. 2008, *ApJ*, 678, 200
Chu, Y.-H., Gruendl, R. A., Chen, C.-H. R., *et al.* 2005, *ApJ*, 634, L189
Cooper, R. L., Guerrero, M. A., Chu, Y.-H., Chen, C.-H. R., & Dunne, B. C. 2004, *ApJ*, 605, 751
Danforth, C. W., Howk, J. C., Fullerton, A. W., Blair, W. P., & Sembach, K. R. 2002, *ApJS*, 139, 81
Dennerl, K., Haberl, F., Aschenbach, B., *et al.* 2001, *A&A*, 365, L202
Dixon, W. V. D. & Sankrit, R. 2008, *ApJ*, 686, 1162
Dunne, B. C., Points, S. D., & Chu, Y.-H. 2001, *ApJS*, 136, 119
Fukui, Y., Mizuno, N., Yamaguchi, R., Mizuno, A., & Onishi, T. 2001, *PASJ*, 53, L41
Fukui, Y., Kawamura, A., Minamidani, T. *et al.* 2008, *PASJ*, 178, 56
Gibson, B. K., Giroux, M. L., Penton, S. V., Putman, M. E., Stocke, J. T., & Shull, J. M. 2000, *AJ*, 120, 1830
Gouliermis, D. A., Chu, Y.-H., Henning, T., Brandner, W., Gruendl, R. A., Hennekemper, E., & Hormuth, F. 2008, *ApJ*, 688, 1050
Gruendl, R. A. & Chu, Y.-Y. 2008, *ApJS*, submitted
Howk, J. C., Sembach, K. R., Savage, B. D., Massa, D., Friedman, S. D., & Fullerton, A. W. 2002, *ApJ*, 569, 214
Kim, S., Staveley-Smith, L., Dopita, M. A., Sault, R. J., Freeman, K. C., Lee, Y., & Chu, Y.-H. 2003, *ApJS*, 148, 473
Koo, B.-C., McKee, C. F., Lee, J.-J., *et al.* 2008, *ApJ*, 673, L147
Lehner, N., Howk, J. C., Keenan, F. P., & Smoker, J. V. 2008, *ApJ*, 678, 219
Leroy, A., Bolatto, A., Stanimirovíc, S., Mizuno, N., Israel, F., & Bot, C. 2007, *ApJ*, 658, 1027
Maddox, L. A., Williams, R. M., Dunne, B. C., & Chu, Y.-H. 2008, *ApJ*, submitted
Meixner, M., Gordon, K. D., Indebetouw, R., *et al.* 2006, *AJ*, 132, 2268
Mizuno, N., Rubio, M., Mizuno, A., Yamaguchi, R., Onishi, T., & Fukui, Y. 2001a, *PASJ*, 53, L45
Mizuno, N., Yamaguchi, R., Mizuno, A., *et al.* 2001b, *PASJ*, 53, 971

Mizuno, N., Muller, E., Maeda, H., Kawamura, A., Minamidani, T., Onishi, T., Mizuno, A., & Fukui, Y. 2006, *ApJ*, 643, L107

Muller, E., Staveley-Smith, L., & Zealey, W. J. 2003a, *MNRAS*, 338, 609

Muller, E., Staveley-Smith, L., Zealey, W., & Stanimirović, S. 2003b, *MNRAS*, 339, 105

Muller, E., Stanimirović, S., Rosolowsky, E., & Staveley-Smith, L. 2004, *ApJ*, 616, 845

Points, S. D., Chu, Y.-H., Kim, S., Smith, R. C., Snowden, S. L., Brandner, W., & Gruendl, R. A. 1999, *ApJ*, 518, 298

Points, S. D., Chu, Y.-H., Snowden, S. L., & Staveley-Smith, L. 2000, *ApJ*, 545, 827

Points, S. D., Chu, Y.-H., Snowden, S. L., & Smith, R. C. 2001, *ApJS*, 136, 99

Putman, M. E., Gibson, B. K., Staveley-Smith, L, *et al.* 1998, *Nature*, 394, 752

Sankrit, R. & Dixon, W. V. D. 2007, *PASP*, 119, 284

Sasaki, M., Haberl, F., & Pietsch, W. 2002, *A&A*, 392, 103

Sembach, K. R., Howk, J. C., Savage, B. D., & Shull, J. M. 2001, *AJ*, 121, 992

Simon, J. D., Bolatto, A. D., Whitney, B. A., *et al.* 2007, *ApJ*, 669, 327

Smith, D. A. & Wang, Q. D. 2004, *ApJ*, 611, 881

Smith, R. C. The MCELS Team 1999, in *New Views of the Magellanic Clouds*, 190, 28

Stanimirović, S., Staveley-Smith, L., Dickey, J. M., & Sault, R. J. 1999, *MNRAS*, 302, 417

Stanimirović, S., Staveley-Smith, L., & Jones, P. A. 2004, *ApJ*, 604, 176

Staveley-Smith, L., Kim, S., Calabretta, M. R., Haynes, R. F., & Kesteven, M. J. 2003, *MNRAS*, 339, 87

Townsley, L. K., Broos, P. S., Feigelson, E. D., Brandl, B. R., Chu, Y.-H., Garmire, G. P., & Pavlov, G. G. 2006, *AJ*, 131, 2140

Whitney, B. A., Sewilo, M., Indebetouw, R., *et al.* 2008, *AJ*, 136, 18

Keele Hall by night.

The Magellanic System: Stars, Gas, and Galaxies
Proceedings IAU Symposium No. 256, 2008
Jacco Th. van Loon & Joana M. Oliveira, eds.

The magnetic field structure of the Small Magellanic Cloud

**Antonio Mário Magalhães[1], Aiara Lobo Gomes[1],
Aline de Almeida Vidotto[1], Cláudia Vilega Rodrigues[2],
Antonio Pereyra[3], John Wisniewski[4], Karen Bjorkman[5],
Jon Bjorkman[5], Marilyn Meade[6] and Brian L. Babler[6]**

[1] Depto. de Astronomia, IAG, Universidade de São Paulo, Rua do Matão 1226, São Paulo,
SP 05508-900, Brazil
email: `mario@astro.iag.usp.br`

[2] INPE/DAS, S.J. dos Campos, SP, Brazil

[3] ON/CNPq, Brazil

[4] University of Washington, USA

[5] University of Toledo, USA

[6] University of Wisconsin-Madison, USA

Abstract. We describe two studies of the interstellar magnetic field in regions of the Small Magellanic Cloud (SMC), including those affected by the interaction with the Large Magellanic Cloud (LMC). We use optical polarization data from aligned grains in the interstellar medium of the SMC in order to map the sky-projected direction of the magnetic field and determine characteristics of the SMC and Pan-Magellanic field structures. The earlier, photoelectric data are reanalyzed and they provide values for the average projected magnetic field intensity (1.7×10^{-6} G) and the random field component intensity (3.5×10^{-6} G). Another on-going program uses imaging data and, when concluded, will allow more local estimates of the field intensity in the SMC NE/Wing regions. Additional goals include cross-correlating our field mapping results with those of point sources and structures found by the Spitzer Space Telescope in the SMC between 3.6 and 8 μm.

Keywords. polarization, techniques: polarimetric, galaxies: individual (SMC), Magellanic Clouds, galaxies: magnetic fields

1. Introduction

The Magellanic Clouds form with the Galaxy a triple system in continuous interaction. When of the last approach, for instance, the SMC reached 2−5 kpc from the LMC, about 0.2 Gyr ago (Westerlund 1993). These interactions have had great influence on the Clouds. For example, the SMC may have been "stretched" in its Northeast and Wing sections towards the LMC during this last collision. The H I bridge and Hα nebulosities are other effects of such past collisions (Sofue 1994).

Magnetic fields are thought to be strong enough to influence the gas dynamics in galaxies (Zweibel & Heiles 1997; Beck 2008). Our on-going project aims at studying the interstellar magnetic field in these SMC regions affected by the collisions. We employ the technique of optical imaging polarimetry, detecting the polarization caused by magnetically aligned grains in the SMC.

2. Interstellar polarization

Interstellar grains can be partially aligned in the magnetic field of the interstellar medium (Davis & Greenstein 1951; Hoang & Lazarian 2008). The detailed mechanism is still being actively discussed, with radiative torque receiving support lately (Hoang & Lazarian 2008). Regardless of the theory, the mechanism is expected to first align the small axis of the grain with its angular momentum, or spin, axis and then align the spin axis with the magnetic field. The end result has the polarizing grains rotating with their longest profile perpendicular to the magnetic field. Starlight shining through such grains will then have the transmitted electric field vibrations predominantly along the sky-projected magnetic field direction.

Starlight polarization is hence a direct tracer of the ISM magnetic field projected on the sky. This technique has been used in the optical to map Galactic fields in dark clouds (e.g., Pereyra & Magalhães 2004) and in shocked regions (Pereyra & Magalhães 2007). It can also be used to probe the magnetic field correlation length in the Galaxy's ISM (Magalhães *et al.* 2005). On a Galactic scale, Heiles (1996) and Fosalba *et al.* (2002) have used the compiled data for the Galaxy in order to probe from the field structure in the Solar neighborhood to the ratio of random to uniform components of the field.

Observed linear polarization values towards the general ISM in the Galaxy are typically of a few percent, although higher values ($\sim 10-15\%$ or so) may be observed towards regions where the magnetic field geometry is favourable (Pereyra & Magalhães 2004, 2007). In the SMC case however, in part because of the small dust-to-gas ratio and in part to our incomplete knowledge, the reported values of interstellar polarization are in general considerably smaller (Rodrigues *et al.* 1997; Wisniewski *et al.* 2007a; section 3).

3. Interstellar polarization in the SMC

Mathewson & Ford (1970) and Schmidt (1976) provided the first measurements of starlight polarization for SMC members. These works and the first photoelectric measurements of Magalhães *et al.* (1990) have suggested a magnetic field aligned with the H I bridge (McGee & Newton 1981, 1986). More recently, Mao *et al.* (2008) have reanalyzed the Mathewson & Ford (1970) data and, together with radio Faraday rotation data, concluded that the overall magnetic field of the SMC may indeed lie along the line connecting the centers of each Cloud.

Here we re-discuss the Magalhães *et al.* (1990) data, the details of which we expect to publish shortly (Magalhães *et al.*, in preparation). We also follow the practise of separating the SMC main body and NE/Wing into 5 sections (I through V; Schmidt 1976; Rodrigues *et al.* 1997). In particular, we use the corresponding five Galactic foreground correction values of Rodrigues *et al.* (1997), in their study of SMC dust (their Table 2). Since the SMC interstellar polarization values are rather small (typically $\sim 1\%$ or smaller), this correction is crucial. The observational uncertainty in each data value is added quadratically to that of the foreground correction.

The total sample has 132 SMC stars distributed along its main body and NE/Wing sections. After correcting for the Galactic foreground values and selecting only objects with SMC polarization P and error σ_P such that $P/\sigma_P \geqslant 2$, 66 objects remain. We then transform their equatorial position angle to a "Magellanic" position angle, Θ_M, such that this new angle is zero along the SMC-LMC direction (Schmidt 1976, Mao *et al.* 2008).

Fig. 1 shows the distribution of the Θ_M values for the whole sample. Clustering around two position angle values, $\Theta_M \sim 0°$ and $\sim 60°$, can be noted. In other words, one of the tendencies is for the field to indeed lie along the SMC-LMC direction, and another

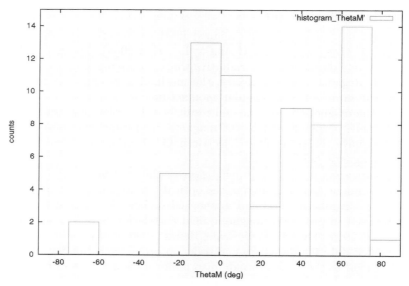

Figure 1. Histogram of position angles of optical polarization vectors across the SMC (main body and NE/Wing sections combined), using the data of Magalhães *et al.* (1990). $\Theta_M = 0$ along the SMC-LMC direction.

one represent a tendency of the field to roughly lie along the N-S direction. The latter is because the average correction from the equatorial to the Magellanic position angle for the main body is $\sim 113°$, giving almost a N-S direction (i.e., $0° \equiv 180°$).

These two tendencies are confirmed if we examine separately the histograms of Θ_M for regions I through III (main body) and IV-V (NE and Wing sections). Fig. 2 shows the histogram across the main body and the values again cluster around $\Theta_M \sim 0°$ and $\sim 60°$, albeit with a less marked tendency. Fig. 3, for the NE+Wing sections, shows again the two tendencies but this time more clearly. This may be understood if we remember that lines of sight through the main body are more likely to cross higher density regions that those in the NE/Wing sections; higher densities will tend to drag the field around compared to less dense regions and there should be a broader dispersion of polarization values along the line of sight. This is consistent with the fact that the SMC is more distant than the LMC and the bridge is seen 'pointing' to the observer; the observed polarization is mostly produced in these layers lying in front of the SMC (Magalhães *et al.* 1990).

The distribution of position angles allows us to estimate the component of the magnetic field intensity in the plane of the sky using the formulation of Falceta-Gonçalves *et al.* (2008). We first estimate the random component of the magnetic field, δB, using their eq. 7, which equates the random thermal and magnetic energies. Using the SMC H I velocity dispersion (22 ± 2 km s^{-1}; Stanimirović *et al.* 2004) and density ($n_H \sim 0.1$ cm^{-3}, Mao *et al.* 2008), we find $\delta B \approx 3.5 \times 10^{-5}$ G. This values compares well with the estimate of Mao *et al.* (2008) using synchrotron emission.

Now, eq. 9 of Falceta-Gonçalves *et al.* (2008) provides a modified form for the Chandrasekhar-Fermi relation between the dispersion in position angles and the total field intensity projected on the sky, $B_{\text{sky}}^u + \delta B$ (which they call $B_{\text{sky}}^{\text{ext}} + \delta B$). Essentially, for a given random velocity dispersion a narrow dispersion in Θ_M implies a larger sky-projected, uniform field intensity, as physically expected. Using this equation, we obtain $B_{\text{sky}}^u \approx 1.7 \times 10^{-6}$ G. This is to be compared to $(1.6 \pm 0.4) \times 10^{-6}$ G obtained by Mao

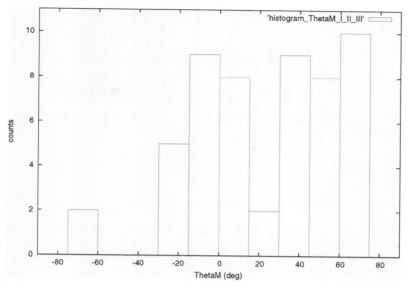

Figure 2. Same as Fig. 1 but for the SMC main body only (regions I–III of Schmidt 1976).

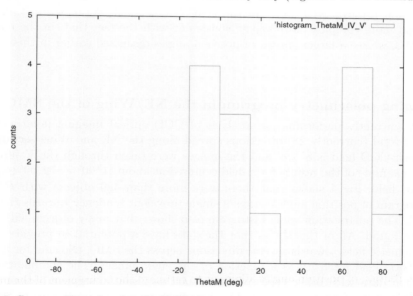

Figure 3. Same as Fig. 1 but for the SMC NE and Wing sections only (regions IV and V of Schmidt 1976).

et al. (2008) using the Mathewson & Ford (1970) sample. The formal uncertainty in our result is not too bad ($\sim 20\%$) but the real answer, given the uncertainty in the density, for instance, is probably within a factor of a few.

The above discussion concerned the overall magnetic field of the SMC. We (Wisniewski *et al.* 2007a,b) have observed 6 LMC and 6 SMC open clusters for studying Be star disks in low metallicity environments. These observations were used to derive the foreground values of the interstellar polarization within each Cloud (after Galactic foreground subtraction). The SMC interstellar polarization values and their position angles towards the six directions observed are consistent with the sample discussed above. However, in the

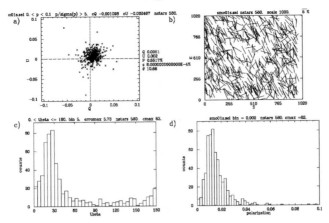

Figure 4. Foreground corrected, V-band imaging polarization data for a $8' \times 8'$ field towards $[1^{\mathrm{h}}00^{\mathrm{m}}, -72°00'$ (eq. 1975)], in the SMC NE region. Stars for which $P/\sigma_P \geqslant 7$ have been selected. From left to right and top to bottom: Q-U diagram, position angle vector diagram, histogram of equatorial position angles and histogram of polarization values.

denser region towards NGC 2100, in the LMC 2 Supershell, Wisniewski *et al.* (2007b) have found more complex morphology, consistent with the fact that, in the main body of the SMC, we see a larger spread of position angles discussed earlier in this section.

4. Imaging polarimetry program in the NE/Wing of the SMC

We are currently performing an analysis of CCD optical imaging polarimetry data towards several (currently 28) directions spread along the NE and Wing sections of the SMC. Each CCD field is $8' \times 8'$ and the images were taken through the V filter. Fig. 4 shows an example of the results for a field centered at about $[1^{\mathrm{h}}00^{\mathrm{m}}, -72°00'$ (eq. 1975)].

For this field, Fig. 4 shows that there were more than 480 objects with $P/\sigma_P \geqslant 7$. The histogram of position angles shows a single prevalent tendency towards that part of the SMC. The polarization value histogram also shows that many objects with $P \gtrsim 2\%$ have been found. All of this shows that the data have a potential to provide estimates of the magnetic field towards several directions across the SMC, allowing for a study of the magnetic field intensity as a function of the environment. The combination of the polarization data with other types of data (such as visual absorption) should provide further quantities of interest, such as $P_{\mathrm{V}}/A_{\mathrm{V}}$, which measures the alignment efficiency.

The *Spitzer Space Telescope* is providing data on the SMC that will have a direct bearing on such polarization data. The programs S^3MC (Bolatto *et al.* 2007) and SAGE-SMC (Gordon *et al.*, this volume) are very good examples. For instance, the study of how the magnetic field structures correlates with the wealth of structures seen by Spitzer should provide information on the physical processes in the SMC interstellar medium. Also, *Spitzer* data on the dust along the SMC Wing will help interpret the polarization data in that region, where the SMC polarization is expected to be very small, and will provide a check on the accuracy of the Galactic foreground correction. Finally, point source catalogs from *Spitzer* imagery, in association with our optical polarization, should eventually provide detection of significant intrinsic polarization for some sources, indicative of non-spherically symmetric material around such stars.

5. Conclusions

Starlight polarimetry is a good tool for mapping the magnetic field across the SMC, despite the overall low dust content of that galaxy. The general magnetic field structure of the SMC presents a Pan-Magellanic component towards the Bridge and a main body component, roughly in the N-S direction. The estimated field intensities are $\delta B \approx 3.5 \times 10^{-5}$ G for the random component and $B_{sky}^{u} \approx 1.7 \times 10^{-6}$ G for the uniform, sky-projected component. The on-going analysis of optical imaging polarimetry across the SMC NE and Wing sections should provide a much better understanding of the behaviour of the magnetic field throughout the SMC and its relationship to the ISM of the galaxy.

Acknowledgements

Research on polarimetry at the Astronomy Department, IAG-USP, is supported by São Paulo state funding agency FAPESP. AMM is also supported by CNPq.

References

Beck, R. 2008, in F. A. Aharonian, W. Hofmann, & F. M. Rieger (eds.), *High Energy Gamma-Ray Astronomy, AIP Conf. Proc.*, 1085, p. 83

Bolatto, A. D., Simon, J. D., Stanimirović, S., *et al.* 2007, *ApJ*, 655, 212

Davis, L. J. & Greenstein, J. L. 1951, *ApJ*, 114, 206

Falceta-Gonçalves, D., Lazarian, A., & Kowal, G. 2008, *ApJ*, 679, 537

Fosalba, P., Lazarian, A., Prunet, S., & Tauber, J. A. 2002, *ApJ*, 564, 762

Heiles, C. 1996, *ApJ*, 462, 316

Hoang, T. & Lazarian, A. 2008, *MNRAS* 388, 117

Magalhães, A. M., Loiseau, N., Rodrigues, C. V., & Piirola, V. 1990 in *Galactic and intergalactic magnetic fields*, IAU Symposium 140 (Dordrecht, The Netherlands: Kluwer Academic Publishers), p. 255

Magalhães, A. M., Pereyra, A., Melgarejo, R., *et al.* 2005, in A. Adamson, C. Aspin, C. J. Davis, & T. Fujiyoshi (eds), *Astronomical Polarimetry: Current Status and Future Directions*, ASP Conference Series 343, p. 305

Mao, S. A., Gaensler, B. M., Stanimirović, S., Haverkorn, M., McClure-Griffiths, N. M., Staveley-Smith, L., & Dickey, J. M. 2008, *ApJ*, in press, *ApJ*, 688, 1029

Mathewson, D. S. & Ford, V. L. 1970, *ApJ*, 160, L43

McGee, R. X. & Newton, L. M. 1981, *Proceedings of the Astronomical Society of Australia*, 4, 189

McGee, R. X. & Newton, L. M. 1986, *Proceedings of the Astronomical Society of Australia*, 6, 471

Pereyra, A. & Magalhães, A. M. 2004, *ApJ*, 603, 584

Pereyra, A. & Magalhães, A. M. 2007, *ApJ*, 662, 1014

Rodrigues, C. V., Magalhães, A. M., Coyne, G. V., & Piirola, V. 1997, *ApJ*, 485, 618

Schmidt, T. 1976, *A&AS*, 24, 357

Sofue, Y. 1994, *PASJ*, 46, 431

Stanimirović, S., Staveley-Smith, L., & Jones, P. A. 2004, *ApJ*, 604, 176

Westerlund, B. E. 1993, in B. Bacheck (ed.), *New Aspects of Magellanic Cloud Research*, 416, p. 7

Wisniewski, J. P., Bjorkman, K. S., Magalhães, A. M., Bjorkman, J. E., Meade, M. R., & Pereyra, A. 2007a, *ApJ*, 671, 2040

Wisniewski, J. P., Bjorkman, K. S., Magalhães, A. M., & Pereyra, A. 2007b, *ApJ*, 664, 296

Zweibel, E. G. & Heiles, C. 1997, *Nature*, 385, 131

The Magellanic System: Stars, Gas, and Galaxies
Proceedings IAU Symposium No. 256, 2008
Jacco Th. van Loon & Joana M. Oliveira, eds.

Early results from the
SAGE-SMC *Spitzer* legacy

Karl D. Gordon[1]**, M. Meixner, R. D. Blum, W. Reach, B. A.
Whitney, J. Harris, R. Indebetouw, A. D. Bolatto, J.-P. Bernard, M.
Sewilo, B. L. Babler, M. Block, C. Bot, S. Bracker, L. Carlson, E.
Churchwell, G. C. Clayton, M. Cohen, C. W. Engelbracht, Y. Fukui,
V. Gorjian, S. Hony, J. L. Hora, F. Israel, A. Kawamura, A. K. Leroy,
A. Li, S. Madden, A. R. Marble, F. Markwick-Kemper, M. Meade,
K. A. Misselt, A. Mizuno, N. Mizuno, E. Muller, J. M. Oliveira,
K. Olsen, T. Onishi, R. Paladini, D. Paradis, S. Points, T. Robitaille,
D. Rubin, K. M. Sandstrom, S. Sato, H. Shibai, J. D. Simon, L. J.
Smith, S. Srinivasan, A. G. G. M. Tielens, U. P. Vijh, S. van Dyk,
J. Th. van Loon, K. Volk and D. Zaritsky**

[1]Space Telescope Science Institute, 3700 San Martin Drive, Baltimore, MD 21218, USA
email: `kgordon@stsci.edu`

Abstract. Early results from the SAGE-SMC (Surveying the Agents of Galaxy Evolution in the tidally-disrupted, low-metallicity Small Magellanic Cloud) *Spitzer* legacy program are presented. These early results concentrate on the SAGE-SMC MIPS observations of the SMC Tail region. This region is the high H I column density portion of the Magellanic Bridge adjacent to the SMC Wing. We detect infrared dust emission and measure the gas-to-dust ratio in the SMC Tail and find it similar to that of the SMC Body. In addition, we find two embedded cluster regions that are resolved into multiple sources at all MIPS wavelengths.

Keywords. galaxies: individual (SMC), galaxies: ISM, Magellanic Clouds, infrared: ISM

1. Introduction

SAGE-SMC is a *Spitzer* legacy program (cycle 4, 285 hours) to map the entire SMC (Bar, Wing, and Tail) with IRAC and MIPS. The SAGE-SMC observations cover ~ 30 deg^2, greatly expanding on the S^{3}MC pathfinder survey (Bolatto *et al.* 2007) which covered the inner ~ 3 deg^2 of the SMC. The main SAGE-SMC goal is to study the evolution of a single galaxy in detail. As the SMC is close ($d \sim 60$ kpc), we can investigate the cycle of star formation and dust by studying the injection of material into the interstellar medium (ISM) from evolved stars, the contents of the present day ISM, and how the ISM is consumed in regions of star formation. The SMC is a unique target for such studies as it is nearby, low metallicity (1/5 Z$_\odot$), and tidally disrupted. The comparison of the SAGE-SMC observations with similar observations of the LMC (SAGE-LMC, Meixner *et al.* 2006) and the Milky Way (e.g., GLIMPSE & MIPSGAL) will provide a solid understanding of galaxy evolution over a wide range of metallicities and star formation histories.

2. Status of observations

The SAGE-SMC observations are taken at 2 epochs with instrumental field-of-views rotated by $\sim 90°$ to help suppress residual instrumental signatures in both MIPS and

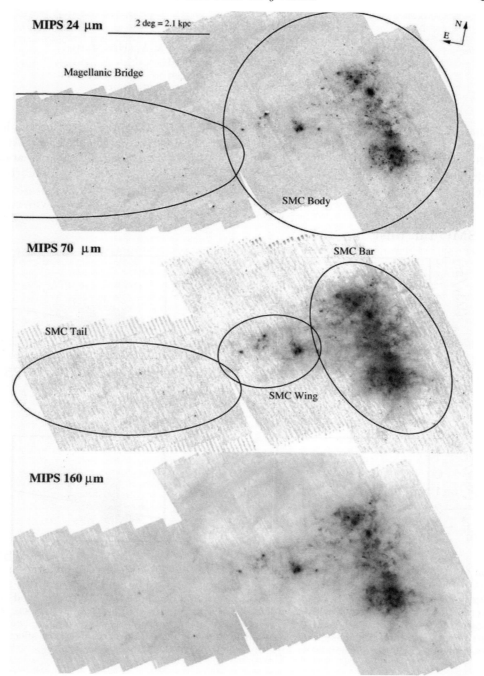

Figure 1. The MIPS 24, 70, and 160 μm mosaics of the SMC are shown.

IRAC and allow for studies of variable sources. The observation dates and status of each epoch of observations are given in Table 2. The MIPS 1st epoch observations are the subject of this paper and are displayed in Fig. 1. The S[3]MC MIPS 70 & 160 μm data have been added to the SAGE-SMC observations to suppress the residual instrumental

Name	Observation Date	Status
MIPS 1st epoch	Sep 2007	Reduced & Analysis started
MIPS 2nd epoch	Jun 2008	Reduced
IRAC 1st epoch	Jun 2008	Reduced
IRAC 2nd epoch	Sep 2008	Reductions Started

signatures in the region of overlap (most of the SMC body and wing). Note that residual baseline drifts in the S^{3}MC 70 μm data were subtracted by comparison with the SAGE-SMC 70 μm data before mosaicking. The epoch 1 MIPS 24 μm point source catalog includes 13,974 high reliability ($>5\sigma$) sources. The similar catalog for the LMC (from SAGE-LMC) includes 39,019 sources.

3. SMC tail dust

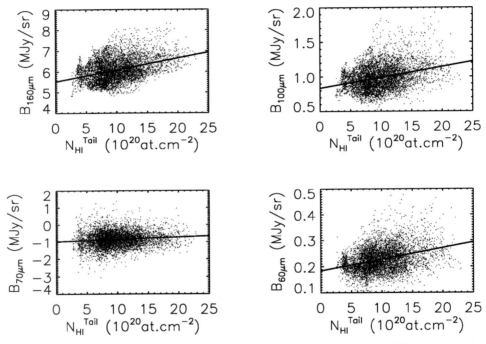

Figure 2. The correlations between the SMC H I column densities and MW foreground subtracted IRAS 25, IRAS 60, MIPS 70, and MIPS 160 μm surface brightnesses are shown. The negative surface brightnesses at MIPS 70 μm are a result of the subtraction of the time varying instrumental baseline that has clearly resulted in an oversubtraction. This does not affect our measurement of the correlation as we are only interested in the slope.

The Magellanic Bridge connects the SMC and LMC with a bridge of H I. Directly adjacent to the SMC, there is a high H I density portion which may be related to the SMC. This region likely represents the closest example of tidally stripped material with recent star formation and no old stars (Harris 2007). One of the goals of the SAGE-SMC observations is to measure the dust content of this region, which we are calling the SMC Tail. The MW cirrus foreground was removed using predictions based on MW H I foreground measurements. The residual IR emission was correlated with the SMC

H I and shown in Fig. 2. Preliminary calculations give an atomic gas-to-dust ratio of ~ 1000. This is similar to the SMC Body gas-to-dust ratio (Bot *et al.* 2004) and indicates that the SMC Tail material has been recently stripped from the SMC Body. This agrees with the measured metallicities of stars in the SMC Tail (Lee et al. 2005) and numerical simulations (Connors et al. 2006).

4. SMC tail young, embedded clusters

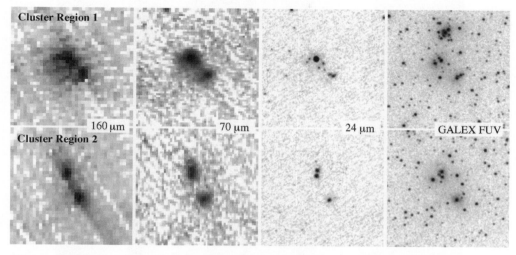

Figure 3. The two cluster regions with resolved sources at all MIPS wavelengths are shown along with the same region observed in the *GALEX* far-UV band.

The SAGE-SMC observations show a number of point sources in the SMC Tail detected at 24, 70, & 160 μm. Two are resolved into multiple sources in all MIPS bands and are associated with young, UV bright clusters of stars. These clusters are prime examples of tidally triggered star formation regions which are still embedded in the natal clouds as seen from the H I images. They provide localized measurements of the atomic gas-to-dust ratio of ~ 200. When combined with existing CO observations, the total gas-to-dust ratios are ~ 250 to 450. The 2nd epoch of MIPS observations will enhance the detections, especially at 160 μm.

References

Bolatto, A. D., Simon, J. D., Stanimirović, S., *et al.* 2007, *ApJ*, 655, 212
Bot, C., Boulanger, F., Lagache, G., Cambrésy, L., & Egret, D. 2004, *A&A*, 423, 567
Connors, T. W., Kawata, D., & Gibson, B. K. 2006, *MNRAS*, 371, 108
Harris, J. 2007, *ApJ*, 658, 345
Lee, J.-K., Rolleston, W. R. J., Dufton, P. L., & Ryans, R. S. I. 2005, *A&A*, 429, 1025
Meixner, M., Gordon, K. D., Indebetouw, R., *et al.* 2006, *AJ*, 132, 2268

Rob Jeffries, live on stage. Before the drama of Othello, thunder, lighting and rain.

Session IV

The star formation process

The Magellanic System: Stars, Gas, and Galaxies
Proceedings IAU Symposium No. 256, 2008
Jacco Th. van Loon & Joana M. Oliveira, eds.

© 2009 International Astronomical Union
doi:10.1017/S1743921308028457

The star formation process in the Magellanic Clouds

J. M. Oliveira

Lennard-Jones Laboratories, School of Physical & Geographical Sciences, Keele University,
Staffordshire ST5 5BG, UK
email: joana@astro.keele.ac.uk

Abstract. The Magellanic Clouds offer unique opportunities to study star formation both on the global scales of an interacting system of gas-rich galaxies, as well as on the scales of individual star-forming clouds. The interstellar media of the Small and Large Magellanic Clouds and their connecting bridge, span a range in (low) metallicities and gas density. This allows us to study star formation near the critical density and gain an understanding of how tidal dwarfs might form; the low metallicity of the SMC in particular is typical of galaxies during the early phases of their assembly, and studies of star formation in the SMC provide a stepping stone to understand star formation at high redshift where these processes can not be directly observed. In this review, I introduce the different environments encountered in the Magellanic System and compare these with the Schmidt-Kennicutt law and the predicted efficiencies of various chemophysical processes. I then concentrate on three aspects that are of particular importance: the chemistry of the embedded stages of star formation, the Initial Mass Function, and feedback effects from massive stars and its ability to trigger further star formation.

Keywords. astrochemistry, stars: formation, stars: luminosity function, mass function, ISM: clouds, H II regions, ISM: molecules, galaxies: evolution, Magellanic Clouds, galaxies: stellar content

1. Introduction

The Magellanic Clouds (MCs) are our closest gas-rich galaxy neighbours. New instrumental advances mean we can now study their resolved stellar populations in great detail. At the same time, in the MCs we can study these populations from outside the galaxies themselves, allowing us a unique view of these populations and how they relate to the gas and dust distributions and the galaxies' structure.

The formation of stars in the early Universe took place in a metal-poor environment, however most of what is currently known about the star formation process is derived from observations in the Milky Way. A great advantage of the MCs is that their Interstellar Media (ISM) are characterised by metallicity significantly lower than that of the Milky Way. Thus the MCs are ideal templates to test whether metallicity significantly influences star formation and thus are unique probes for the environmental conditions more typical of the high-redshift Universe.

Only since relatively recently are we able to study the details of the star formation in the MCs. Firstly, I describe the star formation environment in the Magellanic Clouds. I will then concentrate on three particular facets of the star formation process. I will start by discussing the effects of a lower metal content on the chemistry in molecular clouds. I will then discuss the stellar Initial Mass Function (IMF), as well as massive star feedback and triggered star formation at low metallicity.

2. The star formation environment in the MCs

In the Milky Way star formation is clearly dominated by the spiral arms (see review by Elmegreen 2009). In galaxies with weak or no spiral arms, star formation seems to occur throughout their disks, probably resulting from local gravitational instabilities. This is the case of the MCs, that exhibit no clear spiral arms but are still observed to be forming stars at present†. On the other hand, the MCs are interacting gas-rich galaxies that also interact with the Milky Way. Thus tidal and/or hydrodynamical effects might have an important role in stimulating and regulating star formation in these galaxies. The properties and location of the young stellar populations identified in the SMC tail (the part of the Magellanic Bridge closest to the SMC) suggest that these stars formed within a body of gas that had already been pulled out of the SMC, probably by tidal forces (Harris 2007). Whether these tides were also responsible for triggering the star formation is not clear. It has been proposed that the formation of the giant complex 30 Doradus could be associated with the last Magellanic collision, about 0.2 Gyr ago (Bekki & Chiba 2007), which may have induced a gaseous spiral arm distortion lasting more than an orbital period. The alternative explanation of ram-pressure induced star formation as a result of the LMC ploughing through the hot Galactic halo (de Boer *et al.* 1998) has recently gained support again, explaining also some features of the Magellanic Stream (Nidever, Majewski & Burton 2008).

The star formation efficiency is generally observed to be correlated with the local gas density. This Schmidt law (Schmidt 1959) — and its generalization to include starbursts by Kennicutt (1998) — is usually expressed in terms of the projected star formation rate and gas column density. It is observed that the efficiency of star formation is lower when the gas density is higher, and there appears to be a lower threshold to the gas density that can support star formation (Kennicutt 1998). Typical gas densities for distinct components in the Magellanic System may be estimated from the H I data presented in Nidever *et al.* (2008, see their Fig. 9), where the H I column density was transformed into total gas density assuming that the atomic and molecular gas fractions are equal (see Fig. 4 in Kennicutt 1998), a common average distance of 60 kpc, and ignoring projection effects. These are listed in Table 1 and compared with the Schmidt-Kennicutt law in Fig. 1. Note that the molecular fraction of gas in the Bridge/Stream is likely to be smaller.

Table 1. Relevant properties of different regions in the Magellanic System. Typical gas densities are estimated from Nidever *et al.* (2008).

Region	$Z/Z_\odot$	N(H I) atoms cm^{-2}	Σ(gas) M$_\odot$ pc^{-2}	Star formation
LMC ridge	0.4	$> 3 \times 10^{21}$	> 50	yes/soon
LMC disk	0.4	$1-3 \times 10^{21}$	$\sim 20-50$	yes, scattered
SMC body/wing	0.2	$> 3 \times 10^{21}$	> 50	yes, throughout
SMC tail	0.1	$1(-2) \times 10^{21}$	~ 20	yes, in pockets
Bridge/Stream	?	$< 5 \times 10^{20}$	$\ll 8$	no

The MCs offer a range of gas densities, with the SMC tail being a unique example of a star forming entity close to the critical density, while high-density pockets such as 30 Dor and much of the generally very gas-rich SMC main body support vigorous star formation. The Magellanic Bridge and Stream are apparently too tenuous as they show no

† Although the LMC can be considered a one-armed spiral (see Wilcots, these proceedings), star formation in the LMC is not limited to this particular structure.

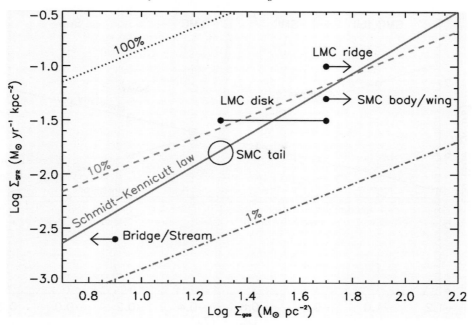

Figure 1. Schmidt-Kennicutt law (Kennicutt 1998) with gas densities for different regions of the Magellanic System indicated (estimated from Nidever *et al.* 2008, see Table 1). Different global star formation efficiencies are also indicated (Kennicutt 1998).

signs of star formation. The transition between quiescent and star-forming ISM within the Magellanic System seems to occur for a gas column density at which Krumholz, McKee & Tumlinson (2008) predict that molecular clouds become self-shielded. It has now become possible to examine the Schmidt-Kennicutt law on pc scales within individual star-forming clouds in the MCs (see Indebetouw, these proceedings), with the aim of identifying deviations from the global law resulting from differences in the physical processes that work on small scales within galaxies.

The Magellanic Clouds have a low metallicity ISM ($0.1-0.4\,Z_\odot$, Table 1). This implies that there is in general less dust in the MCs than in the Milky Way, i.e. the gas-to-dust ratio is higher. As we will see in the next section, this might have important implications for the star formation process. The efficiency of some chemo-physical processes are expected to depend on metallicity. Fig. 2 shows typical timescales for several of these processes that are expected to play an important role in the star formation process and their dependence on metallicity (from Banerji *et al.* 2008). This diagram shows that the relative hierarchy of these processes (i.e. which ones might be more dominant due to shorter timescales) seems to change even in the metallicity range covered by the Magellanic System and the Milky Way. This suggests that metallicity might have an effect on the star formation process, even without having to go back to the conditions prevalent when the very first generations of stars formed.

3. Star formation and molecular cloud chemistry

Why is chemistry important for the early stages of the star formation process? First of all chemistry is essential in determining the cooling in a contracting molecular clump. During the onset of gravitational collapse of a dense cloud, sufficiently dense cores can only develop if the heat produced during the contraction can be effectively dissipated.

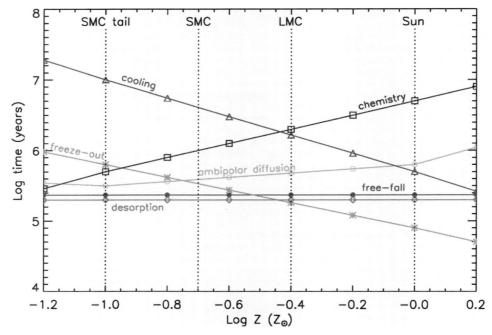

Figure 2. Timescales relevant for the star formation process as a function of metallicity (adapted from Banerji *et al.* 2008). Processes considered are: free-fall collapse (filled circles), freeze-out (crosses; formation of ice coatings on dust grains), ambipolar diffusion (open circles), desorption (diamonds; process by which molecules formed in ice mantles are returned to the gas phase), cooling (triangles) and ion-molecule chemistry (squares; creation of coolant molecules from atomic gas). Some of these processes and their timescales depend on metallicity. In the metallicity range covered by the MCs and the Galaxy, the relative importance of these processes seems to change.

This cooling occurs mainly via fine-structure lines of C or O, for instance via the C$_{\rm II}$ emission line at $158\,\mu$m and also via rotational transitions of abundant molecules like water. At lower metallicity there are obviously fewer C and O atoms and it is expected that this might somehow affect the cooling efficiency (Fig. 2).

Magnetic fields may also play an important role in supporting dense molecular clouds and the cores within them. Magnetic field lines are coupled directly to ions and, via ion-neutral collisions, are also coupled to the neutral material. If this latter coupling is weak, then the neutral material decouples from the magnetic field lines and can then respond more freely to any gravitational perturbation (ambipolar diffusion). With fewer metallic atoms available in low metallicity environments, the ionization state of the gas as a whole could be lower and collapse may proceed closer to free fall. On the other hand, less dust shielding and harsher radiation field (see below) might cause metallic atoms to be ionized to a higher degree. Furthermore, it is not clear how important ambipolar diffusion really is (Crutcher, Hakobian & Troland 2008).

In cold molecular clouds, gas-phase, ice and dust chemistries are strongly inter-linked. Dust grains play an important role by shielding the cold gas in the molecular clouds against ambient UV radiation. Dust grains also provide surfaces onto which chemical reactions occur that would otherwise not be possible. For instance surface chemistry is important in the formation of both H_2 and O_2 molecules. Ice mantles form on the surfaces of dust grains, leading to depletion of molecules from the gas phase.

In the MCs the gas-to-dust ratio is higher. If there is less dust then we can expect that the shielding effect of dust opacity is weaker, compounded by the fact that the ambient UV radiation in MCs is harsher than that in our Galaxy (Welty *et al.* 2006). Chemistry in general might be slower if there is less grain surface available. Extinction curves are different in the diffuse ISM in the MCs (Gordon *et al.* 2003), possibly due to a different grain size distribution. If dust grains were predominantly smaller in molecular clouds in the MCs, they would provide more surface per unit mass, possibly counteracting the effect of a lower total dust mass. In the denser environment of star forming clouds, grains grow and the rate at which this occurs may be different at low metallicity.

Dust composition is possibly also different. It has been suggested that there might be more carbon-rich dust produced by evolved stars in the SMC (Zijlstra *et al.* 2006). It is not clear what effect this would have in dust composition as carbon-rich dust is more easily destroyed (cf. van Loon *et al.* 2008). Nevertheless, if dust composition is different in the MCs, one could expect a different opacity and consequently a different thermal balance. In summary, if the properties of dust grains are intrinsically different in the MCs this could affect chemistry in general and, via cooling, the physics of the star formation process.

3.1. *Ice chemistry*

In cold star forming clouds, ice mantles form on the surface of dust grains. Abundant molecules like water, CO_2 (and to some extent CO) are largely locked into these ice mantles (Bergin *et al.* 1995). The most abundant ice species are water (typically $10^{-5} - 10^{-4}$ with respect to H_2), followed by CO_2 and CO, with combined abundance $10-30\%$ with respect to water ice (van Dishoeck 2004). Ice processing also enriches the gas phase: molecules like methanol, formaldehyde and formic acid are possibly formed via UV and cosmic ray processing of ice mantles before being evaporated into the gas phase. Processing also causes segregation of the different ice species and crystallisation within the ice mantles. Therefore understanding ice chemistry is a powerful route to understand cloud chemistry in general.

Ices have been detected in the envelopes of heavily embedded young stellar objects (YSOs) both in the Milky Way and the MCs. The first evidence of ices in the envelope of a massive YSO in any extra-galactic environment was a serendipitous discovery by van Loon *et al.* (2005). The *Spitzer* IRS and ISAAC/ *VLT* spectrum of IRAS 05328−6827 in the LMC shows clear absorption signatures of water ice at $3.1\,\mu$m, methanol ice at 3.5 and $3.9\,\mu$m and CO_2 ice at $15.2\,\mu$m (Fig. 3). The spectrum also shows a typical silicate dust absorption feature at $10\,\mu$m. Recently Shimonishi *et al.* (2008) identified a few more embedded YSOs with ice signatures in the LMC. van Loon *et al.* (2005) suggests that ice processing observed in IRAS 05328−6827 might be an effect of metallicity, as UV radiation is able to deeper penetrate the less dusty envelopes of metal-poor YSOs. This is a tentative result and a more extensive sample, preferably in the SMC, would be needed to constrain any definitive metallicity trend. Three embedded YSOs have been identified in the SMC (van Loon *et al.* 2008); the ISAAC/ *VLT* spectrum of one of these sources, IRAS 01042−7215, shows clear water ice absorption and hydrogen emission lines, indicating a slightly more evolved object than the LMC example (Fig. 4).

What are the possible effects of low metal content on the chemistry in the envelopes of massive YSOs? Lower carbon and oxygen abundances in the metal-poor ISM of the MCs lead to lower gas-phase CO abundances (Leroy *et al.* 2007). One could therefore also expect a lower CO and CO_2 content of the ice mantles that coat the dust grains. On the other hand a lower dust fraction leads to reduced shielding from an already harder interstellar radiation field. A high degree of ice processing could thus be expected as

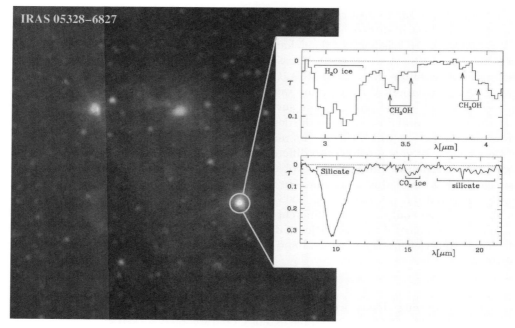

Figure 3. IRAS 05328−6827 in the LMC, the first extra-galactic embedded YSO spectroscopically identified by van Loon *et al.* (2005). The image is a 3-colour IRAC image; the spectra clearly show absorption features due to water and CO$_2$ ices as well as silicate dust.

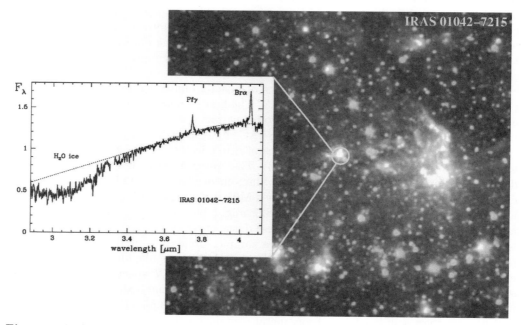

Figure 4. IRAS 01042−7215 in the SMC, a YSO identified in the SMC (van Loon *et al.* 2008). The image is a 3-colour IRAC image; the spectrum clearly show absorption features due to water ice, as well as emission lines, indicating a more evolved object than IRAS 05328−6827 (Fig. 3).

well as a higher degree of crystallinity and ice species segregation. While the first effect could be expected to depress the abundances of all species when compared to Galactic embedded YSOs, the second effect would enhance more complex ice species mainly at the expense of water ice. It is not clear at present which, if any, of these effects dominates. By comparing YSO samples in the SMC, LMC and the Milky Way we can probe metallicities from $0.1 - 1 \, Z_\odot$, allowing us to assess the effect of metallicity on ice chemistry.

Firstly, we will have to construct reliable samples of YSOs in the MCs, selected in a more systematic way, to be compared with Galactic samples. The *Spitzer Space Telescope* imaging surveys of the MCs — SAGE (Meixner *et al.* 2006), S^3MC (Bolatto *et al.* 2007) and SMC-SAGE (Gordon *et al.*, in preparation) — have identified large numbers of YSO candidates (Whitney *et al.* 2008). Follow-up spectroscopic surveys (including SAGE-SPEC and SMC-IRS) target a sizeable YSO sample, that will allow the study of ice features in the mid-IR range. *AKARI* is also being used to investigate ice features on many such objects (Shimonishi *et al.* 2008) in the near-IR, as are groundbased investigations (Oliveira *et al.*, in preparation).

Observations of Galactic YSOs show that the bands of the main ice species are better fitted by laboratory profiles that include admixtures of other ices (Pontoppidan *et al.* 2008). Furthermore, from the previous paragraph it is clear that relative ice abundances, not individual ice species, hold the key to isolate any environmental chemistry effect. Another important point to take into account is that in the MCs we can only expect to study the most massive embedded YSOs, thus one needs to carefully consider the objects' luminosity before comparing the samples. The analysis of the spectral energy distributions of such objects allows to constrain not only the objects' luminosity and evolutionary stage but also dust properties that together with ice chemistry will build a coherent picture of the early star formation process.

One of the major advances since the previous MC Symposium has been the availability of facilities like *Spitzer* which now allow us to study in detail the properties of embedded YSOs outside our Galaxy. We can thus sample star formation environments that are significantly different from those encountered in the Milky Way, with the real possibility of understanding how metallicity influences the early stages of star formation. At the same time, we can now also study global properties of the resulting young stellar populations and investigate for instance how the Initial Mass Function (IMF) reflects their parent environment.

4. From cloud collapse to the stellar IMF

The stellar IMF is determined by a range of inter-linked physical processes: cooling, turbulence, fragmentation, feedback, rotation, magnetic fields etc. The IMF is a rather simplistic snapshot of a very complicated process, and we still struggle to fully understand it and explain its main properties.

At the extreme low metallicities of the early Universe, the IMF is believed to have been top heavy, i.e star formation events gave rise only to extremely massive stars, due mainly to the inability of primordial gas to efficiently cool at low temperatures (Abel, Bryan & Norman 2002). As these populations die and new ones emerge, the Universe gets progressively enriched with copious amount of metals. Once metallicities of the order of 10^{-3} to $10^{-5} \, Z_\odot$ are reached, enough metals are present to allow for more efficient cooling, through fine-structure and molecular transitions, as well as continuum emission from dust produced in supernova explosions. This more efficient cooling allows the gas that will form the next generation of stars to reach lower temperatures, and therefore smaller clumps can be created via fragmentation (e.g., Smith & Sigurdsson 2007). That is to say lower-mass stars can form and the peak of the IMF shifts to lower masses. Thus, once the metal content in the Universe reaches this metallicity threshold stellar populations are characterised by essentially a Salpeter-like IMF with a turn-over or characteristic mass at lower masses, as observed in the solar neighbourhood (Kroupa 2001; Chabrier 2005).

The present day local IMF seems to be universal, but significant variations are observed mainly in more extreme environments, like regions that formed at high redshift (Elmegreen 2008). The IMF can be defined as $\xi(\log M) \propto M^{\Gamma}$, where Γ is the IMF slope. Observationally, the IMF is found to have a slope of $\Gamma \sim -1.35$ (commonly referred to as the "Salpeter slope", Salpeter 1955) down to below a solar mass; it then flattens out somewhat (the so-called IMF plateau) and reaches a peak at about $0.3\,M_{\odot}$, falling sharply into the brown dwarf regime. Theoretical considerations actually struggle to explain the observed IMF constancy in a wide variety of environments (Kroupa 2008). However, this refers mostly to the slope of the IMF at higher masses or to integrated stellar populations; only very recently are we able to probe directly the low-mass IMF in environments that may be distinct to those prevalent in the Milky Way.

In what way could we expect metallicity to influence the lower mass IMF? Based on numerical simulations it has been proposed that the characteristic mass scales with the thermal Jeans mass at the onset of collapse (Bate & Bonnell 2005). One could then intuitively expect the IMF properties to vary with the environmental conditions, via for instance the dependence of the Jeans mass on the temperature and metallicity in the molecular core. However, Elmegreen *et al.* (2008) show that, if grain-gas coupling is taken into account, the thermal Jeans mass should depend only weakly on environmental factors like density, temperature, metallicity and radiation field. In particular, the dependence of the Jeans mass with metallicity could be only as $Z^{-0.3}$. It is also possible that the characteristic mass is not linked to the Jeans mass, depending instead on core sub-fragmentation and protostellar feedback (Elmegreen 2008).

What have we learned so far from constraining the IMF in the MCs? One of the first regions investigated to characterise the high-mass IMF outside our Galaxy was R 136 in the LMC. In the early days, it was thought that R 136 was a single super-massive star (Cassinelli, Mathis & Savage 1981), suggesting an IMF with different properties to those observed in the Galaxy. Modern instrumentation however quickly resolved R 136 into a massive star cluster (Weigelt & Baier 1985). In fact, the most massive stars observed in the MCs have masses of the order of $\sim 150\,M_{\odot}$ (Massey & Hunter 1998), consistent with Galactic massive objects. We should point out that these are observed masses. As discussed in Elmegreen (2008), very massive stars loose a substantial amount of mass extremely quickly. Thus this observed maximum mass is not necessarily the maximum mass as far as the star formation process is concerned.

The IMF of several associations in the MCs is found to have a Salpeter slope down to about $1 - 2\,M_{\odot}$ (Sirianni *et al.* 2000; Selman & Melnick 2005; Kumar, Sagar & Melnick 2008, to name but a few). To study the lower-mass IMF, we need to investigate populations younger than about 10 Myr, as these are less affected by dynamic evolution effects. Young solar-mass stars were firstly identified using near-IR images or Hα emission, for instance in 30 Doradus (Brandl *et al.* 1996) and near SN 1987 A (Panagia *et al.* 2000). More recently a number of *Hubble Space Telescope* (*HST*) surveys have allowed the detection of large numbers of pre-main-sequence (PMS) objects in H II regions in the Clouds, down to below half a solar mass (e.g. Nota *et al.* 2006; Gouliermis *et al.* 2006, 2007).

Two young star forming regions in the SMC have been recently under very close scrutiny (see also contributions by Gouliermis and Sabbi in this volume). NGC 602 in the wing of the SMC shows that star formation can be triggered in low-density environments, probably by compression and turbulence associated with H I shell interactions (Nigra *et al.* 2008). The massive cluster in the center has many PMS stars associated with it (Gouliermis, Quanz & Henning 2007) and embedded ongoing star formation has been detected towards the rim of the H II region (Carlsson *et al.* 2007). NGC 346 on the other hand is the largest OB association in the SMC and it has a more complex gas and dust morphology than NGC 602. A rich PMS has been discovered in this region (Nota *et al.* 2006) and star formation is still ongoing in its denser parts (Simon *et al.* 2007). It has been proposed that several star formation episodes have taken place in NGC 346 (Sabbi *et al.* 2007; Hennekemper *et al.* 2008). For both these young regions the IMF is found to be Salpeter-like down to about a solar mass (Schmalzl *et al.* 2008; Sabbi *et al.* 2008); below that mass incompleteness becomes a serious issue.

Even with *HST* imaging we are unable to fully sample the IMF plateau and determine the characteristic mass of young stellar populations in the MCs. But even when new instrumentation allows us to do so, we might struggle to find conclusive answers. If the characteristic or peak mass depends on metallicity as proposed by Elmegreen *et al.* (2008), in the MCs we could expect it to be about $0.4-0.6\,M_\odot$, where in the Galaxy it typically is $0.3\,M_\odot$. Furthermore, even in the Milky Way we observe variations in the peak mass that seem to be significant (Luhman *et al.* 2007; Oliveira *et al.* 2009) however we are as yet unable to explain these differences. Another possible complication is that the peak mass also depends (albeit weakly) on the density and radiation field, and in the MCs we do sample lower densities and harsher radiation fields. Therefore a metallicity effect on the IMF properties might prove difficult to isolate.

5. Feedback and triggering

Another factor we have to take into account is that star formation has a cyclic component to it: feedback from massive stars can quench/prolong a star formation episode and maybe even trigger new such events. If feedback effects are different at low metallicity, can this affect the outcome of a star formation event?

Infant massive stars start to shape their environments very early in their evolution, by ionizing the gas and heating the dust (Oey & Clarke 2007). Stellar UV radiation and winds create expanding blisters of ionized gas. At the interfaces between these H II regions and the cold molecular cloud, the heating powers shocks that compress the molecular material. Supernova explosions also disrupt severely the neighbouring ISM. In the first instance, massive stars violently destroy their parent molecular clouds. However, observational evidence indicates that at the rims of giant H II regions star formation is ongoing, NGC 6611 and the Eagle Nebula in the Milky Way being one such example (Oliveira 2008). This seems to suggest that under the right conditions, radiative and/or mechanical feedback of massive stars can trigger further star formation in the remnant molecular gas (see review by Chu in this volume).

The observed properties of massive stars in the Magellanic Clouds show some important differences when compared to their galactic counterparts (see review by Evans in this volume). Massive stellar winds are slower and less dense in the low metallicity environment of the MCs (Mokiem *et al.* 2007). For the same stellar mass, massive stars in the Clouds are also hotter than their Galactic counterparts (Massey *et al.* 2005), which could compensate slightly the metallicity effect on the wind as hotter stars drive faster outflows. The lower dust-to-gas ratio in the MCs implies that dust shielding is less efficient and radiative feedback from massive stars may thus be more effective at lower metallicity. All this seems to suggest that massive star feedback might operate differently at low metallicity. But do we know how this affects the triggering of star formation?

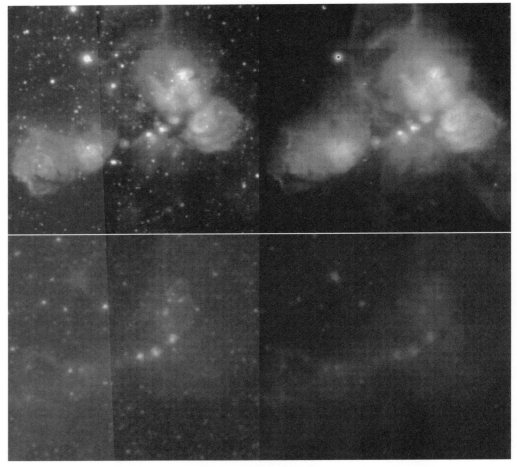

Figure 5. Multi-wavelength images of N 113 in the LMC (Oliveira *et al.* 2006). The images at the top are the Hα+continuum (left) and Hα emission (right), showing the complex structure of the hot gas. The image on the bottom left shows an IRAC/*Spitzer* 3-colour composite, showing the location of the dense dust lane in the center of the region. The image on the bottom right shows a composite of Hα+continuum with IRAC/*Spitzer* 8 μm dust emission. It shows to dramatic effect how the bubbles of ionized gas created by the massive stars are compressing the dense lanes of dust. Star formation is ongoing in the compressed dust, as shown by the presence of maser emission (Oliveira *et al.* 2006).

N 113 is a small H II region in the LMC. Multiple young stellar populations are associated with this region (Bica, Claria & Dottori 1992) and star formation is ongoing within radio continuum sources in the central part of the nebula (Wong *et al.* 2006), as pinpointed by the detection of water and OH maser emission (Lazendic *et al.* 2002; Brooks & Whiteoak 1997). Fig. 5 shows in detail the interplay between the ionized gas and the molecular material in N 113. Oliveira *et al.* (2006) use Hα emission and continuum images together with IRAC/*Spitzer* images to show that the ionized gas bubbles created by the massive stars in the region are compressing the dense lanes of molecular material where star formation is ongoing. Further persuasive evidence for triggered star formation in the MCs is found at the rims and interfaces of LMC supergiant shells where ongoing star formation is occurring (see contribution by Chu, this volume). This paints the picture that star formation triggered by massive stars and their interaction with the ISM also occurs in the Clouds.

One problem with analysing the processes that can trigger star formation is to establish a cause-effect relationship: it is very difficult to distinguish between sequential star formation that occurs without the need for extra triggers and star formation that occurs only due to the direct action of the massive stars. One of the few examples where triggering may have been proved is the superbubble N 51D in the LMC. By measuring the thermal pressure in a dust globule and comparing it to ambient conditions, Chu *et al.* (2005) find that star formation in the globule may have been induced recently by the thermal pressure in the superbubble interior.

As we have seen in Section 2, in the MCs there are other mechanisms that could trigger star formation: ram pressure and/or tidal effects resulting from interactions within the Magellanic System and with the Milky Way. Even though we know that massive stars have different observed properties at low metallicity, it is not clear whether this significantly influences star formation triggering.

6. Final remarks

I reviewed what we know about the star formation environment in the Magellanic Clouds, in particular concerning gas density and metallicity. I described how particular components of the star formation process, namely, the chemistry of the early embedded stages, stellar IMF and stellar feedback and triggering, might operate differently in environments with low metal abundance as in the Magellanic Clouds. The Magellanic System offers an ideal laboratory to study both the details and global properties of star formation at low metallicity. This kind of studies has only recently become possible; with so much new data and exciting facilities coming on, we have very much to learn!

References

Abel, T., Bryan, G. L., & Norman, M. L. 2002, *Sci*, 295, 93
Banerji, M., Viti, S., Williams, D. A., & Rawlings, J. M. C. 2008, *ApJ*, in press, arXiv0810.3662
Bate, M. R. & Bonnell, I. A. 2005, *MNRAS*, 356, 1201
Bekki, K. & Chiba M. 2007, *PASJ*, 24, 21
Bergin, E. A., Langer, W. D., & Goldsmith, P. F. 1995, *ApJ*, 441, 222
Bica, E. L., Claria, J. J., & Dottori, H. 2002, *AJ*, 103, 1859
Bolatto, A. D., Simon, J. D., Stanimirović, S., *et al.* 2007, *ApJ*, 655, 212
Brandl, B., Sams, B. J., Bertoldi, F., *et al.* 1996, *ApJ*, 466, 254
Brooks, K. & Whiteoak, J. B. 1997, *MNRAS*, 291, 395
Carlson, L. R., Sabbi, E., Sirianni, M., *et al.* 2007, *ApJ*, 665, 109
Cassinelli, J. P., Mathis, J. S., & Savage, B. D. 1981, *Sci*, 212, 1497
Chabrier G. 2005, in E. Corbelli, F. Palla, & H. Zinnecker (eds.), *The Initial Mass Function 50 years later*, ApSS Library 327 (Dordrecht: Springer), p. 41
Chu, Y.-H., Gruendl, R. A., Chen, C.-H. R., *et al.* 2005, *ApJ*, 634, 189
Crutcher, R. M., Hakobian, N., & Troland, T. H. 2008, in press, arXiv0807.2862
de Boer, K. S., Braun, J. M., Vallenari, A., & Mebold, U. 1998, *A&A*, 329, L49
Elmegreen, B. G. 2009, in *The Galaxy Disk in Cosmological Context*, IAU Symposium 254, arXiv0810.5406
Elmegreen, B. G. 2008, in *The Evolving ISM in the Milky Way and Nearby Galaxies: Recycling in the Nearby Universe*, 4th Spitzer Science Center Conference, arXiv:0803.3154
Elmegreen, B. G., Klessen, R. S., & Wilson, C. D. 2008, *ApJ*, 681, 365
Gordon, K. D., Clayton, G. C., Misselt, K. A., Landolt, A. U., & Wolff, M. J. 2003, *ApJ*, 594, 279
Gouliermis, D. A., Brandner, W., & Henning, T. 2006, *ApJ*, 636, 33

Gouliermis, D. A., Henning, T., Brandner, W., Dolphin, A. E., Rosa, M., & Brandl, B. 2007, *ApJ*, 665, 27

Gouliermis, D. A., Quanz, S. P., & Henning, T. 2007, *ApJ*, 665, 306

Harris, J. 2007, *ApJ*, 658, 345

Hennekemper, E., Gouliermis, D. A., Henning, T., Brandner, W., & Dolphin, A. E. 2008, *ApJ*, 672, 914

Kennicutt, R. C., Jr. 1998, *ApJ*, 498, 541

Kroupa, P. 2001, *MNRAS*, 322, 231

Kroupa, P. 2008, in G. Israelian & G. Meynet (eds.), *The Metal Rich Universe* (Cambridge, UK: Cambridge University Press), p. 227

Krumholz, M. R., McKee, C. F., & Tumlinson, J. 2008, *ApJ* in press, arXiv:0811.0004

Kumar, B., Sagar, R., & Melnick, J. 2008, *MNRAS*, 386, 1380

Lazendic, J. S., Whiteoak, J. B., Flamer, I., Harbinson, P. D., & Kuiper, T. B. H. 2002, *MNRAS*, 331, 969

Leroy, A., Bolatto, A. D., Stanimirović, S., Mizuno, N., Israel, F., & Bot, C. 2007, *ApJ*, 658, 1027

Luhman, K. L., Joergens, V., Lada, C., Muzerolle, J., Pascucci, I., & White, R., 2007, in B. Reipurth, D. Jewitt, and K. Keil (eds.), *Protostars and Planets V*, (Tucson: University of Arizona Press), p. 443

Massey, P. & Hunter, D. A. 1998, *ApJ*, 493, 180

Massey, P., Puls, J., Pauldrach, A. W. A., Bresolin, F., Kudritzki, R. P., & Simon, T. 2005, *ApJ*, 627, 477

Meixner, M. Gordon, K. D., Indebetouw, R. *et al.* 2006, *AJ*, 132, 2268

Mokiem, M. R., de Koter, A., Vink, J. S., *et al.* 2007, *A&A*, 473, 603

Nidever, D. L., Majewski, S. R., & Burton W. B. 2008, *ApJ*, 679, 432

Nigra, L., Gallagher, J. S., III, Smith, L. J., Stanimirović, S., Nota, A., & Sabbi, E. 2008, *PASP*, 120, 972

Nota, A., Sirianni, M., Sabbi, E., *et al.* 2006, *ApJ*, 640, 29

Oey, M. S. & Clarke, C. J. 2007, in M. Livio & E. Villaver (eds.), *Massive Stars: From Pop III and GRBs to the Milky Way*, (Cambridge, UK: Cambridge University Press), astro-ph/0703036

Oliveira, J. M., van Loon, J.Th., Stanimirović, S., & Zijlstra, A.A. 2006, *MNRAS*, 372, 1509

Oliveira, J. M., Jeffries R. D., & van Loon J. Th. 2008, 2009, *MNRAS*, 392, 1034

Oliveira, J. M. 2008, in B. Reipurth (ed.), *The Handbook of Star Forming regions*, in press, arXiv:0809.3735

Panagia, N., Romaniello, M., Scuderi, S., & Kirshner, R. P. 2000, *ApJ*, 539, 197

Pontoppidan, K. M., Boogert, A. C. A., Fraser, H. J., *et al.* 2008, *ApJ*, 678, 1005

Sabbi, E., Sirianni, M., Nota, A., *et al.* 2008, *AJ*, 135, 173

Sabbi, E., Sirianni, M., Nota, A., *et al.* 2007, *AJ*, 133, 44

Salpeter, E. E. 1955, *ApJ*, 121, 161

Schmalzl, M., Gouliermis, D. A., Dolphin, A. E., & Henning, T. 2008, *ApJ*, 68, 290

Schmidt, M. 1959, *ApJ*, 129, 243

Selman, F. J. & Melnick, J. 2005, *A&A*, 443, 851

Shimonishi, T., Onaka, T., & Kato, D., *et al.* 2008, *ApJ*, 686, L99

Simon, J. D., Bolatto, A, D., & Whitney, B. A., *et al.*, 2007, *ApJ*, 669, 327

Sirianni, M., Nota, A., Leitherer, C., De Marchi, G., & Clampin, M. 2000, *ApJ*, 533, 203

Smith, B. D. & Sigurdsson, S. 2007, *ApJ*, 661, L5

van Dishoeck, E. 2004, *ARAA*, 42, 119

van Loon, J. Th., Oliveira, J. M., Wood , P. R., *et al.* 2005, *MNRAS*, 364, L71

van Loon, J. Th., Cohen, M., Oliveira, J. M., *et al.* 2008, *A&A*, 487, 1055

Weigelt, G. & Baier, G. 1985, *A&A*, 150, 18

Welty, D. E., Federman, S. R., Gredel, R., Thorburn, J. A., & Lambert, D. L. 2006, *ApJS*, 165, 138

Whitney, B. A., Sewilo, M., Indebetouw, R., *et al.* 2008, *AJ*, 136, 18

Wong, T., Whiteoak, J. B., Ott, J., Chin, Y.-N., & Cunningham, M. R. 2006, *ApJ*, 649, 224

Zijlstra, A. A., Matsuura M., Wood P. R., *et al.* 2006, *MNRAS*, 370, 1961

The Magellanic System: Stars, Gas, and Galaxies
Proceedings IAU Symposium No. 256, 2008
Jacco Th. van Loon & Joana M. Oliveira, eds.

© 2009 International Astronomical Union
doi:10.1017/S1743921308028469

The properties of molecular clouds across the Magellanic System

Norikazu Mizuno

Department of Astrophysics, Nagoya University, Furocho Chikusa-ku, Nagoya 464-8602, Japan
email: norikazu@a.phys.nagoya-u.ac.jp

Abstract. Most stars form in Giant Molecular Clouds (GMCs) and regulate the evolution of galaxies in various respects. The formed stars affect the surrounding materials strongly via their UV photons, stellar winds, and supernova explosions, which lead to trigger the formation of next-generations of stars in the GMCs. It is therefore crucial to reveal the distribution and properties of GMCs in a galaxy. The Magellanic System is a unique target to make such detailed comprehensive study of GMCs. This is because it is nearby and the LMC is nearly face-on, making it feasible to unambiguously identify associated young objects within GMCs. Recent millimeter and sub-millimeter observations in the Magellanic System have started to reveal the distribution and properties of the individual GMCs in detail and their relation to star formation activities. From the *NANTEN* CO surveys, three types of GMCs can be classified in terms of star formation activities; Type I is starless, Type II is with H II regions only, and Type III is associated with active star formation indicated by huge H II regions and young star clusters. The further observations to obtain detailed structure of the GMCs by *Mopra* and *SEST* and to search for the dense cores by *ASTE* and *NANTEN2* in higher tansition lines of CO have been carried out with an angular resolution of about 5 to 10 pc. These observations revealed that the differences of the physical properties represent an evolutionary sequence of GMCs in terms of density increase leading to star formation. Type I and II GMCs are at the early phase of star formation where density does not yet become high enough to show active star formation, and Type III GMCs represent the later phase where the average density is increased and the GMCs are forming massive stars.

Keywords. stars: formation, ISM: clouds, Magellanic Clouds, radio lines: ISM, submillimeter

1. Introduction

The Magellanic system including the LMC, SMC, and Bridge is a valuable astrophysical laboratory to study star formation and cloud evolution because of its proximity to the Sun. In particular, the LMC offers the best site because of its unrivaled closeness and of the nearly face-on view to us. Star formation requires a cool and high-density interstellar medium (ISM) in which most of the hydrogen is in molecular form; therefore, studies of distribution and properties of molecular clouds are very important for understanding the star formation process. Emission from the tracer CO molecule has been widely used to estimate the distribution and amount of H_2 in the Galaxy and other galaxies.

The SMC and Bridge are also unique targets for understanding the processes of star formation in a galaxy that is extremely different from the Milky Way, its low metallicity of $1/5^{th}$–$1/10^{th}$ solar, and its low dust-to-gas ratio. Its history of interactions with the Milky Way and LMC, mean that star formation may be predominantly driven by a combination of tidally-induced galaxy-galaxy interactions and shell formation. These conditions approximate those encountered in young, star-forming galaxies at high redshift. In such a low-metal environment, CO may not trace the whole amount of H_2 in the GMC due to deeper CO photodissociation (see contributions by C. Bot; A. Leroy, this volume), but stars form within dense parts of the GMCs which can be traced by CO. For a better understanding of the young massive cluster (*populous cluster*) formation, which is yet poorly understood, we need to reveal the properties of CO clouds and their relation to star formation.

In this contribution, the global distributions and properties of GMCs at 40 pc scale by the *NANTEN* CO survey are presented in sections 2 and 3, and the comparisons with star/clustr formation and H I gas are discussed in sections 4 and 5. The new results of submillimeter observations with *ASTE* and *NANTEN2* are presented in section 6.

2. Giant molecular clouds (GMCs) in the Magellanic System

The LMC. Fig. 1 shows the molecular clouds detected with the 2nd *NANTEN* CO Survey (Fukui *et al.* 2008) on an optical image of the LMC. The mass of the molecular gas in total is $\sim 5 \times 10^6$ M$_\odot$, if we use a CO luminosity to hydrogen column density conversion factor, the X_{CO}-factor, of $\sim 7 \times 10^{20}$ cm^{-2} (K km s^{-1})$^{-1}$ (see section 3). The CO distribution of the LMC is found to be clumpy with several large molecular cloud complexes, unlike the H I gas distribution (Kim *et al.* 1998), which is composed of many filamentary and shell-like structures. The cloud complex south of 30 Doradus is remarkable, stretching in a nearly straight line from north to south, as already noted in the previous CO observations (Cohen *et al.* 1988; Kutner *et al.* 1997; Fukui *et al.* 1999; Mizuno *et al.* 2001a). The current survey shows that the clouds in this molecular cloud complex, "the molecular ridge" are actually connected to one another by low-density molecular gas, and the recent high-resolution CO survey with *Mopra* (MAGMA survey; Ott *et al.* 2008) have confirmed this feature. The arclike distribution of molecular clouds along the south-eastern optical edge of the galaxy (the "CO Arc" in Fukui *et al.* 1999) is also clearly seen. The current sensitive survey confirms that this CO Arc indeed clearly represents an arclike edge of the molecular gas distribution in this eastern boundary of the LMC. Some have speculated that this feature is due to hydrodynamical collision between the LMC and SMC (Fujimoto & Noguchi 1990) or ram pressure pileup of gas due to the motion of the LMC through a halo of hot, diffuse gas (de Boer *et al.* 1998; Kim *et al.* 1998). Supershells may also be playing a role in the formation of GMCs as in the case of LMC 4 (Yamaguchi *et al.* 2001a). A comprehensive comparison between supergiant shells and GMCs shows that only about 1/3rd of the GMCs are located towards supershells, suggesting the effects of supershells are not predominant (Yamaguchi *et al.* 2001b). There is neither an excess nor a deficit of CO associated with the stellar bar, but the bright H II regions are all clearly associated with molecular clouds.

The SMC and Bridge. Fig. 2 shows the GMCs superimposed on a grayscale image made using the 3.6, 4.5, and 8.0 μm bands from the IRAC instrument on the *Spitzer Space Telescope* (Bolatto *et al.* 2007). The CO map is from *NANTEN* (Mizuno *et al.* 2001b). Unlike the LMC, they are not spread throughout the galaxy but appear preferentially on the northern and southern ends of the galaxy. Another grouping is located to the east (left) of the SMC along the H I bridge that connects the LMC and SMC. The dust emission is associated with the molecular gas traced by CO, but appears to be more extended than the CO emission (Rubio *et al.* 2004; Leroy *et al.* 2007; Bot *et al.* 2007). A lower dust content would have an impact on the physical and chemical properties of the gas component, e.g., low shielding enhancing molecular dissociation rates. This has a direct consequence for the chemical state but also for the physical properties of the gas, less extended emission and a higher degree of clumping would be expected.

The first detection of a molecular cloud in the Magellanic Bridge was reported by Muller *et al.* (2003) based on CO observations towards a IRAS 100 μm emission source

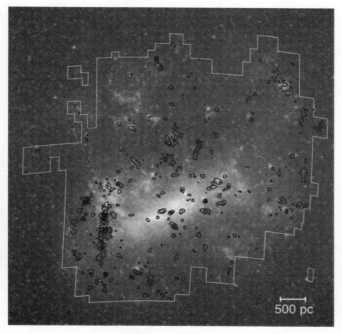

Figure 1. Optical image of the LMC with GMCs mapped with the *NANTEN* telescope (~40 pc resolution) indicated within the boundary of the survey area.

detected in the Bridge as had been the case for SMC CO detections earlier (Rubio *et al.* 1993). Mizuno *et al.* (2006), reported the detection in CO(1–0) emission of seven new and independent sites in the Bridge out of 16 sites investigated which had correlated bright H I emission and 100 μm emission. The molecular clouds detected have a weak CO emission (a factor of 10 weaker than that of the SMC), as expected in low mettalicity systems. Mizuno *et al.* have estimated the molecular mass of the clouds and these range between 1–7 $\times 10^3$ M$_\odot$. These molecular clouds lie at a distance of a few tens of parsec of OB associations and Hα emission nebulae. Such evidences support that the molecular clouds in the Bridge could be the sites of new star formation in the Magellanic Bridge. Some peaks of the molecular clouds contain embedded clusters, indicating that star formtion is still going on (see posters by Rubio *et al.* and Gordon *et al.* in this volume). Its distance of a few kpc to the SMC wing implies that these have to have formed in situ and could be a prototype of star formation induced in a galaxy-galaxy interaction.

3. The properties of GMCs

To study the systematic comparison of GMC properties, individual clouds were identified by the new cloud-finding algorithm, FITSTOPROPS (Rosolowsky & Leroy 2006). Rosolowsky and Leroy developed a method for minimizing the biases that plague such comparisons. For example, measurement of the cloud radius depends on the sensitivity of the measurements, and they suggest a robust method to extrapolate to the expected radius in the limit of infinite sensitivity. All data (LMC, SMC, the Galaxy) were re-analyzed using the same methods to minimize systematics and cloud properties of GMCs, such as size (radius), line width, CO luminosity, virial mass were derived (*for the details*, see Fukui *et al.* 2008). GMC properties measured across the Magellanic Clouds are very much

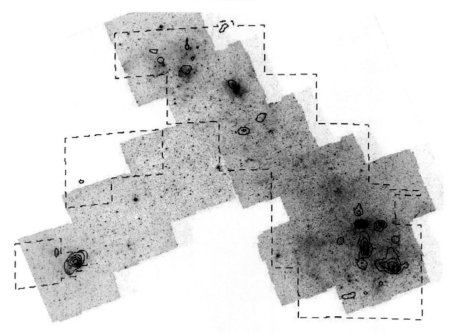

Figure 2. GMCs in the SMC observed with the *NANTEN* telescope overlayed on an a near-infrared image of the galaxy from the *Spitzer Space Telescope* (Bolatto *et al.* 2007). The lines indicate the survey boundary. The CO clouds are clearly associated with regions of transiently heated small grains or PAHs that appear as dark, nebulous regions in the image.

compatible with those in the Galaxy and nearby galaxies (Blitz *et al.* 2007; Bolatto *et al.* 2008).

The mass spectrum (Fig. 3, *left*) of the GMCs in the LMC with $5 \times 10^4 \leqslant M_{\rm CO}/M_\odot \leqslant 10^7$ is well fitted by a power law, $N_{\rm CO}(> M_{\rm CO}) \propto (M_{\rm CO}/M_\odot)^{-0.75\pm0.06}$. This slope is consistent with the previous results obtained from the Galaxy and nearby galaxies (Fukui *et al.* 2008; Blitz *et al.* 2007). The slope of the mass spectrum becomes steeper if we fit only the massive clouds, e.g., $N_{\rm CO}(> M_{\rm CO}) \propto (M_{\rm CO}/M_\odot)^{-1.2\pm0.2}$ for $M_{\rm CO} \geqslant 3 \times 10^5$ $M_\odot$, which suggests mass truncation. This may suggest that the disruption of the molecular clouds is faster in the massive clouds (Fukui *et al.* 2001, 2008). It may also suggest that cloud formation takes place inhomogeneously; the mass spectra in different regions of the galaxy may have different slopes and the truncation of the slope might appear when we sum up all the mass spectra within the galaxy. The reason of the truncation is not yet known but the current results present new information leading to a better knowledge of the cloud formation and disruption. A least-squares fit to the virial mass versus CO luminosity relation in the LMC shows a power law, $\log(M_{\rm vir}/M_\odot) = 26 \log[L_{\rm CO}/({\rm K\ km\ s^{-1}\ pc^2})]^{1.1\pm0.3}$, with Spearman rank correlation of 0.8 (Fig. 3, *right*). This relation suggests that clouds are virialized and the CO luminosity can be a good tracer of mass in the LMC with a quite-constant conversion factor from $L_{\rm CO}$ to mass throughout the mass range $10^4 < (M_{\rm vir}/M_\odot) < 10^7$. The average value of $\log(M_{\rm vir}/L_{\rm CO})$ is 1.2 ± 0.3, corresponding to an $X_{\rm CO}$-factor of $\sim(7 \pm 2) \times 10^{20}$ cm^{-2} (K km s^{-1})$^{-1}$. The CO clouds in the SMC and the Warp region lie along the best-fitting power law of GMCs in the LMC.

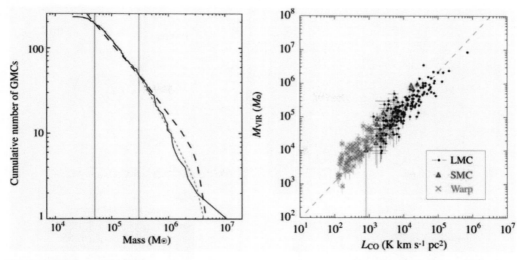

Figure 3. *Left*: Cumulative mass spectra of the M_{CO} of the 230 GMCs in the LMC, *Right*: Virial mass, M_{vir}, of the GMCs as a function of luminosity, L_{CO}. The line presents a best fit to the LMC data (filled circles) with a slope of 1.1 ± 0.1. Triangles show the clouds in the SMC (Mizuno *et al.* 2001b) and crosses those in the Galactic Warp region (Nakagawa *et al.* 2005).

4. Star/cluster formation in the GMCs

The overall distribution of the Hα local peaks and GMCs in the Magellanic Clouds are well coincident to each other, while the extent of the individual Hα emitting regions and the size of the GMCs are different. Almost all the luminous H II regions with high flux densities, like 30 Dor, N 159, N 11 in the LMC and the N 83/84 complex, N 12A, N 24 in the SMC are associated with GMCs, while fainter and diffuse extended Hα emission shows a lower degree of association. Some luminous H II regions, like N 66 in the SMC, are associated with only small molecular clouds, which may indicate molecular gas dissipation after formation of massive H II regions. It is also noted that there are several GMCs not associated with Hα emission (Fig. 4). Such *"starless GMCs"* (starless molecular clouds in the sense that they are not associated with H II regions or young clusters) are apparently very rare in the Solar vicinity (e.g., Maddalena's cloud), where most of the Galactic GMCs are O-star-forming clouds (Fukui *et al.* 1999; Yamaguchi *et al.* 2001b). From the comparisons of the LMC GMCs with H II regions and young clusters, a large number of the young clusters ($\tau < 10$ Myr) and H II regions are found within 100 pc of the GMCs. On the other hand, older clusters show almost no correlation with the GMCs. We have examined the association between the individual GMCs and the H II regions and young clusters and found that about a half of the H II regions and young clusters are associated with the GMCs.

From the *NANTEN* CO surveys, three types of GMCs can be classified in terms of star formation activities (Kawamura *et al.* 2007, 2008); Type I shows no signature of star formation, Type II is associated with relatively small H II region(s) and Type III with both H II region(s) and young stellar cluster(s). It is also found that there is not a significant difference in the distribution of the line widths and sizes of the GMCs among the three types for those with a mass above the completeness limit, 5×10^4 M$_\odot$, while the mass distribution of the Type III GMCs is different from those of Types I and II. The mass distribution of Types I and II shows a peak at $M_{CO} \sim 10^5$ M$_\odot$, while that of the

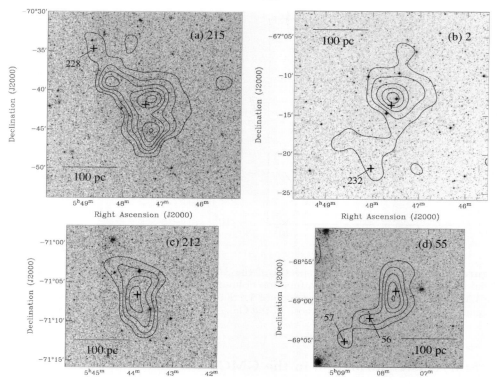

Figure 4. Examples of GMCs without massive star formation (Type I GMCs). *NANTEN* CO maps are shown in contours superimposed on DSS2 images.

Type III is rather flat. We interpret that these types represent the evolutionary sequence (see Fig. 5 and Kawamura *et al.* 2007, 2008); i.e. the youngest phase is Type I followed by Type II and the last phase is Type III where most active star formation takes place leading to cloud dispersal. The number of the three types of GMCs should be proportional to the time scale of each evolutionary stage if a steady state is a good approximation. By adopting the time scale of the youngest stellar clusters, 10 Myr, we roughly estimate the timescales of Types I, II, and III to be 6 Myr, 13 Myr and 6 Myr, respectively for those with a mass above the completeness limit, 5×10^4 M$_\odot$. This corresponds to a lifetime of the GMCs of 20–30 Myr.

Recent surveys of the Magellanic Clouds by the IR satellites, like *Spitzer* (e.g., Meixner *et al.* in this volume, "Surveying the Agents of a Galaxy's Evolution") and *AKARI* have been strong tools to identify younger, and lower mass YSOs (Whitney *et al.* 2008). From the comparisons of these YSOs and the GMCs, the distribution of the YSOs appears to be well correlated with that of CO. Significant enhancement of the YSO density is seen toward some of large H II regions, such as 30 Dor, N 11, and N 44 in the LMC. YSO densities towards GMCs with high-mass star formation (Types II and III) are significantly (a factor of two or three) higher than that towards Type I GMCs. This demonstrates that the Type I GMCs have actually lower star formation activity than the others. It also means that the high-mass star formation coincides with the increase of the number of YSOs.

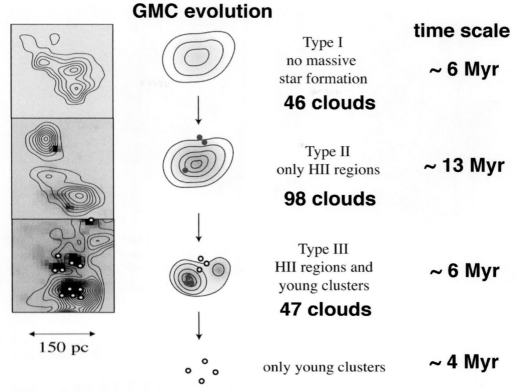

Figure 5. Evolutionary sequence of the GMCs. The left panels are examples of GMC Type I, II, and III from the top panel, respectively. Each panel presents an Hα image from Kim *et al.* (1999) with GMCs identified by *NANTEN* (Fukui *et al.* 2008) in contours. Open circles indicate the positions of young clusters (Bica *et al.* 1996). The middle panels are illustrations for each evolutionary stage. Open circles and filled circles represent young clusters and H II regions, respectively (Kawamura *et al.* 2008).

5. GMCs and H I gas

GMCs are the major sites of star formation and the GMC formation and evolution must be a crucial step in the evolution of a galaxy. To examine the relationship of GMCs to the remainder of the neutral ISM, we plot the CO clouds on top of H I maps of the Magellanic Clouds in Fig. 6. Every GMC in each of the galaxies is found on a bright filament or clump of H I, where $N(\mathrm{HI}) \gtrsim\sim 10^{21}$ cm^{-2}, but the reverse is not true: there are many bright filaments of H I without CO. The Magellanic System comprises H I-rich galaxies, $[M(\mathrm{HI}) \sim 5 \times 10^{8}$ M$_\odot$ (LMC), 4×10^{8} M$_\odot$ (SMC), 5×10^{7} M$_\odot$ (Bridge)] (Brüns *et al.* 2005). Molecular mass traced by CO, M_{CO}, covers only $\sim$10% (LMC), 1% (SMC), 0.05% (Bridge) of $M(\mathrm{HI})$, respectively, in these galaxies. In the LMC, large H I integrated intensities and peak brightnesses are prerequisites for detection of CO, with the detection fraction rising steadily for larger values of both quantities, though never reaching 100%. From the 2-D (2 dimensional) analysis (Fig. 7 *left*), the peak of the histogram is only 33%, so even at the highest H I peak brightnesses, only about 1/3rd of the pixels are associated with CO (see poster by Wong *et al.* this volume). Fukui (2007) made a 3-D (3 dimensional) comparison of CO and H I in the LMC where the 3-D datacube has a velocity axis in addition to the two axes in the sky. Fig. 7 (*right*) shows a histogram of the H I integrated intensity in 3-D and the pixels with significant CO emission (greater

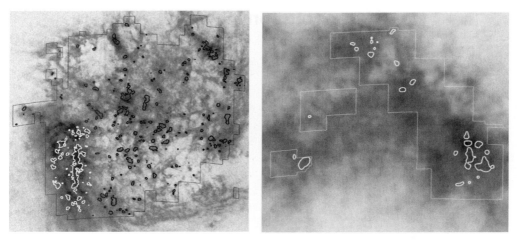

Figure 6. CO emission overlaid on maps of H I emission for the LMC (*left*) and the SMC (*right*). The H I maps are from Kim *et al.* (2003, LMC) and Stanimirović *et al.* (1999, SMC). CO emission (contours) is found exclusively on bright filaments of atomic gas though not every bright H I filament has CO emission.

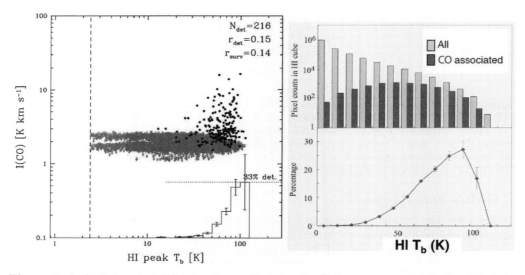

Figure 7. *Left*: Integrated CO intensity vs. H I intensity (2-D comparisons). Black solid circles are CO detections, red arrows represent 3-σ upper limits. *Right*: (*Upper panel*) Distribution of H I integrated intensity in 3-D. Distribution of the H I integrated intensity where the significant CO emissions are detected is also shown in gray. (*Lower panel*) The ratio of the H I pixels with and without CO emission.

than 0.7 K km s^{-1}) are marked. This histogram shows that the CO fraction increases steadily with the H I intensity, and about one third of the pixels exhibit CO emission near T_b(HI) of ~90 K. This is consistent with the 2-D analysis within error. Fig. 8 shows comparisons with CO and H I intensity for the three GMC types. This clearly shows that the H I intensity tends to increase from Type I to Type III. The H I intensity surrounding GMCs becomes greater with the GMC evolution and star formation. The dependence of H I intensity on Types of GMC indicates that GMCs are "dressed" in H I and that the "H I dress" grows in time. The enveloping H I gas is accreting onto GMCs and is converted into

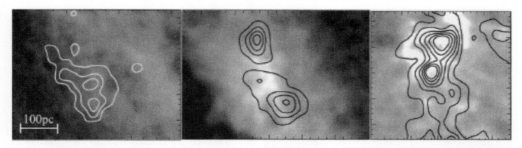

Figure 8. Examples of the H I and CO (contour) distributions of the GMC Type I (*Left*), II (*Middle*), and III (*Right*), respectively. The images are H I integrated intensity maps; velocity is integrated over the range where the significant CO emissions are detected.

H_2 due to increased optical depth. This leads to increase the molecular mass of GMC, i.e. the observed mass increases from Type I to III (see section 5). The timescale of the GMC evolution is ~ 10 Myr and the increased molecular mass is of the order of $\sim 10^6$ $M_\odot$. Hence, a mass accretion rate of $\sim 10^{-1}$ $M_\odot$ yr^{-1} is required. We roughly estimate that this rate is consistent with that calculated for a spherical accretion of the H I gas having $n(\mathrm{HI}) \sim 10$ cm^{-3} at an infall speed of ~ 7 km s^{-1}. In the SMC and Bridge, the H I is so widespread that the CO clouds appear as small, isolated clouds in a vast sea of H I.

6. High (sub)millimeter observations of GMCs

At the resolution of the *NANTEN* telescope (~ 40 pc), many of the star-forming clumps (cores) in the GMCs in the Magellanic System remain unresolved. To more reliably determine their sizes, virial masses, and relationship with star formation and H I structures etc. at 10 pc scale, recently, MAGMA ("The Magellanic Mopra Assessment": see posters by Hughes *et al.* and Muller *et al.* in this volume) have started. This new project will cover all the brightest CO clouds in the LMC and SMC as detected by *NANTEN* at a resolution of $\sim 45''$ (10 pc). Maps are revealing molecular cloud properties across flux-limited samples in both galaxies. They are also being used to address the relationship with star formation and CO–H I correlations, which we are investigating globally using the *NANTEN* data, can be studied on the scales of individual GMCs.

Recent advances in submillimeter observations have allowed us to determine physical parameters of molecular clouds over much larger ranges than in the millimeter region by comparing line intensities between different transitions. High-J CO and C I lines are important to constrain physical conditions ($n(H_2)$, T_{kin}), and heating mechanism (UV heating etc.) in clouds (Minamidani *et al.* in this volume; Minamidani *et al.* 2008). These submillimeter studies were initiated with the *SEST* 15-m telescope in Chile followed by the *AST/RO* 1.6-m telescope in Antarctica. Subsequently, in the 2000s, the development of new instruments at an altitude of 5000 m in the Atacama in northern Chile resulted in a superior capability because of the high altitude and dry characteristics of the site. The instruments installed in the Atacama include the *ASTE* 10 m, *APEX* 12 m, and *NANTEN2* 4 m telescopes (Fig. 9). All these instruments are beginning to take new molecular data with significantly better quality than before in terms of noise level (Fig. 10), as well as angular resolution (Fig. 11). It is also noteworthy that the current frequency coverage extends as high as 800 GHz.

The N 159 region is one of the most interesting star/cluster-forming regions in the LMC. This region has been well studied at mm/submm and infrared wavelengths (see poster by Y. Mizuno *et al.* in this volume). It has been shown that the N 159 complex

Figure 9. *NANTEN2 (left)* and *ASTE (right)* telescopes.

consists of 3 distinct and spatially well separated clumps (Fig. 11). N 159W is a prominent star-forming region and has the highest observed CO(J =1–0, 2–1, 3–2) intensities in the LMC. This clump is interesting to study the origin of young massive (populous) clusters. On the contrary, the southern clump, N 159S, lacks signs of on-going massive star formation (neither emission nebulae nor clusters) and appears to be a cold, more quiescent cloud. This clump is a good candidate of a chemically young GMC and crucial for the origin of starless GMCs.

For the SMC, submillimeter observations of CO(4–3/7–6) and [C I] lines towards the star forming regions are ongoing with *NANTEN2*. From the LVG model calculations, we derived the densities and temperatures of the dense clumps in the GMCs, which show a good correlation with the star formation activities in the GMCs. The GMC associated with N 66 is hot and dense, on the other hand, the GMC associated SMC B1#1-1, which is a quiescent GMC devoid of any massive star forming activity, shows up cold and less dense. The relation between these properties and the evolutionary sequence of GMCs are discussed in the poster by N. Mizuno *et al.* (this volume).

The current study of GMCs in the Magellanic System provides us with promise on what we can learn on GMCs and star formation therein when *ALMA* is used for the Local Group. It would be particularly important to find very high mass cloud cores of $\sim 10^5$–10^6 M$_\odot$, as promising candidates for proto-globular clusters.

Acknowledgements

NANTEN2 is an international collaboration between 10 universities including the Universities of Cologne and Bonn. The original *NANTEN* telescope was operated based on a mutual agreement between Nagoya University and the Carnegie Institution of Washington. We also acknowledge that the operation of *NANTEN* was realized by contributions from many Japanese public donators and companies. This work is financially supported in part by a Grant-in-Aid for Scientific Research (KAKENHI) from the Ministry of Education, Culture, Sports, Science and Technology of Japan (Nos. 15071203 and 18026004) and from JSPS (Nos. 14102003, 20244014, and 18684003). This work is also financially supported in part by core-to-core program of a Grant-in-Aid for Scientific Research from the Ministry of Education, Culture, Sports, Science and Technology of Japan (No. 17004). It is a pleasure to thank my collaborators, Nagoya University (A. Kawamura, Y. Mizuno, E. Muller, T. Onishi, M. Murai, A. Ohama, H. Yamamoto, A. Mizuno, Y. Fukui), the *NANTEN2* consortium (M. Rubio, J. Pineda, F. Bertoldi, U. Klein, J. Stutzki, Y. Yonekura), the *ASTE* team (T. Minamidani, N. Yamaguchi, M. Ikeda, K. Tatematsu, T. Hasegawa), the MAGMA project (T. Wong, A. Hughes, J. Pineda, J. Ott,

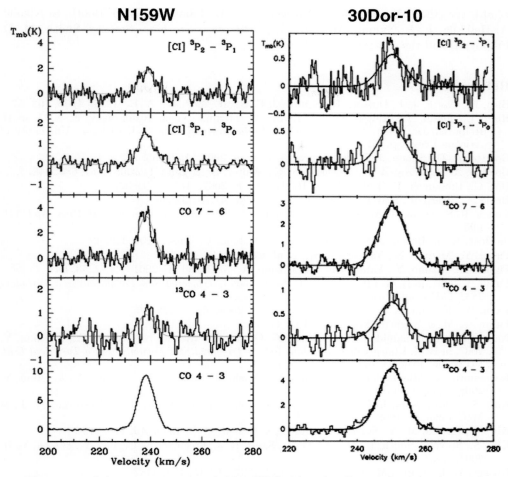

Figure 10. Submm spectra towards N 159W (*left*) and 30 Dor-10 (*right*) taken with *NANTEN2* (Pineda *et al.* 2008).

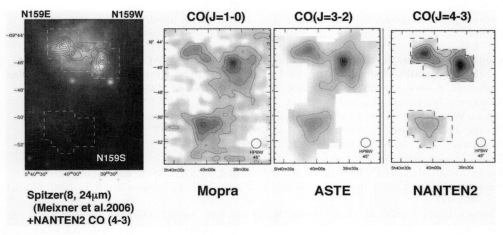

Figure 11. Integrated intensity maps of CO($J =1$–0/3–2/4–3) towards the N 159 region in the LMC. the resolutions of these maps were smoothed to the same beam size of $45''$.

et al.), the *SEST* Nagoya-Onsala project (L. E. B. Johansson, R. S. Booth, S. Nikolić, A. Heikkila, L.-Å. Nyman, M. Lerner), *ATNF* users (S. Kim, Staveley-Smith, M. D. Filipović), and the *Spitzer* SAGE Team.

References

Bica, E., Clariá, J. J., Dottori, H., Santos, J. F. C., Jr., & Piatti, A. E. 1996, *ApJS*, 102, 57

Blitz, L., Fukui, Y., Kawamura, A., Leroy, A., Mizuno, N., & Rosolowsky, E. 2007, in B. Reipurth, D. Jewitt, & K. Keil (eds.), *Protostars and Planets V* (Tucson: University of Arizona Press), p. 951

Bolatto, A. D., Simon, J. D., Stanimirović, S., *et al.* 2007, *ApJ*, 655, 212

Bolatto, A. D., Leroy, A. K., Rosolowsky, E., Walter, F., & Blitz, L. 2008, *ApJ*, 686, 948

Bot, C., Boulanger, F., Rubio, M., & Rantakyro, F. 2007, *A&A*, 471, 103

Brüns, C., Kerp, J., Staveley-Smith, L., *et al.* 2005, *A&A*, 432, 45

Cohen, R. S., Dame, T. M., Garay, G., Montani, J., Rubio, M., & Thaddeus, P. 1988, *ApJ*, 331, L95

de Boer, K. S., Braun, J. M., Vallenari, A., & Mebold, U. 1998, *A&A*, 329, 49

Fukui, Y., Mizuno, N., Yamaguchi, R., *et al.* 1999, *PASJ*, 51, 745

Fukui, Y., Mizuno, N., Yamaguchi, R., Mizuno, A., & Onishi, T. 2001, *PASJ*, 53, L41

Fukui, Y. 2007, in B. G. Elmegreen & J. Palouš (eds.), *Triggered Star Formation in a Turbulent ISM*, IAU Conf. Proc. (Cambridge: CUP), p. 31

Fukui, Y., Kawamura, A., Minamidani, T., *et al.* 2008, *ApJS*, 178, 56

Fujimoto, M. & Noguchi, M. 1990, *PASJ*, 42, 505

Kawamura, A., Minamidani, T., Mizuno, Y., Onishi, T., Mizuno, N., Mizuno, A., & Fukui, Y. 2007, in B. G. Elmegreen & J. Palouš (eds.), *Triggered Star Formation in a Turbulent ISM*, IAU Conf. Proc. (Cambridge: CUP), p. 101

Kawamura, A., Minamidani, T., Mizuno, Y., Onishi, T., Mizuno, N., Mizuno, A., & Fukui, Y. 2008, *ApJS*, submitted

Kim, S., Staveley-Smith, L., Dopita, M. A., Freeman, K. C., Sault, R. J., Kesteven, M. J., & McConnell, D. 1998, *ApJ*, 503, 674

Kim, S., Dopita, M. A., Staveley-Smith, L., & Bessell, M. S. 1999, *AJ*, 118, 2797

Kim, S., Staveley-Smith, L., Dopita, M. A., Sault, R. J., Freeman, K. C., Lee, Y., & Chu, Y.-H. 2003, *ApJS*, 148, 473

Kutner, M. L., Rubio, M., Booth, R. S., *et al.* 1997, *A&AS*, 122, 255

Leroy, A., Bolatto, A. D., Stanimirović, S., Mizuno, N., Israel, F., & Bot, C. 2007, *ApJ*, 658, 1027

Minamidani, T., Mizuno, N., Mizuno, Y., *et al.* 2008, *ApJS*, 175, 485

Mizuno, N., Yamaguchi, R., Mizuno, A., *et al.* 2001a, *PASJ*, 53, 971

Mizuno, N., Rubio, M., Mizuno, A., Yamaguchi, R., Onishi, T., & Fukui, Y. 2001b, *PASJ*, 53, L45

Mizuno, N., Muller, E., Maeda, H., Kawamura, A., Minamidani, T., Onishi, T., Mizuno, A., & Fukui, Y. 2006, *ApJ*, 643, L107

Muller, E., Staveley-Smith, L., & Zealey, W. J. 2003, *MNRAS*, 338, 609

Nakagawa, M., Onishi, T., Mizuno, A., & Fukui, Y. 2005, *PASJ*, 57, 917

Ott J., Wong, T., Pineda, J., *et al.* 2008, *PASA*, 25, 129

Pineda, J. L., Mizuno, N., Stutzki, J., *et al.* 2008, *A&A*, 482, 197

Rosolowsky, E. & Leroy, A. 2006, *PASP*, 118, 590

Rubio, M., Lequeux, J., & Boulanger, F. 1993, *A&A*, 271, 9

Rubio, M., Boulanger, F., Rantakyro, F., & Contursi, A 2004, *A&A*, 425, L1

Stanimirović, S., Staveley-Smith, L., Dickey, J. M., Sault, R. J., & Snowden, S. L. 1999, *MNRAS*, 302, 417

Whitney, B. A., Sewilo, M., Indebetouw, R., *et al.* 2008, *AJ*, 553, 185

Yamaguchi, R., Mizuno, N., Onishi, T., Mizuno, A., & Fukui, Y. 2001a, *ApJ*, 553, L185

Yamaguchi R., Mizuno, N., Onishi, T., Mizuno, A., & Fukui, Y. 2001b, *PASJ* 53, 959

Yamaguchi R., Mizuno, N., Mizuno, A., *et al.* 2001c, *PASJ* 53, 985

The Magellanic System: Stars, Gas, and Galaxies
Proceedings IAU Symposium No. 256, 2008
Jacco Th. van Loon & Joana M. Oliveira, eds.

Properties of star forming regions in the Magellanic Clouds

Mónica Rubio

Departamento de Astronomía, Universidad de Chile, Casilla 36-D, Santiago, Chile
email: mrubio@das.uchile.cl

Abstract. Understanding the process of star formation in low metallicity systems is one of the key studies in the early stages of galaxy evolution. The Magellanic Clouds, being the nearest examples of low metallicity systems, allow us to study in detail their star forming regions. As a consequence of their proximity we can resolve the molecular clouds and the regions of star formation individually. Therefore we can increase our knowledge of the interaction of young luminous stars with their environment. We will present results of multiwavelenghts studies of LMC and SMC massive star forming regions, which includes properties of the cold molecular gas, the embedded young population associated with molecular clouds, and the interaction of newly born stars with the surrounding interstellar medium, based on *ASTE* and *APEX* submillimeter observations complemented high sensitivity NIR groud based observations and *Spitzer* results.

Keywords. stars: formation, ISM: clouds, H II regions, Magellanic Clouds

1. Introduction

The molecular clouds constitute the cold interstellar material where stars are born. The rate of star formation depends on the quantity of molecular gas and on the concentration in the clouds as a result of gravitational bounding. The characterization of the properties of these clouds and of the star formation efficiency is a key factor to understand the evolution of galaxies.

The Small and Large Magellanic Clouds (SMC and LMC) are the two closest galaxies to the Sun that are rich in gas. These are two irregular galaxies where the structure of interstellar matter is largely influenced by the action of new stars on their environment rather than by the large scale dynamics of high mass spiral galaxies such as the Milky Way. The two clouds share another important characteristic of galaxies in the early stage of evolution. The presence of heavy elements (metallicity) is 10 to 4 times weaker than that of the Solar neighbourhood. The Magellanic Clouds are therefore considered as the nearest environments that allow to study interstellar matter, gas and dust, in the early stages of galaxy evolution.

As a consequence of their proximity, the observations resolve the molecular clouds and the regions of star formation individually. Nevertheless, it is difficult to trace the cold gas because H_2 is not directly observable in these conditions. It is necessary to employ tracers. The Magellanic Clouds have been extensively observed in the first rotation transitions of the CO molecule, but these observations leave many uncertainties on the mass and structure of the clouds. The photo-chemical models suggest that a dominant part of the molecular gas in low metallicity galaxies is weak in CO.

Multiple molecular clouds in the Magellanic Clouds have been surveyed based on CO line emission. These observations show that the CO emission in the Magellanic Clouds is very weak. In effect, the CO emission compared with star formation tracers, such as the band B luminosity or the Hα luminosity, is much weaker than in normal galaxies (Leroy *et al.* 2007). Even more, the comparison of the dynamic masses of molecular gas and of the mass derived from CO luminosity gives very disperse values of the conversion factor $X_{CO} = N(H_2)/I_{CO}$, and in average 10 times greater than in our galaxy (Rubio *et al.* 1993; Mizuno *et al.* 2001).

2. Properties of molecular clouds

Giant Molecular Clouds (GMCs) properties have been derived from complete CO surveys in the Magellanic Clouds. The *NANTEN* survey covered the LMC (Fukui *et al.* 1999, 2008) and SMC (Mizuno *et al.* 2001) at 2.4′ resolution, surpassing the 8.8′ resolution survey (Cohen *et al.* 1988; Rubio *et al.* 1991). The *NANTEN* survey has derived the properties of GMCs of typical sizes of about 50 pc and a summary of these properties is given by Mizuno, in these proceedings.

Higher angular resolution observations were possible with the 15m *SEST* telescope but these did not cover the entire galaxies. The *ESO-SEST* Key Programme: "CO in the Magellanic Clouds" is the best available high angular CO(1–0) survey with 20 pc spatial resolution of the Magellanic Clouds. In a series of articles, the properties of the surveyed molecular clouds were given. These can be summarized as follows: The LMC has molecular clouds with sizes between 10 and 40 pc, line widths between 2.5 and 10 km s^{-1} and CO luminosities between 0.8×10^3 and 1.3×10^4. The SMC molecular clouds have sizes between 10 and 20 pc, line widths between 1.6 and 6.4 km s^{-1} and CO luminosities between 0.5×10^3 to 6.4×10^4.

In general, these molecular clouds at *SEST* resolution follow a linewidth–size relationship where $\Delta v \propto R^{0.5}$ (Israel *et al.* 1993; Rubio *et al.* 1993, etc.), typical of gravitationally bound molecular clouds as found in our Galaxy and described by Larson (1981).

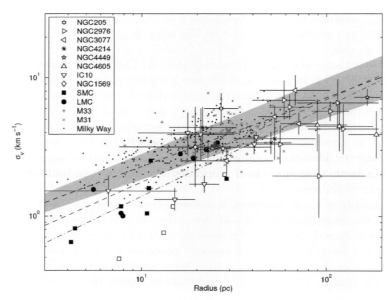

Figure 1. Extragalactic luminosity line width relation for molecular clouds from Bolatto *et al.* (2008, their figure 2). Measurements for different galaxies are indicated with open and filled symbols depending on whether they were obtained from CO(1–0) or CO(2–1) observations, respectively. The grey bar is the relation for Galactic GMCs (Solomon *et al.* 1987) and the blue dotted line the one obtained considering only the dwarf galaxies.

A mayor difference found in the properties of Magellanic molecular clouds and Galactic molecular clouds is that Magellanic GMCs show CO luminosities a factor of 5 to 10 weaker than those found in our Galaxy for similar linewidths. This result is still valid for CO clouds observed at 20 pc resolution but the factor decreases to 3 to 5 times less luminous for the LMC and SMC, respectively. This difference has been an important result in revealing that the CO emission may not be tracing the total molecular mass in low metallicity molecular clouds. Thus, it has become essential for the study of star formation in these galaxies to determine the total molecular gas mass of these clouds.

2.1. *Molecular mass*

To determine the amount of molecular gas is difficult as the cold H_2 gas is almost impossible to observe. The most used method to derive the mass in molecular gas is through the CO luminosity of the molecular cloud. In our Galaxy, an empirical relationship has been determined between the column density of H_2 and the CO velocity-integrated emission of the CO cloud. This relationship, calibrated by gamma-ray measurements, virial mass comparison, and extinction measurements (Bloemen *et al.* 1986) gives consistent results and a calibration factor, known as the X factor, allows to deduce the cloud molecular mass from its CO emission.

In the Magellanic Clouds, all the CO observations have consistently given a lower molecular mass of the clouds when determined by their CO luminosity and compared to virial mass determination using their CO linewidths and size. This result is also found even when we have had a ten-fold spatial resolution improvement in the observations as achieved by *SEST* CO(2–1) observations.

The explanation of this result has been argued as follows: the CO molecule in the Magellanic Clouds, and hence in low metallicty systems, is photodissociated and only traces part of the molecular cloud, keeping large envelopes of H_2 hidden from the observations. The CO molecule survives only when it is shielded from the UV radiation field. In the LMC and SMC, this shielding occurs at column densities of molecular H_2 gas about 0.1 times the column densities of molecular clouds in our Galaxy of $N(H_2) \approx 1 - 1.5 \times 10^{21}$ cm^{-2}. The molecular hydrogen molecule self-shields at a column density lower than CO, the UV photon penetration is much larger, and CO only traces the dense core. As a consequence the molecular mass in molecular clouds in low metallicity systems is mainly H_2 and not CO.

Bolatto *et al.* (2008) have made a comparison of molecular cloud properties in extragalactic systems, including the LMC and SMC, and found that extragalactic GMCs follow approximately similar size–linewidth, luminosity–size, and luminosity–linewidth relations as Galactic molecular clouds. But the SMC molecular cloud properties deviate from typical Galactic molecular cloud properties much more than LMC molecular clouds (Figure 1). The SMC shows lower CO luminosity for similar size as Galactic clouds but smaller linewidths. Thus, these clouds may not be supported gravitationally and other mechanisms might be playing a role, e.g., magnetic fields and/or turbulence. The LMC with only a 1/4 $Z_\odot$ metallicity does not show this deviation. The impact of the metallicity on these results is still to be explored as the range of metallicities of his sample is in average 1/5 that of the Milky Way. Observationally this is a difficult problem as low metallicity systems have very weak CO emission and thus require large amounts of telescope time to detect its emission. The SMC, due to its proximity, is therefore an excellent nearby system to study and determine the molecular cloud properties in low metallicity systems.

An alternative method to determine the molecular mass of molecular clouds is by their dust emission. The dust emission peaks in the submillimeter spectral range and the

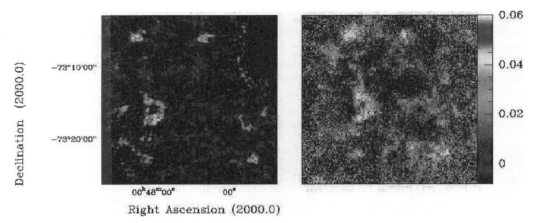

Figure 2. The continuum emission of the SW region of the SMC, with in the left panel the 1.2-mm SIMBA image and in the right one the 850 μm LABOCA image. The bright sources, observed also in CO, are seen at both frequencies while at 850 μm extended emission is clearly seen in the region. The scale is mJy beam^{-1}.

installation of mm and submm bolometer cameras in Chile provide a unique opportunity to measure the continuum dust emission from LMC and SMC molecular clouds. These observations can be critical to resolve the discrepancy between the virial mass and the CO mass determination.

The first 1.2-mm continuum observations of molecular clouds in the Magellanic Clouds were done with the SIMBA bolometer at the *SEST* telescope in La Silla, Chile. In the SMC, Rubio *et al.* (2004) reported the results for a cold quiescent cloud in which they found that the gas mass derived from the continuum dust emission was larger than the virial mass determination of the cloud. This result was also found in studies of several molecular clouds in the SW region of the SMC Bar by Bot *et al.* (2007) as well as in the active star forming region N 66 (Perez 2006). The analysis of observations of the SMC has shown for many clouds in the same region that the masses of gas deduced from the dust emission were ten times higher than those deduced by the application of the virial theorem using the size of the clouds and the gas velocity dispersion measured in CO (Rubio *et al.* 2004; Bot *et al.* 2007; fig. 1 of Bot *et al.*, these proceedings). This large difference in mass is surprising. It is larger than the uncertainty on the emissivity of grains and of the gas-to-dust mass ratio. It is known that for weak metallicity the CO molecule does not trace but the densest regions of molecular clouds (Lequeux *et al.* 1994). It is possible that the dynamics of these regions does not trace the gravitational potential of the cloud. Bot *et al.* (2007) have proposed that a significant part of the support of molecular clouds against gravitation could be due to a magnetic field. The importance of the magnetic field depends on the mechanism of cloud formation. Its role must be more important where the clouds are created by unidirectional compression on the surfaces of the bubbles blown by star clusters rather than by the isotropic action of gravity.

New observations done at 0.8 μm (350 GHz) using the LABOCA bolometer at the *APEX* telescope confirm that in the SW region of the SMC the molecular mass derived from the dust emission is larger than the virial mass derived from CO(2–1) observations (Bot *et al.*, these proceedings). Figure 2 shows the 1.2-mm SIMBA and the 0.8-mm LABOCA images of the SW SMC Bar. The bright sources have been mapped in CO. The 0.8 μm LABOCA image shows diffuse continuum emission almost all over the observed area. Bot *et al.* (these proceedings) have found that after subtracting the free-free emission and CO line contribution, and using a range of possible dust emissivities, analyzing different temperature dependencies, and using the gas-to-dust ratios as determined for the SMC, the gas mass derived from the dust emission continues to be consistently larger than the virial mass determination, being a factor of 2 to 10 larger. Analysis is under way but the confirmation that dust emission gives larger gas masses is consistent with the fact that SMC molecular clouds must have large envelopes of H_2.

These studies are based on a certain number of hypothesis on the dust properties in the millimeter domain. Little is known about dust outside the Galaxy, particularly the impact of metallicity on its composition and the ratio in gas mass over dust (Reach *et al.* 2000; Weingartner & Draine 2001; Bot *et al.* 2004). Some studies of dust emission in low metallicity galaxies farther than the Magellanic Clouds have shown variations in dust properties (Galliano *et al.* 2003, 2005) but the molecular clouds in these galaxies are not resolved and the variations are not currently well understood. In the Galaxy, variations of dust millimeter emissivity have been observed between cold molecular clouds, regions of hot-star formation and the diffuse medium (Dupac *et al.* 2003). These variations have been associated with the evolution in the composition and the distribution in size of dust by the formation of mantles and grain coagulation (see Boulanger, these proceedings).

There have been indirect methods to determine the amount of molecular mass in the molecular clouds by using the mid-IR emission from gas and dust.

In the SMC, Leroy *et al.* (2007), using the 160 μm *Spitzer* images (Bolatto *et al.* 2007) plus the H I gas emission determined molecular hydrogen columnn densities which they compared to the molecular clouds as mapped by the *NANTEN* CO(1–0) survey. They obtain that the 160 μm emission has a spatial extension larger than the volume where CO emission has been detected, consistent with molecular clouds having a large envelope of H_2 molecular gas and dense but smaller CO cores. The mid-IR images show large dusty envelopes suggesting that low metallicity dense clouds have low-density and CO-free H_2 envelopes. A large fraction of the molecular mass could be contained in such lower density envelopes (Lequeux *et al.* 1994).

This result is also confirmed by a recent a study of one SMC molecular cloud (N 83) (Leroy *et al.*, in preparation) in which they have combined the 160 μm and 70 μm *Spitzer* data, the H I emission, and the high resolution *SEST* CO (2–1) data with 10 parsec resolution. Leroy and collaborators derive from the H I and 160 μm opacities a gas-to-dust ratio 14 times larger than that of the diffuse gas in our Galaxy. They find that the H_2 densities of molecular gas in this cloud are larger than 350 $M_\odot$ pc^{-2} while in the molecular ring in the Milky Way $N(H_2) \sim 8$ $M_\odot$ pc^{-2}. The CO emission comes from $A_V \sim 1.6$ mag, consistent with the PDR models of SMC molecular clouds where CO selfshields much deeper than the molecular hydrogen as modeled by Lequeux *et al.* (1994) using SMC gas conditions, i.e., a 1/10 lower dust content, a lower abundance of C and O, and a higher, about 1000 larger UV radiation field, than the ISM conditions of the Galaxy. Leroy *et al.* (these proceedings) suggest that the sizes of the dense molecular clumps are 1/10 of those of the dense clumps in Galactic molecular clouds. The small size of the dense clumps would explain that the conversion factor in the SMC would be larger than that determined in the Galaxy implying that the mass determination from CO luminosities has to be lower than in Galacic molecular clouds. For example, the typical mass of a molecular cloud in the SMC is $M = 10^4$ $M_\odot$ while in the Galaxy a low mass star forming region such as ρ Ophiuchus has a mass of $M \sim 5 \times 10^5$ $M_\odot$ for the principal CO clump. Thus, star formation can be quite different in these low metallicity molecular clouds.

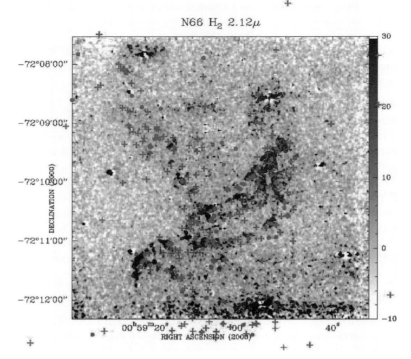

Figure 3. The 2.12 μm H_2 image of N 66. The *Spitzer* YSO list of Simon *et al.* (2007) is plotted with crosses and the near-IR sources with $J - K_s > 1$ mag as filled circles. Young massive objects follow the gas emission and concentrate on the H_2 knots.

2.2. *Star formation*

Stellar associations in the Magellanic Clouds contain the richest sample of young bright stars and thus our knowledge of their young massive stars has been collected from studies of such stellar systems (e.g., Massey *et al.* 2002). Almost every young association coincides with one or more H II regions as catalogued by Henize (1956) and Davies *et al.* (1976). These young associations also host large numbers of PMS stars (Nota *et al.* 2006; Gouliermis *et al.* 2006). These young OB associations correlate with the distribution of molecular clouds. In the LMC, young associations with ages less than one million years are found associated with GMCs (Yamaguchi *et al.* 2001; Fukui *et al.* 2008).

Observationally the study of star formation in the Magellanic Clouds can be adressed combining ground-based observations with IR capabilities of 8m-class telescopes operating in the southern hemisphere (Chile), and space-based NICMOS/*HST* and *Spitzer* observations. Several studies have investigated massive star forming regions and their molecular clouds.

In the LMC, these studies include the bright Giant H II region 30 Doradus and its massive super star cluster R 136 (Rubio *et al.* 1998; Walborn *et al.* 1999; Brandner *et al.* 2001; Oliveira *et al.* 2006), the second largest H II region N 11 (Barbá *et al.* 2003), the young H II region N 159 (Jones *et al.* 2005; Rubio *et al.* 2008), and a compact H II region N 4 (Contursi *et al.* 2007).

In the SMC, these studies have concentrated on the brightest H II region N 66 and its central cluster NGC 346 (Contursi *et al.* 2000; Rubio *et al.* 2000), the H II regions N 27 (LIRS 49) and N 12 (LIRS 36), and NGC 602/N 90 (Gouliermis *et al.* 2008).

Deep near-IR JHK imaging has been obtained towards these regions to study the embedded population in molecular clouds. Infrared color–color and color–magnitude diagrams have shown the existence of infrared sources that have colors consistent with massive YSOs. Their colors imply important IR excess and/or large extinction.

In N 66 in the SMC, Nota *et al.* (2006) and Sabbi *et al.* (2008), using *HST* observations. determined the spatial distribution of PMS stars and found a very good correlation of PMS and young stars with CO emission peaks. A comparison of their catalogue and the near-IR photometry of ISAAC/*VLT* images of IR sources with $(K_s - H) > 1.5$ mag shows a strong association of the PMS star location to the H_2 2.12 μm gas emission with a large concentration towards molecular knots in N 66 (Rubio *et al.* 2000; Rubio 2007).

Using the available *Spitzer* LMC (Meixner *et al.* 2006) and SMC (Bolatto *et al.* 2007) mid IR imaging, complete point source catalogues have been made and used to select embedded and/or young stellar objects based on their IRAC color diagrams. In the SMC, Simon *et al.* (2007) produced a point source catalogue and determined YSO candidates from SED fits. These YSO candidates show a very strong spatial correlation with the molecular cloud peaks and H_2 2.12 μm dense knots as seen in Figure 3. The H_2 dense knots are most probably PDRs formed by the embedded massive stars which are photo-dissociating their natal cloud. IR spectroscopy of three IR embedded sources associated with H_2 knots in N 66 confirm their massive YSO's nature (Rubio & Barbá 2008). Similarly, NGC 602/N 90 hosts several *Spitzer* YSO candidates associated with this massive star forming region (Gouliermis *et al.* 2008).

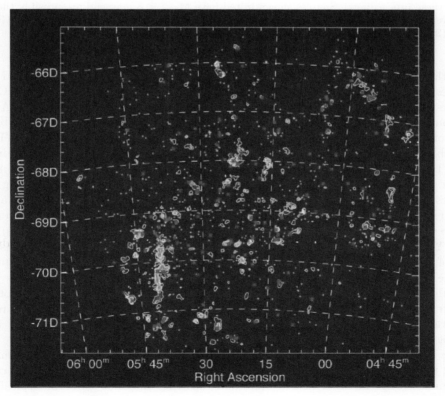

Figure 4. The *NANTEN* CO(1–0) map of the LMC. The distribution of the *Spitzer* YSO candidates (Whitney *et al.* 2008) are plotted as yellow crosses. The young OB association with ages less tha 10 Myr are show in red dots.

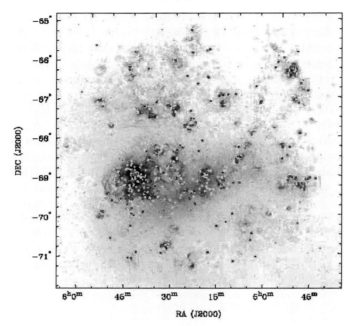

Figure 5. MCELS Hα image of the LMC overlaid with positions of *Spitzer* YSO candidates (Chu, these proceedings). Sources shown in red have $M_{8\mu m} < 8$ mag and are possibly the most massive candidates while those in solid circles as considered as definitive YSOs following Gruendl *et al.* (2008).

In the LMC, Whitney *et al.* (2008), and Gruendl *et al.* (2008), have produced point sources catalogues from *Spitzer* IRAC images. The distribution of massive YSO candidates is well correlated to the molecular clouds as can be seen in Figure 4. A comparison between the *Spitzer* catalogues and the existing near-IR ground-based observations of massive YSOs in specific regions do not show a one-to-one correlation. This is probably due to the difficulty in recovering the point sources from the *Spitzer* images when these suffer fron strong contamination from the bright nebular emission in the region. Different selection criteria for analyzing the IRAC images and extracting point sources also contribute to this result. For example, Gruendl *et al.* (2008) has a YSO list for the LMC which differs from that of Whitney *et al.* (2008). We display their catalogue over an Hα image in Figure 5. In N 11, Smith *et al.* (these proceedings) show more YSO candidates than the Whitney *et al.* list. In Figure 6, we show the distribution of the molecular gas and the YSOs as identified in the Whitney *et al.* catalogue for the N 11 nebula and interestingly the well studied N 11A and N 11B regions where several near-IR embedded sources have been detected (Barbá *et al.* 2003) do not have *Spitzer* YSO counterparts. A similar situation is found in the massive star forming region, 30 Doradus, where the embedded YSO sources are not listed in the *Spitzer* point source catalogues.

The IRAC images have a resolution 6″ at 8 μm and thus cannot disentangle massive young stars which are born in compact groups. *Spitzer* colors of point sources can be contaminated by the other nearby stars and thus they may not show the typical YSO colors. Even for near-IR ground-based observations obtained with typical 0.6″ seeing or better, 10 times better than the resolution of *Spitzer* images at 8 μm, multiple systems are not resolved. Only *HST* observations can resolve the multiple systems at the Magellanic

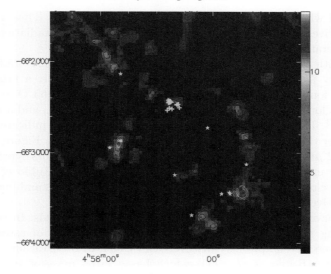

Figure 6. CO(1–0) image of N 11. Overlaid are the positions of the YSO candidates (Whitney *et al.* 2008) as (yellow) stars. The N 11A and N 11B ground-based YSO positions (Barbá *et al.* 2003) are indicated as (yellow) crosses.

Clouds' distance. In 30 Doradus, the NICMOS/*HST* observations showed that several of the bright IR sources ($K_s \sim 12$) detected in ground-based photometry were compact multiple systems with 2 or 3 bright IR sources (Walborn *et al.* 1999), confirming that massive star formation occurs in compact groups. Thus, studies of these regions will require better algorithms to extract the point sources from the *Spitzer* images.

We know that massive star forming regions undergo several episodes of star formation. There is observational evidence of residual molecular clouds with dense clumps that could undergo new star formation. In 30 Doradus, N 11, and N 66 a second generation of star formation is seen in the border of molecular clouds near the massive stars already formed. A particular case is that of N 66 and its ionizing cluster, in which spectroscopically confirmed massive YSOs have been discovered located near the optical O stars, all of

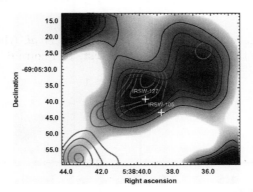

Figure 7. Dense molecular cloud at $20''$ from R 136 in 30 Doradus. The image shows the CO(2–1) integrated emission and superimposed in red are the H_2 2.12 μm emission contours. Two crosses indicate the position of IRS W-105 and IRS W-127 (Rubio *et al.* 1998), embedded sources with typical YSO IR colors.

them associated with a strong H_2 molecular knot and CO peak. Thus, dense molecular clumps can survive the interaction of strong winds and high UV radiation field produced by the first generation of massive stars.

In a region with extreme physical ambient conditions such as 30 Doradus in the LMC, we have recently found a dense CO clump at only 20″ NE (i.e. a projected distance of 5 pc) of the ionizing compact cluster R 136 which contains 65 O and n WR stars. This molecular cloud has a radius of 3 pc, a mass $\sim 1.5 \times 10^4$ M$_\odot$ and a mean density of $\sim 2.5 \times 10^3$ cm^{-3}. The detection of CS(2–1) towards the CO peak indicates a density even higher, i.e. larger than 10^6 cm^{-3}. The molecular cloud, shown in Figure 7, is associated with an H_2 2.12 μm emission knot and two IR sources with colors typical of a massive YSOs, probably embedded in the molecular cloud, are found (Rubio et al. 2008). This dense molecular cloud has survived the action of the strong winds and high UV radiation field of the massive stars in R 136.

Few studies have been done towards quiescent molecular clouds. In contrast to massive star forming regions these quiescent clouds show no sign of associated star formation. They were serendipitously discovered in CO(1–0) observations. The best studied is SMC B1#1 (Rubio et al. 1993) which has a virial mass of 10^4 M$_\odot$ and a gas mass as determined from continuum dust emission which is a factor of 10 larger (Rubio et al. 2004). The cloud is illuminated externally by a B-type star (Reach et al. 2000) and the 24 μm Spitzer image shows a larger spatial emission than the corresponding CO emitting volume. This cloud has been compared to have characteristics of a low-mass star forming regions such as ρ Ophiuchus in our Galaxy (Boulanger, these proceedings).

3. The Magellanic Bridge: a lower metallicity system

The Magellanic Bridge has a metallicity even lower than the SMC, about 1/20 of the Galaxy. In such extreme ISM conditions, CO(1–0) emission was detected using the NANTEN telescope towards seven H I clouds with column densities exceeding $N(\text{HI}) > 10^{21}$ and bright 100 μm emission, $I_{100\mu m} > 2.6$ MJy sr^{-1} (Mizuno et al. 2006). The properties of the CO clouds were determined from one single CO spectrum and these

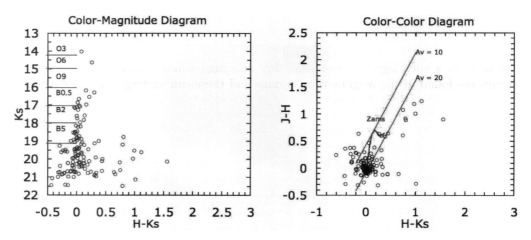

Figure 8. Color–Color and Color–Magnitude diagram of NGC 796 in the Magellanic Bridge molecular cloud. Several sources have color corresponding to Herbig Ae/Be stars while others are only seen in K.

have CO luminosities of several $\sim 10^3$ $L_\odot$ and their CO linewidths were in average smaller than the SMC molecular clouds CO linewidths and molecular masses of several 10^3 $M_\odot$, determined from their CO lumninosities and using the *NANTEN* SMC conversion factor (Mizuno *et al.* 2001).

IR studies done by Nishiyama *et al.* (2007) towards the CO cloud associated to the NGC 796 cluster classified several IR sources as Herbig Ae/Be stars. We have done deep near-IR observations towards these CO clouds to investigate their embedded population using deep JHK_s imaging with the near-IR camera ISAAC on the *VLT* at Paranal Observatory last December 2007. Towards the molecular cloud associated with the cluster NGC 796, the IR color–color and color–magnitude diagrams, shown in Figure 8, show several IR sources that can be identified as Herbig Ae/Be stars or highly reddened O stars from their IR colors and position in the J vs. $J - K$ diagram (Bik *et al.* 2006) Several of the brighter stars classified by Nishiyama *et al.* (2007) as Herbig Ae/Be stars are found to be multiple systems in our images. Several other weak sources are only seen in K_s ($K \sim 18$ mag) have been identified. (Rubio *et al.*, these proceedings). A detailed study of this region is underway.

4. Summary

The Magellanic Clouds and in particular the SMC are ideal systems to study the process of star formation in a low metallicity ISM. Molecular clouds in these systems seem to be in virial equilibrium as they follow, in general, Larson's law. Nevertheless, there is evidence that SMC molecular clouds could be magnetically supported or in pressure bound equilibrium (Bot *et al.*, these proceedings; Bolatto *et al.* 2008).

The molecular mass in these molecular clouds may not be traced by its CO emission as in spiral galaxies with Galactic metallicities and dust content. The CO molecule, used as standard tracer of the H_2 is highly affected by the low dust content and the low C and O abundances permitting a high penetration of UV radiation into the interior of the molecular clouds and thus shielding of the CO molecule occurs at higher column densities of gas. In the SMC, where the metallicity and dust content are a factor of 10 lower than in the Milky Way, this column density is several 10^{22} mol cm^{-2}.

The massive star forming regions in the Magellanic Clouds show a clear association to molecular clouds with masses $\sim 10^4$ $M_\odot$. These regions show on-going star formation and this is confirmed by a clear association between ground-based and *Spitzer* YSO candidates and the molecular clouds. Stars are formed in these metal-poor molecular clouds and also second generation of star formation occurs. Interestingly in regions of massive star formation with high UV radiation fiels and high winds residual dense molecular clumps are found in the neighbouring regions and these can undergo star formation. In many of the cases, the massive stars formed are found in compact clusters or multiple systems. Thus, the process of how stars form seems to be independent of the local conditons once a critical column density for the clouds is achieved.

The study of the IMF towards the molecular clouds in the Magellanic Clouds can be adressed with 8m ground-based telescopes. At present, we are able to detect stars with masses as low as 1 $M_\odot$. *Spitzer* data could lower this limit but the spatial resolution is not adequate to resolve compact systems. The detailed study of the dense molecular cores at the LMC and SMC will be possible when *ALMA* starts operation.

Acknowledgements

M.R wishes to acknowledge support from FONDECYT (CHILE) Grant No 12345608. She is supported by the Chilean *Center for Astrophysics* FONDAP No. 15010003.

References

Barbá, R. H., Rubio, M., Roth, M, *et al.* 2003, *AJ*, 125, 1940

Bik, A., Kaper, L., Waters, L. B. F. M., *et al.* 2006, *A&A*, 455, 561

Bloemen, J. B. G. M., Strong, A. W., Mayer-Hasselwander, H. A., *et al.* 1986, *A&A*, 154, 25

Bolatto, A. D., Simon, J. D., Stanimirović, S., *et al.* 2007, *ApJ*, 655, 212

Bolatto, A, Leroy, A., Rosolowsky, E., *et al.* 2008, *ApJ*, 686, 948

Bot, C., Boulanger, F., Lagache, G., Cambrésy, L., & Egret, D. 2004, *A&A*, 423, 567

Bot, C, Boulanger, F., Rubio, M., & Rantakyrö, F. 2006, *A&A*, 471, 103

Brandner, W., Grebel, E., Barbá, R., *et al.* 2001, *AJ*, 122, 858

Cohen, R. S, Dame, T. M, Garay,, G., *et al.* 1988, *ApJ*, 331, L95

Contursi, A., Lequeux, J., Cesarsky, D., *et al.* 2000, *A&A*, 362, 310

Contursi, A, Rubio, M., Sauvage, M., *et al.* 2007, *A&A*, 469, 539

Davies, R. D., Elliot, K. H., & Meaburn, J. 1976, *MemRAS*, 81, 89

Dupac, X., del Burgo, C., Bernard, J.-P., *et al.* 2003, *MNRAS*, 344, 105

Fukui, Y, Mizuno, N., Yamaguchi, R., *et al.* 1999, *PASJ*, 51, 745

Fukui, Y, Kawamura, A., Minamidani, T., *et al.* 2008, *ApJS*, 178, 56

Galliano, F., Madden, S. C., Jones, A. P., Wilson, C. D., Bernard, J. -P., & Le Peintre, F. 2003, *A&A*, 407, 159

Galliano, F., Madden, S. C., Jones, A. P., Wilson, C. D., & Bernard, J. -P. 2005, *A&A*, 434, 867

Gouliermis, D., Brandner, W., & Henning, Th. 2006, *ApJ*, 636, L133

Gouliermis, D. A., Chu, Y. -H., Henning, Th., *et al.* 2008, *ApJ*, arXiv:0710.1352

Gruendl, R. A., Chu, Y., *et al.* 2008, submitted

Henize, K. H. 1956, *ApJS*, 2, 315

Israel, F. P., Johansson, L. E. B., Lequeux, J., *et al.* 1993, *A&A*, 276, 25

Jones, T. J., Woodward, C. E., Boyer, M. L., *et al.* 2005, *ApJ*, 620, 731

Larson, R. B. 1981, *MNRAS*, 194, 809

Lequeux, J., Le Bourlot, J., Pineau des Forêts, G., Roueff, E., Boulanger, F., & Rubio, M. 1994, *A&A*, 292, 371

Leroy, A., Bolatto, A., Stanimirović, S., Mizuno, N., Israel, F., & Bot, C. 2007, *ApJ*, 658, 1027

Massey, P. 2002, *ApJS*, 141, 81

Meixner, M., Gordon, K., Indebetouw, R., *et al.* 2006, *AJ*, 133, 2268

Mizuno, N, Rubio, M., Mizuno, A., Yamaguchi, R., Onishi, T., & Fukui, Y. 2001, *PASJ*, 53, 971

Mizuno, N., Muller, E., Maeda, H., *et al.* 2006, *ApJ*, 643, 107

Nishiyama, S., Haba, Y., Kato, D., *et al.* 2007, *ApJ*, 658, 358

Nota, A., Sirianni, M., Sabbi, E., *et al.* 2006, *ApJ*, 640, L29

Oliveira, J. M., van Loon, J. Th., Stanimirović, S., & Zijlstra, A.A. 2006, *MNRAS*, 373, 75

Perez, L. 2005, MSc thesis, Universidad de Chile, Santiago, Chile

Reach, W. T., Boulanger, F., Contursi, A, *et al.* 2000, *A&A*, 361, 895

Rubio, M., Garay, G, Montani, J., *et al.* 1991, *ApJ*, 368, 173

Rubio, M., Lequeux, J., Boulanger, F., *et al.* 1993, *A&A*, 271, 1

Rubio, M., Barbá, R. H., Walborn, N., *et al.* 1998, *AJ* 116, 1708

Rubio, M., Contursi, A., Lequeux, J., *et al.* 2000, *A&A*, 359, 1139

Rubio, M., Boulanger, F., Rantakyrö, F., & Contursi, A., 2004, *A&A*, 425, L1

Rubio, M. 2007, in B. G. Elmegreen & J. Palouš (eds.), *Triggered star formation in a turbulent ISM*, IAU Conf.Proc. 237 (Cambridge: CUP), p. 40

Rubio, M., & Barbá, R. H. 2008, *ApJL*, submitted

Rubio, M., Paron, S., & Dubner, G. 2008, *A&A*, submitted

Sabbi, E., Siriani, M, Nota, A., *et al.* 2008, *AJ*, 135, 173

Simon, J. D., Bolatto, A. D., Whitney, B. A., *et al.* 2007, *ApJ*, 669, 327

Solomon, P. M., Rivolo, A. R., Barrett, J., & Yahil, A. 1987, *ApJ*, 319, 730

Walborn, N. R., Barbá, R. H., Brandner, W., *et al.* 1999, *AJ*, 117, 225

Weingartner, J. C., & Draine, B. T. 2001, *ApJ*, 548, 296

Whitney, B. A., Sewilo, M., Indebetouw, R., *et al.* 2008, *AJ*, 136, 18

Yamaguchi, R., Mizuno, N., Mizuno, A., *et al.* 2001, *PASJ*, 53, 985

The Magellanic System: Stars, Gas, and Galaxies
Proceedings IAU Symposium No. 256, 2008
Jacco Th. van Loon & Joana M. Oliveira, eds.

Star-formation masers in the Magellanic Clouds: A multibeam survey with new detections and maser abundance estimates

J. A. Green[1,2]†, **J. L. Caswell**[2], **G. A. Fuller**[1], **A. Avison**[1],
S. L. Breen[2,3], **K. Brooks**[2], **M. G. Burton**[4], **A. Chrysostomou**[5],
J. Cox[6], **P. J. Diamond**[1], **S. P. Ellingsen**[3], **M. D. Gray**[1], **M. G. Hoare**[7],
M. R. W. Masheder[8], **N. M. McClure-Griffiths**[2], **M. Pestalozzi**[5,11],
C. Phillips[2], **L. Quinn**[1], **M. A. Thompson**[5], **M. A. Voronkov**[2],
A. Walsh[9], **D. Ward-Thompson**[6], **D. Wong-McSweeney**[1], **J. A. Yates**[10]
and R. J. Cohen[1]‡

[1]Jodrell Bank Centre for Astrophysics, Alan Turing Building, University of Manchester, Manchester, M13 9PL, UK

[2]Australia Telescope National Facility, CSIRO, PO Box 76, Epping, NSW 2121, Australia

[3]School of Mathematics and Physics, University of Tasmania, Private Bag 37, Hobart, TAS 7001, Australia

[4]School of Physics, University of New South Wales, Sydney, NSW 2052, Australia

[5]Centre for Astrophysics Research, Science and Technology Research Institute, University of Hertfordshire, College Lane, Hatfield, AL10 9AB, UK

[6]Department of Physics and Astronomy, Cardiff University, 5 The Parade, Cardiff, CF24 3YB, UK

[7]School of Physics and Astronomy, University of Leeds, Leeds, LS2 9JT, UK

[8]Astrophysics Group, Department of Physics, Bristol University, Tyndall Avenue, Bristol, BS8 1TL, UK

[9]School of Maths, Physics and IT, James Cook University, Townsville, QLD 4811, Australia

[10]University College London, Department of Physics and Astronomy, Gower Street, London, WC1E 6BT, UK

[11]Göteborgs Universitet Insitutionen för Fysik, Göteborg, Sweden

Abstract. The results of the first complete survey for 6668-MHz CH_3OH and 6035-MHz excited-state OH masers in the Small and Large Magellanic Clouds are presented. A new 6668-MHz CH_3OH maser in the Large Magellanic Cloud has been detected towards the star-forming region N 160a, together with a new 6035-MHz excited-state OH maser detected towards N 157a. We also re-observed the previously known 6668-MHz CH_3OH masers and the single known 6035-MHz OH maser. Neither maser transition was detected above ~ 0.13 Jy in the Small Magellanic Cloud. All observations were initially made using the CH_3OH Multibeam (MMB) survey receiver on the 64-m *Parkes* radio telescope as part of the overall MMB project. Accurate positions were measured with the *Australia Telescope Compact Array (ATCA)*. In a comparison of the star formation maser populations in the Magellanic Clouds and our Galaxy, the LMC maser populations are demonstrated to be smaller than their Milky Way counterparts. CH_3OH masers are under-abundant by a factor of ~ 50, whilst OH and H_2O masers are a factor of ~ 10 less abundant than our Galaxy.

Keywords. masers, surveys, stars: formation, Magellanic Clouds

† E-mail:james.green@csiro.au
‡ Deceased 2006 November 1.

1. Introduction

The Magellanic Clouds exhibit maser emission across a number of molecular transitions from both regions where stars are forming (e.g., UCHIIs) and regions where they are dying (e.g., red supergiants and asymptotic giant branch stars). The latter, collectively known as circumstellar masers, and example transitions being 22-GHz water and 43, 86 and 123-GHz silicon oxide, are explored in detail elsewhere (van Loon *et al.* 1998; van Loon *et al.* 2001). Also dealt with elsewhere are the Magellanic Cloud masers associated with supernova remnants (Brogan *et al.* 2004; Roberts & Yusef-Zadeh 2005). The current work focuses purely on masers associated with star-formation, specifically water (H_2O), hydroxyl (OH) and methanol (CH_3OH).

Only three extragalactic 6668-MHz CH_3OH masers were known to exist before the current survey, and all were seen in the LMC. They were detected towards the nebulae N 105a/MC23 (Sinclair *et al.* 1992) and N 11/MC18 (Ellingsen *et al.* 1994), and towards the IRAS source 05011-6815 (Beasley *et al.* 1996). One 6035-MHz maser was known in the Magellanic clouds, detected towards the nebula N 160a/MC76 (Caswell 1995). The first systematic survey for 6668-MHz CH_3OH and 6035-MHz OH masers in the Magellanic clouds has been completed with the *Parkes* 64-m radio telescope. We report on the new detections of the survey together with maser abundance estimates.

2. Observations

Both the LMC and the SMC were surveyed by the Methanol Multibeam (MMB) project (Cohen *et al.* 2007) in the sidereal time ranges when the Galactic plane was not visible from the *Parkes* radio telescope. The observational setup was the same as for the main MMB survey, and as such is described in detail in Green *et al.* (2008). In addition to scanning the Magellanic Clouds in the same way as the Galactic plane survey, we also conducted targeted observations of known maser and star-formation regions. The scanning survey regions were chosen to fully sample the CO and HI distributions of Fukui (2001) and Staveley-Smith (2003). The LMC region surveyed is shown in Fig. 1, along with the CO clouds contained within the region. For the SMC the MMB team mapped $299° < l < 305°$, $-42° < b < -46°$, taking two passes and resulting in an rms noise of ~ 0.13 Jy. The LMC was scanned across $2°$ in longitude at a rate of $0.08°$ per min. These scans were separated by alternating latitude steps of 1.07 arcmin and 15 arcmin. This fully samples a $2° \times 7°$ block of the LMC in 56 scans. The SMC was scanned across $3°$ in longitude at a rate of $0.15°$ per min, fully sampling a $3° \times 4°$ block in 32 scans. Data processing for the scanning technique is the same as for the MMB Galactic survey Green *et al.* (2008). The scanning observations were made over the period 2006 January to 2007 November, concurrently with the Galactic survey, making use of the complementary LST range.

Targeted observations were performed toward 11 star-formation regions in the LMC and two known 22-GHz H_2O masers (Scalise & Braz 1982) in the SMC (see Table 1). These points in the sky were tracked with the pointing centre cycling through each of the seven receiver beams. This means each of the seven receiver beams was on source for 10 minutes (with the exception of N 157a where 20 minute integrations were taken). Data processing for the targeted pointings used the package ASAP. The bandpass reference for each beam was estimated using the median spectrum of the six "off-source" positions, and each median reference then combined with the corresponding total power spectrum "on-source" to determine a baseline- and gain-corrected quotient spectrum. The seven spectra were then combined to give a final best spectrum with an effective 70 min integration time (140 min for N 157a) and a typical rms of ~ 25 mJy in the total intensity spectrum. Flux densities were calibrated using observations of the continuum source PKS 1934−638. The

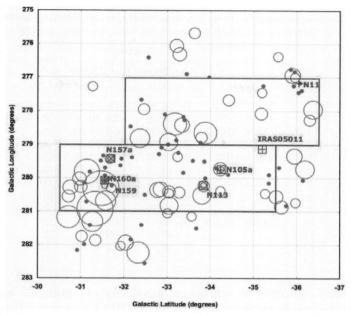

Figure 1. Map of the survey region of the LMC with small and large CO (J = 1−0) clouds from the data of Fukui *et al.* (2001) with overlaid maser positions. Small CO clouds are represented by dots and large clouds by circles scaled proportionally to the total cloud surface area. The labelled symbols represent the masers with squares the ground-state OH, circles 22-GHz H_2O, crosses 6035-MHz excited-state OH, and pluses 6668-MHz CH_3OH. One pass was conducted over the full region, $275° < l < 283°$, $−30° < b < −37°$ with an rms noise of ~ 0.22 Jy. All regions with detected CO received a second pass and then as the known CH_3OH masers were all within the middle two blocks, the MMB team chose to concentrate on these and thus conducted a further two passes over $277° < l < 281°$, $−30° < b < −37°$, resulting in an rms noise of ~ 0.09 Jy. We then further concentrated on the regions $279° < l < 281°$, $−30.5° < b < −35.5°$ and $277° < l < 279°$, $−32° < b < −36.5°$. These regions contain all the known masers, the largest CO clouds and over 70 % of the small CO clouds. A further four passes were conducted on these two regions (solid line outlined region), resulting in an rms noise of ~ 0.06 Jy.

targeted *Parkes* observations were taken in 2006 June (N 11, N 105a, IRAS 05011−6815), 2006 September 3 (N 160a) and 2007 January 25 (N 157a) and are listed with their rms noise levels in Table 1.

In addition to the targeted *Parkes* observations, the *Australia Telescope Compact Array* (*ATCA*) was used 2007 July 21−22 to obtain precise positions for the new CH_3OH maser and for the two excited-state OH masers. A loop of the three targets and the phase calibrator 0530−727 was repeated 22 times over a 10-hour time span and five times on the following day. Within each loop, the three targets had integration times of 15, 2 and 4 minutes respectively. The array was in a 6 km configuration (6C), yielding a FWHM beamsize of 2.8 × 1.6 arcsec. The bandwidth was 4 MHz, spread across 2048 frequency channels, the same as for the *Parkes* spectra.

3. Results

In addition to detecting the three known CH_3OH masers at N 11, N 105a and IRAS 05011−6815 (Fig. 1), a new CH_3OH maser was detected ($> 5\sigma$ in the targeted observations) in association with the known excited-state OH maser in N 160a. The maser had

Table 1. *Parkes* pointing positions for targeted regions in the Magellanic Clouds, together with rms noise levels for the total intensity spectra.

Position	Name	RA(J2000) h m s	Dec(J2000) ° ′ ″	rms noise (mJy)
1	N 11/MC18	04 56 47	−66 24 35	23
2	IRAS 05011	05 01 02	−68 10 28	26
3	N 105a/MC23	05 09 52	−68 53 29	26
4	″	05 09 59	−68 54 34	26
5	N 113/MC24	05 13 25	−69 22 46	26
6	″	05 13 18	−69 22 21	26
7	N 157a/MC74	05 38 47	−69 04 46	20
8	″	05 38 45	−69 05 07	20
9	N 159	05 39 29	−69 47 19	26
10	N 160a/MC76	05 39 44	−69 38 34	26
11	″	05 39 39	−69 39 11	26
12	S7	00 46 39	−72 40 49	22
13	S9	00 47 31	−73 08 20	21

a peak flux density of 0.20 Jy at a heliocentric velocity of 248 km s^{-1} and is spatially coincident with the previously known excited-state OH. A new excited-state OH maser was also detected ($> 5\sigma$ in the targeted observations), through a pointing at N 157a. The maser had a peak flux density of 0.22 Jy at a heliocentric velocity of 258 km s^{-1}. Nothing was detected towards the SMC pointings or towards an additional H$_2$O maser (recently discovered by J. Lovell, private communication).

4. The abundance of star-formation masers in the Magellanic Clouds

Through adopting a "luminosity" parameter of the peak flux density multiplied by the square of the distance, with a distance to the LMC of 50 kpc (Feast 1999), the maser populations of our Galaxy and the LMC can be compared to provide insight into how maser emission is affected by environmental parameters, such as metallicity and the ambient UV field, together with tracing different rates of star-formation. The lowest luminosity maser observed in the Magellanic Clouds for each species is taken as a "cut-off", and for sources above this cut-off an empirical ratio of the size of the maser population is estimated. The Galactic populations are taken as the estimated number of masers above the equivalent Galactic "cut-off". In cases where the Galactic population is well understood, but the Magellanic Cloud population is incomplete (even above the cut-off), then the ratio will be an upper limit. The results are discussed below by species and are summarised in Table 2.

1665-MHz OH. The Brooks & Whiteoak (1997) measurements list four sites of maser emission in the LMC, with strongest peaks ranging from 221 mJy to 580 mJy. Their measurements could have detected sources as weak as 50 mJy in favourable cases, but 200 mJy is treated as the cut-off, i.e. 500 Jy kpc^2. Adopting the ground-state OH luminosity function of Caswell & Haynes (1987), but scaling to a Sun−Galactic Centre distance of 8.5 kpc, the Galactic population is estimated to be 60 sources above the corrected cut-off and a ratio of $\leqslant 15$. The LMC results do not represent a complete survey, so the ratio is taken as a lower limit.

6035-MHz OH. The two LMC detections have "luminosities" of ~ 625 Jy kpc^2 and ~ 1000 Jy kpc^2 and so for this transition also, we adopt an approximate cut-off of 500 Jy kpc^2. From the discussion of Caswell & Vaile (1995), updated to include information from Caswell (2001) and Caswell (2003), only two sources in the Galaxy are found to

Table 2. Comparison of maser populations. References: [a]Caswell & Haynes (1987); [b]Greenhill *et al.* (1990); [c]Sinclair *et al.* (1992); [d]Ellingsen *et al.* (1994); [e]Caswell & Vaile (1995); [f]Beasley *et al.* (1996); [g]Brooks & Whiteoak (1997); [h]Lazendic *et al.* (2002) and van Loon & Zijlstra (2001); [i]Current Work; [†]Although the Galactic ground-state OH population is unlikely to change significantly, both the CH₃OH and excited-state OH Galactic populations will be better defined once the MMB is completed and the excited-state OH Galactic population could conceivably increase. Therefore the 1:1 ratio for excited-state OH should be considered with this in mind. Luminosities for the LMC are based on a distance of 50 kpc and Galactic luminosities assume that the Galactic Centre is 8.5 kpc from the Sun. *See §5.4.1.

Molecule	Transition (MHz)	Luminosity Cut-off ($Jy\,kpc^2$)	LMC Pop.	Galactic Pop.	Pop. Ratio (Gal/LMC)
OH	1665	500	$\geqslant 4^g$	60^a	$\leqslant 15$
OH	6035	500	$\geqslant 2^{e,i}$	$2,20^e$	$\leqslant 1^\dagger$, $\leqslant 10^*$
CH₃OH	6668	500	$\geqslant 4^{c,d,f,i}$	214^i	$\leqslant 53^\dagger$
H₂O	22235	2500	$\geqslant 8^h$	88^b	$\leqslant 11$

be above this threshold. However, these are clearly very small number statistics, and this problem may be countered by using information on the whole Galactic population (exceeding 100). A general comparison with ground-state OH suggests that, to similar luminosity limits, the 6035-MHz masers are one-third as common (Caswell 2001). That would suggest a Galactic population of 20 above the cut-off, and a ratio of ~ 10.

6668-MHz CH₃OH. The sources in N 11, N 105a and N 160a would have "luminosities" between ~ 500 and $\sim 750\,Jy\,kpc^2$, giving a luminosity threshold of $500\,Jy\,kpc^2$. From the preliminary results of the MMB survey, resolving kinematic distances where possible and assuming the remaining are at the near distance gives a Galactic population of 214 (which is likely to be a lower limit as it is possible a larger proportion than assumed could be at the far distance). This results in a difference of populations by a factor of ~ 50. As the CH₃OH results are based on a complete survey of the LMC, the population ratio for this maser species should be the most reliable. It therefore seems significant that, compared to our own Galaxy, the CH₃OH masers in the LMC are under-abundant relative to OH, which is perhaps suggestive that the LMC environment is less conducive for CH₃OH emission, than OH.

22-GHz H₂O. The small beams at this frequency make both the Galactic and Magellanic populations likely to be severely incomplete. However, using the known detections as described in Lazendic *et al.* (2002) coupled with the detection by van Loon & Zijlstra (2001), we have a minimum flux density of $\sim 1\,$Jy (when smoothing the 1.6 Jy weakest maser to a linewidth of $1\,km\,s^{-1}$). At the LMC distance this gives a "luminosity" threshold of $2500\,Jy\,kpc^2$ (a factor of ~ 5 greater than the previous similar study of Brunthaler *et al.* 2006). Applying the derived Galactic luminosity function of Greenhill *et al.* (1990) this gives a Galactic population of 88 and a ratio of ~ 11.

5. Conclusions

A complete survey of the Magellanic Clouds has been conducted as part of the MMB project and supplemented by a targeted search of known star-formation regions in the Large Magellanic Cloud. This has detected two new extragalactic masers: a fourth 6668-MHz CH₃OH maser towards the star-forming region N 160a in the LMC; and a second extragalactic 6035-MHz OH maser towards N 157a. The MMB team have also re-observed the three previously known 6668-MHz CH₃OH masers and the single 6035-MHz OH maser in the LMC. All these masers have had their positions determined with the *ATCA* and exhibit stability in spectral structure and intensity, even over the ~ 10 year separation between the current and previous observations. The current lower limits on the LMC maser populations are demonstrated to be up to a factor of ~ 50 smaller than the Milky Way, with the CH_3OH showing the largest discrepancy. With the LMC star formation rate believed to be about a tenth of our Galaxy (Israel 1980), H_2O and OH masers are broadly compatible, but the CH_3OH masers in the LMC are still under-abundant by a factor of ~ 5. This remaining disparity may be due to the lower oxygen and carbon abundances in the LMC, in agreement with the speculation of Beasley *et al.* (1996).

Acknowledgements

JAG, AA, JCox and DW-McS acknowledge the support of a Science and Technology Facilities Council (STFC) studentship. LQ acknowledges the support of the EU Framework 6 Marie Curie Early Stage Training programme under contract number MEST-CT-2005-19669 "ESTRELA". The Parkes Observatory and the Australia Telescope Compact Array are part of the Australia Telescope which is funded by the Commonwealth of Australia for operation as a National Facility managed by CSIRO. The authors dedicate this paper to the memory of R. J. Cohen.

References

Beasley, A. J., Ellingsen, S. P., Claussen, M. J., & Wilcots, E. 1996, *ApJ*, 459, 600
Brogan, C. L., Goss, W. M., Lazendic, J. S., & Green, A. J. 2004, *ApJ*, 128, 700
Brooks, K. J. & Whiteoak, J. B. 1997, *MNRAS*, 291, 395
Brunthaler, A., Henkel, C., de Blok, W. J. G., Reid, M. J., Greenhill, L. J., & Falcke, H. 2006, *A&A*, 457, 109
Caswell, J. L. 1995, *MNRAS*, 272, L31
Caswell, J. L. & Vaile, R. A. 1995, *MNRAS*, 273, 328
Caswell, J. L. 2001, *MNRAS*, 326, 805
Caswell, J. L. 2003, *MNRAS*, 341, 551
Caswell, J. L. & Haynes, R. F. 1987, *Aust. J. Phys.*, 40, 855
Cohen, R. J., Caswell, J. L., Brooks, K., *et al.* 2007, *IAUS*, 237, 403
Ellingsen, S. P., Whiteoak, J. B., Norris, R. P., Caswell, J. L., & Vaile, R. A. 1994, *MNRAS*, 269, 1019
Feast, M. 1999, *PASP*, 111, 775
Fukui, Y., Mizuno, N., Yamaguchi, R., Mizuno, A., & Onishi, T. 2001, *PASJ*, 53, 41
Green, J. A., Caswell, J. L., Fuller, G. A., *et al.*, 2008, *MNRAS*, 385, 948
Greenhill, L. J., Moran, J. M., Reid, M. J., *et al.* 1990, *ApJ*, 364, 513
Israel, F. P. 1980, *A&A*, 90, 246
Lazendic, J. S., Whiteoak, J. B., Klamer, I., Harbison, P. D., & Kuiper, T.B.H. 2002, *MNRAS*, 331, 969
Pestalozzi, M. R, Minier, V., & Booth, R. S. 2005, *A&A*, 432, 737
Pestalozzi, M. R, Chrysostomou, A., Collett, J. L., Minier, V., Conway, J., & Booth, R. S. 2007, *A&A*, 463, 1009
Roberts, D. A. & Yusef-Zadeh, F. 2005, *ApJ*, 129, 805
Scalise, E. & Braz, M. A. 1982, *ApJ*, 87, 528
Sinclair, M. W., Carrad, G. J., Caswell, J. L., Norris, R. P., & Whiteoak, J. B. 1992, *MNRAS*, 256, 33
Staveley-Smith, L., Kim, S., Calabretta, M. R., Haynes, R. F., & Kesteven, M. J. 2003, *MNRAS*, 339, 87
van Loon, J. T., te Lintel Hekkert, P., Bujarrabal, V., Zijlstra, A. A., & Nyman, L.-Å. 1998, *A&A*, 337, 141
van Loon, J. T. & Zijlstra, A. A. 2001, *ApJ*, 547, L61
van Loon, J. T., Zijlstra, A A., Bujarrabal, V., & Nyman, L.-Å. 2001, *A&A*, 368, 950

The Magellanic System: Stars, Gas, and Galaxies
Proceedings IAU Symposium No. 256, 2008
Jacco Th. van Loon & Joana M. Oliveira, eds.

AKARI near-infrared spectroscopy: Detection of H_2O and CO_2 ices toward young stellar objects in the Large Magellanic Cloud

Takashi Shimonishi[1], Takashi Onaka[1], Daisuke Kato[1], Itsuki Sakon[1], Yoshifusa Ita[2], Akiko Kawamura[3] and Hidehiro Kaneda[4]

[1]Department of Astronomy, Graduate School of Science, The University of Tokyo, 7-3-1 Hongo, Bunkyo-ku, Tokyo 113-0003, Japan, email: shimonishi@astron.s.u-tokyo.ac.jp

[2]National Astronomical Observatory of Japan, Japan

[3]Department of Astrophysics, Nagoya University, Japan

[4]Institute of Space and Astronautical Science, Japan Aerospace Exploration Agency, Japan

Abstract. We present the first results of the *AKARI* Infrared Camera near-infrared spectroscopic survey of the Large Magellanic Cloud (LMC). The circumstellar material of young stellar objects (YSOs) are affected by galactic environments such as a metallicity or radiation field. Ices control the chemical balance of circumstellar environments of embedded YSOs. We detected absorption features of the H_2O ice 3.05 μm and the CO_2 ice 4.27 μm stretching mode toward seven massive YSOs in the LMC. This is the first detection of the 4.27 μm CO_2 ice feature toward extragalactic YSOs. The present samples are for the first time spectroscopically confirmed to be YSOs. We used a curve-of-growth method to evaluate the column densities of the ices and derived the CO_2/H_2O ratio to be 0.45±0.17. This is clearly higher than that seen in Galactic massive YSOs (0.17±0.03). We suggest that the strong ultraviolet radiation field and/or the high dust temperature in the LMC may be responsible for the observed high CO_2 ice abundance.

Keywords. circumstellar matter, stars: pre-main-sequence, ISM: molecules, galaxies: individual (LMC), Magellanic Clouds

1. Introduction

Properties of extragalactic young stellar objects (YSOs) provide us important information on the understanding of the diversity of YSOs in different galactic environments. The Large Magellanic Cloud (LMC), the nearest irregular galaxy to our Galaxy (∼50 kpc; Alves 2004), offers an ideal environment for this study since it holds a unique metal-poor environment (Luck *et al.* 1998). Because of its proximity and nearly face-on geometry, various types of surveys have been performed toward the LMC (e.g., Zaritsky *et al.* 2004; Meixner *et al.* 2006; Kato *et al.* 2007, and references therein).

An infrared spectrum of a YSO shows absorption features of various ices which are thought to be an important reservoir of heavy elements and complex molecules in a cold environment such as a dense molecular cloud or an envelope of a YSO (e.g., Chiar *et al.* 1998; Nummelin *et al.* 2001; Whittet *et al.* 2007; Boogert *et al.* 2008). These ices are thought to be taken into planets and comets as a result of subsequent planetary formation activity (Ehrenfreund & Schutte 2000). Studying the compositions of ices as functions of physical environments is crucial to understand the chemical evolution in circumstellar environments of YSOs and is a key topic of astrophysics. H_2O and CO_2 ices are ubiquitous and are major components of interstellar ices (Boogert & Ehrenfreund 2004). Since the absorption profile of the ices is sensitive to the chemical composition

Table 1. Observation parameters and column densities of ices.

No.	AKARI ID	Obs. Date	Other Name	RA (J2000.0)	DEC (J2000.0)	$N(H_2O)$ (10^{17} cm^{-2})	$N(CO_2)$ (10^{17} cm^{-2})
ST1	J053931−701216	2007 Apr 12	05393117−7012166[a]	5:39:31.15	−70:12:16.8	$9.6^{+1.9}_{-1.9}$	$6.7^{+4.8}_{-3.6}$
ST2	J052212−675832	2006 Nov 24	NGC 1936	5:22:12.56	−67:58:32.2	$11.1^{+1.6}_{-1.4}$	$3.1^{+1.7}_{-1.5}$
ST3	J052546−661411	2007 Jun 22	......[b]	5:25:46.69	−66:14:11.3	$29.7^{+4.5}_{-4.6}$	$15.5^{+13.8}_{-9.4}$
ST4	J051449−671221	2006 Jun 6	IRAS F05148−6715	5:14:49.41	−67:12:21.5	$18.7^{+2.3}_{-2.3}$	$8.2^{+4.5}_{-3.7}$
ST5	J053054−683428	2007 Mar 13	IRAS 05311−6836	5:30:54.27	−68:34:28.2	$31.7^{+4.7}_{-4.5}$	$12.4^{+9.1}_{-6.8}$
ST6	J053941−692916	2006 Oct 25	05394112−6929166[a]	5:39:41.08	−69:29:16.8	$59.1^{+38.4}_{-25.0}$	...
ST7	J052351−680712	2006 Nov 29	IRAS 05240−6809	5:23:51.15	−68:07:12.2	...	$55.3^{+40.3}_{-30.3}$

Notes: [a] 2MASS ID; [b] The source is in a cluster.

of icy grain mantles and the thermal history of local environments, the ices are important tracers to investigate the properties of YSOs. However, our knowledge about the ices around extragalactic YSOs is limited because few observations have been performed toward extragalactic YSOs (e.g., van Loon *et al.* 2005). Therefore infrared spectroscopic observations toward YSOs in the LMC are important if we are to improve our understanding of the influence of galactic environments on the properties of YSOs and ices (Shimonishi *et al.* 2008).

AKARI is the first Japanese satellite dedicated to infrared astronomy launched in February 2006 (Murakami *et al.* 2007). We have performed a near infrared spectroscopic survey of the LMC using the powerful spectroscopic survey capability of the Infrared Camera (IRC; Onaka *et al.* 2007) on board *AKARI*. In this paper, we present 2.5–5 μm spectra of newly confirmed YSOs in the LMC with our survey, and discuss the abundances of H_2O and CO_2 ice.

2. Observations and data reduction

The observations reported here were obtained as a part of the *AKARI* IRC survey of the LMC (Ita *et al.* 2008). An unbiased slit-less prism spectroscopic survey of the LMC has been performed since May 2006. In this survey, the IRC02b *AKARI* astronomical observing template (AOT) with the NP spectroscopy mode was used to obtain low-resolution spectra ($R \sim 20$) between 2.5 and 5 μm.

The spectral analysis was performed using the standard IDL package prepared for the reduction of *AKARI* IRC spectra (Ohyama *et al.* 2007). The wavelength calibration accuracy is estimated to be about $\sim$0.01 μm (Ohyama *et al.* 2007).

3. The selection of YSOs

We select infrared-bright objects from the point-source catalog of the *Spitzer* SAGE project (Meixner *et al.* 2006) with the following selection criteria: (1) [3.6] − [4.5] > 0.3 mag and [5.8] − [8.0] > 0.6 mag, and (2) [3.6] < 12 mag and [4.5] < 11.5 mag, where [wavelength] represents the photometric value in magnitude at each wavelength in μm. The criterion (1) refers to the YSO model of Whitney *et al.* (2004), and the criterion (2) comes from the detection limit of the *AKARI* IRC NP mode. This rough selection is applied to the sources located in the survey area of *AKARI* IRC, and about 300 sources are selected. These photometrically selected sources include not only massive YSOs, but also a large number of dusty evolved stars since their infrared spectral energy distribution (SED) are similar to each other. For the accurate selection of YSOs, we select the sources that show absorption features of the 3.05 μm H_2O ice and the 4.27 μm CO_2 ice stretching mode in their NIR spectra taken by the present spectroscopic

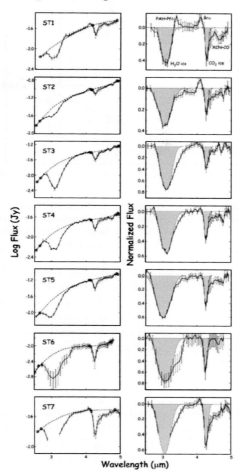

Figure 1. *AKARI* IRC 2.5–5μm spectra of YSOs in the LMC. Left: Plots of $\log_{10}$ flux(Jy) vs. wavelength(μm). Open circles represent the points used for the continuum determination. Dashed lines represent the derived continuum. Right: Plots of normalized flux $(1 - F/F_c$; F and F_c represent observed flux and continuum flux, respectively) vs. wavelength(μm). Shaded regions around 3.0 and 4.3 μm represent fitted Gaussians for the absorption features of H_2O and CO_2 ices, and the areas correspond to the equivalent widths. The positions of H_2O, CO_2, XCN, CO ice absorption bands, PAH emission bands, and hydrogen recombination lines are shown.

survey. The presence of CO_2 ice is strong evidence of YSOs since the detection of CO_2 ice toward dusty evolved stars has not been reported (Sylvester *et al.* 1999). Spectral overlapping with other sources located in the dispersion direction is a serious problem for slit-less spectroscopy, which makes it difficult to obtain reliable spectra. We check the overlapping contamination by visual inspection, and we only use the sources without such contamination in the following analysis.

As a result, we spectroscopically confirmed seven massive YSOs in the LMC for the first time. The sources are listed in Table 1 with the observation parameters. Six of the seven sources are included in the recent YSO candidates catalog (Whitney *et al.* 2008), and one source (ST6) is a newly found YSO. The spectra of these sources are shown in Fig. 1 together with the results of spectral fitting (see §4 for details). The absorption features of H_2O and CO_2 ices are rather broadened due to the low spectral resolution

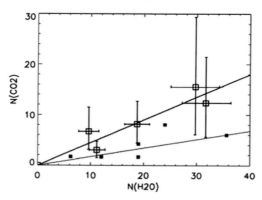

Figure 2. CO_2 ice vs. H_2O ice column density in units of 10^{17} cm^{-2}. Open squares with error bars represent the results of this study. Filled squares represent those of Galactic massive YSOs (Gibb *et al.* 2004). Upper and lower solid lines represent $CO_2/H_2O \sim 0.45$ and 0.17, respectively. The sources ST6 and ST7 are not plotted due to their large errors.

of the *AKARI* IRC NP spectroscopy mode but clearly detected. This is the first clear detection of the 4.27 μm stretching mode of CO_2 ice toward extragalactic YSOs. In addition, unresolved emission of PAHs and the hydrogen recombination line Pfδ around 3.3 μm, the Brα line at 4.05 μm and blended absorption features of 4.62 μm XCN and 4.67 μm CO ices around 4.65 μm are detected toward several sources. However, it is difficult to evaluate the column densities of XCN and CO ices accurately with the present low spectral resolution data.

4. Spectral fitting

We fit a polynomial of the second to fourth order to the continuum regions and divide the spectra by the fitted continuum (Fig. 1). The wavelength regions for the continuum are set to be 2.5–2.7 μm, 3.6–3.7 μm, 4.0–4.15 μm, and 4.9–5.0 μm (Gibb *et al.* 2004).

Due to the low spectral resolution of the IRC NP spectroscopy mode, direct comparison of the observed spectra with the absorption profiles of laboratory ices is difficult. However, the equivalent width of absorption does not depend on the spectral resolution of the spectrum. Therefore we used a curve-of-growth method to derive the column densities of the ices. A Gaussian profile with the fixed central wavelength is fitted to the absorption bands to derive the equivalent width (Fig. 1). The vertical axis of the plot is shown in units of the normalized flux because it is difficult to estimate the optical depth directly from the present low resolution data. We use the laboratory absorption profiles of H_2O and CO_2 ices taken from the Leiden Molecular Astrophysics database (Ehrenfreund *et al.* 1996) to calculate the curve-of-growth. The profiles of pure H_2O ice and mixture of H_2O:CO_2 (100:14) ice both at 10 K are used for the calculation since these compositions are typical in the interstellar ices (Nummelin *et al.* 2001; Gibb *et al.* 2004). The present spectrum cannot resolve the polar and apolar CO_2 ice features, and the present analysis assumes the polar CO_2 ice only. However, contribution of the apolar ice is generally small toward YSOs (Gerakines *et al.* 1995, 1999). We adopt the band strengths of H_2O and CO_2 ices to be 2.0×10^{-16} and 7.6×10^{-17} cm molecule^{-1} (Gerakines *et al.* 1995), respectively. The derived column densities are listed in Table 1.

5. Results and discussion

The obtained column densities of H_2O and CO_2 ices are plotted in Fig. 2. The error bars become larger for the larger column density due to the saturation effect of the curve-of-growth. A linear fit to the data points indicates that the CO_2/H_2O ice column density ratio in the LMC is 0.45±0.17. The large uncertainty mainly comes from the errors in the curve-of-growth analysis. For comparison, column densities of Galactic massive YSOs taken from Gibb *et al.* (2004) and their CO_2/H_2O ice column density ratio of 0.17±0.03 (Gerakines *et al.* 1999) are also plotted in Fig. 2. A similar CO_2/H_2O ratio of 0.18±0.04 is also observed toward a Galactic quiescent dark cloud (Whittet *et al.* 2007), while a relatively high CO_2/H_2O ratio of 0.32±0.02 is observed toward Galactic low- and intermediate-mass YSOs, and some of them reach ∼0.4 (Pontoppidan *et al.* 2008). Although the uncertainty is large, it is clear from the present results that the CO_2/H_2O ice ratio in the LMC is higher than the typical ratios of the Galactic objects. Since the distribution range of the H_2O ice column density in the LMC is comparable to that of the massive Galactic YSOs, it can be concluded that the abundance of the CO_2 ice is higher in the LMC. The present results suggest that the different galactic environment of the LMC is responsible for the high CO_2 abundance.

The formation mechanism of CO_2 ice in circumstellar environments of YSOs is not understood, but a number of scenarios have been proposed. Several laboratory experiments indicate that CO_2 ice is efficiently produced by UV photon irradiation of H_2O-CO binary ice mixtures (e.g., Watanabe *et al.* 2007). The LMC has an order-of-magnitude stronger UV radiation field than our Galaxy due to its active massive star formation (Israel *et al.* 1986), which could lead to higher CO_2/H_2O ratios in the LMC. A high CO_2/H_2O ratio toward a YSO in the LMC is also reported in van Loon *et al.* (2005), who suggest that a different radiation environment in the LMC is one of the reasons for the high CO_2 abundance. On the other hand, models of diffusive surface chemistry suggests that high abundance of CO_2 ice can be produced at relatively high dust temperatures (Ruffle & Herbst 2001). Several studies have reported that the dust temperature in the LMC is generally higher than in our Galaxy based on far-infrared to submillimeter observations of diffuse emission (e.g., Sakon *et al.* 2006). Therefore the high dust temperature may also have an effect on the high CO_2 ice abundance in the LMC.

6. Summary and future works

We performed a near infrared spectroscopic survey of the LMC with *AKARI* IRC. We spectroscopically confirmed seven massive YSOs that show absorption features of H_2O and CO_2 ices. This is the first detection of the 4.27 μm CO_2 ice feature toward extragalactic YSOs. The derived ice column densities indicate that the abundance of CO_2 ice is clearly higher in the LMC than in our Galaxy. The relatively strong UV radiation field and/or high dust temperature in the LMC may be responsible for the observed high abundance of CO_2 ice. Our study shows the difference in the chemical composition around extragalactic YSOs, suggesting that extragalactic YSOs hold quite different environments from Galactic ones.

With our low resolution NIR spectra, it is difficult to separate the effect of the UV radiation field and the dust temperature on the high abundance of CO_2 ice in the LMC. The 4.62 μm XCN feature is known to be indicative of strong UV irradiation (e.g., Bernstein, Sandford, & Allamandola 2000). The presence of the CO ice which has a narrow absorption feature at 4.67 μm will constrain the dust temperature due to its low sublimation temperature. Furthermore, detailed profile analysis of the 3.05 μm H_2O ice stretching mode and the 15.2 μm CO_2 ice bending mode should reveal the temperatures and compositions of the ices (Ehrenfreund *et al.* 1996; Öberg *et al.* 2007). Near- to mid-infrared future observations with sufficient wavelength resolution are necessary to investigate the properties of extragalactic YSOs.

On the other hand, we need to increase samples of extragalactic YSOs which show absorption features of ices. The present spectroscopic survey is expanding the survey area, and also is performing a survey toward the Small Magellanic Cloud. These observations are expected to increase the samples of the Magellanic Clouds' YSOs, and will contribute to the study of extragalactic YSOs.

References

Alves, D. R. 2004, *New Astron. Revs*, 48, 659

Bernstein, M. P., Sandford, S. A., & Allamandola, L. J. 2000, *ApJ*, 542, 892

Boogert, A. C. A. & Ehrenfreund, P. 2004, in A. N. Witt, G. C. Clayton, & B. T. Draine (eds.), *Astrophysics of Dust*, ASP-CS, 309, p. 547

Boogert, A. C. A., Pontoppidan, K. M., Knez, C., *et al.* 2008, *ApJ*, 678, 985

Chiar, J. E., Gerakines, P. A., Whittet, D. C. B., Pendleton, Y. J., Tielens, A. G. G. M., Adamson, A. J., & Boogert A. C. A. 1998, *ApJ*, 498, 716

Ehrenfreund, P., Boogert, A. C. A., Gerakines, P. A., Jansen, D. J., Schutte, W. A., Tielens, A. G. G. M., & van Dishoeck, E. F. 1996, *A&A*, 315, L341

Ehrenfreund, P. & Schutte, W. A. 2000, *Adv. Sp. Res.*, 25, 2177

Gerakines, P. A., Schutte, W. A., Greenberg, J. M., & van Dishoeck, E. F. 1995, *A&A*, 296, 810

Gerakines, P. A., Whittet, D. C. B., Ehrenfreund, P., *et al.* 1999, *ApJ*, 522, 357

Gibb, E. L., Whittet, D. C. B, Schutte, W. A., *et al.* 2000, *ApJ*, 536, 347

Gibb, E. L., Whittet, D. C. B., Boogert, A. C. A., & Tielens, A. G. G. M. 2004, *ApJS*, 151, 35

Israel, F. P., de Graauw, Th., van de Stadt, D., & de Vries, C. P. 1986, *ApJ*, 303, 186

Ita, Y., Onaka, T., Kato, D., *et al.* 2008, *PASJ*, 60, S435

Kato, D., Nagashima, C., Nagayama, T., *et al.* 2007, *PASJ*, 59, 615

Luck, R. E., Moffett, T. J., Barnes III, T. G., & Gieren, P. W. 1998, *AJ*, 115, 605

Meixner, M., Gordon, K. D., Indebetouw, R., *et al.* 2006, *AJ*, 132, 2268

Murakami, H., Baba, H., Barthel, P., *et al.* 2007, *PASJ*, 59, S369

Nummelin, A., Whittet, D. C. B., Gibb, E. L., Gerakines, P. A., & Chiar, J. E. 2001, *ApJ*, 558, 185

Öberg, K. I., Fraser, H. J., Boogert, A. C. A., Bisschop, S. E., Fuchs, G. W., van Dishoeck, E. F., & Linnartz, H. 2007, *A&A*, 462, 1187

Ohyama, Y., Onaka, T., Matsuhara, H., *et al.* 2007, *PASJ*, 59, S411

Onaka, T., Matsuhara, H., Wada, T., *et al.* 2007, *PASJ*, 59, S401

Pontoppidan, K. M., Boogert, A. C. A., Fraser, H. J., *et al.* 2008, *ApJ*, 678, 1005

Ruffle, D. P. & Herbst, E. 2001, *MNRAS*, 324, 1054

Sakon, I., Onaka, T., Kaneda, H., *et al.* 2006, *ApJ*, 651, 174

Shimonishi, T., Onaka, T., Kato, D., Sakon, I., Ita, Y., Kawamura, A., & Kaneda, H. 2008, *ApJ*, 686, L99

Sylvester, R. J., Kemper, F., Barlow, M. J., de Jong, T., Waters, L. B. F. M., Tielens, A. G. G. M., & Omont, A. 1999, *A&A*, 352, 587

van Loon, J. Th., Oliveira, J. M., Wood, P. R., *et al.* 2005, *MNRAS*, 364, L71

Watanabe, N., Mouri, O., Nagaoka, A., Chigai, T., Kouchi, A., & Pironello, V. 2007, *ApJ*, 668, 1001

Whitney, B. A., Indebetouw, R., Bjorkman, J. E., & Wood, K. 2004, *ApJ*, 617, 1177

Whitney, B. A., Sewilo, M., Indebetouw, R., *et al.* 2008, *AJ*, 136, 18

Whittet, D. C. B., Shenoy, S. S., Bergin, E. A., Chiar, J. E., Gerakines, P. A., Gibb, E. L., Melnick, G. J., & Neufeld, D. A. 2007, *ApJ*, 655, 332

Zaritsky, D., Harris, J., Thompson, I. B., & Grebel, E. K. 2004, *AJ*, 128, 1606

The Magellanic System: Stars, Gas, and Galaxies
Proceedings IAU Symposium No. 256, 2008
Jacco Th. van Loon & Joana M. Oliveira, eds.

Insights from *Spitzer* on massive star formation in the LMC

Rémy Indebetouw[1], Barbara A. Whitney[2], Marta Sewilo[3], Thomas Robitaille[4], Margaret Meixner[3] and the (rest of the) SAGE team

[1]University of Virginia and National Radio Astronomy Observatory, P.O. Box 400325, Charlottesville, VA, 22904 USA, email: remy@virginia.edu

[2]Space Science Institute, Boulder, CO, USA

[3]Space Telescope Science Institute, Baltimore, MD, USA

[4]Smithsonian Astrophysical Observatory, Cambridge, MA, USA

Abstract. *Spitzer*'s sensitive mid-IR photometric surveys of the Magellanic Clouds provide a relatively extinction-free census of star formation activity, and sub-parsec resolution permits the study of individual massive protostars and small clusters. Using the SAGE survey of the LMC, we identify over 1000 massive YSO candidates by their MIR colors. Analysis of their spectral energy distributions (SEDs) constrains the stellar content and evolutionary state, beginning to realize for the first time the unique potential of the Clouds to study an entire galaxy's population of individual protostars. We probe the physics underlying the Schmidt-Kennicutt scaling law by analyzing how it begins to break down at $10-100\,\mathrm{pc}$ spatial scales. MIR spectroscopic surveys currently underway like SAGE-SPEC will enable us to couple the circumprotostellar dust distribution (the evolutionary state reflected in the SED) with the physical state of the gas, dust and ice.

Keywords. surveys, stars: formation, ISM: clouds, H II regions, galaxies: individual (LMC), Magellanic Clouds

1. Introduction

Surveying the Agents of a Galaxy's Evolution (SAGE, Meixner *et al.* 2006)) is a *Spitzer Space Telescope* legacy program to image the entire Large Magellanic Cloud (LMC) with the IRAC (3.6, 4.5, 5.8, & 8.0 μm, Fazio *et al.* 2004) and MIPS (24, 70, & 160 μm, Rieke *et al.* 2004) cameras. SAGE provides a complete census of stars and the interstellar medium in the entire galaxy down to the confusion limit, except in bright H II regions where fainter point sources are lost in bright diffuse emission. SAGE can study the life cycle of dust — formation, processing, and destruction — in the ISM down to column densities of 1.2×10^{21} cm^{-2}. SAGE can also find and characterize all massive young stellar objects in the entire galaxy. The SAGE LMC survey has been followed up by a large-scale photometric survey of the Small Magellanic Cloud and part of the Bridge (SAGE-SMC; PI Gordon), and several mid-infrared spectroscopic surveys. These include a complete spectral cube of 30 Doradus (Indebetouw *et al.* 2009), spectra of 200 Young Stellar Object (YSO) candidates (PI Looney), and of several hundred young and evolved stars (SAGE-SPEC, PI Tielens).

These mid-infrared observations are greatly improving our understanding of star formation, massive (proto)stellar feedback, and physical conditions in the ISM of the Magellanic system. In particular, we can address aspects of star formation that take place on meso-scales, between 0.1 and 1000pc. The very small scales of star formation are best studied in our own Galaxy, and for low-mass (solar or less) stars, these detailed nearby studies have resulted in a fairly complete theory. Individual high-mass star formation still holds many puzzles at high spatial resolution, but if we assume that there is consistency in how a *cluster* of stars forms from a molecular clump (parsec-scale), then we can use the SAGE data to connect those individual small clusters and clumps to the kiloparsec and larger-scale star formation, scaling relations commonly used for galaxies as a whole. We can in particular address:

- physical conditions in clouds and cores;
- stability/collapse/triggering of clouds;
- distribution of SF regions in space/time;
- feedback from protostars on clouds.

2. YSO candidates: selection and population analysis

Figure 1 shows one of the many color—magnitude diagrams that can be constructed using SAGE point source photometry. Various known populations are marked. YSOs occupy the bright red part of mid-IR color—magnitude space. Main sequence stars have nearly zero infrared color, evolved stars have modestly red colors except for a relatively rarer population of extreme AGB stars and proto-planetary nebula which can be bright and quite red. Thus a careful set of color—magnitude cuts can isolate YSO candidates reasonably reliably (spectroscopic confirmation is required for certainty of course). As described in detail in Whitney *et al.* (2008), we can select over 1000 massive YSO candidates, greatly increasing the statistical sample relative to the $\sim$ dozen massive YSOs in the LMC previously described in the literature. At fainter magnitudes, reddened and redshifted background galaxies begin to contaminate any sample, effectively limiting this analysis to greater than $5-10\,M_\odot$ (SAGE sensitivity provides photometry of isolated protostars down to $\sim 3\,M_\odot$ depending on evolutionary state).

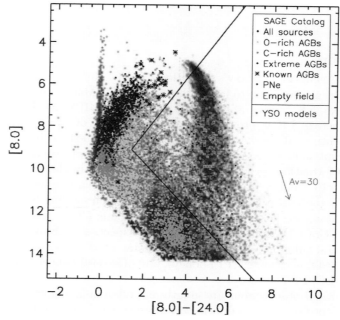

Figure 1. A SAGE color—magnitude diagram showing the locations of known object populations, modeled young stellar objects (YSOs), and a selection cut used to isolate YSO candidates.

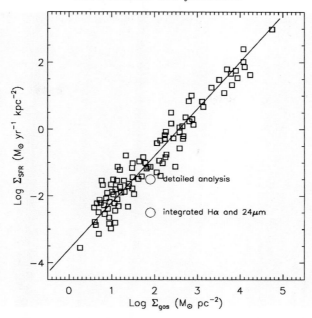

Figure 2. The molecular ridge south of 30 Doradus falls below the Schmidt-Kennicutt relation (Kennicutt 1998) in total emission because each of its regions is a poor cluster, not fully sampling the hot end of the stellar initial mass function. When the individual regions are analyzed, their mass does sum to something consistent with SK.

The YSO catalog permits many different analyses. We find that the YSOs are highly correlated with the gas surface density and are relatively strongly clustered with each other (Whitney *et al.* 2008). The spectral energy distribution of a YSO can be compared against a grid of dust radiative transfer models (Robitaille *et al.* 2007), constraining the central mass and luminosity (spectral class given a set of pre-main-sequence evolutionary tracks), and the circumstellar dust distribution (disk vs. envelope) and total mass (corresponding to a relative evolutionary state from younger strongly accreting sources to older sources with remnant disks). We can also analyze the stability of individual molecular clouds and clumps; for example in the molecular ridge south of 30 Doradus we find that the clouds with a higher ratio of CO luminosity (or molecular gas mass assuming a constant X-factor conversion) to virial mass have greater $24\,\mu$m and total infrared luminosities. Considering the entire galaxy, Yang *et al.* (2007) find that the YSO candidates are found in Toomre-unstable regions, but only if one includes the gravitational potential of the stars as well as the gas in the instability criterion.

We can use SED analysis of YSOs and small H II regions to dissect the Schmidt-Kennicutt (SK) scaling relation at unprecedented small scales (few tens of parsecs). The relation states that star formation rate (measured by infrared plus Hα or UV emission, Calzetti *et al.* 2005; Kennicutt *et al.* 2007) is proportional to the gas surface density to the 1.4 power. It is commonly used in extragalactic work and holds tightly above kiloparsec scales across Hubble type. However, we expect it to break down around the Giant Molecular Cloud scale (100 pc or so), where local conditions may not be sufficiently averaged. In Indebetouw *et al.* (2008) we analyze the strangely faint southern molecular ridge of the LMC, south of 30 Doradus. The total 24 μm and Hα emission from the ridge places it below the SK relation (Figure 2), but if one analyzes each region individually, one finds that they are mostly poor clusters that are not fully sampling the initial mass

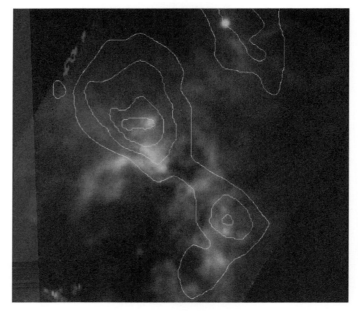

Figure 3. 30 Doradus in the mid-infrared showing the ionization structure of the bubble H II region. blue=[Ne III], green=[Ne II], red=8 μm PAH feature (continuum subtracted) contours=^{12}CO(1–0) (Johansson *et al.* 2008).

function (IMF). With a few 1000 stars and most massive stars only of B type, they do not have the total luminosity that one would expect for a fully sampled IMF. Their mass does add up to a total consistent with SK, but the mode of star formation is non-typical and doesn't include rich bright clusters.

3. Spectroscopy and feedback

Mid-infrared spectra contain several pairs of atomic fine-structure lines which can be used to constrain the ionization parameter and ionizing field in an H II region (e.g., Morisset *et al.* 2004). We apply this analysis to 30 Doradus, shown in Figure 3. The ionization gradient from inside the bubble out into the molecular cloud is already clear from the transition from [Ne III] (blue) to [Ne II] (green). Photoionization modeling reveals a similar structure, of a hot bubble ionized by a relatively hard radiation field (especially a the eastern end which is dominated by a Wolf-Rayet star). Local features from individual hot stars and the remnant molecular material are much stronger than any radial gradient from R 136 (Indebetouw *et al.* 2009).

As described by K. Sandstrom at this meeting, *Spitzer* spectra also contain aromatic dust emission features (usually attributed to polycyclic aromatic hydrocarbons, PAHs) which allow one to determine the ionization state of small dust in photo-dissociation regions around YSOs and H II regions, and to study the destruction of dust by massive (proto)stars. As described by J. Oliveira and T. Shimonishi (elsewhere in this volume), we can also make detailed analyses of the circumstellar dust and ices around protostars, leading to much finer constraints on their evolutionary state and the early stages of the heating and processing of their natal material.

In conclusion, we have only just begun to learn from the massive mid-infrared datasets of the Magellanic system provided by the *Spitzer Space Telescope*, especially the SAGE,

SAGE-SMC, and SAGE-SPEC surveys. We have already increased the number of known massive protostellar objects by several orders of magnitude in the LMC, characterized their population and evolutionary state, determined stability conditions locally and globally, and will continue to examine the physics behind extragalactic star formation scaling laws. Spectra of H II regions and YSOs in the Clouds will yield even more detail on their physical conditions, and the formation and feedback of massive stars as a function of metallicity and galactic environment.

References

Calzetti, D., Kennicutt, R. C., Bianchi, L., *et al.* 2005, *ApJS*, 633, 871
Fazio, G. G., Hora, J. L., Allen, L. E., *et al.* 2004, *ApJS*, 154, 10
Indebetouw, R., De Messieres, G., *et al.* 2009, *AJ*, submitted
Indebetouw, R., Whitney, B. A., Kawamura, A., *et al.* 2008, *AJ*, 136, 1442
Johansson, L. E. B., Greve, A., Booth, R. S., *et al.* 1998, *A&A*, 331, 857
Kennicutt, R. C., Jr. 1998, *ApJ*, 498, 541
Kennicutt, R. C., Jr., Calzetti, D., Walter, F., *et al.* 2007, *ApJ*, 671, 333
Meixner, M., Gordon, K. D., Indebetouw, R., *et al.* 2006, *AJ*, 132, 2268
Morisset, C., Schaerer, D., Bouret, J.-C., & Martins, F. 2004, *A&A*, 415, 577
Rieke, G. H., Young E. T., Engelbracht C. W., *et al.* 2004, *ApJS*, 154, 25
Robitaille, T. P., Whitney, B. A., Indebetouw, R., & Wood, K. 2007, *ApJS*, 169, 328
Whitney, B. A., Sewilo, M., Indebetouw, R., *et al.* 2008, *AJ*, 136, 18
Yang, C.-C., Gruendl, R. A., Chu, Y.-H., Mac Low, M.-M., & Fukui, Y. 2007, *ApJ*, 671, 374

The Magellanic System: Stars, Gas, and Galaxies
Proceedings IAU Symposium No. 256, 2008
Jacco Th. van Loon & Joana M. Oliveira, eds.

© 2009 International Astronomical Union
doi:10.1017/S1743921308028512

Time resolved star formation in the SMC: the youngest star clusters

Elena Sabbi[1], **Linda J. Smith**[1,2], **Lynn R. Carlson**[3], **Antonella Nota**[1,4], **Monica Tosi**[5], **Michele Cignoni**[5], **Jay S. Gallagher III**[6], **Marco Sirianni**[1,4] **and Margaret Meixner**[1]

[1]Space Telescope Science Institute
3700 San Martin Drive, Baltimore, MD, 21218, USA
email: sabbi@stsci.edu

[2]University College London, London, UK

[3]Johns Hopkins University, Baltimore, MD, USA

[4]European Space Agency, Research and Scientific Support Department, Baltimore, MD, USA

[5]INAF-Osservatorio Astronomico di Bologna, Bologna, Italy

[6]University of Wisconsin, Madison, WI, USA

Abstract. The two young clusters NGC 346 and NGC 602 in the Small Magellanic Cloud provide us with the opportunity to study and the efficiency of feedback mechanism at low metallicity, as well as the impact of local and global conditions in cluster formation and evolution. I describe the latest results from a multi-wavelength, large-scale study of these two clusters. *HST*/ACS images reveal that the clusters have very different structures: NGC 346 is composed by a number of sub-clusters which appear coeval with ages of 3 ± 1 Myr, strongly suggesting formation by the hierarchical fragmentation of a turbulent molecular cloud (Nota *et al.* 2006; Sabbi *et al.* 2007a). NGC 602, on the contrary, appears as a single small cluster of OB stars surrounded by pre-main sequence stars. For both clusters high-resolution spectroscopy of the ionized gas shows little evidence for gas motions. This suggests that at the low SMC metallicity, the winds from the hottest stars are not powerful enough to sweep away the residual gas. Instead we find that stellar radiation is the dominant process shaping the interstellar environment of NGC 346 and NGC 602.

Keywords. stars: formation, stars: luminosity function, mass function, stars: pre-main-sequence, galaxies: individual (SMC), Magellanic Clouds, galaxies: star clusters

1. Introduction

As the closest star forming dwarf galaxy ($\sim$60 kpc), the Small Magellanic Cloud (SMC) represents an ideal laboratory for detailed studies of resolved stellar populations in this extremely common class of objects. Its present day metallicity ($Z = 0.004$) and low dust content (1/30 of the Milky Way) make the SMC the best local analog to the vast majority of late-type dwarfs. Deep images acquired with the Advanced Camera for Survey (ACS) on board of the *Hubble Space Telescope* (*HST*) provide excellent photometry well below the turn-off (TO) of its oldest stellar population, allowing us to infer an accurate star formation history (SFH) over the entire Hubble time.

The high spatial resolution of the Advanced Camera for Surveys (ACS) allows us to spatially resolve even the densest star cluster in the SMC to probe the cluster formation and evolution in late-type dwarf galaxies. Furthermore the availability of multi-wavelength surveys from the radio band to the far-IR, combined with the simple kinematics of the galaxy allow us to identify the possible triggers of star formation (SF).

At solar metallicities, during the first Myr of a star cluster evolution, the powerful winds from the massive stars remove the gas left over from star formation. If star formation efficiency is sufficiently low, the gas dispersion can unbind the cluster (e.g., Bastian & Goodwin 2006). However the reduced stellar wind power at low metallicity can modify early evolution of clusters.

With the aim to understand the impact of global and local conditions on the early star cluster evolution we recently investigated the stellar content of two of the youngest star clusters in the SMC (NGC 346 and NGC 602 respectively) using deep ACS/WFC images in the filters F555W (∼V) and F814W (∼I). We used *Spitzer Space Telescope (SST)* IRAC images to investigate if star formation is still ongoing and high resolution spectra to study the kinematics of the ionized gas still associated to these clusters.

2. NGC 346

NGC 346 represents the most active star-forming region in the SMC. It is located towards the northern end of the bar of the SMC and contains almost half of the known O stars of the entire galaxy (Massey *et al.* 1989) which ionized N 66, the largest H II region in the SMC.

HST/ACS images (Plates 1 and 2 in Nota *et al.* 2006) show that even if NGC 346 is ∼3 Myr old (Bouret *et al.* 2003; Nota *et al.* 2006; Sabbi *et al.* 2007a), it still contains some of its residual gas. This suggests that supernova explosions have not occurred yet in the central region of this cluster. This conclusion nicely agrees also with the low diffuse X-ray flux (Nazé *et al.* 2002), and with the neutral gas and CO maps (Rubio *et al.* 2000; Contursi *et al.* 2000).

The photometric analysis of NGC 346 based on *HST*/ACS F555W and F814W deep images (Nota *et al.* 2006; Sabbi *et al.* 2007a) indicates that stars in NGC 346 have an age of 3 ± 1 Myr. The inspection of the m_{F555W} vs. $m_{F555W} - m_{F814W}$ color—magnitude diagram (CMD) reveals the presence of a rich population of faint and red objects (Fig. 1, left panel), consistent with low mass (0.6–3 $M_\odot$) pre-main sequence (preMS) stars, likely coeval with the cluster.

Stars in NGC 346 are not uniformly distributed within the ionized nebula, but are organized in at least 16 sub-clusters that differ in size and stellar densities. 15 of the sub-clusters are still embedded in nebulosities, appear to be connected by filaments of dusts and gas, and are, within the uncertainties of isochrones fitting, likely coeval (Sabbi *et al.* 2007a). The 16th sub-cluster is ∼ 10–15 Myr older, and probably not associated to NGC 346.

From the analysis of *Spitzer* data, Simon *et al.* (2007) found 111 embedded young stellar objects (YSOs) in NGC 346. They found that all but one of the 15 sub-clusters identified in Sabbi *et al.* (2007a) contain YSOs, showing that star formation is still ongoing in the entire region.

The quality of *HST*/ACS data allowed us to derive the present day mass function (PDMF) of NGC 346: the PDMF is Salpeter over the mass range of 0.8–60 $M_\odot$ (Fig. 2, open circles). However the PDMF varies as a function of the radial distance from the cluster center, indicating that the cluster is, probably primordially, mass segregated (Sabbi *et al.* 2007b). Simon *et al.* (2007) noted a similar distribution for the YSOs, with the most massive objects located in the central regions.

Assuming a sound-speed of 10 km s^{-1}, the crossing-time from the cluster center to the periphery is ∼ 2 Myr, which means that NGC 346 is about one crossing-time old.

The observed sub-clustered structure and the coevality of the sub-clusters strongly suggests that NGC 346 was formed by the collapse, and subsequent hierarchical

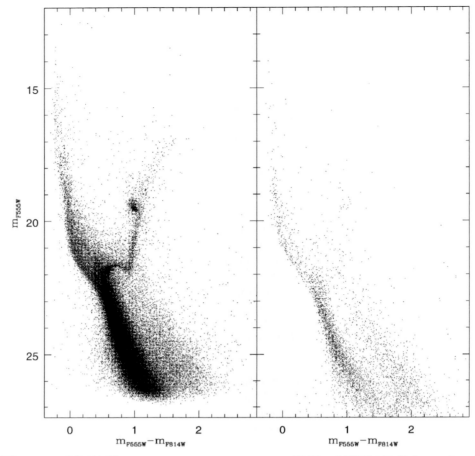

Figure 1. *HST*/ACS m_{F555W} vs. $m_{\mathrm{F555W}} - m_{\mathrm{F814W}}$ CMDs of NGC 346 (*left panel*) and NGC 602 (*right panel*).

fragmentation of a giant molecular cloud (GMC) into multiple seeds of star formation (Elmegreen 2000; Klessen & Burkert 2000; Bonnell & Bate 2002). In this model, the fragmentation of the GMC is due to supersonic turbulent motions in the gas. The turbulence induces shocks, and causes the formation of filamentary structures, and local density enhancements. High-density regions that become self-gravitating can collapse to form stars, and this occurs simultaneously at different locations within the cloud (Bonnell *et al.* 2003).

The formation of shocks in the gas, due to the initial supersonic turbulence, rapidly removes kinetic energy from the gas (Ostriker *et al.* 2001). To test if this is the case, we obtained high-resolution spectra with the University College London Echelle Spectrograph (UCLES) at the *Anglo-Australian Telescope* (*AAT*). We observed a number of sub-clusters in N 66. Hα and [O III] emission lines show that the ionized gas is quiescent, with no evidence for large-scale gas motions. The Hα profiles are single, with a velocity dispersion of $\sim$14 km s^{-1} and a constant velocity along the length of each slit position (1 arc min). Even at the center of NGC 346, where the most massive stars are located, we detect no significant ionized gas motions, further supporting the idea that NGC 346 is a good observational counterpart of the hierarchical fragmentation models of a GMC.

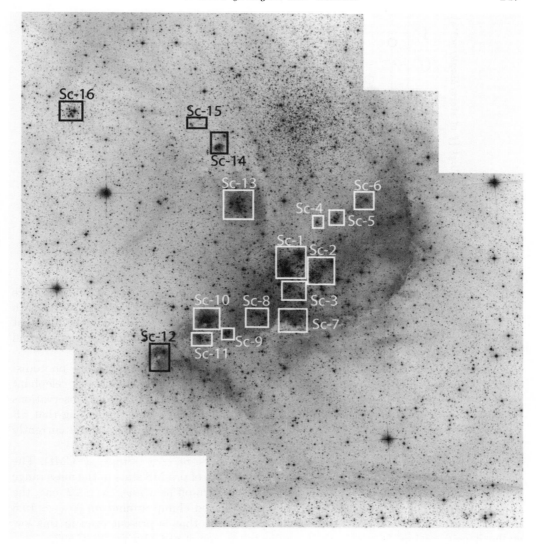

Figure 2. *HST*/ACS F555W image NGC 346 showing the 16 identified sub-clusters.

The lack of large-scale motions also strongly suggests that at the low metallicity of the SMC, stellar winds are much reduced, and the dominant form of interaction is via stellar radiation, rather than by winds, in agreement with Bouret *et al.* (2003) who measured the mass-loss rates of six O stars in the center of NGC 346, and found that their winds are considerably weaker than their Galactic analogs.

3. NGC 602

NGC 602 is a small and bright star forming region, located in the Wing of the SMC. This is the relatively diffuse southeastern region that connects the SMC to the Large Magellanic Cloud via the Magellanic Bridge. NGC 602 is an example of massive star formation in a region with diffuse interstellar medium (ISM) without any apparent direct kinetic trigger, such as a recent nearby supernova.

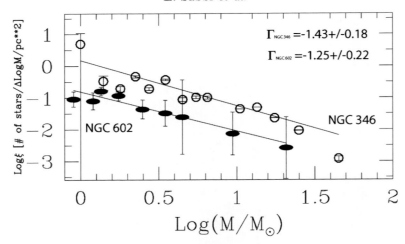

Figure 3. PDMFs of NGC 346 (open circles) between 0.8 and 60 $M_\odot$ and NGC 602 (filled ellipses) between 0.8 and 30 $M_\odot$.

Its isolation alone makes it a great laboratory for the study of star formation since it lacks the complexities introduced by kinetic interactions and optical confusion with other star clusters. This, combined with the low metallicity and the relatively quiescent environment, also makes it a candidate for comparisons to theoretical works on primordial H II regions driven by Population III stars (e.g., Abel *et al.* 2007).

Two ridges of dust and gaseous filaments surround the central cluster. Many photodissociation regions (PDRs) are visible along the ridges, as well as magnificent "elephant trunks" and "pillars of creation" (see Fig. 1 by Carlson *et al.* 2007). *SST* observations revealed numerous class 0.5 and I YSOs on the verge of the pillars, indicating that SF propagated from the center of NGC 602, and a second generation of stars is currently forming in the periphery of the cluster (Carlson *et al.* 2007).

Figure 1–right panel shows the NGC 602 m_{F555W} vs. $m_{F555W} - m_{F814W}$ CMD. The most striking feature of this CMD is the rich population of pre-MS stars in the mass range 0.6–3 $M_\odot$ (Carlson *et al.* 2007). The main-sequence turn-off near $m_{F555W} \simeq 22$ mag, the sub-giant curve also at $m_{F555W} \simeq 22$ mag, and the red clump around $m_{F555W} \simeq 19.5$ mag belong likely to the SMC field stellar population that is present even in this low stellar density region. Isochrone fitting indicates an age of ~ 4 Myr for NGC 602.

We derived the PDMF of NGC 602 in the mass range 0.8–30 $M_\odot$ (Fig. 2–filled ellipses). From the weighted least mean squares fit of the data we derived a slope $\Gamma = -1.25 \pm 0.22$ (Cignoni *et al.* 2009) in excellent agreement with what we found in NGC 346.

Even if the morphology of this association is reminiscent of an expanding bubble of gas, UCLES high-resolution spectra show very low velocity dispersion, excluding large-scale motions (Nigra *et al.* 2008). Another possible explanation for the NGC 602 morphology is that we are observing a cavity eroded by the OB UV stellar radiation.

NGC 602 and its associated H II region, N 90, formed in a relatively isolated and diffuse environment. Its isolation from other regions of massive star formation and the relatively simple surrounding H I shell structure encouraged us to try to constrain the processes that may have led to its formation. Using the shell catalog derived from the 21 cm neutral hydrogen (H I) spectrum survey data (Staveley-Smith *et al.* 1997; Stanimirović *et al.* 1999) we identified a distinct H I cloud component that is likely the progenitor cloud of the cluster and the H II region which probably formed in blister fashion from the cloud's periphery. A comparison between H I and H II kinematics suggests that star formation in NGC 602 was triggered by compression and turbulence associated with H I shell interaction $\sim$ 7 Myr ago (Nigra *et al.* 2008).

4. Conclusions

We have recently analyzed the two very young SMC star clusters NGC 346 and NGC 602, which are located in two regions characterized by very different gas and stellar densities. We derived the PDMF of the clusters over two orders of magnitude, further confirming its universality.

Both clusters contain noticeable populations of pre-MS stars and YSOs, indicating that in both cases SF is still ongoing and residual gas is still present. The ionized gas is quiescent, and we did not find evidence of stellar wind interaction, confirming the hypothesis that mechanical feedback is reduced at low metallicities.

Our multi-wavelength approach allowed us to infer very different origins for the clusters, suggesting that different local condition might affect deeply affect the formation and the evolution of star clusters since the earliest phases.

References

Abel, T., Wise, J. H., & Bryan, G. L. 2007, *ApJ*, 659, L87

Bastian, N. & Goodwin, S. P. 2006, *MNRAS*, 369, L6

Bonnell, I. A. & Bate, M. R. 2002, *MNRAS*, 336, 659

Bonnell, I. A., Bate, M. R., & Vine, S. 2003, *MNRAS*, 343, 413

Bouret, J.-C., Lanz, T., Hillier, D. J., Heap, S. R., Hubeny, I., Lennon, D. J., Smith, L. J., & Evans, C. J. 2003, *ApJ*, 595, 1182

Carlson, L. R., Sabbi, E., Sirianni, M., *et al.* 2007, *ApJ*, 665, L109

Cignoni, M., Sabbi, E., Nota, A., *et al.* 2009, *AJ*, in press

Contursi, A., Lequeux, J., Cesarsky, D., *et al.* 2000, *A&A*, 362, 310

Elmegreen, B. G. 2000, *AJ*, 530, 227

Klessen, R. S. & Burkert, A. 2000, *ApJS*, 128, 287

Massey, P., Parker, J. W., & Garmany, C. D. 1989, *AJ*, 98, 1305

Nazé, Y., Hartwell, J. M., Stevens, I. R., Corcoran, M. F., Chu, Y.-H., Koenigsberger, G., Moffat, A. F. J., & Niemela, V. S. 2002, *ApJ*, 580, 225

Nigra, L., Gallagher, J. S., III, Smith, L. J., Stanimirović, S., Nota, A. & Sabbi, E. 2008, *PASP*, 120, 972

Nota, A., Sirianni, M., Sabbi, E., *et al.* 2006, *ApJ*, 640, L29

Ostriker, E. C., Stone, J. M., & Gammie, C. F. 2001, *ApJ*, 546, 980

Rubio, M., Contursi, A., Lequeux, J., Probst, R., Barbá, R., Boulanger, F., Cesarsky, D., & Maoli, R. 2000, *A&A*, 359, 1139

Sabbi, E., Sirianni, M., Nota, A., *et al.* 2007a, *AJ*, 133, 44

Sabbi, E., Sirianni, M., Nota, A., Gallagher, J., Tosi, M., Smith, L. J., Angeretti, L., & Meixner, M. 2007b, *AJ*, 133, 44

Simon, J. D., Bolatto, A. D., Whitney, B. A., *et al.* 2007, *ApJ*, 670, 313

Stanimirović, S., Staveley-Smith, L., Dickey, J. M., Sault, R. J., & Snowden, S. L. 1999, *MNRAS*, 302, 417

Staveley-Smith, L., Sault, R. J., Hatzidimitriou, D., Kesteven, M. J., & McConnell, D. 1997, *MNRAS*, 289, 225

The Magellanic System: Stars, Gas, and Galaxies
Proceedings IAU Symposium No. 256, 2008
Jacco Th. van Loon & Joana M. Oliveira, eds.

The sub-solar initial mass function in the Large Magellanic Cloud†

Dimitrios A. Gouliermis

Max-Planck-Institut für Astronomie, Königstuhl 17, 69117 Heidelberg, Germany
email: dgoulier@mpia.de

Abstract. The Magellanic Clouds offer a unique variety of star forming regions seen as bright nebulae of ionized gas, related to bright young stellar associations. Nowadays, observations with the high resolving efficiency of the *Hubble Space Telescope* allow the detection of the faintest infant stars, and a more complete picture of clustered star formation in our dwarf neighbors has emerged. I present results from our studies of the Magellanic Clouds, with emphasis in the young low-mass pre-main sequence populations. Our data include imaging with the Advanced Camera for Surveys of the association LH 95 in the Large Magellanic Cloud, the deepest observations ever taken with *HST* of this galaxy. I discuss our findings in terms of the initial mass function, which we constructed with an unprecedented completeness down to the sub-solar regime, as the outcome of star formation in the low-metallicity environment of the LMC.

Keywords. Magellanic Clouds, galaxies: star clusters, open clusters and associations: individual (NGC 346, NGC 602, LH 52, LH 95), stars: evolution, stars: pre–main-sequence, stars: luminosity function, mass function

1. Introduction

The conversion of gas to stars is determined by the star formation process, the outcome of which are stars with a variety of masses. The distribution of stellar masses in a given volume of space at the time of their formation is known as the Initial Mass Function (IMF), given that all stars were born simultaneously. The IMF dictates the evolution and fate of stellar systems, as well as of whole galaxies. The evolution of a stellar system is driven by the relative initial numbers of stars of various masses, from the short-lived high-mass stars ($M \gtrsim 8 \ M_\odot$), which enrich the ISM with elements heavier than H and He, to the low-mass stars ($M \lesssim 1 \ M_\odot$), which lock large amounts of mass over long timescales. It is therefore of much importance to quantify the relative numbers of stars in different mass ranges and to identify systematic variations of the IMF with different star-forming conditions, which will allow us to understand the physics involved in assembling each of the mass ranges.

There are various parameterizations of the IMF (see Kroupa 2002 and Chabrier 2003 for reviews), of which a commonly used is that of a power law of the form $\xi(\log M) \propto M^\Gamma$, or alternatively $\xi(M) \propto M^\alpha$. The IMF, thus, is characterized by the derivatives

$$\Gamma = \frac{d \log \xi(\log M)}{d \log M} \quad \text{or} \quad \alpha = \frac{d \log \xi(M)}{d \log M},$$

depending on whether stars are distributed according to their masses in logarithmic or linear scales. The above derivatives, which relate to each other as $\alpha = \Gamma - 1$, correspond to the so-called slope of the IMF. A reference value for the IMF slope, as found by

† Based on observations made with the NASA/ESA *Hubble Space Telescope*, obtained at the Space Telescope Science Institute, which is operated by the Association of Universities for Research in Astronomy, Inc. under NASA contract NAS 5-26555.

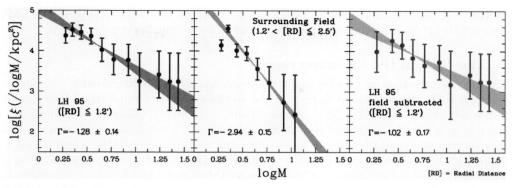

Figure 1. The main-sequence MF of the association LH 95 from observations with the 1-m telescope at Siding Spring Observatory. *Left*: The MF of the system (radial distance $r \lesssim 1.'2$ from its center) with the field population included. *Middle*: The main-sequence MF of the considered surrounding field within radial distance $1.'2 \lesssim r \lesssim 2.'5$. *Right*: The main-sequence MF of the association after the field contribution has been subtracted, which accounts for the IMF of the system. The lack of stars with masses $\gtrsim 10$ $M_\odot$ in the surrounding field introduces a steep MF for its stellar population. The shaded regions represent the uncertainties of the MF. The corresponding MF slopes Γ are given for stars up to the higher observed mass, which in the region of the system is ~ 28 $M_\odot$. The slope of the IMF of LH 95 is a bit shallower than a Salpeter (1955) IMF because of its high-mass content. Data from Gouliermis *et al.* (2002).

Salpeter (1955) in the solar neighborhood for stars with masses between 0.4 and 10 $M_\odot$, is $\Gamma = -1.35$ (or $\alpha = -2.35$). In general, although the IMF appears relatively uniform when averaged over whole clusters or large regions of galaxies (Chabrier 2003), the measured IMF shows local spatial variations, which could be the result of physical differences, or even purely statistical in nature (Elmegreen 2004).

The Large Magellanic Cloud (LMC) is the nearest undisrupted neighboring dwarf galaxy to our own. It has a spatially varying sub-solar metallicity of $Z \sim 0.3$–0.5 $Z_\odot$, while its star formation rate is almost the same as the Milky Way (Westerlund 1997). It demonstrates an energetic star formation activity with its H I shells (Kim *et al.* 1999), H II regions (Davies *et al.* 1976), and molecular clouds (Fukui *et al.* 1999), all linked to ongoing star formation. A wide variety of young stellar systems, the stellar associations (Gouliermis *et al.* 2003), located at regions of recent star formation in the LMC form a complete sample of targets with various characteristics for the study of the stellar IMF in this galaxy. Therefore, the LMC, being so close to us, has provided an ideal alternative environment for the study of extragalactic star formation and the derived IMF.

Photometric and spectroscopic investigations of young stellar associations in the LMC, limited so far to ground-based observations, revealed that the IMF of the high-mass stars in these systems is consistent from one system to the other. All measured IMF slopes are found to be clustered around the Salpeter value, not very different from that of OB associations of our Galaxy (Massey 2006), suggesting that the massive IMF appears more or less to be universal with a typical Salpeter slope (Massey 2003). It should be noted that there are small but observable differences in the constructed IMF and its slope between different systems, which may imply that the IMF is probably determined by the local physical conditions (Hill *et al.* 1995), or that the variability in the slope of the IMF could be the result of different star formation processes (Parker *et al.* 1998). However, taking into account the constraints in the construction of the IMF, one may argue that the IMF variations are possibly observational and/or statistical in nature (Kroupa 2001). An example of the spatial dependence of the IMF and the effect of the contamination by the field population is demonstrated in Fig. 1, where the massive IMF in the LMC association LH 95 constructed earlier by the author and collaborators is shown.

2. The IMF in the low-mass regime

Complete investigations concerning the low-mass stars are so far available only for the Galactic IMF, which is found to become flat in the sub-solar regime (Reid 1998), down to the detection limit of 0.1 $M_\odot$ or lower (Lada *et al.* 1998). Observed variations from one region of the Galaxy to the next in the numbers of low-mass stars and brown dwarfs over intermediate- and high-mass stars affect the corresponding IMF, which seems to depend on the position (e.g., Luhman *et al.* 2000; Briceño *et al.* 2002; Preibisch *et al.* 2003). Explanations suggested for the low-mass IMF variations include stochasticity in the ages and ejection rates of proto-stars from dense clusters (e.g., Bate *et al.* 2002), differences in the photo-evaporation rate from high-mass neighboring stars (e.g., Kroupa & Bouvier 2003), or a dependence of the Jeans mass on column density (Briceño *et al.* 2002) or Mach number (Padoan & Nordlund 2002), and they may also be affected by variations in the binary fraction (Malkov & Zinnecker 2001).

While the discussion over the low-mass part of the Galactic IMF continues, more information on the IMF toward smaller masses in other galaxies is required for the understanding of its dependence on the global properties of the host-galaxy and for addressing the issue of its universality. The LMC, due to its proximity, can serve as the best proxy for the study of extragalactic young low-mass stellar populations, and due to its global differences from the Milky Way, it provides a convenient alternative environment for the investigation of the IMF variability. Under these circumstances, issues that should be addressed can be summarized to *what is the low-mass stellar population of young stellar associations in the LMC, what is the shape of the corresponding IMF*, and *how it compares to that of the Milky Way*. A substantial amount of such information in both the Magellanic Clouds (MCs) was thus far lacking, mostly due to observational limitations. However, recent observations of the MCs with the *Hubble Space Telescope* (*HST*) changed dramatically our view of star formation in these galaxies.

3. Low-mass PMS populations in the Magellanic Clouds

In the Galaxy previous observations have confirmed the existence of faint pre-main sequence (PMS) stars as the low-mass population of nearby OB associations and star-forming regions (e.g., Hillenbrand *et al.* 1993; Brandl *et al.* 1999; Preibisch *et al.* 2002; Sherry *et al.* 2004). The only extragalactic places where such stars could be resolved are the MCs, and recent imaging with *HST* revealed the PMS populations of their stellar associations, and allowed their study. Our investigation of the LMC association LH 52 with WFPC2 observations extended the stellar membership of MCs associations to their PMS populations for the first time (Gouliermis *et al.* 2006a). Subsequent photometry with the *Advanced Camera for Surveys* (ACS) of the association NGC 346 in the Small Magellanic Cloud (SMC) led to the discovery of an extraordinary number of low-mass PMS stars in its vicinity (Nota *et al.* 2006; Gouliermis *et al.* 2006b), providing the required statistical sample for the investigation of the clustering behavior (Hennekemper *et al.* 2008) and the IMF (Sabbi *et al.* 2008) of these stars.

Our photometric study of another young SMC star cluster, NGC 602, with *HST*/ACS revealed a coherent sample of PMS stars, ideal for the study of the low-mass IMF in the SMC and the complications in its construction (Schmalzl *et al.* 2008). The IMF of the PMS population of NGC 602 was constructed by counting the stars between evolutionary tracks on the CMD (Fig. 2 - *left*). While this method is independent of any age-gradient among the stars, it still depends on the selected models. As a consequence, the shape of

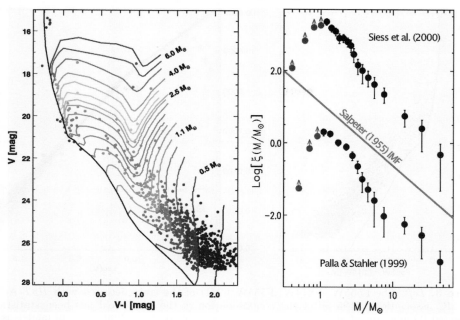

Figure 2. Construction of the IMF of the young SMC cluster NGC 602. *Left*: The CMD of the stars of the system after the contamination by the field population is removed, with PMS evolutionary tracks from the models by Siess *et al.* (2000) overlaid. The ZAMS from the models by Girardi *et al.* (2002) is also shown. The IMF of NGC 602 for the PMS stars is constructed by counting stars between evolutionary tracks using both Palla & Stahler (1999) and Siess *et al.* (2000) PMS tracks, while the ZAMS is used for the construction of a mass-luminosity relation for the bright MS stars with $M \gtrsim 6$ M$_\odot$ for the derivation of their masses. *Right*: The derived IMF, as it is constructed with use of both the models by Siess *et al.* (2000) (top) and Palla & Stahler (1999) (bottom). The grey line corresponds to a Salpeter (1955) IMF with $\alpha \simeq -2.3$. Stellar numbers are corrected to a bin-size of 1 M$_\odot$. Data from Schmalzl *et al.* (2008).

the derived IMF for stars with $M \lesssim 2$ M$_\odot$ appears somewhat different with the use of different PMS tracks (Fig. 2 - *right*). Although this IMF seems to flatten for masses close to solar, no important change in the slope is identified. We found that, in general, a single-power law with a slope of $\alpha \simeq -2.2 \pm 0.3$ represents well the average IMF of the system as it is constructed with the use of all considered grids of models (Schmalzl *et al.* 2008). This result is in line with that of Sabbi *et al.* (2008) for NGC 346. The IMF of both NGC 346 and NGC 602 is limited to stars of 1 M$_\odot$ due to the observations, which did not allow the detection of statistically significant numbers of sub-solar PMS stars.

4. The sub-solar IMF in the Large Magellanic Cloud

The first complete sample of extragalactic sub-solar PMS stars is detected by the author and collaborators by utilizing the *high sensitivity and spatial resolving power* of ACS in combination with its *large field of view* in the LMC association LH 95 (Gouliermis *et al.* 2007). Two pointings, one on LH 95 and another on the general field, were observed in the filters F555W ($\approx V$) and F814W ($\approx I$) with the longest exposures ever taken with *HST* of the LMC (in total 5,000 sec per filter per field). These state-of-the-art observations allowed us the construction of the CMD of the association in unprecedented detail and to decontaminate it for the average LMC stellar population (Fig. 3 - *left*). Although LH 95 represents a rather modest star-forming region, our photometry, with a detection limit of $V \lesssim 28$ mag (at 50% completeness), revealed more than 2,500 PMS stars with masses down to ~ 0.3 M$_\odot$.

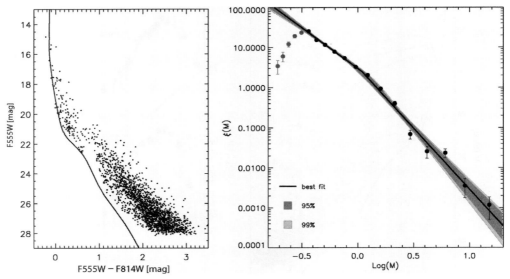

Figure 3. *Left*: The F555W−F814W, F555W CMD of the stars detected with *HST*/ACS in the LMC association LH 95, after the contamination of the LMC field has been statistically subtracted (Gouliermis *et al.* 2007). The ZAMS from Girardi *et al.* (2002) grid of models is also plotted. This is the most complete CMD of an extragalactic star-forming region ever constructed. *Right*: The IMF of the association LH 95 constructed from our ACS photometry with the use of a new observable plane for the PMS models by Siess *et al.* (2000) designed by us for the metallicity of the LMC and the photometric system of ACS. The best two-part power-law fit is drawn with solid line, while the shaded areas represent the 95% and 99% confidence uncertainties in the slope determination and the break point (the "knee"). Units of the IMF, $\xi(M)$, are logarithmic number of stars per solar mass per pc^2. Data from Da Rio *et al.* 2009.

The subsequent interpretation of these observations by Da Rio, Gouliermis & Henning 2009 led to the introduction of a new set of observational evolutionary models, derived from the theoretical calculations by Siess *et al.* (2000). We converted luminosities from these evolutionary tracks into observable magnitudes in the ACS photometric system by taking into account the parameters T_{eff}, $\log(g)$ and $[M/H]$ for a large variety of stellar objects. We used these models with the observations of LH 95 to derive the IMF of the system by assigning a mass value to each PMS star. This IMF, shown in Fig. 3 (*right*), is reliably constructed for stars with masses down to ~ 0.3 M$_\odot$, the lowest mass ever observed within reasonable completeness in the MCs. Consequently, its construction offers *an outstanding improvement in our understanding of the low-mass star formation in the LMC*. We verified statistically that the field-subtracted completeness-corrected IMF of LH 95 has a definite change in its slope for masses $M \lesssim 1$ M$_\odot$, where it becomes more shallow (Da Rio *et al.* 2009). In general, the shape of this IMF agrees very well with a multiple power-law, as the typical Galactic IMF down to the sub-solar regime. A definite change is identified, though, in the slope of the IMF (the "knee of the IMF") at ~ 1 M$_\odot$, a higher mass-limit than that of the average Galactic IMF derived in various previous investigations (e.g., Kroupa 2001, 2002). As far as the slope of the IMF of LH 95 is concerned, it is found to be systematically steeper than the classical Galactic IMF (e.g., Scalo 1998; Kroupa 2002) in both low- and high-mass regimes. No significant differences in the shape of the average IMF of LH 95 from that of each of the three individual PMS sub-clusters of the association is also found. This clearly suggests that the IMF of LH 95 is not subject to local variability.

Acknowledgements

The author kindly acknowledges the support of the German Research Foundation (DFG) through the individual grant GO 1659/1-1.

References

Bate, M. R., Bonnell, I. A., & Bromm, V. 2002, *MNRAS*, 332, L65
Brandl, B., Brandner, W., Eisenhauer, F., *et al.* 1999, *A&A*, 352, L69
Briceño, C., Luhman, K. L., Hartmann, L., Stauffer, J. R., & Kirkpatrick, J. D. 2002, *ApJ*, 580, 317
Chabrier, G. 2003, *PASP*, 115, 763
Davies, R. D., Elliot, K. H., & Meaburn, J. 1976, *MemRAS*, 81, 89
Da Rio, N., Gouliermis, D., & Henning, T. 2009, *ApJ*, in press
Elmegreen, B. G. 2004, *MNRAS*, 354, 367
Fukui, Y., Mizuno, N., Yamaguchi, R., *et al.* 1999, *PASJ*, 51, 745
Girardi, L., Bertelli, G., Bressan, A., Chiosi, C., Groenewegen, M. A. T., Marigo, P., Salasnich, B., & Weiss, A. 2002, *A&A*, 391, 195
Gouliermis, D., Keller, S. C., de Boer, K. S., Kontizas, M., & Kontizas, E. 2002, *A&A*, 381, 862
Gouliermis, D., Kontizas, M., Kontizas, E., & Korakitis, R. 2003, *A&A*, 405, 111
Gouliermis, D., Brandner, W., & Henning, T. 2006a, *ApJ*, 636, L133
Gouliermis, D. A., Dolphin, A. E., Brandner, W., & Henning, T. 2006b, *ApJS*, 166, 549
Gouliermis, D. A., Henning, T., Brandner, W., *et al.* 2007, *ApJ*, 665, L27
Hennekemper, E., Gouliermis, D. A., Henning, T., *et al.* 2008, *ApJ*, 672, 914
Hill, R. S., Cheng, K.-P., Bohlin, R. C., O'Connell, R. W., Roberts, M. S., Smith, A. M., & Stecher, T. P. 1995, *ApJ*, 446, 622
Hillenbrand, L. A., Massey, P., Strom, S. E., & Merrill, K. M. 1993, *AJ*, 106, 1906
Kim, S., Dopita, M. A., Staveley-Smith, L., Bessell, M. S. 1999, *AJ*, 118, 2797
Kroupa, P. 2001, *MNRAS*, 322, 231
Kroupa, P. 2002, *Science*, 295, 82
Kroupa, P. & Bouvier, J. 2003, *MNRAS*, 346, 369
Lada, E. A., Lada, C. J., & Muench, A. 1998, in G. Gilmore & D. Howell, *The Stellar Initial Mass Function*, ASP Conf.Ser. 142, p 107
Luhman, K. L., Rieke, G. H., Young, E. T., Cotera, A. S., Chen, H., Rieke, M. J., Schneider, G., & Thompson, R. I. 2000, *ApJ*, 540, 1016
Malkov, O. & Zinnecker, H. 2001, *MNRAS*, 321, 149
Massey, P. 2003, *ARAA*, 41, 15
Massey, P. 2006, in M. Livio & T. M. Brown, *The Local Group as an Astrophysical Laboratory* (Cambridge, UK: Cambridge University Press), p. 164
Nota, A., Sirianni, M., Sabbi, E., *et al.* 2006, *ApJ*, 640, L29
Padoan, P. & Nordlund, Å. 2002, *ApJ*, 576, 870
Palla, F. & Stahler, S. W. 1999, *ApJ*, 525, 772
Parker, J. Wm., Hill, J. K., Cornett, R. H., *et al.* 1998, *AJ*, 116, 180
Preibisch, T., Brown, A. G. A., Bridges, T., Guenther, E., & Zinnecker, H. 2002, *AJ*, 124, 404
Preibisch, T., Stanke, T., & Zinnecker, H. 2003, *A&A*, 409, 147
Reid, N. 1998, in G. Gilmore & D. Howell, *The Stellar Initial Mass Function*, ASP Conf. Ser. 142, p. 121
Sabbi, E., Sirianni, M., Nota, A., *et al.* 2008, *AJ*, 135, 173
Salpeter, E. E. 1955, *ApJ*, 121, 161
Scalo, J. 1998, in G. Gilmore & D. Howell, *The Stellar Initial Mass Function*, ASP Conf.Ser. 142, p. 201
Schmalzl, M., Gouliermis, D. A., Dolphin, A. E., & Henning, T. 2008, *ApJ*, 681, 290
Sherry, W. H., Walter, F. M., & Wolk, S. J. 2004, *AJ*, 128, 2316
Siess, L., Dufour, E., & Forestini, M. 2000, *A&A*, 358, 593
Westerlund, B. E. 1997, *The Magellanic Clouds* (Cambridge, UK: Cambridge Univ. Press)

The Magellanic System: Stars, Gas, and Galaxies
Proceedings IAU Symposium No. 256, 2008
Jacco Th. van Loon & Joana M. Oliveira, eds.

An observational study of GMCs in the Magellanic Clouds with the *ASTE* telescope

**Tetsuhiro Minamidani[1], Norikazu Mizuno[2], Yoji Mizuno[2],
Akiko Kawamura[2], Toshikazu Onishi[2], Ken'ichi Tatematsu[3],
Tetsuo Hasegawa[3], Masafumi Ikeda[4] and Yasuo Fukui[2]**

[1] Department of Physics, Faculty of Science, Hokkaido University N10W8, Kita-ku,
Sapporo, 060-0810, Japan
email: tetsu@astro1.sci.hokudai.ac.jp

[2] Department of Astrophysics, Nagoya University Furo-cho, Chikusa-ku,
Nagoya 464-8602, Japan

[3] National Astronomical Observatory of Japan Mitaka, Tokyo 181-8588, Japan

[4] Research Center for the Early Universe and Department of Physics, University of Tokyo
Tokyo 113-0033, Japan

Abstract. We report the results of the submillimeter observations with the *ASTE* 10 m telescope toward the giant molecular clouds (GMCs) in the Magellanic Clouds to reveal the physical properties of dense molecular gas, the principle sites of star and cluster formation. Six GMCs in the Large Magellanic Cloud have been mapped in the $^{12}CO(J = 3 - 2)$ transition and 32 clumps are identified in these GMCs at a resolution of 5 pc. These data are combined with $^{12}CO(J = 1 - 0)$ and $^{13}CO(J = 1 - 0)$ results and compared with LVG calculations to derive the density and temperature of clumps. The derived density and temperature are distributed in wide ranges. We have made small mapping observations in the $^{13}CO(J = 3 - 2)$ transition toward 9 representative peak positions of clumps to determine the density and temperature of clumps. These physical properties are constrained well and there are differences in density and temperature among clumps. We suggest that these differences of clump properties represent an evolutionary sequence of GMCs in terms of density increase leading to star formation.

Keywords. ISM: clouds, ISM: molecules, galaxies: individual (LMC), Magellanic Clouds, radio lines: ISM, submillimeter

1. Introduction

Giant molecular clouds (GMCs) are important as sites of star and cluster formation. GMCs consist mainly of molecular hydrogen (H_2), which does not have a permanent electric dipole moment. This means a lack of appropriate emission lines excited under the typical condition of GMCs. Carbon monoxide, CO, is the second most abundant molecule and has rotational transitions excited under the typical condition in GMCs. These CO transitions are mainly used to trace molecular components in the interstellar medium.

The observations of *NANTEN* 4-m telescope in 2.6 mm $^{12}CO(J = 1 - 0)$ emission reveal the distribution of GMCs in the whole Large Magellanic Cloud (LMC), at 40 pc resolution (e.g., Fukui *et al.* 2008). Fukui *et al.* (2008) identified 272 GMCs. These GMCs are classified into 3 types in terms of massive star formation activity; Type I shows no signs of massive star formation, Type II is associated with H II regions, and Type III is associated with both H II regions and young clusters (e.g., Kawamura *et al.* 2008).

Kawamura *et al.* (2008) suggest that these GMC types correspond to the evolutionary stages of GMCs.

Observations of higher transitions ($J = 2 - 1$, $J = 3 - 2$, $J = 4 - 3$, $J = 7 - 6$) of CO spectra of GMCs suggested that there is either warm gas or dense gas, or both in GMCs (e.g., Sorai *et al.* 2001; Johansson *et al.* 1998; Heikkilä *et al.* 1999; Bolatto *et al.* 2005; Israel *et al.* 2003; Kim *et al.* 2004; Kim 2006; Pineda *et al.* 2008; Minamidani *et al.* 2008). We aim to obtain submillimeter molecular data at better signal-to-noise ratios than in previous studies to estimate temperatures and densities of molecular gas in the GMCs. We combine the ^{12}CO($J = 3 - 2$) and ^{13}CO($J = 3 - 2$) data observed with the *ASTE* telescope, and the ^{12}CO($J = 1 - 0$) and ^{13}CO($J = 1 - 0$) data obtained with *SEST* and *MOPRA* telescopes.

2. Observations and results

Observations of ^{12}CO($J = 3 - 2$) and ^{13}CO($J = 3 - 2$) transitions at 345 GHz and 330 GHz, respectively, were made with the *ASTE* 10-m telescope at Pampa la Bola in Chile. The details of ^{12}CO($J = 3 - 2$) observations and results are described by Minamidani *et al.* (2008). ^{13}CO($J = 3 - 2$) observations were made using the single "cartridge type" SIS receiver (SC345), and the spectrometer was XF-type digital autocorrelator. The half-power beam width was 23″ at 330 GHz, and this corresponds to 5.6 pc at the distance of the LMC, 50 kpc. We have made 3×3 points mapping observations with 20″ grid toward 9 representative peaks of ^{12}CO($J = 3 - 2$) clumps observed by Minamidani *et al.* (2008). They were 30 Dor No.1, No.2, No.3, No.4, N 159 No.1, No.2, No.4, N 206D No.1, and GMC 225 No.1. The clumps in the 30 Dor and the N 159 regions are in Type III GMCs, the clump in the N 206D region is in Type II GMC, and the clump in the GMC 225 region is in Type I GMC. Figure 1 shows the results of the N 159 region.

3. LVG analysis

To estimate the physical properties of the molecular gas, we have performed the LVG (Large Velocity Gradient) analysis (Goldreich & Kwan 1974) of the CO rotational transitions. The LVG radiative transfer code simulates a spherically symmetric cloud of constant density, temperature, and velocity gradient. It solves the equations of statistical equilibrium for the fractional population of CO rotational levels at given density and temperature. It includes the lowest 40 rotational levels and uses the Einstein's A and collisional coefficients of Schöier *et al.* (2005). Calculated ranges are 5–200 K in kinetic temperature and 10–10^6 cm^{-3} in density of molecular hydrogen.

Figure 2 shows the results of LVG analysis for reference. This indicates that 2 intensity ratios, ^{13}CO($J = 3 - 2$) to ^{13}CO($J = 1 - 0$) and ^{12}CO($J = 3 - 2$) to ^{13}CO($J = 3 - 2$), are orthogonal in wide ranges of density and temperature as compared with the ratios of ^{12}CO($J = 3 - 2$) to ^{12}CO($J = 1 - 0$) and ^{12}CO($J = 1 - 0$) to ^{13}CO($J = 1 - 0$) (See Figure 19 of Minamidani *et al.* 2008). Figure 2 also shows the results of LVG analysis of representative 3 clumps, 30 Dor No.1, N 206D No.1, and GMC 225 No.1.

The results of LVG analysis for studied clumps are presented in Figure 3. The physical properties of these clumps are well constrained as compared with previous analysis using the ratios of ^{12}CO($J = 3-2$) to ^{12}CO($J = 1-0$) and ^{12}CO($J = 1-0$) to ^{13}CO($J = 1-0$) (See Figure 21 of Minamidani *et al.* 2008).

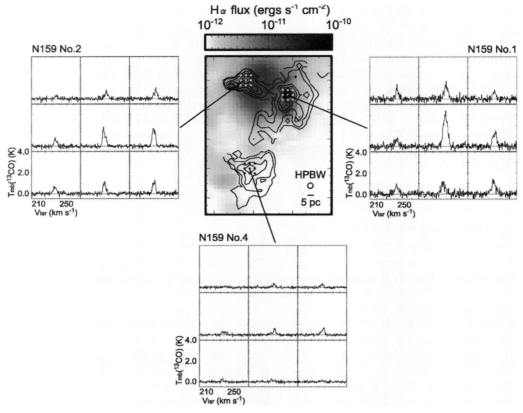

Figure 1. $^{13}CO(J = 3 - 2)$ spectra in the N 159 region. Black and white image is Hα, contours are $^{12}CO(J = 3 - 2)$ integrated intensity, and red circles indicate positions of young clusters (<10 Myr, SWB0, Minamidani *et al.* 2008). Positions observed in $^{13}CO(J = 3 - 2)$ are indicated by yellow and blue dots.

4. Evolution of GMCs

The results of the LVG analysis indicate that clumps are distributed $\sim 7 \times 10^2$– 1.5×10^5 cm^{-3} in density and $15 - 200$ K in temperature. These differences of clump density and temperature generally correspond to the types of GMCs; a clump in the Type I GMC is cool and less dense, and clumps in the Type III GMCs are warm and dense. We should note that the properties of clumps are different even in the same GMC. It seems that the properties of clumps depend on the stage of evolution and are affected by the local environments.

Acknowledgements

A part of this study was financially supported by MEXT Grant-in-Aid for Scientific Research on Priority Area (No. 15071202 and No. 15071203) and by JSPS (No. 14102003, core-to-core program 17004, and No. 18684003). The *ASTE* project is driven by Nobeyama Radio Observatory (NRO), a branch of National Astronomical Observatory of Japan (NAOJ), in collaboration with University of Chile, and Japanese institutes including University of Tokyo, Nagoya University, Osaka Prefecture University, Ibaraki University, Kobe University, and Hokkaido University. We are grateful to all the members of *ASTE* team.

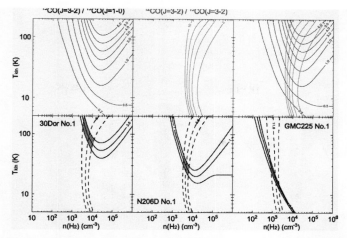

Figure 2. Contour plots of LVG analysis for (*upper panels*) reference and (*lower panels*) representative 3 clumps. The vertical axis is kinetic temperature, T_{kin}, and the horizontal axis is molecular hydrogen density, $n(H_2)$. Solid lines are the ratios of $^{13}CO(J = 3 - 2)$ to $^{13}CO(J = 3 - 2)$ and dashed lines are the ratios of $^{12}CO(J = 3 - 2)$ to $^{13}CO(J = 3 - 2)$. Hatched areas are the regions in which these two ratios overlap within errors.

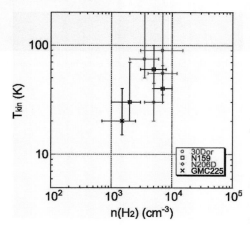

Figure 3. Plot of LVG results. The vertical axis is kinetic temperature, T_{kin}, and the horizontal axis is molecular hydrogen density, $n(H_2)$.

References

Bolatto, A. D., Israel, F. P., & Martin, C. L. 2005, *ApJ*, 633, 210

Fukui, Y., Kawamura, A., Minamidani, T., *et al.* 2008, *ApJS*, 178, 56

Goldreich, P. & Kwan, J. 1974, *ApJ*, 189, 441

Heikkilä, A., Johansson, L. E. B., & Olofsson, H. 1999, *A&A*, 344, 817

Israel, F. P., Johansson, L. E. B., Rubio, M., *et al.* 2003, *A&A*, 406, 817

Johansson, L. E. B., Greve, A., Booth, R. S., *et al.* 1998, *A&A*, 331, 857

Kawamura, A., *et al.* 2008, *ApJS*, submitted

Kim, S., Walsh, W., & Xiao, K. 2004, *ApJ*, 616, 865

Kim, S. 2006, *PASP*, 118, 94

Minamidani, T., Mizuno, N., Mizuno, Y., *et al.* 2008, *ApJS*, 175, 485

Pineda, J. L., Mizuno, N., Stutzki, J., *et al.* 2008, *A&A*, 482, 197

Schöier, F. L., van der Tak, F. F. S., van Dichoeck, E. F., & Black, J. H. 2005, *A&A*, 432, 369

Sorai, K., Hasegawa, T., Booth, R. S., *et al.* 2001, *ApJ*, 551, 794

Poster area, frequented by the curious and those enjoying a conversation — and coffee.

Session V

The star formation history and chemical evolution

The Magellanic System: Stars, Gas, and Galaxies
Proceedings IAU Symposium No. 256, 2008
Jacco Th. van Loon & Joana M. Oliveira, eds.

Breaking the age–metallicity degeneracy: The metallicity distribution and star formation history of the Large Magellanic Cloud

Andrew A. Cole[1], Aaron J. Grocholski[2], Doug Geisler[3], Ata Sarajedini[4], Verne V. Smith[5] and Eline Tolstoy[6]

[1]School of Maths & Physics, University of Tasmania, Private Bag 37, Hobart, Tasmania 7005, Australia, email: andrew.cole@utas.edu.au

[2]Space Telescope Science Institute, 3700 San Martin Drive, Baltimore, MD, 21218 USA
[3]Dept. of Astronomy, University of Florida, P.O.Box 112055, Gainesville, FL, 32611 USA
[4]Departamento de Fisica, Universidad de Concepción, Casilla 160-C, Concepción, Chile
[5]Gemini Project, NOAO, Tucson, AZ, 85719 USA [6]Kapteyn Astronomical Institute, University of Groningen, P.O. Box 800, 9700AV Groningen, The Netherlands

Abstract. We have obtained metallicities from near-infrared calcium triplet spectroscopy for nearly a thousand red giants in 28 fields spanning a range of radial distances from the center of the bar to near the tidal radius. We have used these data to investigate the radius-metallicity and age-metallicity relations. A powerful application of these data is in conjunction with the analysis of deep *HST* color–magnitude diagrams (CMDs). Most of the power in determining a robust star-formation history from a CMD comes from the main-sequence turnoff and subgiant branches. The age-metallicity degeneracy that results is largely broken by the red giant branch color, but theoretical model RGB colors remain uncertain. By incorporating the observed metallicity distribution function into the modelling process, a star-formation history with massively increased precision and accuracy can be derived. We incorporate the observed metallicity distribution of the LMC bar into a maximum-likelihood analysis of the bar CMD, and present a new star formation history and age–metallicity relation for the bar. The bar is certainly younger than the disk as a whole, and the most reliable estimates of its age are in the $5-6$ Gyr range, when the mean gas abundance of the LMC had already increased to $[Fe/H] \gtrsim -0.6$. There is no obvious metallicity gradient among the old stars in the LMC disk out to a distance of $8-10$ kpc, but the bar is more metal-rich than the disk by $\approx 0.1-0.2$ dex. This is likely to be the result of the bar's younger average age. In both disk and bar, 95% of the red giants are more metal-rich than $[Fe/H] = -1.2$.

Keywords. techniques: spectroscopic, stars: abundances, stars: evolution, galaxies: abundances, galaxies: evolution, galaxies: individual (LMC), Magellanic Clouds, galaxies: stellar content, galaxies: structure

---

In the decade since the last IAU Symposium on the Magellanic Clouds, study of the star-formation history and chemical evolution of the field stars has become a major path toward understanding the history of the clouds. This has become important because of the growing realization that the field stars and the clusters have experienced significantly different star-formation histories (e.g., Holtzman *et al.* 1999; Smecker-Hane *et al.* 2002). Great progress has come about due to the enormously successful synergy between extremely deep *Hubble Space Telescope* imaging and wide-field photometric surveys spanning most of the central regions of both clouds. A concomitant explosion of data has occurred on the spectroscopic front, owing to the emergence of the 8-meter class telescopes and massively multiplexed spectrographs. Between the pioneering abundance survey of LMC clusters by Olszewski *et al.* (1991) and the field red giant spectroscopy of 39 disk stars in Cole, Smecker-Hane & Gallagher (2000) just one spectroscopic study of the *old* field star population appeared, the conference paper on the LMC outer disk/halo by Olszewski (1993). Since then, the amount of data has multiplied literally a hundredfold, with studies including Cole *et al.* (2005) and Carrera *et al.* (2008) using the calcium triplet, and higher-dispersion ground being broken by Smith *et al.* (2002) and Pompéia *et al.* (2008).

Our group has commenced measuring the chemical abundances of field stars in the LMC disk and bar to fill in the parameter space spanning the cluster age gap, to measure the variation of metallicity with position across the LMC, and to attempt to discern the nature of the bar as a subpopulation of the disk or a distinct entity. This requires the measurement of old stars, spanning the entire ≈ 13 Gyr history of the LMC. The most common bright stars with this range of ages are red giants, which top out in the LMC at a magnitude I ≈ 14.8. At these levels, the most reliable easy way to obtain large spectroscopic samples in a moderate amount of time is to use the near-infrared Ca II triplet at $\lambda \approx 8600$ Å. This method was calibrated with an extensive set of globular cluster data by Rutledge *et al.* (1997), and the calibration extended into the open cluster regime by Cole *et al.* (2004). The Ca triplet had previously been applied to Magellanic Cloud clusters by Olszewski *et al.* (1991) and Da Costa & Hatzidimitriou (1998). In addition to the immediate significance of metallicity information, and the radial velocity data that comes along with the spectra, this effort pays an enormous dividend in giving added value to very deep photometric studies.

The evolutionary history of a galaxy is encoded in the distributions of age and metallicity of its long-lived stellar populations. The most powerful way to recover the information from these populations is by obtaining deep color–magnitude diagrams of the galaxy and using them to derive the variation in star-formation rate as a function of time. An example of the variations that star-formation history imposes on the color-magnitude diagram is the progressive fading of the brightest main-sequence turnoff stars with population age. This can create very obvious features in a differential comparison of luminosity functions (e.g., Butcher 1977) or color–magnitude diagrams (e.g., Smecker-Hane *et al.* 2002) when the fields compared differ in mean age. Comparisons between LMC bar and disk fields in Holtzman *et al.* (1999) and in Smecker-Hane *et al.* (2002) have consistently shown an excess of stars in the bar corresponding to the main-sequence turnoff of a dominant stellar population that is much younger than a Hubble time. However, the inferred age, metallicity, and mass fraction in the younger population are model dependent — for a vivid example of the systematic effects introduced by these dependencies, see the excellent writeup of the 2001 Coimbra experiment on *HST* data taken at the center of the LMC bar, by Skillman & Gallart (2002).

Age-metallicity degeneracy is the chief source of non-uniqueness in obtaining the star formation rate as a function of time (star formation history, SFH) from deep color–magnitude diagrams (CMDs). The age–metallicity degeneracy of a simple stellar population is explicitly illustrated in Figure 1, where a tripling of the age of an isochrone is countered by a halving of the metallicity in order to create a nearly identical red giant branch (RGB). The isochrones have been chosen to be representative of the upper envelope of LMC metallicities, and have been shifted by a distance modulus $(m - M)_0 = 18.5$ mag for ease of comparison to real LMC data. In the presence of any realistic photometric error, the two giant branches in the left panel of Fig. 1 will be indistinguishable. However, CMDs that reach the subgiant branch and main sequence turnoff (shaded area) can break the degeneracy — the age and metallicity tradeoff is different for stars that are core

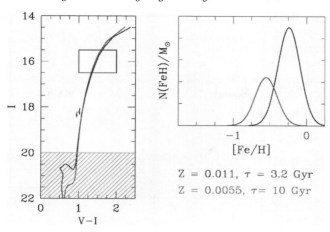

Figure 1. Illustration showing how spectroscopic metallicity measurements can break the age-metallicity degeneracy. *Left panel:* Isochrones from the Padua group (Girardi, private communication) showing how two red giant branches can have the same color if a factor of three in age is compensated by a simultaneous factor of one-half in metallicity. In this simplified example, the degeneracy is broken by observations that reach the main-sequence turnoff (shaded area), but this can be ambiguous in real, complex stellar populations. *Right panel:* The metallicity distribution functions that would be obtained from spectra of red giants selected at random from within the box in the left panel. The isochrone metallicities have been converted to [Fe/H], and convolved with the observational errors reported in Cole *et al.* (2005). Knowledge of the metallicity distribution function prevents an over-reliance on RGB colors that are notoriously difficult to compute accurately. When the photometry is deep enough, the metallicity distribution allows a more precise disentangling of age, metallicity, and reddening in complex stellar populations than provided by the CMD alone.

hydrogen-burning and not fully convective (e.g., Worthey 1999). In this case, sufficiently deep observations can break the age–metallicity degeneracy without ambiguity.

However, for a complex stellar population such as the LMC, this is not as easily done. The deep CMDs show a continuous distribution of stars corresponding to an extremely wide range of possible combinations of age and metallicity. Smearing by photometric error and differential reddening further complicate the picture, such that even with observations as deep as shown in Fig. 1, the interpretation of CMDs is challenging. Comparison of the CMDs to suites of synthetic models via χ^2 minimization or maximum likelihood techniques can find the statistically *most likely* resolution of the degeneracy, but in effect all such comparisons rely on the RGB color to supply the additional information necessary to break the degeneracy. When the ages are older than a few Gyr and the metallicities are low this is likely to be reasonable (e.g., Gallart, Zoccali & Aparicio 2005), but this is often not the case, and is certainly untrue for the Magellanic Clouds. Such an approach necessarily introduces a large degree of systematic uncertainty that will not be reflected in the reported errorbars on the solution, because of the inherent uncertainty in modelling the RGB color (e.g., Salaris 2002). Reliance on the RGB color to break the age-metallicity degeneracy may provide a satisfactory statistical fit to the CMD, but it necessarily imprecise, and incorporates a high degree of model dependence into the solution.

The right panel of Fig. 1 shows an alternative based on incorporating spectroscopic information directly into the maximum-likelihood fits to the CMD data. If a box is drawn around an area of the CMD that is accessible to spectroscopic observations (in Fig. 1a, the box corresponds to the selection area from Cole *et al.* 2005), and the metallicities of

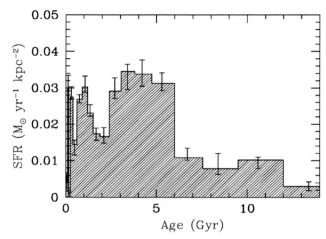

Figure 2. Star formation history of a field at the center of the LMC bar. The epoch of bar formation can be reliably dated to ≈ 6 Gyr ago, when the metallicity of the gas had already risen to [Fe/H] ≳ −0.6. The star-formation history shown here, combined with the metallicity distribution function of Cole *et al.* (2005), closely tracks the "bursting" chemical evolution model presented in Pagel & Tautvaišienė (1998), but with the burst epoch pushed back to ≈ 6 Gyr ago.

the two overlapping isochrones are plotted, the degeneracy is clearly broken: even with the metallicities convolved with the ±0.15 dex errorbar typical of modern calcium triplet studies, there is an obvious difference in the resulting metallicity distribution functions (MDF). In the extremely simple case shown, two distinct stellar populations with very different properties can be identified in a vanishingly narrow RGB even without recourse to main-sequence photometry.

Note that the total number of red giants sampled per unit *created mass* of the stellar population decreases with increasing age and decreasing metallicity — there are fewer old, metal-poor stars in the MDF of Fig. 1, even though the 3.2 Gyr and 10 Gyr populations were born in equal numbers. This bias arises because of the observational selection: the number of stars sampled is effectively proportional to the integral over the initial mass function (IMF) from the bottom to the top of the selection box along the RGB. The higher the age of the population, the smaller the mass range sampled by the selection box, and for stellar masses below ≈ 1.3 M$_\odot$, the IMF is not weighted strongly enough to lower mass stars (older red giants) to compensate.

This raises the possibility that the MDF and the CMD can be *simultaneously* fit in order to derive the SFH and age–metallicity relation without the necessity of assuming a chemical evolution model or relying on uncertain RGB colors. The LMC is an ideal test case for this method, because both very deep CMDs that reach far below the oldest main-sequence turnoffs and large numbers of spectroscopic metallicities for stars spanning a wide range of ages are available.

New results: SFH of the bar, metallicity of the disk

Cole *et al.* (2004, 2005) measured [Fe/H] for 373 red giants in the LMC bar, at the location of the field analyzed in Smecker-Hane *et al.* (2002) and examined in Skillman & Gallart (2002). They found a mean metallicity [Fe/H] = −0.45 that was well-fit by a Gaussian with a 10% tail down to [Fe/H] < −2. They noted the very wide range in RGB color at given metallicity and concluded that the age-metallicity relation must have been

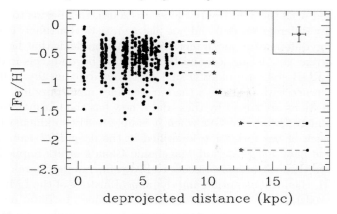

Figure 3. Metallicities of 511 red giants in the LMC disk and bar. The locations of the fields are given in Grocholski *et al.* (2006); radial distances are found assuming the fields lie in the plane of the LMC disk as described in van der Marel (2001). The errorbar shows a representative average 1σ error of $\pm\,0.13$ dex and $\pm\,1$ kpc in distance. For stars near globular clusters projected beyond 7.5°, a second point is plotted for each star, showing the effect on the derived radius if these *field* stars are assumed to lie at the same location along the line of sight as the nearby clusters (open stars).

extremely flat over most of the LMC history. Confirming the earlier result of Cole *et al.* (2000), they found no sign of the cluster metallicity/age gap. By finding the model star-formation + chemical evolution history that simultaneously reproduces both the *HST* CMD and the MDF from Cole *et al.* (2005), we derive the SFH anew, shown in Fig. 2. We also derive the age-metallicity relation, and find excellent agreement with the results in Cole *et al.* (2005) and with the bursting models of Pagel & Tautvaišienė (1998), if the models are modified to cause the burst to occur at 6 Gyr instead of at 3 Gyr. The large star formation rate at an age of 1 Gyr is reflected in the large number of carbon stars observed in the LMC bar.

We have also obtained several hundred new spectra in and around clusters scattered across the face of the LMC (Grocholski *et al.* 2006). There are 511 field red giants in this sample, which probes 27 regions from the bar to 15° away. The shapes of the disk and bar MDFs are broadly similar: both distributions cut off sharply on the high-metallicity end, and both contain an identical 5% fraction of stars with $[Fe/H] \leqslant -1.2$. However, the disk shows a far higher fraction of stars between $-1.2 \leqslant [Fe/H] \leqslant -0.7$: this range contains 20% of the disk stars, but just 5% of those in the bar. Detailed modelling in conjunction with the disk CMDs will be necessary to confirm this, but it seems likely that this is due to the higher number of young stars in the bar field: the CMD analysis requires a younger bar, and Figure 1 shows that an observationally "unbiased" sample will tend to be dominated by young, metal-rich RGB stars.

Figure 3 shows the metallicities of every field star in our *disk* sample, along with its deprojected radius (according to the disk model of van der Marel 2001). There is no clear radial gradient in metallicity out to at least 8–10 kpc deprojected distance. Virtually all of the stars at larger radii turned out to be Milky Way foreground dwarfs or cluster members (by radial velocity), so the details of the metallicity gradient beyond this point are not well constrained in this data. The small number of stars at radii of 10−18 kpc in this sample mean that the significance of any inferred gradient is strongly tied to the field/cluster discrimination. Only one of our disk fields has a similar metallicity distribution to the bar: the innermost field is projected against the bar, and is the highest metallicity field observed; combined with the field from Cole *et al.* (2005), this begins to suggest that the disk-bar metallicity difference is real and not just a statistical artifact. It is important to note that we see no difference in mean metallicity between fields that lie within the $\approx 4°$ limit of active current star formation and those that lie beyond; this may be related to the fact that the RGB stars are older than 1 Gyr, and dynamically well-mixed, but in that case the difference with respect to the bar becomes more puzzling.

Several new questions are raised by these data: how has the bar maintained its apparently separate identity for several Gyr when it is apparently so kinematically unimportant? Can the epoch of bar formation identified in the deep *HST*+abundance data be reconciled with the new, longer-period Magellanic Clouds orbits implied by the recent proper motion data (Kallivayalil *et al.* 2006; Kallivayalil, van der Marel & Alcock 2006)? Although the SFH, chemical evolution and dynamical history of the LMC since the time corresponding to redshift $z = 1$ is becoming well-constrained by data, putting the pieces together into a coherent understanding is proving to be a challenge. At the other end of time, the earliest 1–2 Gyr of LMC history remain very poorly constrained. This is a critical time for galaxy evolution. Comparing the state of our knowledge at the last Clouds meeting (IAU 190) to our understanding today, it is possible to optimistically look forward another decade to a complete description of the halo SFH and metallicity and the detailed chemical abundances of dozens of stars with $[\mathrm{Fe/H}] \lesssim -3$ (another factor of five below the most metal-poor LMC stars currently known).

Partially based on data obtained at the *Very Large Telescope* of ESO's Paranal Observatory, under programmes 70.B-0398 and 74.B-0417, and on observations made with the NASA/ESA *Hubble Space Telescope*, obtained at the Space Telescope Science Institute, which is operated by the Association of Universities for Research in Astronomy, Inc., under NASA contract 5-26555.

References

Butcher, H. 1977, *ApJ*, 216, 372

Carrera, R., Gallart, C., Hardy, E., Aparicio, A., & Zinn, R. 2008, *AJ*, 135, 836

Cole, A. A., Smecker-Hane, T. A., & Gallagher, J. S. 2000, *AJ*, 120, 1808

Cole, A. A., Smecker-Hane, T. A., Tolstoy, E., Bosler, T. L., & Gallagher, J. S. 2004, *MNRAS*, 347, 367

Cole, A. A., Gallagher, J. S., Tolstoy, E., & Smecker-Hane, T. A. 2005, *AJ*, 129, 1465

Da Costa, G. S. & Hatzidimitriou, D. 1998, *AJ*, 115, 1934

Gallart, C., Zoccali, M., & Aparicio, A. 2005, *ARAA*, 43, 387

Grocholski, A. J., Cole, A. A., Sarajedini, A., Geisler, D., & Smith V. V. 2006, *AJ*, 132, 1630

Holtzman, J. A., Gallagher, J. S., III, Cole, A. A., *et al.* 1999, *AJ*, 118, 2262

Kallivayalil, N., van der Marel, R. P., Alcock, C., *et al.* 2006, *ApJ*, 638, 772

Kallivayalil, N., van der Marel, R. P., & Alcock, C. 2006, *ApJ*, 652, 1213

Olszewski, E. W. 1993, *ASPC*, 48, 351

Olszewski, E. W., Schommer, R. A., Suntzeff, N. B., & Harris, H. C. 1991, *AJ*, 101, 515

Pagel, B. E. J. & Tautvaišienė, G. 1998, *MNRAS*, 299, 535

Pompéia, L., Hill, V., Spite, M., *et al.* 2008, *A&A*, 380, 379

Rutledge, G. A., Hesser, J. E., & Stetson, P. B. 1997, *PASP*, 109, 883

Salaris, M. 2002, *ASPC*, 274, 50

Skillman, E. D. & Gallart C. 2002, *ASPC*, 274, 535

Smecker-Hane, T. A., Cole, A. A., Gallagher, J. S., & Stetson, P. B. 2002, *ApJ*, 566, 239

Smith, V. V., Hinkle, K. H., Cunha, K., *et al.* 2002, *AJ*, 124, 3241

van der Marel, R. P. 2001, *AJ*, 122, 1827

Worthey, G. 1999, *ASPC*, 192, 283

The Magellanic System: Stars, Gas, and Galaxies
Proceedings IAU Symposium No. 256, 2008
Jacco Th. van Loon & Joana M. Oliveira, eds.

© 2009 International Astronomical Union
doi:10.1017/S1743921308028561

The star formation history in 12 SMC fields

Noelia E. D. Noël[1], Antonio Aparicio[1], Carme Gallart[1], Sebastián L. Hidalgo[2], Edgardo Costa[3] and René A. Méndez[3]

[1] Instituto de Astrofísica de Canarias, Spain

[2] University of Minnesota, Department of Astronomy, USA

[3] Universidad de Chile, Departamento de Astronomía, Chile

Abstract. We present a quantitative analysis of the star formation history (SFH) of 12 fields in the Small Magellanic Cloud (SMC) based on unprecedented deep [(B–R),R] color—magnitude diagrams (CMDs) from Noël *et al.* (2007). Our fields reach down to the oldest main sequence (MS) turnoff with high photometric accuracy, which is vital for obtaining accurate SFHs. We use the IAC-pop code (Aparicio & Hidalgo 2009) to obtain the SFH, using a single CMD generated using IAC-star (Aparicio & Gallart 2004). We find that there are three main periods of enhancement of star formation: a young one peaked at ∼0.2–0.5 Gyr old, only present in the eastern and in the central-most fields; one at intermediate ages, peaked at ∼4–5 Gyr old in all fields; and an old one, peaked at ∼10 Gyr in all the fields but the western ones, in which this old enhancement splits into two, peaked at ∼8 Gyr old and at ∼12 Gyr old. This "two-enhancement" zone seems to be a robust feature since it is unaffected when using different stellar evolutionary libraries, implying that stars in the SMC take a Hubble time or more to mix. This indicates that there was a global enhancement in $\psi(t)$ at ∼4–5 Gyr ago in the SMC. We also find that the age of the old population is similar at all radii and at all azimuth and we constrain the age of this oldest population to be older than ∼11.5 Gyr old. The intermediate-age population, in turn, presents variations with both, radii and azimuth. Theoretical studies based on results from larger spatial areas are needed to understand the origin of the young gradient. This young component is highly affected by interactions between Milky Way/LMC/SMC. We do not find yet a region dominated by an old, Milky Way-like, halo at 4.5 kpc from the SMC center, indicating either that this old stellar halo does not exist in the SMC or that its contribution to the stellar populations, at the galactocentric distances of our outermost field, is negligible.

Keywords. stars: formation, galaxies: evolution, galaxies: individual (SMC), galaxies: interactions, Magellanic Clouds, galaxies: stellar content, galaxies: structure

1. Introduction

Containing stars born over the whole lifetime of a galaxy, the color—magnitude diagram (CMD) is a fossil record of the SFH. For the Milky Way satellites, it is possible to obtain accurate SFHs, from CMDs reaching the oldest main-sequence (MS) turnoffs, using ground-based telescopes. Reaching the oldest MS turnoffs is vital for breaking the age-metallicity degeneracy and properly characterising the intermediate-age and old population (see Gallart *et al.* 2005). The Magellanic Clouds (MCs), our nearest irregular satellites, provide an ideal environment for this work. In this paper, we will focus on the Small Magellanic Cloud (SMC). The SMC has been historically neglected in favor of its larger neighbor, the Large Magellanic Cloud (LMC). However, there has recently been growing interest in the SMC, related to the discovery from new proper motion measurements — which give constraints to the past orbital motions of the MCs (Kallivayalil *et al.* 2006; Piatek *et al.* 2008) — that it may have a different origin to the LMC (see, e.g., Bekki *et al.* 2004). If true, this would imply that its SFH, evolution and structure could differ significantly from that of the LMC.

2. The SFH of the SMC fields

For each field we obtained three different solutions for the SFH, $\psi(t, z)$, using three different age binning sets in order to reduce sampling problems associated with binning.

The results for the three different sets were then combined by fitting a cubic spline. The spline fit is the final $\psi(t)$ solution we adopted for each field and, together with the results for the 3 age binning sets, is shown in figure 1. The best-fit SFHs for the binning sets are represented by a different symbol and color in figure 1: red triangles are for age-1, blue squares are for age-2, and green circles are for age-3. Each point carries its vertical error bar that is the formal error from IAC-pop, calculated as the dispersion of 20 solutions with $\chi_\nu^2 = \chi_{\nu,\min}^2 + 1$, where $\chi_{\nu,\min}^2$ is that of the solution shown in the figure (see Aparicio & Hidalgo 2008). Horizontal tracks are not error bars but show the age interval associated to each point. We do not have a constraint on $\psi(t)$ at 13 Gyr old and so the end point of our spline fit is arbitrary. Choosing zero for the end point gives good agreement between the integrated SFH under our spline fit and those of our measured SFHs for the three age binnings.

2.1. *Main characteristics of the $\psi(t)$ solutions for our SMC fields*

As can be seen from figure 1, the eastern fields and the central-most field, smc0057 — located in the south — show recent star formation. In particular, the eastern fields show a recent enhancement from ∼1.5–2 Gyr ago until the present, while smc0057 shows a recent peak of star formation ∼1 Gyr ago, which seems to be mostly extinguished at the present time. These $\psi(t)$ peaks are quantified in figure 2, in which the intensity of each $\psi(t)$ enhancement as a function of radius and age for all the fields are represented together with the pericenter passages of the SMC (see figure caption for details). The intensity of each enhancement is the area under a gaussian function fitted to the elevation in the spline fit. The $\psi(t)$ enhancements at young ages in the Eastern fields and in smc0057, peaked between ∼0.2–0.5 Gyr ago, are not seen in other fields located at similar galactocentric distances. The age of these young peaks is in agreement with the age found in the Magellanic Bridge by Harris (2007) and by other authors in the wing area of the SMC (see, for example, McCumber *et al.* 2005). The three eastern fields — the only ones presently forming stars — are located in regions of large amounts of H I, unlike the rest of our fields, including smc0057. A conspicuous intermediate-age enhancement peakes between ∼4 and ∼5 Gyr old in all fields. In addition, there is a small enhancement at ∼2–2.5 Gyr old in the southern and in the western fields. This enhancement is shifted toward younger ages, at ∼1.5–2 Gyr old in the eastern fields.

Finally, an old-age enhancement occurs at ∼10 Gyr in the eastern and southern fields, which seems to be "split" into two, at ∼8 and ∼12 Gyr old, in the western fields.

2.2. *Global bursts of star formation in the SMC*

Phase mixing in a galaxy occurs when stars initially close in space — for example stars formed in a star forming region — spread out over time because they have slightly different energies and angular momenta. Stars are said to be fully phase mixed if there is no memory left that they were born close together. The rate at which stars phase mix depends on the gravitational potential, on the initial proximity of the stars, and on their orbits. As a consequence of the latter, perfectly circular orbits will never mix in radius, while perfectly radial orbits never mix in angle.

The presence of the $\psi(t)$ enhancement at ∼4–5 Gyr old among all the SMC fields, together with the large variations found for ages younger than ∼4 Gyr old, would suggest that the phase mixing time in the SMC is of the order of ∼4 Gyr. However, we find also evidences for spatial variations at older ages: the western fields present two $\psi(t)$

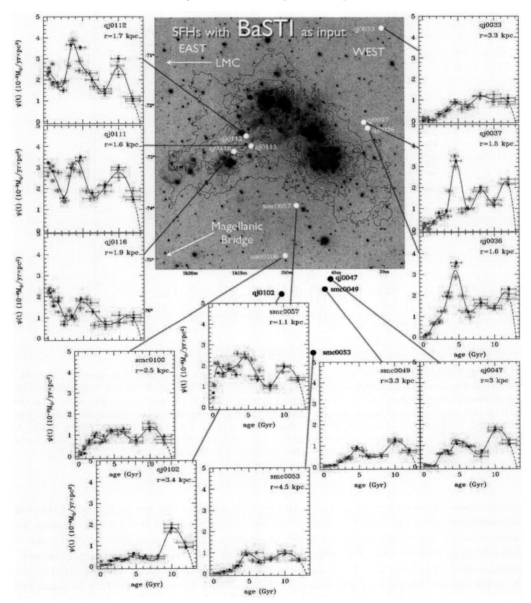

Figure 1. The derived SFHs of our SMC fields. BaSTI stellar evolution library was used as input of IAC-star. Each solution shows the SFH obtained for 3 three age binning schemes (red triangles: age-1, blue squares age-2, and green circles age-3). Each point carries its vertical error bar that is the formal error from IAC-pop, calculated as the dispersion of 20 solutions with $\chi_\nu^2 = \chi_{\nu,\min}^2 + 1$. Horizontal tracks are not error bars, but show the age interval associated to each point. The solid line shows the results of a cubic spline fit to the results. We do not have a constraint on the $\psi(t)$ at 13 Gyr old and the end point of our spline fit was chosen to be zero arbitrarily (dashed lines between 12 and 13 Gyr ago in the spline fit). See text for details.

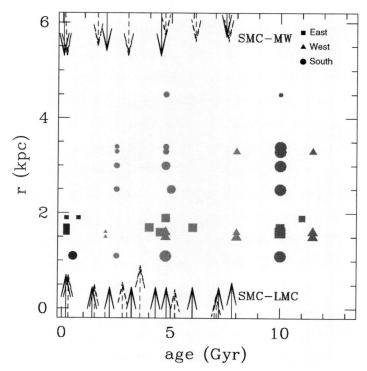

Figure 2. Intensity of the $\psi(t)$ enhancements together with pericenter passages of the SMC. We fitted a gaussian function to the elevations in figure 1. The size of the symbols depends on the intensity of the enhancement. The bottom arrows indicate the pericenter passages with the LMC while the top arrows show the encounters with the Milky Way (solid-lined arrow represent data from Kallivayalil *et al.* (2006) and dashed-lined ones are data obtained from Bekki & Chiba 2005). The size of the arrows represent the intensity of the encounter. Note that some enhancements are hidden behind larger ones. See text for details.

enhancements at ∼8 Gyr old and at ∼12 Gyr old, while in the rest of the fields this old enhancement occurs at ∼10 Gyr old. Time resolution of our SFH solutions is worst for older ages and the results could be model-dependent. However, the spliting of the oldest population in the western fields seems a robust result. On the one hand the two "bumps" remain when sampling the old population with different age binnings. On the other hand, these bumps are impervious when using different stellar evolution libraries. Besides, the formal errors from IAC-pop are significantly smaller than the bumps. In summary, the different behavior of the oldest population in the western fields would imply that stars in the SMC take a Hubble time or more to phase mix. An alternative would be that these variations in older age star distributions might be the result of globular clusters recently dissolved. The following estimative reasoning should be enough to show that this is very unlikely. At least two globulars would be necessary to cause the ∼8 and the ∼12 Gyr old bumps. The mass converted into stars under each bump in each field is of the order of ∼2–5×10^4 M$_\odot$. Considering that their traces are extended over a spatial area larger than our three western fields and that not all the stars in the bumps would be expected to originate in the clusters, their initial masses would have been well above 10^5 M$_\odot$, perhaps close to ∼10^6 M$_\odot$. Since the evaporation times for such massive clusters are over a Hubble time, the remnants should still be observable, which is not the case.

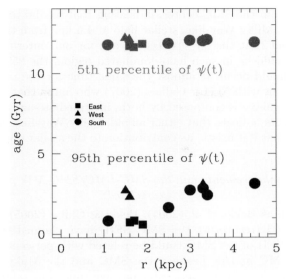

Figure 3. The age at the 5th and at the 95th percentiles of $\psi(t)$ for each of our SMC fields are represented as a function of radius, position angle, and age. See text for details.

We therefore favor the fact that the enhancement of star formation at $\sim$4–5 Gyr old would be a general burst in the SMC and not a sign of phase mixing over these timescales in this galaxy. However, to confirm our findings more studies are needed with the same depth as the ones presented here but covering larger spatial areas.

2.3. *Spatial distribution of the stellar populations in our SMC fields*

One of the most intriguing issues regarding the SMC evolution is the age and distribution of its oldest stars. In order to shed light into this, we calculated the age at the 5th percentile of $\psi(t)$ in each of our SMC fields, i.e., the population age by which 5% of the total stars were formed in each field. The results are presented in figure 3, in which the 5th percentile as a function of radius and age for all the fields is shown. The 5th percentile age in all fields presents a flat distribution at $\sim$11.5 Gyr. This shows that the age of the old population in all our SMC fields is essentially the same, independently of the galactocentric distance or the position angle. In addition, we constrain the age of the oldest population to be older than $\sim$11.5 Gyr old.

The asymmetric shape of the SMC main body is known to be exclusively due to its young populations. A way to quantify how this young component is contributing to the irregular appearance, is to find the age at the 95th percentile of $\psi(t)$ in all the SMC fields, which is also presented in figure 3. The dependence of the 95th percentile age with the position angle is strong: for the southern fields located at $r \gtrsim 3$ kpc it occured $\sim$2.5–3 Gyr ago, at around the same time as in the western fields located at 1.5, 1.6, and 3.3 kpc. The three eastern fields and the central-most field, smc0057, are the only ones for which the 95th percentile age is less than 1 Gyr. These facts show that younger star formation episodes did not occur across the entire galaxy at the same time, but rather toward its central southern part and in the wing area.

Another important — and controversial — fact regarding Local Group dwarf galaxies in general and the SMC in particular, is the composition of the outer extended stellar populations. The 95th percentile age for all the SMC fields shows a smooth gradient while going further away from the SMC center, indicating that we did not yet reach a region dominated by an old, Milky Way-like, stellar halo at 4.5 kpc from the SMC center. This is stressed by the fact that the 95th percentile age for our outermost field occured at ∼3 Gyr ago. If we would be in such halo-dominated region, the 95th percentile and the 5th percentile age should occur at almost the same time for the outermost fields. Our results are in agreement with Noël & Gallart (2007) who found that up to ∼6.5 kpc from the SMC center, the galaxy is composed by both, intermediate-age and old populations. In summary, our results indicate that either an old, Milky Way-like, stellar halo does not exist in the SMC or that if it exists, its contribution to the stellar population is negligible at ∼4.5 kpc.

2.4. Are the $\psi(t)$ enhancements and the SMC-LMC/SMC-MW pericenter passages correlated?

Several authors, such as Bekki *et al.* (2004), Bekki & Chiba (2005) (through numerical simulations) and HZ04 (recovering the SMC SFH) have claimed that the episodes of enhancement in the SFH of the SMC could be related with pericenter passages between the SMC and the LMC and/or between the SMC and the Milky Way. We compare the enhancements in $\psi(t)$ and the pericenter passage times but taking into account the intensity of the $\psi(t)$ enhancements. The intensity of each enhancement is the area under a gaussian function fitted to the elevation in the spline fit. The results are shown in figure 2. The young enhancements in the eastern fields and in smc0057 between at ∼0.2–0.5 Gyr old are not seen in the other fields at the same galactocentric distances. The strongest enhancement at intermediate ages, common to most of the fields (independently of the location in terms of position angle and radius) peaks at ∼4.75 Gyr old. In field qj0112, this enhancement is unfolded into two, peaked at ∼4 Gyr old and at ∼6 Gyr old, respectively. The most conspicuous old enhancement is the one at ∼10 Gyr old which is present in all fields except for the western ones, for which the enhancements at ∼8 and at ∼12 Gyr old, mentioned above, are instead very noticeable.

The SMC-LMC and SMC-Milky Way pericenter passages according to Kallivayalil *et al.* (2006) and Bekki & Chiba (2005) are shown by the arrows in figure 2. There seems to be a correlation between SMC-Milky Way encounters given by Kallivayalil *et al.* (2006) and the enhancements in $\psi(t)$ we found at ∼2.5 Gyr ago, at ∼4.75 Gyr ago and at ∼8 Gyr ago.

References

Aparicio, A. & Gallart, C. 2004, *AJ*, 128, 1465
Aparicio, A. & Hidalgo, S. L. 2009, *AJ*, submitted
Bekki, K., Couch, W. J., Beasley, M. A., Forbes, D. A., Chiba, M., & Da Costa, G. S. 2004, *ApJ*, 610, L93
Bekki, K. & Chiba, M. 2005, *ApJ*, 625, L107
Gallart, C., Zoccali, M., & Aparicio, A. 2005, *ARAA*, 43, 10
Harris, J. 2007, *ApJ* 658, 345
Kallivayalil, N., van der Marel, R. P., & Alcock, C. 2006, *ApJ*, 652, 1213
McCumber, M. P., Garnett, D. R., & Dufour, R. J. 2005, *AJ*, 130, 1083
Noël, N. E. D., Gallart, C., Costa, E., & Méndez, R. A. 2007, *AJ*, 133, 2037
Noël, N. E. D. & Gallart, C. 2007, *ApJ*, 665, L23
Piatek S., Pryor, C., & Olszewski, E. W. 2008, *AJ*, 135, 1024

The Magellanic System: Stars, Gas, and Galaxies
Proceedings IAU Symposium No. 256, 2008
Jacco Th. van Loon & Joana M. Oliveira, eds.

The chemical enrichment history of the Magellanic Clouds field populations

**Ricardo Carrera[1], Carme Gallart[2], Antonio Aparicio[2],
Edgardo Costa[3], Eduardo Hardy[3,4], Rene A. Méndez[3],
Noelia E. D. Noël[2] and Robert Zinn[5]**

[1] Osservatorio Astronomico di Bologna, Via Ranzani 1, I-40127 Bologna, Italy
email: ricardo.carrera@oabo.inaf.it

[2] Instituto de Astrofísica de Canarias, Vía Láctea sn, E-38200 La Laguna, Spain

[3] Departamento de Astronomía, Universidad de Chile, Casilla 36–D, Santiago, Chile

[4] National Radio Astronomy Observatory, Casilla El Golf 16-10, Las Condes, Santiago, Chile

[5] Department of Astronomy, Yale University, New Haven, USA

Abstract. We report the results of our project devoted to study the chemical enrichment history of the field population in the Magellanic Clouds using Ca II triplet spectroscopy.

Keywords. galaxies: abundances, Magellanic Clouds, galaxies: stellar content

1. Introduction

The study of the ages and metallicities of the resolved stars in a nearby galaxy provides very detailed information on its evolutionary history. The Magellanic Clouds are examples of galaxies where this method can be applied very successfully. However, their vastness and our limitations in observing sizable samples of stars imply that there are particularly large gaps of knowledge in this area, compared with others. For example, their chemical enrichment histories have mainly been studied from their cluster systems. However, these objects have some drawbacks, like the age-gap of the LMC clusters or the single old cluster known in the SMC. Although the SMC is more metal-poor than the LMC, the chemical evolution of the cluster systems has been qualitatively similar in both galaxies. They show a first episode of chemical enrichment followed by a period of slow chemical evolution until around 4 Gyrs ago. Then, both galaxies took off again (Olszewski *et al.* 1991; Piatti *et al.* 2001). However, as mentioned above, there are some epochs in which the lack of clusters makes it difficult to extract definite conclusions. The chemical evolution of the Magellanic Clouds has also been investigated from their planetary nebulae (Dopita *et al.* 1991; Idiart *et al.* 2007), which show a behaviour similar to the clusters one. Finally, Cole *et al.* (2005) have studied the chemical evolution of the LMC bar based on a sample of red giant branch (RGB) stars. The age—metallicity relationship (AMR) of the bar stars is similar to the cluster's one, particularly for the older ages. However, the increase of metallicity observed in the clusters in the last 3 Gyrs is not observed in the bar.

What about the field populations in the LMC disk and in the SMC? Do they share the cluster and planetary nebulae behaviour in each galaxy? Does chemical evolution show a global pattern in each galaxy or on the contrary, does it depend on the position? To address these questions, and as part of a large project devoted to study the stellar content of the Magellanic Clouds, we have obtained metallicities and ages for stars in 4 large LMC and 13 SMC fields. In this paper we will discuss the procedure used to derive metallicities and ages, and describe our main conclusions on the chemical enrichment history of both Magellanic Clouds.

2. Deriving stellar metallicities and ages

Although the best way to derive chemical abundances is high-resolution spectroscopy, this technique needs huge amounts of telescope time to measure a significant number of stars even in the nearest galaxies. The alternative is low-resolution spectroscopy which, together with the modern multi-object spectrographs, allows us to observe large samples in a reasonable time. However, in external galaxies, even low-resolution spectroscopy is only possible for the brightest objects, which in most cases are RGB stars. A powerful index to derive metallicities in these stars is the infrared Ca II triplet (CaT). This index is linearly correlated with metallicity in the range $-2.2 \leqslant [Fe/H] \leqslant +0.47$, with no measurable influence of age in the interval $13 \leqslant Age/Gyr \leqslant 0.25$ (Carrera *et al.* 2007). Tough this index may overestimate metallicities for values below -2.5 dex (Koch *et al.* 2008), the range of ages and metallicities in which it shows a linear behaviour agrees with the ones expected in the Magellanic Clouds.

The position of the RGB on the color—magnitude diagram (CMD) suffers from age–metallicity degeneracy. However, when the metallicity has been obtained in an alternative way, as in this case from spectroscopy, this age—metallicity degeneracy can be broken up, and stellar ages can be estimated from the position of the stars in the CMD. To do that, we computed a polynomial relationship to derive stellar ages from metallicities and positions of stars in the CMD. For that purpose, a synthetic CMD computed with IAC-STAR (Aparicio & Gallart 2004) with the overshooting BaSTI stellar stellar evolution models (Pietrinferni *et al.* 2004) as input was used. Details of this procedure can be found in Carrera *et al.* (2008a) and Carrera *et al.* (2008b).

3. LMC chemical enrichment history

In the LMC, we have studied four fields located at $2°3$, $4°0$, $5°5$ and $7°1$ northward of the bar. The CMDs of these fields have been presented by Gallart *et al.* (2008). We are obtaining detailed SFHs for them (see Gallart *et al.* 2008, and Meschin *et al.* in this volume) using the synthetic CMD technique. In each field we observed spectroscopically more than 100 stars in the upper part of the RGB. The results of this work can be found in Carrera *et al.* (2008a). Table 1 summarizes the mean metallicities and their dispersions in each field. The inner fields have a similar mean metallicity, while it is a factor of two more metal-poor in the outermost field.

Table 1. Mean values of metallicity distributions of LMC fields.

Field	$\langle [Fe/H] \rangle$	$\sigma_{[Fe/H]}$
Bar	-0.39	0.19
$2°3$	-0.47	0.31
$4°0$	-0.50	0.37
$5°5$	-0.45	0.31
$7°1$	-0.79	0.44

To understand why the outermost field is more metal-poor, we obtained the AMR of each field (Fig. 1). The age error in each age interval is indicated in the top panel. The age distributions for each field have been plotted in inset panels taking into account (*solid line*) and not (*histogram*) the age determination uncertainties (see Carrera *et al.* 2008a for details).

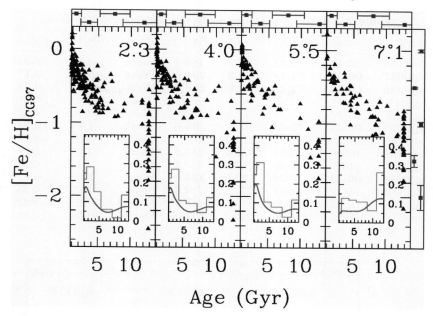

Figure 1. Age–metallicity relationships for the four LMC fields in our sample. Inset panels show the age distribution computed taking into account (*solid line*) and not (*histogram*) the age determination uncertainties. The top panel show the age error in each interval. The left panels show the metallicity error.

The AMR is, within the uncertainties, very similar for all fields, and also similar to the cluster ones. As expected, the most metal-poor stars in each field are also the oldest ones. A rapid chemical enrichment at a very early epoch is followed by a period of very slow metallicity evolution until around 3 Gyr ago, when the galaxy started another period of chemical enrichment that is still ongoing. Furthermore, the age histograms for the three innermost fields are similar, although the total number of stars decreases when moving away from the centre. The outermost field has a lower fraction of "young" (1–4 Gyr) intermediate-age stars. This indicates that its lower mean metallicity is related to the lower fraction of intermediate-age, more metal-rich stars rather than to a different chemical enrichment history (for example, a slower metal enrichment).

4. SMC chemical enrichment history

In the SMC we secured spectroscopy of stars in 13 fields spread about the galaxy body. The corresponding photometry has been published by Noël *et al.* (2007) and their detailed SFHs are presented in this volume by Noël *et al.* (2008). The positions of these fields, together with their mean metallicities are listed in Table 2. The fields are ordered by their distance to the center, which is shown in column 4. Fields in different regions are indicated by different font types: eastern fields in normal, western fields in boldface and southern fields in italics. Mean metallicities are very close to [Fe/H]~ -1 in all fields within $r \lesssim 2°.5$ from the SMC center. A similar value is observed for the southern fields up to $r \lesssim 3°$ (qj0047 and smc0049). For the outermost fields, qj0033 in the West, and qj0102 and qj0053 in the South, the mean value is clearly more metal-poor than in the others.

Table 2. SMC observed fields, mean metallicities and dispersion.

Field	α_{2000}	δ_{2000}	r(')	PA (°)	Zone	$\langle[\text{Fe/H}]\rangle$	$\sigma_{[Fe/H]}$
smc0057	*00:57*	*−73:53*	*65.7*	*164.4*	*South*	*−1.01*	*0.33*
qj0037	**00:37**	**−72:18**	**78.5**	**294.0**	**West**	**−0.95**	**0.17**
qj0036	**00:36**	**−72:25**	**79.8**	**288.0**	**West**	**−0.98**	**0.25**
qj0111	*01:11*	*−72:49*	*80.9*	*89.5*	*East*	*−1.08*	*0.21*
qj0112	*01:12*	*−72:36*	*87.4*	*81.0*	*East*	*−1.16*	*0.32*
qj0035	**00:35**	**−72:01**	**95.5**	**300.6**	**West**	**−1.09**	**0.24**
qj0116	*01:16*	*−72:59*	*102.5*	*95.2*	*East*	*−0.96*	*0.26*
smc0100	*01:00*	*−74:57*	*130.4*	*167.5*	*South*	*−1.07*	*0.28*
qj0047	*00:47*	*−75:30*	*161.7*	*187.7*	*South*	*−1.15*	*0.27*
qj0033	**00:33**	**−70:28**	**172.9**	**325.0**	**West**	**−1.58**	**0.57**
smc0049	*00:49*	*−75:44*	*174.8*	*184.6*	*South*	*−1.00*	*0.28*
qj0102	*01:02*	*−74:46*	*179.5*	*169.4*	*South*	*−1.29*	*0.42*
smc0053	*00:53*	*−76:46*	*236.3*	*179.4*	*South*	*−1.64*	*0.50*

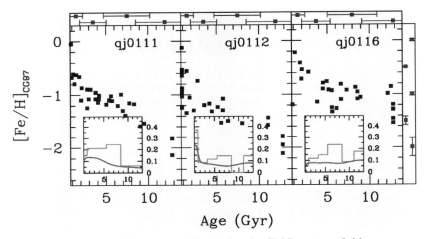

Figure 2. The same as Fig. 1, for the SMC eastern fields.

The fact that the mean metallicity decreases when moving away from the center implies that there is a metallicity gradient in the SMC. This is the first time that a spectroscopic metallicity gradient has been reported in SMC stellar populations. The detection of this gradient has been possible because we have covered a large radius range, up to 4° from the SMC center.

To investigate the nature of the gradient, we have calculated the AMR for each field. They are plotted in Fig. 2, 3 and 4 for fields situated to the East, West and South respectively. Inset panels, as in Fig. 1, show the age distribution of the observed stars, with and without taking into account the age uncertainty (*solid line and histogram*, respectively).

All the AMR plotted in Fig. 2, 3 and 4 show a rapid chemical enrichment at a very early epoch. Even though in some fields we have not observed enough old stars to sample this part of the AMR, note that 12 Gyr ago all fields have reached [Fe/H]$\simeq$ −1.4 to −1.0. This initial chemical enrichment was followed by a period of very slow metallicity evolution until around 3 Gyr ago. Then, the galaxy started another period of chemical enrichment, which is observed in the innermost fields, which are, however, the only ones where we observed enough young stars to sample this part of the AMR. In all cases,

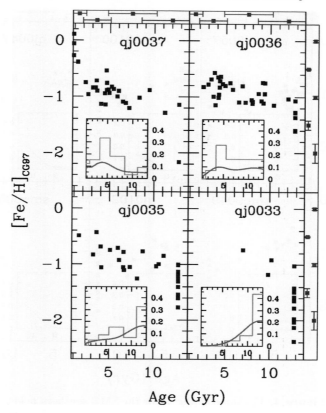

Figure 3. The same as Fig. 1, for the SMC western fields.

the mean metallicity is similar to that of the other fields at similar galactocentric radii. The field AMRs obtained in this work are similar to those for clusters (the reader should take into account that differences in the metallicity scales exist among different works), although there is only one cluster older than 10 Gyr (e.g., Piatti *et al.* 2001), and for planetary nebulae, with the exception that in these objects it is not observed the chemical enrichment episode at a very early epoch (Idiart *et al.* 2007).

For eastern fields, located in the wing, most of the observed stars have ages younger than 8 Gyr, but there is also a significant number of objects older than 10 Gyr. At a given galactocentric distance, eastern fields show a large number of young stars ($\leqslant 3$ Gyr) in comparison to the western ones. For the western and southern fields, the fraction of intermediate-age stars, which are also more metal-rich, decreases as we move away from the center, although the average metallicity in each age bin is similar. This indicates the presence of an age gradient in the galaxy, which may be the origin of the metallicity one. It is noticeable that for the most external fields, qj0033 and qj0053, we find a predominantly old and metal-poor stellar population.

5. Future work

We have investigated the chemical evolution of the field populations in both Magellanic Clouds. However, there are still some points which should be investigated. For example, in the LMC, we have studied 4 fields at different galactocentric distances northward of the bar. Now it is necessary to investigate what happens in other positions such as facing

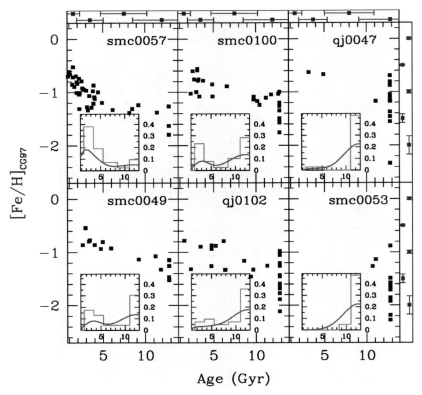

Figure 4. The same as Fig. 1, for the SMC southern fields.

the SMC and in the opposite direction. This could give us clues about the effects of the interactions with the SMC in the LMC field populations. Also, in both galaxies it is necessary to investigate the behaviour of the abundances of different chemical species as a function of position in the galaxy.

References

Aparicio, A. & Gallart, C. 2004, *AJ*, 128, 1465
Carrera, R., Gallart, C., Pancino, E., & Zinn, R. 2007, *AJ*, 134, 1298
Carrera, R., Gallart, C., Hardy, E., Aparicio, A., & Zinn, R. 2008, *AJ*, 135, 836
Carrera, R., Gallart, C., Aparicio, A., Costa, E., Méndez, R. A., & Nöel, N. E. D. 2008, *AJ*, 136, 1039
Cole, A. A., Tolstoy, E., Gallagher III, J. S., & Smecker-Hane, T. A. 2005, *AJ*, 129, 1465
Dopita, M. A., Vassiliadis, E., Wood, P. R., *et al.* 1997, *ApJ*, 474, 188
Gallart, C., Stetson, P. B., Meschin, I. P., Pont, F., & Hardy, E. 2008, *ApJ*, 682, 89
Idiart, T. P., Maciel, W. J., & Costa, R. D. D. 2007, *A&A*, 472, 101
Koch, A., Grebel, E. K., Gilmore, G. F., *et al.* 2008, *AJ*, 135, 1580
Nöel, N. E. D., Gallart, C., Costa, E., & Méndez, R. A. 2007, *AJ*, 133, 2037
Olszewski, E. W., Schommer, R. A., Suntzeff, N. B., & Harris, H. C. 1991, *AJ*, 101, 515
Piatti, A. E., Santos, J. F. C., Clariá, J. J., *et al.* 2001, *MNRAS*, 325, 792
Pietrinferni, A., Cassisi, S., Salaris, M., & Castelli, F. 2004, *ApJ*, 612, 168

The Magellanic System: Stars, Gas, and Galaxies
Proceedings IAU Symposium No. 256, 2008
Jacco Th. van Loon & Joana M. Oliveira, eds.

© 2009 International Astronomical Union
doi:10.1017/S1743921308028585

Spatial variations in the star formation history of the Large Magellanic Cloud

Carme Gallart[1], Ingrid Meschin[1], Antonio Aparicio[1], Peter B. Stetson[2] and Sebastián L. Hidalgo[1]

[1]Instituto de Astrofísica de Canarias. 38200 La Laguna. Tenerife, Spain
email:`carme, imeschin, antapaj, shidalgo@iac.es`

[2]Herzberg Institute of Astrophysics, National Research Council, Victoria, BC, Canada V9E 2E7
email: `Peter.Stetson@nrc-cnrc.gc.ca`

Abstract. Based on the quantitative analysis of a set of wide-field color—magnitude diagrams reaching the old main sequence-turnoffs, we present new LMC star-formation histories, and their variation with galactocentric distance. Some coherent features are found, together with systematic variations of the star-formation history among the three fields analyzed. We find two main episodes of star formation in all three fields, from 1 to 4 and 7 to 13 Gyr ago, with relatively low star formation around $\simeq$ 4−7 Gyr ago. The youngest age in each field gradually increases with galactocentric radius; in the innermost field, LMC 0514−6503, an additional star formation event younger than 1 Gyr is detected, with star formation declining, however, in the last $\simeq$ 200 Myr. The population is found to be older on average toward the outer part of the galaxy, although star formation in all fields seems to have started around 13 Gyr ago.

Keywords. galaxies: evolution, galaxies: formation, galaxies: individual (LMC), Magellanic Clouds

1. Introduction

Modern hydrodynamical simulations of galaxy formation and evolution are beginning to be able to predict both temporally and spatially resolved structure in galaxies (e.g., Abadi *et al.* 2003; Governato *et al.* 2007; Roškar *et al.* 2008), and thus, it is timely to offer detailed observational counterparts for comparison with the models.

From the observational side, late-type spiral and irregular galaxies are found to become on average redder with increasing distance from the center (e.g., Taylor *et al.* 2005). On another hand, Local Group dwarf irregular galaxies show stellar population gradients such that young populations disappear in the outer part (e.g., Bernard *et al.* 2007). This has sometimes been interpreted (Minniti & Zijlstra 1996) as the presence of an old stellar halo in these galaxies. However, with typical groundbased observations of galaxies at $\simeq$ 1 Mpc, such as those of Minniti & Zijlstra, it is not possible to ascertain the actual age composition, after the last few hundred Myr, as a function of position. Carbon stars indicate the presence of intermediate-age populations at large radius. For example, in IC 1613 Albert *et al.* (2000) found 24 C stars beyond R $\simeq$ 10′, where the color—magnitude diagrams (CMD) of Bernard *et al.* (2007) show few or no stars on the young main sequence (MS).

A quantitative determination of the star-formation history (SFH) as a function of radius can be obtained from CMDs reaching down to the oldest MS turnoffs (Gallart *et al.* 1999; Gallart *et al.* 1999, 2005; Cole *et al.* 2007). These can be obtained using ground-based facilities for all the Milky Way satellites. In addition, an important fraction of the Local Group galaxies can be observed down to the same absolute magnitude limit using the ACS on board the *HST* (e.g., Bernard *et al.* 2008; Cole *et al.* 2007; Brown *et al.* 2007).

Thanks to their proximity, the Magellanic Clouds are in principle ideal candidates for detailed studies of their SFHs, and can help to shed some light on the actual origin of the stellar-population gradients observed in late-type spiral and irregular galaxies. Their huge size in the sky, however, makes it challenging to observe areas large enough to contain representative samples of stars. Most studies to date that do reach the oldest MS turnoffs of field LMC stars have used the WFPC2 on the *HST* to observe relatively small fields in the bar or at relatively small galactocentric distances (single WFPC2 fields in the case of Holtzman et al. 1999; mosaics of a few fields in the case of Smecker-Hane *et al.* 2002). In other cases, the field stars have been studied in images taken for the purpose of observing clusters (e.g., Olsen 1999; Javiel *et al.* 2005). The situation is similar in the case of the SMC (Dolphin *et al.* 2001; McCumber *et al.* 2005; Chiosi & Vallenari 2007; Noël *et al.* 2007).

In this paper, we will discuss some new results on the SFH of the LMC and its variation as a function of galactocentric distance, based on a quantitative analysis of a set of four wide-field CMDs reaching the old MS turnoffs, and ranging from 2.3° to 7.1°. We will compare these results with others on late-type and irregular galaxies in order to shed light on some of the issues raised above.

2. The data

We obtained V- and I-band images of four LMC fields with the Mosaic II camera on the CTIO Blanco 4 m telescope in December 1999 and January 2001. Fields were chosen to span a range of galactocentric distances, from $\simeq 2.3°$ to 7.1° (2.0 to 6.2 kpc) northward from the kinematic center of the LMC. We will name the fields according to their RA and DEC (J2000.0) as LMC 0512−6648, LMC 0514−6503, LMC 0513−6333 and LMC 0513−6159, in order of increasing galactocentric distance. Profile-fitting photometry was obtained with the DAOPHOT/ALLFRAME suite of codes (Stetson 1994) and calibrated to the standard system using observations of several Landolt (1992) fields obtained in the same runs. Finally, a large number of artificial star tests were performed in each frame following the procedure described in Gallart *et al.* (1999); these are used both to derive completeness factors and to model photometric errors in the synthetic CMDs.

3. The star formation history and its spatial variations

Figure 1 shows the $M_I \lesssim 4$ mag CMDs of the four LMC fields, with isochrones from the overshooting set of the BaSTI library (Pietrinferni *et al.* 2004)† superimposed. The metallicities have been chosen to approximately reproduce the common chemical enrichment law for the same fields derived by Carrera *et al.* (2008), using Ca II triplet spectroscopy. Note how well this combination of ages and metallicities reproduces the position and shape of the RGB. The number of stars observed with good quality photometry down to $M_I \lesssim 4$ in each field — in order of increasing galactocentric distances — are 30 0000, 21 4000, 86 000 and 39 000 respectively. All the CMDs reach the oldest MS turnoff ($M_I \simeq 3.0$ mag) with good photometric precision and completeness fractions over 75% (except for the innermost field, in which crowding is very severe). The two innermost fields show CMDs with a prominent, bright MS and a well populated red clump

† The new 2008 version of the BaSTI stellar evolution models (see http://www.oa-teramo.inaf.it/BASTI) is used through the paper. This set shows a much better agreement with other stellar evolution models than the older one.

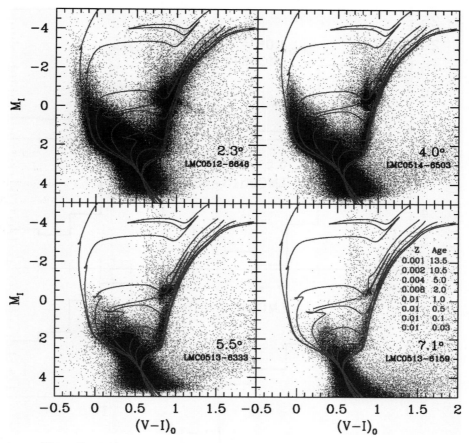

Figure 1. $[(V - I)_0, M_I]$ CMDs for the four fields. Isochrones with ages and metallicities as labelled, and a zero-age horizontal-branch of $Z = 0.001$ by Pietrinferni *et al.* (2004) have been superimposed. A distance modulus of $(m - M)_0 = 18.5$ and $E(B - V) = 0.10, 0.05, 0.037$ and 0.026 magnitudes, respectively, have been assumed.

typical of a population which has had ongoing star formation from $\simeq 13$ Gyr ago to the present time. The two outermost fields clearly show a fainter MS termination, indicating truncated or sharply decreasing star formation in the last few hundred Myr or few Gyr, respectively (see below). No extended horizontal branch is observed in any of the fields, but all fields host a number of stars redder than the RGB tip (and redder than the color interval shown in the figure), which are candidate AGB stars.

Gallart *et al.* (2008) discussed the LMC disk stellar populations and their gradients, based on a comparison of the observed CMD with the BaSTI isochrones, as well as a comparison of the observed color functions (CF) with synthetic CFs. They concluded the following: (a) the area around the old MS turnoff is well populated in all four CMDs, indicating that star formation started at about the same time, $\simeq 13$ Gyr ago, in all four fields, or that old stars have been able to migrate out to the outermost galactocentric radii observed here; (b) the bulk of the star formation may be truncated in the outermost fields (LMC 0513−6159 and LMC 0513−6333) at ages $\simeq 1.5$ and 0.8 Gyr respectively; and (c) there are indications of enhanced star formation (as compared to a constant star formation rate) in the innermost fields LMC 0514−6503 and LMC 0512−6648 in the same time range, likely having started even earlier, ~ 4 Gyr ago. Enhanced star formation

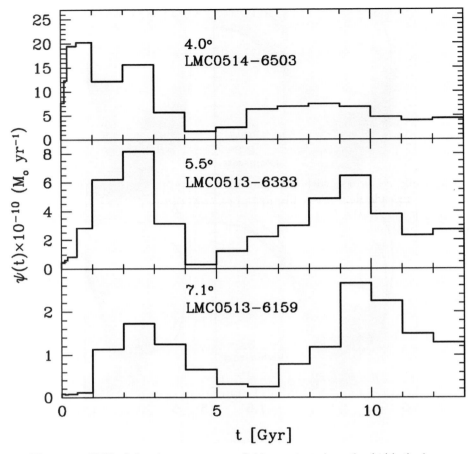

Figure 2. SFH of the three outermost fields, projected on the $(\psi(t), t)$ plane.

starting around 4 Gyr ago possibly extended also to field LMC 0513−6333; however in this case it would have been followed by the truncation discussed above.

To obtain a quantitative derivation of the SFH in the observed fields, we have used the IAC-STAR, IAC-POP and MINNIAC codes (Aparicio & Gallart 2004; Aparicio & Hidalgo 2008; Hidalgo *et al.* 2008, in preparation) to compute synthetic CMDs and compare the distribution of stars in the observed and synthetic CMDs, obtaining the best solution through χ^2 minimization (see the references above and Meschin *et al.* 2008, in these proceedings, for details on the synthetic CMD technique and our particular implementation of it). Here, and in Meschin *et al.* (2008), we show preliminary solutions for the three outermost fields, LMC 0513−6159, LMC 0513−6333 and LMC 0514−6503. We are in the process of testing the robustness of the solutions and their dependence on variations of some of the input parameters, such as the binary fraction or the IMF, or on the distance modulus and reddening adopted. We believe, however, that the present solutions provide a reasonable description of the main features characterizing the LMC SFH at different galactocentric radii. We have assumed the Kroupa (2001) IMF, a binary fraction of 0.4 with mass ratios $q \geqslant 0.5$, a distance modulus $(m − M)_0 = 18.5$ and reddenings $E(B − V) = 0.05$, 0.037 and 0.026 magnitudes, for the innermost to the outermost field, respectively. The BaSTI stellar evolution library has been used as input to IAC-STAR to compute the synthetic CMDs. Figure 2 shows the projection of the SFHs on the $(\psi(t), t)$ plane (see the poster by Meschin *et al.* 2008 for a 3-D representation of the SFH on the $(\psi(t, z), t, z)$ space).

Some coherent features, together with systematic variations of the SFH among fields, are noticeable in Figure 2. In all fields, local maxima of $\psi(t)$ are found around $\simeq$3–4 and 8–10 Gyr ago, with relatively low star formation activity around $\simeq 4-7$ Gyr ago. The ratio between the amount of star formation younger and older than the age of the minimun $\psi(t)$ decreases toward the outer part of the galaxy, thus the population is on average older there. Star formation in all fields seems to have started around 13 Gyr ago. These features are consistent with the age distribution of star clusters in the LMC. The trends regarding the young and young intermediate-age SFH that have been found through the comparison with isochrones and the use of the CF (Gallart *et al.* 2008; see above) are fully confirmed with the quantitative analysis presented in this paper: the youngest age in each field gradually increases with galactocentric radius, from 2.3° to 7.1°. In particular, in the field at 4.0° an additional star formation event younger than 1 Gyr is detected, with star formation declining, however, in the last $\simeq 200$ Myr.

4. Discussion

Since no plausible galaxy-formation theory predicts positive metallicity gradients, Taylor *et al.* (2005) offered two possible explanations for the fact that the late-type galaxies in their sample become redder outwards: a change in the mean age of stellar population, or a change in the dust characteristics. The present work shows that the youngest, bluest stars are progressively missing outwards in the LMC disk, and that the fraction of old and intermediate-age population relative to the younger population increases at increasing galactocentric radius. Therefore, a reddening of the intrinsic integrated color is expected. A similar trend in found in the dwarf galaxies observed in the framework of the LCID project (Gallart *et al.* 2007), and in particular in the case of IC 1613 and Phoenix: the population gets gradually older outwards, though a sizable amount of intermediate-age population is found at all radii observed (so far). Other irregular galaxies (and M33) are not close enough for such a detailed picture to be obtained, but shallower CMDs hint at a similar situation. In addition, in the case of the LMC, the surface brightness profile remains exponential to the largest galactocentric radius observed in this work, and shows no evidence of disk truncation. Combining the information on surface brightness and stellar population, Gallart *et al.* (2004) concluded that the LMC disk extends — and dominates over a possible stellar halo — out to a distance of at least 6 kpc.

The age gradient in the youngest LMC population is correlated with the H I column density as measured by Staveley-Smith *et al.* (2003): the two innermost fields are located at $\simeq 0.7$ kpc on either side of $R_{H\alpha}$, LMC 0512−6648 on the local maximum of the azimuthally averaged H I column density with $\simeq 1.63 \times 10^{21}$ cm^{-2} (close to the H I threshold for star formation; Skillman 1987) and LMC 0514−6503 where the azimuthally averaged H I column density is only $\simeq 5 \times 10^{20}$ cm^{-2}. Finally, the two outermost fields are close to the H I radius considered by Staveley-Smith *et al.* to be at an H I density of 10^{20} cm^{-2}. The outermost field, LMC 0513−6159, is approximately halfway to the tidal radius (van der Marel *et al.* 2002). If the youngest stars in each field were formed *in situ*, in the LMC we are observing an outside-in quenching of the star formation at recent times ($\simeq 1.5$ Gyr), possibly implying a decrease in size of the H I disk able to form stars. Alternatively, star formation may have been confined to the central $\simeq 3$–4 kpc, where gas resides (or is accreted to), and stars then migrate outwards (e.g., Roškar *et al.* 2008). In fact, it is expected that both star-formation sites and stars migrate across the LMC disk due to tidal interactions with the Milky Way and the SMC (e.g., Bekki & Chiba 2005). Of course, a combination of the two scenarios is also possible.

Acknowledgements

C.G., A.A., I.M. & S.L.H acknowledge the support from the IAC and the Spanish MEC (AYA2004-06343).

References

Abadi, M. G., Navarro, J. F., Steinmetz, M., & Eke, V. R. 2003, *ApJ*, 591, 499
Albert, L., Demers, S., & Kunkel, W. E. 2000, *AJ*, 119, 2780
Aparicio, A. & Gallart, C. 2004, *AJ*, 128, 1465
Aparicio, A. & Hidalgo, S. L. 2008, *AJ*, submitted
Bekki, K. & Chiba, M. 2005, *MNRAS*, 356, 680
Bernard, E. J., Aparicio, A., Gallart, C., Padilla-Torres, C. P., & Panniello, M. 2007, *AJ*, 134, 1124
Bernard, E. J., Gallart, C., & Monelli, M., *et al.* 2008, *ApJ*, 678, L21
Brown, T. M., Smith, E., & Ferguson, H. C., *et al.* 2007, *ApJ*, 658, L95
Carrera, R., Gallart, C., Hardy, E., Aparicio, A., & Zinn, R. 2008, *AJ*, 135, 836
Chiosi, E. & Vallenari, A. 2007, *A&A*, 466, 165
Cole, A. A., Skillman, E. D., Tolstoy, E., *et al.* 2007, *ApJ*, 659, L17
Dolphin, A. E., Walker, A. R., Hodge, P. W., Mateo, M., Olszewski, E. W., Schommer, R. A., & Suntzeff, N. B. 2001, *ApJ*, 562, 303
Gallart, C., Freedman, W. L., Aparicio, A., Bertelli, G., & Chiosi, C. 1999, *AJ*, 118, 2245
Gallart, C., Stetson, P. B., Hardy, E., Pont, F., & Zinn, R. 2004, *ApJ*, 614, L109
Gallart, C., Zoccali, M., & Aparicio, A. 2005, *ARAA*, 43, 387
Gallart, C. and the LCID Team. 2007, in A. Vazdekis & R. F. Peletier (eds.), *Stellar Populations as Building Blocks of Galaxies*, Proc. IAU Symp. No. 241 (San Francisco: ASP), p. 290
Gallart, C., Stetson, P. B., Meschin, I. P., Pont, F., & Hardy, E. 2008, *ApJ*, 682, 89
Governato, F., Willman, B., Mayer, L., Brooks, A., Stinson, G., Valenzuela, O., Wadsley, J., & Quinn, T. 2007, *MNRAS*, 374, 1479
Holtzman, J. A., Gallagher, J. S., III, & Cole, A. A., *et al.* 1999, *AJ*, 118, 2262
Javiel, S. C., Santiago, B. X., & Kerber, L. O. 2005, *A&A*, 431, 73
Kroupa, P. 2001, *MNRAS*, 322, 231
Landolt, A. U. 1992, *AJ*, 104, 340
McCumber, M. P., Garnett, D, R., & Dufour, R. J. 2005, *AJ*, 130, 1083
Minniti, D. & Zijlstra, A. 1996, *ApJ*, 467, L13
Noël, N., Gallart, C., Costa, E., & Méndez, R. 2007, *AJ*, 133, 2037
Olsen, K. A. G. 1999, *AJ*, 117, 2244
Pietrinferni, A., Cassisi, S., Salaris, M., & Castelli, F. 2004, *ApJ*, 612, 168
Roškar, R., Debattista, V. P., Stinson, G. S., Quinn, T. R., Kaufmann, T., & Wadsley, J. 2008, *ApJ*, 675, L65
Skillman, E. 1987, in C. J. Lonsdale Persson (ed.), *Star Formation in Galaxies* (NASA CP-2466; Washington: NASA), p. 263
Smecker-Hane, T. A., Cole, A. A., Gallagher, J. S., III, & Stetson, P. B. 2002, *ApJ*, 566, 239
Staveley-Smith, L., Kim, S., Calabretta, M. R., Haynes, R. F., & Kesteven, M. J. 2003, *MNRAS*, 339, 87
Stetson, P. B. 1994, *PASP*, 106, 250
Taylor, V. A., Jansen, R. A., Windhorst, R. A., Odewahn, S. C., & Hibbard, J. E. 2005, *ApJ*, 630, 784
van der Marel, R. P., Alves, D. R., Hardy, E., & Suntzeff, N. B. 2002, *AJ*, 124, 2639

The Magellanic System: Stars, Gas, and Galaxies
Proceedings IAU Symposium No. 256, 2008
Jacco Th. van Loon & Joana M. Oliveira, eds.

Metallicity and kinematics of a large sample of LMC and SMC clusters

A. J. Grocholski[1], M. C. Parisi[2], D. Geisler[3], A. Sarajedini[4], A. A. Cole[5], J. J. Clariá[2] and V. V. Smith[6]

[1] Space Telescope Science Institute, USA

[2] Observatorio Astronómico, Universidad Nacional de Córdoba, Argentina

[3] Departamento de Astronomía, Universidad de Concepción, Chile

[4] Department of Astronomy, University of Florida, USA

[5] University of Tasmania, Australia

[6] Gemini Project, National Optical Astronomy Observatory, USA

Abstract. We have carried out a large-scale investigation of the metallicity and kinematics for a number of LMC and SMC star clusters using Ca II triplet spectra obtained at the *VLT*. Our sample includes 28 LMC and 16 SMC clusters, covering a wide range of ages and spatial extent of the host galaxy. We determine mean cluster velocities to about 2 km s^{-1} and metallicities to 0.05 dex (random error), from about 7 members per cluster. Herein we present the main results for this study for the cluster metallicity distributions, metallicity gradients, age-metallicity relations and kinematics.

Keywords. stars: kinematics, Magellanic Clouds, galaxies: star clusters

1. Introduction

Star clusters in the LMC and SMC are of fundamental importance for a variety of reasons. On a cosmological scale, star clusters in the Magellanic Clouds are necessary for the understanding of stellar populations in distant galaxies since they occupy regions of the age-metallicity plane that are devoid of Milky Way clusters. Thus, accurate knowledge of their properties is needed to properly employ LMC and SMC clusters as tests of stellar evolution models as well as empirical templates of simple stellar populations. For the LMC and SMC specifically, their populous star clusters, which preserve a record of their host galaxy's chemical abundances at the time of their formation, provide the most straightforward means of determining the star formation and chemical enrichment histories of these two galaxies.

Despite their utility, the LMC and SMC cluster systems have surprisingly little up-to-date information available. The only previous large scale study of cluster abundances and velocities in the LMC was performed by Olszewski *et al.* (1991); they observed the Ca II triplet (CaT) lines in RGB stars in ∼80 LMC clusters. However, the accuracy of their results was limited by the technology (4 m class telescopes with single-slit spectrographs) available at the time. For the SMC, the situation is worse, with only two clusters having abundances based on high-resolution spectra (NGC 330 - Hill 1999; NGC 121 - Johnson *et al.* 2004) and an additional six with CaT based velocities and metallicities (Da Costa & Hatzidimitriou 1998, DH98). To improve upon our knowledge of the abundances and kinematics of star clusters in the Magellanic Clouds, we have taken advantage of the multiplexing capability of modern spectrographs and obtained CaT spectra for hundreds of individual RGB stars spread across 28 LMC and 16 SMC clusters. Herein we provide a brief description of our work, while the full details can be found in Grocholski *et al.* (2006, hereafter G06) and Parisi *et al.* (2008, hereafter P08) for the LMC and SMC, respectively.

2. Data

Spectroscopic observations of our LMC clusters were carried out in December 2004, and our SMC targets were observed in November 2005. In both cases we used the FORS2 spectrograph in mask exchange unit mode on the Antu (*VLT*-UT1) 8.2 m telescope at *ESO*'s Paranal Observatory. Our spectra were centered on the CaT lines ($\sim$8500–8700 Å), with a resolution of 2–3 Å. Data processing was carried out within the IRAF environment. Using standard IRAF tasks, we flat-fielded the images, fixed bad pixels, corrected the images for distortions, and extracted and normalized the stellar spectra. The resulting S/Ns were typically 25–50 pixel^{-1}. Radial velocities for all stars were determined via cross-correlation with 30 template stars using the IRAF task *fxcor* (Tonry & Davis 1979). We find a good agreement amongst the template-derived velocities, with a standard deviation of typically $\sim$6 km s^{-1} for each star.

We measure the equivalent width of each CaT line as follows. Since the near-infrared portion of a star's spectrum can be contaminated by weak metal lines (and possibly weak molecular bands), measuring the true equivalent width of the CaT lines at moderate resolution is virtually impossible. Instead, we follow the method of Armandroff & Zinn (1998) by defining continuum bandpasses on each side of each CaT line and then measuring the "pseudocontinuum" as a linear fit to the mean value in each pair of continuum windows. The "pseudo-equivalent width" is then determined by fitting the sum of a Gaussian and a Lorentzian to each CaT line with respect to the pseudocontinuum. Note that, for some low S/N stars in SMC clusters, we have fit only a Gaussian (see P08 for more details). We adopt the same definition for the summed equivalent width, ΣW, as Cole *et al.* (2004), namely,

$$\Sigma W \equiv EW_{8498} + EW_{8542} + EW_{8662}. \tag{2.1}$$

While both theoretical (Jørgensen, Carlsson & Johnson 1992) and empirical (Cenarro *et al.* 2002) studies have shown that effective temperature, surface gravity, and metallicity all play significant roles in determining CaT line strengths, it is well established that there is a linear relationship between a star's absolute magnitude and ΣW for red giants of a given metallicity. Similar to previous authors, we remove the effects of luminosity and temperature on ΣW by defining a reduced equivalent width, W', as

$$W' \equiv \Sigma W + \beta(V - V_{HB}), \tag{2.2}$$

where the introduction of the brightness of the cluster's horizontal branch, V_{HB}, removes any dependence on cluster distance or reddening as well. For intermediate age clusters without fully formed HBs, we adopt the median value of their core helium burning red clump stars in place of V_{HB}. Since our target clusters span a range of ages and metallicities similar to the calibration clusters used by Cole *et al.* (2004), we adopt their value of $\beta = 0.73$. Finally, as shown by Rutledge *et al.* (1997) for Milky Way globular clusters, there is a linear relationship between a cluster's reduced equivalent width and its abundance on the Carretta & Gratton (1997) metallicity scale. Cole *et al.* (2004) extended this relation to a larger range of abundances and ages, and so we adopt their relationship,

$$[Fe/H] = (-2.966 \pm 0.032) + (0.362 \pm 0.014)W'. \tag{2.3}$$

Cluster members are isolated from field stars using a combination of three criteria: distance from the cluster center, radial velocity, and metallicity. For the LMC, we identify an average of eight members per cluster and determine mean cluster velocities to typically 1.6 km s^{-1} and mean metallicities to 0.04 dex (random error). Similarly, for the SMC we find an average of 6.4 cluster members, and calculate mean cluster velocities and metallicities to 2.7 km s^{-1} and 0.05 dex, respectively. For eight of our LMC clusters and all of our SMC clusters, we report the first spectroscopically derived metallicity and radial velocity values based on individual stars within these clusters.

We note that, while for the LMC our results are based solely on our CaT data sample, for the SMC we have enlarged our cluster sample size by including the abundances and velocities derived from the CaT by DH98 for 6 clusters and photometric abundances for 5 clusters obtained by Piatti *et al.* (2001, 2007; see P08 for more details).

3. Results

Kinematics. The LMC is known to have primarily a disk-like structure, with little evidence of a halo component. This has been show through studies of the LMC's H I gas (Kim *et al.* 1998), field stars (e.g., van der Marel & Cioni 2001) and populous clusters (Grocholski *et al.* 2007). Our cluster kinematics (see Fig. 3) are in excellent agreement with the work of Schommer *et al.* (1992) and confirm that all of the clusters so far studied, including our eight "new" targets, have disk-like kinematics with no obvious signature of the existence of a pressure-supported halo.

In contrast, the SMC appears to have a much more complex structure. Distances derived from photometry of star clusters show that the SMC has a large line-of-sight depth that may vary across the face of the galaxy (Crowl *et al.* 2001). Kinematics in the SMC, however, give mixed results. Stanimirović *et al.* (2004) found differential rotation in the SMC's H I gas and suggest that the inner 3 kpc may correspond to a disk-like structure left over from when the SMC was rotationally supported. On the other hand, stellar kinematics indicate that the SMC is likely a pressure supported system (e.g., Harris & Zartisky 2006). While our clusters show a hint of possible rotation (see Fig. 15 in P08), the velocity amplitude (~ 10 km s^{-1}) is considerably smaller than the dispersion (~ 21 km s^{-1}), and thus our results agree with previous work in that the stellar component of the SMC is in a pressure-supported state.

Metallicity. The metallicity distribution, whether derived from individual stars or clusters, is an important diagnostic of the global chemical enrichment of a galaxy, and provides a straightforward way of comparing stellar populations in different galaxies. In Fig. 2 we plot the metallicity distribution of both our LMC (top panel) and SMC (bottom panel) cluster samples. With the LMC, we see a few of the well-known old metal-poor clusters, and a very tight metallicity distribution for the more metal-rich clusters, with a mean [Fe/H] = -0.48 and $\sigma = 0.09$. On the other hand, the entire SMC cluster sample shows a very broad distribution of abundances, with mean [Fe/H] = -1.00 and $\sigma = 0.20$, and the hint of a possible bimodal distribution, although a larger sample is needed to explore this further. There are also two other key differences between the metallicity distribution in the LMC and SMC. First is the fact that the most metal-poor cluster in the SMC, NGC 121, is > 0.5 dex more metal-rich than the most metal-poor clusters in the LMC. And the second is that the most metal-rich clusters in the LMC are ~ 0.3 dex more metal-rich than the ones in the SMC.

Since we are dealing with clusters, for which ages are readily determined, we can further explore the metallicity distribution by creating age-metallicity relations (AMRs) for both the LMC and SMC. In the top panel of Fig. 3 we plot the AMR for the LMC, where we have assumed that the old metal-poor clusters are all coeval (13 Gyr), and the ages of the remaining clusters are obtained from main sequence turnoff fitting (Grocholski

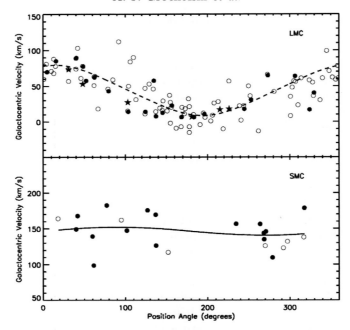

Figure 1. Radial velocity plotted as a function of position for the LMC and SMC. In the top panel, open circles are the data from Schommer *et al.* (1992), while the filed symbols are from G06. The dashed curve represents the best-fit rotation curve from Schommer *et al.* (1992). In the bottom panel, filled symbols are data from P08 while the open symbols are clusters from DH98. The best fit to the SMC data is shown as a solid curve.

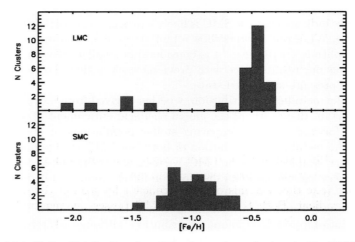

Figure 2. Metallicity distribution function of populous clusters in the LMC and SMC.

et al. these proceedings). The dashed line represents the smooth chemical enrichment model of Pagel & Tautvaišienė (1998) and the solid line is their bursting model. While the well known age gap in the LMC makes it difficult to identify the best fit model for ages older than ~4 Gyr, the intermediate age clusters suggest that the bursting model is the best representation of the LMC's chemical enrichment history. In the SMC's AMR (bottom panel of Fig. 3), we see a more complicated situation. The oldest cluster in the SMC, NGC 121, is only slightly younger (12 Gyr) than the oldest clusters in the

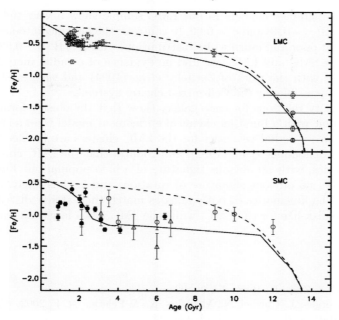

Figure 3. Age—metallicity relations for both the LMC and SMC. The dashed and solid lines represent the smooth and bursting chemical evolution models of Pagel & Tautvaišienė (1998), respectively. For the SMC, we have supplemented our CaT data (filled symbols) with data from DH98 (open circles) and Piatti *et al.* (2001, 2007; open triangles).

LMC, yet is has an abundance that is roughly 1 dex more metal-rich than the metal-poor LMC clusters, which indicates that the SMC was "pre-enriched" prior to the start of cluster formation. From the formation of NGC 121 up until about 4 Gyr ago, SMC clusters showed very little enrichment, and while the clusters younger than ~ 4 Gyr are generally more consistent with a bursting chemical enrichment history (solid line), there is still considerable scatter amongst these clusters, with a mean that is only a few tenths of a dex higher than for the oldest clusters. Our LMC and SMC results reinforce the mysterious situation where the "LMC managed to make metals but no clusters during the age gap while the SMC managed to make clusters but no metals" (Da Costa 1991).

Zaritsky *et al.* (1994) studied H II region oxygen abundances in 39 galaxies and found that disk abundance gradients are ubiquitous in spiral galaxies. However, the presence of a classical bar (one that extends over a significant fraction of the disk length) tends to weaken the gradient. Both the LMC and SMC possess strong stellar bars and, as expected, neither galaxy's cluster system exhibits an abundance gradient (see Fig. 10 in G06 and Fig. 13 in P08).

4. Summary

We have used the multi-object spectrograph, FORS2, on the *VLT* to obtain moderate-resolution near-infrared spectra of the Ca II triplet lines in individual red giant branch stars in a sample of LMC and SMC clusters. These data have allowed us to derive radial velocities and abundances for ~ 7 members per cluster with small random errors (2 km s^{-1} in velocity and 0.05 dex in metallicity). The main results of our study are as follows:

1. The intermediate-age clusters in our LMC sample show a very tight distribution, with a mean [Fe/H] $= -0.48$ and $\sigma = 0.09$. In contrast, the intermediate-age SMC clusters have a more metal-poor and much wider distribution, with [Fe/H] $= -0.88$ and $\sigma = 0.17$.

2. For both the SMC and LMC, we find no evidence of a radial metallicity gradient. This is consistent with the work of Zaritsky *et al.* (1994) and suggests that the bar in each galaxy is responsible for the well-mixed cluster systems.

3. Age–metallicity relations for each galaxy show that the cluster formation histories are more consistent with a bursting chemical enrichment model than a smooth one.

4. We find that our radial velocities for the LMC clusters are in excellent agreement with the work of Schommer *et al.* (1992), and confirm that the LMC cluster system has disk-like kinematics, with no obvious signature of a halo population. The SMC, on the other hand, shows no obvious signature of rotation, and is thus likely a pressure supported system. The kinematics of both galaxies match their stellar distributions, which are known to be disk-like for the LMC, while the SMC has a large (and varying) line-of-sight depth.

References

Armandroff, T. E. & Zinn, R. 1988, *AJ*, 96, 92

Cenarro, A. J., Gorgas, J., Cardiel, N., Vazdekis, A., & Peletier, R. F. 2002, *MNRAS*, 329, 863

Carretta, E. & Gratton, R. G. 1997, *A&A*, 121, 95

Cole, A. A., Smecker-Hane, T. A., Tolstoy, E., Bosler, T. L., & Gallagher, J. S. 2004, *MNRAS*, 347, 367

Crowl, H. H., Sarajedini, A., Piatti, A. E., Geisler, D., Bica, E., Clariá, J. J., & Santos, J. F. C. Jr. 2001, *AJ*, 122, 220

Da Costa, G. 1991, in R. Haynes & D. Milne (eds.), *IAUS 148, The Magellanic Clouds*, p. 143

Da Costa, G. S. & Hatzidimitriou, D. 1998, *AJ*, 115, 1934 (DH98)

Grocholski, A. J., Cole, A. A., Sarajedini, A., Geisler, D., & Smith, V. V. 2006, *AJ*, 132, 1630 (G06)

Grocholski, A. J., Sarajedini, A., Olsen, K. A. G., Tiede, G. P., & Mancone, C. L. 2007, *AJ*, 134, 680

Harris, J. & Zaritsky, D. 2006, *AJ*, 131, 2514

Hill, V. 1999, *A&A*, 345, 430

Johnson, J. A., Bolte, M., Hesser, J. E., Ivans, I. I., & Stetson, P. B. 2004, in A. Mc William & M. Rauch (eds.), *Carnegie Observatories Astrophysics Series 4*, p. 29

Jørgensen, U. G., Carlsson, M., & Johnson, H. R. 1992, *A&A*, 254, 258

Kim, S., Staveley-Smith, L., Dopita, M. A., Freeman, K. C., Sault, R. J., Kesteven, M. J., & McConnell, D. 1998, *ApJ*, 503, 674

Olszewski, E. W., Schommer, R. A., Suntzeff, N. B., & Harris, H. C. 1991, *AJ*, 101, 515

Pagel, B. E. J. & Tautvaišienė, G. 1998, *MNRAS*, 299, 535

Parisi, M. C., Grocholski, A. J., Geisler, D., Sarajedini, A., & Clariá, J. J. 2008, *AJ*, submitted [arXiv:0808.0018] (P08)

Piatti, A. E., Santos, J. F. C., Jr., Clariá, J. J., Bica, E., Sarajedini, A., & Geisler, D. 2001, *MNRAS*, 325, 792

Piatti, A. E., Sarajedini, A., Geisler, D., Gallart, C., & Wischnjewsky, M. 2007, *MNRAS*, 381, L84

Rutledge, G. A., Hesser, J. E., Stetson, P. B., Mateo, M., Simard, L., Bolte, M., Friel, E. D., & Copin, Y. 1997, *PASP*, 109, 883

Schommer, R. A., Olszewski, E. W., Suntzeff, N. B., & Harris, H. C. 1992, *AJ*, 103, 447

Stanimirović, S., Staveley-Smith, L., & Jones, P. A. 2004, *ApJ*, 604, 176

Tonry, J., & Davis, M. 1979, *AJ*, 84, 1511

van der Marel, R. P. & Cioni, M.-R. L. 2001, *AJ*, 122, 1807

Zaritsky, D., Kennicutt, R. C., Jr. & Huchra, J. P. 1994, *ApJ*, 420, 87

The Magellanic System: Stars, Gas, and Galaxies
Proceedings IAU Symposium No. 256, 2008
Jacco Th. van Loon & Joana M. Oliveira, eds.

© 2009 International Astronomical Union
doi:10.1017/S1743921308028603

Integrated spectral properties of star clusters in the Magellanic Clouds

Andrea V. Ahumada[1,2,3], M. L. Talavera[4], J. J. Clariá[2,3],
J. F. C. Santos Jr.[5], E. Bica[6], M. C. Parisi[2,3] and M. C. Torres[2]

[1]European Southern Observatory - ESO, Chile, email: aahumada@eso.org

[2]Observatorio Astronómico, Universidad Nacional de Córdoba, Argentina

[3]Consejo Nacional de Investigaciones Científicas y Técnicas, CONICET, Argentina

[4]Observatorio Astronómico Centroamericano de Suyapa, UNAH, Honduras

[5]Departamento de Física, ICEx, UFMG, Belo Horizonte, Brazil

[6]Departamento de Astronomia, UFRGS, Porto Alegre, Brazil

Abstract. We present flux-calibrated integrated spectra in the optical spectral range of concentrated star clusters in the Large and Small Magellanic Clouds (LMC-SMC), approximately half of which constitute unstudied objects. We have mainly estimated ages and foreground interstellar reddening values from the comparison of the line strengths and continuum distribution of the cluster spectra with those of template spectra with known parameters. Also reddening values were estimated by interpolation between the extinction maps of Burstein & Heiles (1982) (BH). A good agreement between ages and reddenings derived through the different procedures was found. The ages of the 27 LMC star clusters range from 5 to 125 Myr, while those of the 13 SMC vary from 4 to 350 Myr.

Keywords. techniques: spectroscopic, Magellanic Clouds, galaxies: star clusters

1. Introduction

The study of extragalactic stellar systems provides relevant information on the star formation and chemical histories of the host galaxies. Despite the multiple observational as well as theoretical projects undertaken in the last few years, our currently existing knowledge of both the stellar formation processes and chemical evolution of galaxies is, in general, incomplete. Even for the galaxies in the Local Group, our present understanding is definitely limited. In this state of affairs, the stellar cluster systems of the Magellanic Clouds (MCs), on account of their proximity, richness, and variety, may furnish us with the ideal ground to conduct a detailed examination of the processes mentioned before. Efforts to create reference spectra of star clusters and grids of their properties to be used as templates for different ages and metallicities in the study of composite stellar populations were made by different authors, e.g., Bica & Alloin (1986a), Bica (1988), Santos Jr. *et al.* (1995), and Piatti *et al.* (2002). The goal of the present study is to collect and analyse a large sample of MCs clusters in view of studying the integrated light properties of such metal deficient clusters; and making them available as template spectra for studies of star clusters in more distant dwarf galaxies.

2. Cluster sample and observations

The determination of MCs cluster parameters, particularly age, is fundamental to understand the structure and evolution of these galaxies. Concentrated clusters, with small angular diameter are certainly the most suitable to carry out integrated spectroscopy observations. This is because the cluster as well as the surrounding background regions are

Table 1. The SMC cluster sample. Cluster identifications are from Lindsay (1958) (L), Kron (1956) (K), Lauberts (1982)(ESO), Westerlund & Glaspey (1971)(WG), Hodge & Wright (1974) (HW), Bruck (1976), and Pietrzyński *et al.* (1998)(OGLE).

Cluster	α_{2000}	δ_{2000}
NGC 242, K 22, L 29, SMC_OGLE 18	00:43:38	−73:26:38
NGC 256, K 23, L 30, ESO29-SC 11, SMC_OGLE 32	00:45:54	−73:30:24
NGC 265, K 24, L 34, ESO29-SC 14, SMC_OGLE 39	00:47:12	−73:28:38
B 50	00:49:03	−73:22:00
K 34, L 53, SMC_OGLE 104	00:55:33	−72:49:58
IC 1611, K 40, SMC_OGLE 118, L 61, ESO29-SC 27	00:59:48	−72:20:02
IC 1626, K 53, L 77, ESO29-SC 30	01:06:14	−73:17:51
IC 1641, HW 62, ESO51-SC 21	01:09:40	−71:46:03
L 95	01:15:00	−71:20:00
B 164	01:29:30	−73:32:00
HW 85	01:42:00	−71:17:00
WG 1	01:42:53	−73:20:00
NGC 796, L 115, ESO30-SC 6	01:54:45	−74:13:00

Table 2. The LMC cluster sample. Cluster identifications are from Lauberts (1982) (ESO), Shapley & Lindsay (1963) (SL), Kontizas *et al.* (1990) (KMHK), Pietrzyński *et al.* (1999) (LMC_OGLE), Hodge & Sexton (1966) (HS), and Lyngå & Westerlund (1963) (LW).

Cluster	α_{2000}	δ_{2000}	SWB
SL 14, KMHK 28	04:40:28	−69:39:00	II
NGC 1695, KMHK 101, SL 40	04:47:44	−69:22:00	III
SL 56, KMHK 142	04:50:32	−70:04:00	II
SL 58, KMHK 153	04:50:59	−69:38:00	III
SL 79, KMHK 213	04:52:53	−71:39:00	III
SL 76, KMHK 206	04:53:09	−68:12:00	III
NGC 1732, KMHK 209, SL 77	04:53:11	−68:39:00	II
SL 116, KMHK 315	04:56:24	−68:48:00	II
SL 168, KMHK 418	05:00:44	−65:27:00	III
NGC 1822, KMHK 513, SL 210	05:05:08	−66:12:00	II
HS 109, LMC_OGLE 82	05:05:37	−68:43:06	II
SL 234, LMC_OGLE 113	05:06:54	−68:43:08	II
SL 255, KMHK 573	05:07:55	−70:03:00	II
NGC 1887, KMHK 700, SL 343	05:16:05	−66:19:00	II
SL 364, KMHK 736	05:17:41	−71:03:00	II
SL 360, LMC_OGLE 328	05:18:11	−69:13:06	0
SL 386, KMHK 770	05:19:50	−65:23:00	II
NGC 1944, KMHK 836, SL 426, ESO33-SC 7	05:21:57	−72:29:00	III
SL 463, KMHK 889, LW 213	05:26:15	−66:03:00	II
SL 477, KMHK 911	05:26:23	−71:41:00	II
NGC 1972, LMC_OGLE 481, SL 480	05:26:48	−69:50:17	II
NGC 2000, KMHK 932, SL 493	05:27:30	−71:52:00	II
NGC 1986, LMC_OGLE 496, SL 489	05:27:38	−69:58:14	II
SL 566, KMHK 1061	05:32:50	−70:47:00	II
NGC 2053, KMHK 1154, SL 623	05:37:40	−67:24:00	II
SL 763, KMHK 1448	05:52:53	−69:47:00	II
NGC 2140, KMHK 1511, SL 773	05:54:17	−68:36:00	II

well sampled along the slit. Besides, the angular diameter requirement results from the fact that the cluster integrated spectrum must reflect the synthesis of its stellar content. In this study we have selected relatively populous and compact MCs clusters to allow good star sampling in the integrated spectra. The observed SMC star cluster sample is given in Table 1, where their designations in different catalogues are provided, while Table 2 shows the LMC cluster sample including also the SWB type (Searle *et al.* 1980).

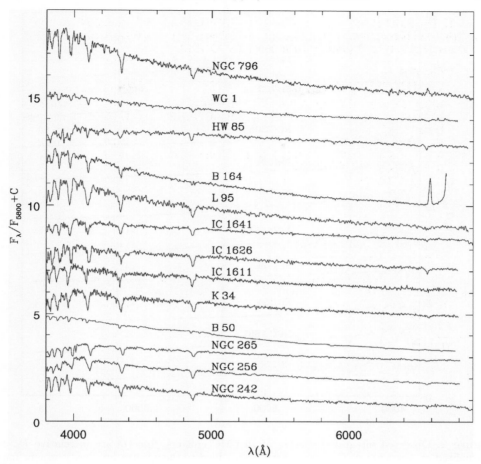

Figure 1. Observed integrated spectra of 13 SMC clusters. Spectra are in relative F_λ units normalised at $\lambda \sim 5800$ Å. Constants have been added to the spectra for clarity, except for the bottom one.

All the observations were carried out with the '*Jorge Sahade*' 2.15 m telescope at the *Complejo Astronómico El Leoncito* (*CASLEO*, San Juan, Argentina) in several runs. We employed a CCD camera containing a Tektronix chip of 1024×1024 pixels attached to a REOSC spectrograph (simple mode). The slit was oriented in the east-west direction and the observations were performed scanning the slit across the objects in the north-south direction in order to get a proper sampling of cluster stars. A grating of 300 grooves/mm was used. The spectral coverage was the visible range: $\sim (3800 - 6900)$ Å, with an average dispersion in the observed region of ~ 140 Å/mm (3.46 Å/pix). The slit width was 4.2″, resulting in a resolution of ~ 14 Å, as measured by the mean full width half maximum of the comparison lines. At least two exposures of 20 minutes of each object were taken, depending on the star concentration of the cluster. Standard stars were also observed for flux calibrations, and comparison lamp exposures were taken for wavelength calibration.

In Figure 1 we present the flux-calibrated integrated spectra of the SMC observed clusters, while Figure 2 shows part of the 27 LMC integrated spectra. The whole LMC sample will be presented elsewhere. All the spectra are in relative flux units, normalised to $F_\lambda = 1$ at $\lambda \sim 5800$ Å. The spectral lines and different slopes of the continuum energy distributions in both figures are primarily the result of age effects.

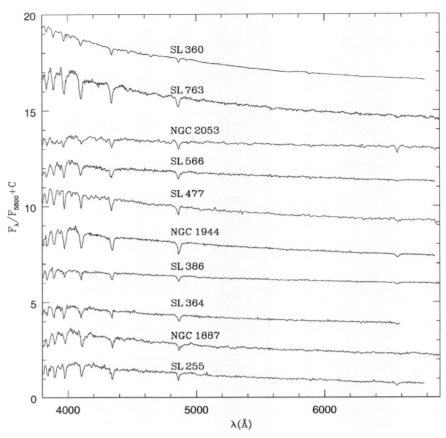

Figure 2. Observed integrated spectra of 10 LMC clusters. Spectra are in relative F_λ units normalised at $\lambda \sim 5800$ Å. Constants have been added to the spectra for clarity, except for the bottom one.

3. Determination of cluster fundamental parameters

Age and foreground reddening values of the clusters were simultaneously derived by means of a template matching method. This was done by achieving the best possible match between the continuum and lines of the cluster's integrated spectrum and those from an integrated spectrum with known properties (e.g., Talavera *et al.* 2006). A direct reddening-independent age estimate was first obtained from EWs of the Balmer absorption lines by interpolating these values in the calibration of Bica & Alloin (1986b). The diagnostic diagrams involving the sum of EWs of selected spectral lines denoted S_h and S_m were employed together with the calibrations with age and metallicity by Santos & Piatti (2004). S_m corresponds to the sum of three metallic lines (K Ca II, G band and Mg I) and S_h corresponds to the sum of three Balmer lines (Hβ, Hγ and Hδ). Foreground reddening E(B–V) values were estimated from the template matching method and also from the interstellar extinction maps by BH. Then, we selected an appropriate set of template spectra according to the age provided by the mentioned EWs and varied reddening and template to get the best match of continuum, Balmer and metal lines of the observed spectrum to that of the template that most resembles it. Reddening corrections were performed employing the interstellar absorption law by Seaton (1979).

Table 3. Age and reddening determinations for SMC clusters.

Cluster	E(B–V)	E(B–V) (BH)	Balmer age (Myr)	S_h, S_m age (Myr)	Template age (Myr)	Adopted age (Myr)
NGC 242	0.08±0.03	0.03	50	27	10−20; 35−65	40±20
NGC 256	0.03±0.02	0.03	50	36	200	150±50
NGC 265	0.03±0.02	0.03	50−500	41	50−110	80±40
B 50	0.00±0.02	0.03	< 10	5	3−5	4±2
K 34	0.08±0.02	0.03	50	95	100−150	200±100
IC 1611	0.10±0.02	0.06	50−100	19	100−150	130±30
IC 1626	0.11±0.02	0.03	300	57	200−350	250±50
IC 1641	0.04±0.01	0.03	500	350	350	350±100
L 95	0.12±0.02	0.03	50−100	57	40	50±20
B 164	0.00±0.02	0.03	∼ 10	7	3−6; 12−35	10±5
HW 85	0.01±0.01	0.03	10−50	82	10−20	20±10
WG 1	0.12±0.02	0.03	< 10	14	3−5	4±2
NGC 796	0.06±0.02	0.03	10-50	22	20	20±10

Table 4. Age and reddening determinations for LMC clusters.

Cluster	E(B–V)	E(B–V) (BH)	Balmer age (Myr)	S_h, S_m age (Myr)	Template age (Myr)	Adopted age (Myr)
SL 14	0.18±0.02	0.08	⩽ 10	8	10−20	10±5
NGC 1695	0.16±0.02	0.06	∼ 50	70	50−110	70±10
SL 56	0.05±0.02	0.10	10−50	18	12−40	40±20
SL 58	0.13±0.03	0.10	∼ 50	50	35−65; 50−110	65±30
SL 79	0.06±0.02	0.03	∼ 100	115	100	100±10
SL 76	0.08±0.02	0.03	50−70	28	12−40	50±30
	0.06±0.02				35−110	
NGC 1732	0.00±0.01	0.03	50	20	35−110	60±10
SL 116	0.00±0.01	0.06	50−70	34	35−65	50±20
SL 168	0.01±0.01	0.03	100	23	35−65	60±20
NGC 1822	0.05±0.02	0.04	50−100	12	100−150	125±25
HS 109	0.08±0.02	0.06	50−100	110	35−65	70±20
SL 234	0.00±0.01	0.06	50	24	50	60±20
SL 255	0.10±0.02	0.10	10−100	22	45−75	60±10
NGC 1887	0.05±0.02	0.04	30−50	18	45−75	70±20
SL 364	0.02±0.01	0.09	∼ 50	19	40	40±10
SL 360	0.10±0.02	0.07	∼ 10	05	3−6	5±2
SL 386	0.17±0.02	0.03	30−50	16	60	70±20
NGC 1944	0.07±0.02	0.07	50−100	22	45−75	60±10
SL 463	0.00±0.01	0.06	10−50	22	12−40	50±10
	0.01±0.02			22	35−65	
SL 477	0.03±0.01	0.07	10−50	14	35−65	40±20
NGC 1972	0.00±0.01	0.07	30−50	20	70	60±10
NGC 2000	0.02±0.01	0.07	50−100	42	40	50±10
NGC 1986	0.10±0.02	0.07	30−50	23	45−75	50±20
SL 566	0.15±0.02	0.09	10−50	22	45−75	50±10
NGC 2053	0.08±0.02	0.06	∼ 50	21	50−110	70±30
SL 763	0.04±0.02	0.08	50−100	41	40	70±20
NGC 2140	0.04±0.01	0.06	50−100	60	50−110; 12−40	60±20

4. Age and reddening values

The parameters determined for the SMC clusters are shown in Table 3. The colour excesses derived for the whole sample range from 0.00 (B 50 and B 164) to 0.12 (WG 1 and L 95), while the ages vary from 4 Myr (B 50 and WG 1) to 350 Myr (IC 1641). Four of the 13 clusters presented here were not previously studied. Table 4 shows ages and colour excesses determined for the selected 27 LMC stellar clusters. The reddening values range between 0.00 (NGC 1732, SL 116, SL 234 and NGC 1972) and 0.18 (SL 14), while the ages range from 5 Myr (SL 360) to 125 Myr (NGC 1822). In this sample, 17 of the 27 stellar clusters do not show previous studies, so we presented here new parameters for 63% of the sample. Within the expected uncertainties, the ages derived in the present work agree with those given in the literature.

5. Summary

We have estimated cluster ages and foreground interstellar reddening values from the comparison of the line strengths and continuum distribution of the cluster spectra with those of template cluster spectra with well-determined physical properties. Reddening values were also estimated by interpolation between the extinction maps of BH. A good agreement between ages and reddening values derived from both procedures was found. The ages of the LMC studied clusters range from 5 to 125 Myr while those of the SMC cluster sample range from 4 to 350 Myr. The present data constitute part of the elements to enhance the spectral libraries at the metallicity levels of the SMC and LMC star clusters.

References

Bica, E. 1988, *A&A*, 195, 76
Bica, E. & Alloin, D. 1986a, *A&A*, 162, 21
Bica, E. & Alloin, D. 1986b, *A&AS*, 66, 171
Bruck, M. T. 1976, *Occasional Reports R. Obs.*, 1, 1
Burstein, D. & Heiles, C. 1982, *AJ*, 87, 1165 (BH)
Hodge, P. W. & Wright, F. W. 1974, *AJ*, 79, 858
Hodge, P. W. & Sexton, J. A. 1966, *AJ*, 71, 363
Kontizas, M., Morgan, D. H., Hatzidimitriou, D., & Kontizas, E. 1990, *A&AS*, 84, 527
Kron, G. E. 1956, *PASP*, 68, 125
Lauberts, A. 1982, *The ESO/Uppsala Survey of the ESO (B) Atlas*, (European Southern Observatory)
Lindsay, E. M. 1958, *MNRAS*, 118, 172
Lyngå, G. & Westerlund, B. E. 1963, *MNRAS*, 127, 31
Piatti, A. E., Bica, E., Clariá, J. J., Santos Jr., J. F. C., & Ahumada, A. V. 2002, *MNRAS*, 335, 233
Pietrzyński, G., Udalski, A., Kubiak, M., Szymański, M., Woźniak, P., & Żebruń, K. 1998, *AcA*, 48, 175
Pietrzyński, G., Udalski, A., Kubiak, M., Szymański, M., Woźniak, P., & Żebruń, K. 1999, *AcA*, 49, 521
Santos Jr., J. F. C., Bica, E., Clariá, J. J., Piatti, A. E., Girardi, L. A., & Dottori, H. 1995, *MNRAS*, 276, 1155
Santos Jr., J. F. C. & Piatti, A. E. 2004, *A&A*, 428, 79
Searle, L., Wilkinson, A., & Bagnuolo, W. G. 1980, *ApJ*, 239, 803
Seaton, M. J. 1979, *MNRAS*, 187, 73p
Shapley, H. & Lindsay, E. M. 1963, *Irish Astron. J.*, 6, 74
Talavera, M. L., Ahumada, A. V., Clariá, J. J., Parisi, M. C., Santos Jr., J. F. C., & Bica, E. 2006, *BAAA*, 49, 311
Westerlund, B. E. & Glaspey, J. 1971, *A&A*, 10, 1

The Magellanic System: Stars, Gas, and Galaxies
Proceedings IAU Symposium No. 256, 2008
Jacco Th. van Loon & Joana M. Oliveira, eds.

© 2009 International Astronomical Union
doi:10.1017/S1743921308028615

The chemical signatures of the Large Magellanic Cloud globular clusters

Alessio Mucciarelli[1]

[1]Dipartimento di Astronomia, Università degli Studi di Bologna, Bologna, Italy
email: alessio.mucciarelli@studio.unibo.it

Abstract. We present the first results of a long-term project based on the analysis of high-resolution optical spectra for a sample of Large Magellanic Cloud globular clusters. The final aim is to build a new, reliable metallicity scale for this cluster system and shed some light on the role played by the different chemical contributors (AGB, SN II and SN Ia), in order to understand the chemical enrichment history of the Large Magellanic Cloud. The analysis of 6 young and intermediate-age clusters and 11 field stars, observed with the UVES@FLAMES spectrograph, provides crucial information about the chemical composition of the dominant stellar population of the LMC. All these stars are metal-rich ([Fe/H]~ -0.4 dex), with solar-scaled [α/Fe] ratios, that point toward an enrichment from SN Ia. Moreover, we observed a general depletion for the [Al/Fe] ratio (indication of a chemical enrichment by metal-poor SN II) and a strong enhancement of the [Ba/Y] ratio (likely due to the enrichment by metal-poor AGB stars).

Keywords. techniques: spectroscopic, stars: abundances, galaxies: evolution, galaxies: Individual (LMC), Magellanic Clouds, galaxies: star clusters

1. The chemistry of the Large Magellanic Cloud: state of the art

The Large Magellanic Cloud (LMC) is the nearest galaxy with a present-day star-formation activity and represents a fundamental laboratory to study the stellar populations. One of its most important peculiarities is the wide age and metallicity distribution covered by its stellar clusters. Several age families for these objects have been recognized: an old population, with clusters coeval with the Galactic ones, an intermediate-age population, including clusters with ages between ~ 1 and ~ 3 Gyr, and finally numerous clusters younger than ~ 1 Gyr. A lot of information about the metallicity of the stellar content of the LMC derived by Ca II triplet surveys was presented by several authors in the last two decades, e.g., Olszewski *et al.* (1991), Cole *et al.* (2005), Grocholski *et al.* (2006), and Carrera *et al.* (2007).

One of the most severe limitations in our knowledge of the LMC stellar populations is the lack of homogeneous age and metallicity scales, fundamental tools to derive a reliable age–metallicity relation and well understand the chemical evolution of this environment. Few works based on high-resolution spectra are actually available for the LMC stellar content, and our knowledge of the chemical signatures of these stellar populations is yet partial. Hill *et al.* (2000) discussed the first measurements from high-resolution spectroscopy for some LMC clusters, showing the abundance patterns for Fe, O and Al. A detailed screening of the chemistry in 4 old LMC clusters has been presented by Johnson *et al.* (2006). Recently, Pompéia *et al.* (2008) find some peculiarities in the chemical composition of LMC field stars, such as the systematic depletion of Ni and iron-peak elements and peculiar and complex abundance pattern of the neutron-capture elements.

2. The global project: towards an age–metallicity relation

We present the first results of a long-term spectro-photometric project devoted to obtain accurate and reliable metallicity and age estimates for a sample of *template* clusters. The whole spectroscopic database obtained by using the *ESO* facility FLAMES@*VLT* includes 9 stellar clusters, spanning the entire age range covered by the LMC cluster system. We have selected 2 young clusters with ages $\lesssim 1$ Gyr (namely NGC 2157 and 2108, Mucciarelli *et al.* (2009)), 4 intermediate-age clusters (NGC 1651, 1783, 1978 and 2173, Ferraro *et al.* (2006) and Mucciarelli *et al.* (2008)) and 3 old clusters (NGC 1786, 2210 and 2257, in preparation). Moreover, 11 giant stars turned out not to be cluster members but to belong to the LMC field and we have analysed also these stars in order to compare the abundance patterns inferred from cluster and field stars.

In the following we summarize the principal aims of this project:

• The definition of a new, homogeneous metallicity scale for a representative sample of LMC *template* clusters, based on the latest generation of high-resolution spectrographs.

• The detailed study of the abundance patterns for the main elemental groups, as iron-peak, light Z-odd, α and neutron-captures elements, in order to well describe the chemical enrichment history of the stellar populations in the LMC.

• The combination of the information about the overall metallicity, [M/H], of each target cluster with the high-resolution, photometric database available in the *HST* archive will provide accurate estimates of the age for these objects. The first results of this procedure have been discussed in Mucciarelli *et al.* (2007a,b). In Figs. 1 and 2 we report the Color–Magnitude Diagram of the two intermediate-age clusters NGC 1978 and 1783, obtained with ACS@*HST* photometry. For these two clusters we have derived an age of 1.9 and 1.4 Gyr, respectively.

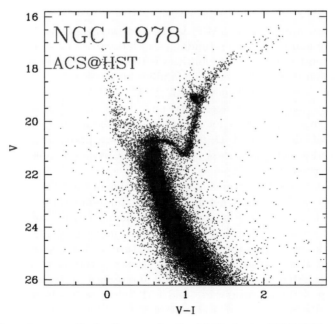

Figure 1. The color–magnitude diagram for the LMC cluster NGC 1978, obtained with ACS@*HST*.

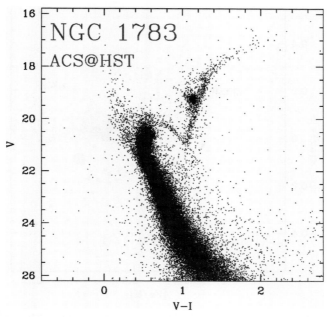

Figure 2. The color–magnitude diagram for the LMC cluster NGC 1783, obtained with ACS@*HST*.

3. The chemical composition

All 6 LMC clusters and 11 field stars analysed here turn out to be metal-rich, with an average value for the clusters of [Fe/H]= −0.35 dex (σ = 0.11 dex), while the field stars exhibit an average iron content of [Fe/H]= −0.49 dex (σ = 0.10 dex). This finding appears to be consistent with the metallicity distribution for the LMC field inferred by different works, in particular the recent studies based on the Ca II triplet by Cole *et al.* (2005) ([Fe/H]= −0.37 ± 0.15 dex), Grocholski *et al.* (2006) ([Fe/H]= −0.48 ± 0.09 dex) and Carrera *et al.* (2007) ([Fe/H]∼ −0.5 dex).

3.1. α-elements

The α-elements are produced mainly by massive stars (and ejected via the SN II events), with a smaller contribution by SN Ia, while the iron is produced mainly by SN Ia. For this reason, the [α/Fe] ratio turns out to be a powerful diagnostic of the relative contribution of SN II to SN Ia. Our stars show sub-solar ratios for [O/Fe] and [Ca/Fe], while [Mg/Fe], [Si/Fe] and [Ti/Fe] match the Galactic patterns. Fig. 3 reports the average value of [Mg/Fe], [Ca/Fe] and [Ti/Fe] of our stars (the red circles are the average values for the LMC clusters and the red asterisks the individual LMC field stars) in comparison with other high-resolution spectroscopic databases: grey points indicate the Galactic stars (Venn *et al.* 2004; Reddy *et al.* 2006) and blue points the Sgr dSph giant stars (Sbordone *et al.* 2007). The LMC stars seem to fall between the envelope described by the Galactic stars and the one from the Sgr dSph. The solar [α/Fe] ratios observed in these LMC stars indicates an enrichment from SN Ia.

3.2. Na and Al

There are two main channels in order to produce Na and Al. The first one is linked to the SN II, in order to provide the extra neutron present in the dominant isotope of both

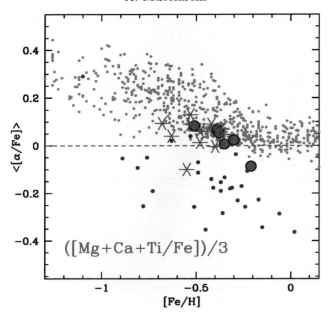

Figure 3. The behaviour of [α/Fe] ratio (defined as the average between [Mg/Fe], [Ca/Fe] and [Ti/Fe]) as a function of [Fe/H]. Red points are the average values for the target LMC clusters, red asterisks are the LMC field stars, grey points indicate the Galactic stars by Venn *et al.* (2004) and Reddy *et al.* (2006) and the blue points are the giant stars of the Sgr dSph by Sbordone *et al.* (2007).

element (^{23}Na and ^{27}Al). The Na and Al production is controlled by the neutron excess and in this way the yields for these elements becomes metallicity-dependent. Another possible channel to produce Na and Al is linked to the AGB stars, in which proton capture reactions can occur through the NeNa and MgAl cycles. This mechanism is invoked in order to explain the so-called *anticorrelations* in the observed Galactic globular clusters. In our cluster sample we find a very high degree of homogeneity for Na, Al, O and Mg, without hints of possible intrinsic chemical anomalies. For this reason, we can consider the AGB channel not very efficient and the Na and Al production attributed to the SN II events. Our target stars, both cluster and field stars, exhibit a general depletion of these two abundance ratios (see Fig. 4), with a strong depletion in the [Al/Fe] ratio (~ -0.40 dex). This finding seems to indicate that the interstellar medium from which these stars were born had been enriched with the ejecta of low-metallicity SN II.

3.3. *Neutron-capture elements*

The LMC stars show a peculiar and dichotomic behaviour for the s-process elements. The elements belonging to the first peak (such as Y and Zr) show a general depletion with respect to the solar value (~ -0.4 dex), while the elements of the second peak (such as Ba, La and Nd) turn out to be enhanced ($\sim +0.5$ dex). The [Ba/Y] ratio (indicated by Venn *et al.* (2004) as an important diagnostic to study this family of elements) turns out to be strongly enhanced, with values of [Ba/Y]$\sim +0.80$ dex, similar to the observed pattern in Sgr dSph but in evident disagreement with the Galactic trend, that exhibits a solar-scaled pattern at the same metallicity level. The interpretation of these abundance patterns is complicated by the multiplicity of the nucleosynthetic sites.

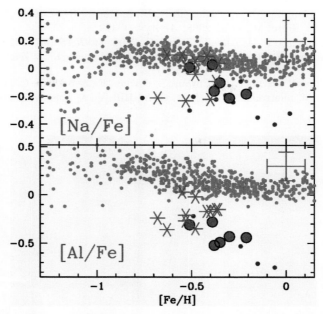

Figure 4. The behaviour of [Na/Fe] (upper panel) and [Al/Fe] (lower panel) ratios as a function of [Fe/H]. Same symbols of Fig. 3.

Theoretical predictions by Travaglio *et al.* (2004) indicate that the AGB yields for this kind of elements are metallicity-dependent. A high value of the [Ba/Y] ratio could suggest a major pollution of the gas by low-metallicity AGB stars.

4. Conclusions

The analysis of high-resolution spectra of 6 young-intermediate-age clusters and 11 field stars of the LMC has revealed some chemical signatures of the dominant stellar populations of this galaxy. (1) All these stars, both of field and cluster, are metal-rich, with an iron content that confirms the previous analysis based on the Ca II triplet; (2) the LMC stars exhibit solar [α/Fe] ratios, indication of the chemical contribution from the ejecta of SN Ia; (3) all the target stars show low values of [Na/Fe] and [Al/Fe] ratios when compared with the Galactic Disk stars; (4) the [Ba/Y] ratio turns out to be enhanced (+0.5 / +1.0 dex), pointing toward an enrichment by metal-poor AGB stars.

References

Carrera, R., Gallart, C., Hardy, E., Aparicio, A., & Zinn, R. 2008, *AJ*, 135, 836
Cole, A. A., Tolstoy, E., Gallagher III, J. S., & Smecker-Hane, T.A. 2005, *AJ*, 129, 1465
Ferraro, F. R., Mucciarelli, A., Carretta, E., & Origlia, L. 2006, *ApJ*, 133, L3
Grocholski, A. J., Cole, A. A., Sarajedini, A., Geisler, D., & Smith, V. V. 2006, *AJ*, 132, 1630
Hill, V., François, P., Spite, M., Primas, F., & Spite, F. 2000, *A&AS*, 364, 19
Johnson, J. A., Ivans, I., & Stetson, P. 2006, *ApJ*, 640, 801
Mucciarelli, A., Ferraro, F. R., Origlia, L., & Fusi Pecci, F. 2007a, *AJ*, 133, 2053
Mucciarelli, A., Origlia, L., & Ferraro, F. R. 2007b, *AJ*, 134, 1813
Mucciarelli, A., Carretta, E., Origlia, L., & Ferraro, F. R. 2008, *AJ*, 136, 375
Mucciarelli, A., Origlia, L., & Ferraro, F. R. 2009, *AJ*, submitted

304 A. Mucciarelli

Olszewski, E. W., Schommer, R. A., Suntzeff, N. B., & Harris, H. C. 1991, *AJ*, 101, 515

Pompéia, L., Hill, V., Spite, M., *et al.* 2008, *A&A*, 480, 379

Reddy, B. E., Lambert, D. L., & Allende Prieto, C. 2006, *MNRAS*, 367, 1329

Sbordone, L., Bonifacio, P., Buonanno, R., Marconi, G., Monaco, L., & Zaggia, S. 2007, *A&A*, 465, 815

Travaglio, C., Gallino, R., Arnone, E., Cowan, J., Jordan, F., & Sneden, C. 2006, *ApJ*, 601, 864

Venn, K. A., Irwin, M. Shetrone, M. D., Tout, C. A., Hill, V., & Tolstoy, E. 2004, *AJ*, 128, 1177

Arrival at Wrenbury Hall, after a leisurely ride through the Cheshire countryside.

The Magellanic System: Stars, Gas, and Galaxies
Proceedings IAU Symposium No. 256, 2008
Jacco Th. van Loon & Joana M. Oliveira, eds.

Multiple stellar populations in massive LMC star clusters

A. D. Mackey, P. Broby Nielsen, A. M. N. Ferguson and J. C. Richardson

Institute for Astronomy, University of Edinburgh, Royal Observatory, Blackford Hill,
Edinburgh, EH9 3HJ, UK, email: dmy@roe.ac.uk

Abstract. The recent discovery of multiple stellar populations in massive Galactic globular clusters poses a serious challenge for models of star cluster formation and evolution. A new angle on this problem is being provided by rich intermediate-age clusters in the Magellanic Clouds. In this contribution we describe the discovery of three such LMC clusters with peculiar main-sequence turn-off morphologies. The simplest interpretation of our observations is that each of these three clusters is comprised of two or more stellar populations spanning an age interval of ∼300 Myr. Surprisingly, such features may not be unusual in this type of cluster.

Keywords. globular clusters: general, galaxies: individual (LMC), Magellanic Clouds, galaxies: star clusters

1. Introduction

It is a long-held astrophysical tenet that rich star clusters, including globular clusters, are composed of single stellar populations. In many cases there is an abundance of excellent supporting evidence — see, for example, the spectacular colour–magnitude diagrams (CMDs) presented recently for NGC 6397 by Richer *et al.* (2008).

New observations, however, are challenging this accepted picture. It has recently been discovered that several of the most massive Galactic globular clusters harbour multiple stellar populations with a wide variety of unexpected characteristics (e.g., Piotto *et al.* 2008). The most strikingly unusual CMD belongs to ω Centauri, which exhibits at least four different populations covering a wide spread in metal abundance and age across the sub-giant branch (SGB) (Villanova *et al.* 2007), as well as a main sequence bifurcation which can only be explained by the presence of a vastly helium-enriched population (Piotto *et al.* 2005). NGC 2808 possesses a triple main sequence split, again implying the presence of helium-enriched populations (Piotto *et al.* 2007); while NGC 1851 and NGC 6388 show clear splits in their SGBs, possibly indicative of populations with ages ∼1 Gyr apart, or of sizeable intra-cluster variations in chemical composition (Milone *et al.* 2008a; Piotto *et al.* 2008; Salaris *et al.* 2008). Even the much lower mass Galactic globular cluster M 4, which does not display a markedly unusual CMD, apparently possesses two stellar populations — one comprised of Na–rich CN-strong stars and one of Na–poor CN-weak stars (Marino *et al.* 2008). Overall, these observed properties pose serious challenges for conventional models of globular cluster formation and evolution.

Rich star clusters in the Magellanic Clouds have the potential to open a new angle on this problem. These objects have masses comparable to Galactic globulars below the peak of the luminosity function, but are typically much younger and hence possess CMDs far more sensitive to internal dispersions in age, for example. In this contribution we report on the recent discovery of several rich intermediate-age LMC clusters

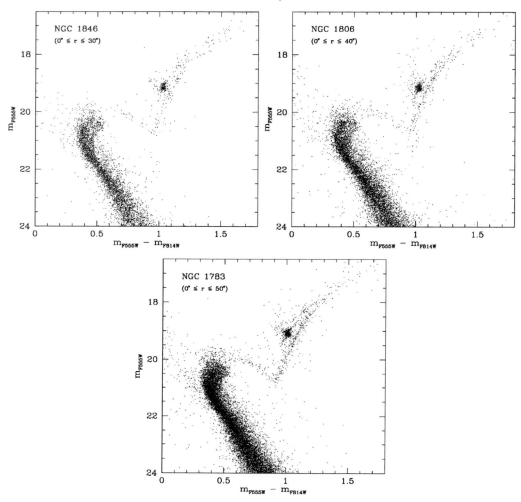

Figure 1. *HST*/ACS CMDs from our *HST* program #9891 snapshot imaging for the three strongest candidates for possessing peculiar main-sequence turn-off morphologies in that sample. The double turn-off in NGC 1846 is clearly visible (Mackey & Broby Nielsen 2007), while the CMD for NGC 1806 shows a hint of a double turn-off. The turn-off for NGC 1783 has no distinct branches but exhibits a rather large spread in colour. Each object possesses a narrow RGB and a compact, well-defined red clump.

with peculiar main-sequence turn-off morphologies, inconsistent with being single stellar populations.

2. Results and analysis

The first indication that some intermediate-age LMC clusters might possess unusual CMDs arose from data obtained with the Advanced Camera for Surveys (ACS) on-board the *Hubble Space Telescope* as part of our snapshot survey of rich star clusters in the Magellanic Clouds (*HST* Program #9891). A careful photometric analysis revealed several objects showing unexpected features around their main-sequence turn-offs.

CMDs for the three strongest candidates in our sample — NGC 1846, 1806, and 1783 — are displayed in Fig. 1. Each exhibits an unusual turn-off morphology. NGC 1846

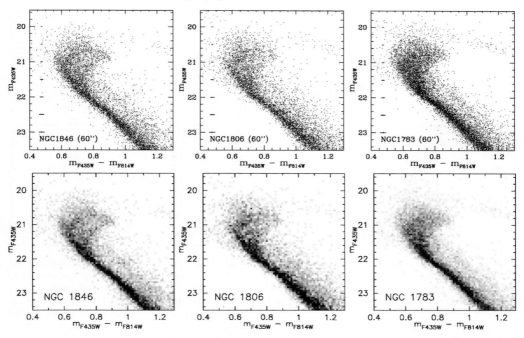

Figure 2. *HST*/ACS CMDs (upper panels) and Hess diagrams (lower panels) constructed from imaging taken under *HST* program #10595, obtained from the public archive. These deep CMDs confirm the peculiar nature of the three clusters identified in our *HST* #9891 snapshot imaging (see Fig. 1). NGC 1846 and 1806 clearly display bifurcated main-sequence turn-offs, while NGC 1783 has a turn-off exhibiting a large spread in colour. Each cluster has a narrow main sequence and a significant population of unresolved binary stars.

possesses two clear turn-off branches, while NGC 1806 shows a hint of two turn-off branches. The turn-off for NGC 1783 has no distinct branches but rather exhibits a spread in colour which is larger than the observational uncertainties would suggest. Apart from these features, the CMDs are as expected for intermediate-age clusters; in particular, each object possesses a narrow RGB and a compact, well-defined red clump. A detailed analysis of the CMD for NGC 1846 is presented by Mackey & Broby Nielsen (2007).

Motivated by our discoveries, we searched the *HST* public archive for additional imaging of these three clusters. All were observed with ACS as part of *HST* Program #10595 between September 2005 — January 2006. These data allowed the construction of higher S/N CMDs than was possible from our snapshot imaging. These archival CMDs are displayed in Fig. 2, and unambiguously show that both NGC 1846 and 1806 possess two main-sequence turn-off branches, while NGC 1783 possesses a turn-off covering a spread in colour much larger than can be explained by the photometric uncertainties (note the very narrow upper main sequence). This feature may represent a bifurcated turn-off in which the branches are unresolved on the CMD, or it may represent a smooth spread of stars. More information on these clusters and their CMDs is presented by Mackey *et al.* (2008); see also Goudfrooij *et al.* (2008) and Milone *et al.* (2008b).

Apart from the peculiar main-sequence turn-off morphologies, each cluster CMD exhibits a very narrow RGB and main sequence. This implies that the turn-off features are not due to significant line-of-sight depth or differential reddening in these clusters. The narrow sequences further suggest minimal internal dispersions in [Fe/H] in each of the three clusters; however we note they place no constraints on the possibility of internal

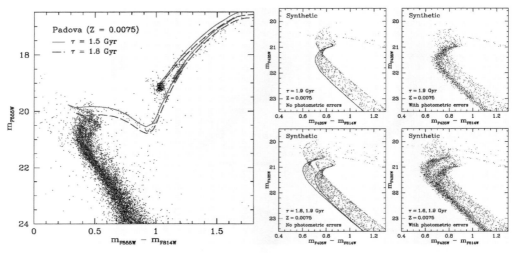

Figure 3. *Left:* Isochrone fit to the *HST* #9891 CMD for NGC 1846. Under the assumption of chemical homogeneity and a uniform distance and foreground extinction, an age spread of ~ 300 Myr within this ~ 1.8 Gyr old cluster can closely reproduce the observed CMD.
Right: Synthetic CMDs aimed at reproducing the *HST* #10595 CMD for NGC 1846. Each includes a population of 40% unresolved binaries. The upper panels include only a single stellar population, while the lower panels include two stellar populations with ages as indicated. Unresolved binaries cannot alone reproduce the observed bifurcated turn-off.

variations in other chemical abundances — for example, CN, O, Na, or $[\alpha/\mathrm{Fe}]$ — as are observed for several of the peculiar massive Galactic globular clusters (e.g., Piotto *et al.* 2008). Mucciarelli *et al.* (2008b) (see also the contribution by Mucciarelli in these proceedings) obtained high resolution spectra for 6 RGB stars in NGC 1783 and found no significant star-to-star dispersion in $[\alpha/\mathrm{Fe}]$; however similar measurements are not presently available for NGC 1846 or 1806. In the absence of this information, we assume complete chemical homogeneity in each of the three systems.

In this scenario, the simplest interpretation of the observed CMDs is that each cluster possesses at least two stellar populations of differing ages. To quantify this, we fit isochrones from the Padova and BaSTI groups (Marigo *et al.* 2008; Pietrinferni *et al.* 2004) to the CMDs. Details of the fitting procedure may be found in Mackey & Broby Nielsen (2007) and Mackey *et al.* (2008). An example may be seen in the left panel of Fig. 3, which shows two Padova isochrones with $Z = 0.0075$ and ages 1.5 Gyr and 1.8 Gyr fit to our *HST* #9891 snapshot CMD for NGC 1846. These models closely reproduce the observed features of the CMD, suggesting that NGC 1846 may consist of two populations with an age difference of ~ 300 Myr. We obtained identical results for NGC 1806 and 1783. For this latter, we fit isochrones to the upper and lower envelopes of the turn-off, to indicate the maximum expected age spread implied by this feature.

Evident on each CMD (especially in Fig. 2) is a spread of stars above and to the red of the main sequence. These are due to unresolved binaries, and imply non-negligible populations of such objects in the three clusters. In fact, in the central cluster regions, counts of stellar detections lying along the main and binary sequences suggest that the ratio of the number of unresolved binary systems with mass ratios $q \gtrsim 0.5$ to the total number of detected objects may locally be as high as $\sim$30–40%. We constructed synthetic CMDs to investigate the role played by unresolved binary stars around the turn-off regions, and in particular investigate whether these objects might reproduce the observed turn-off structures. Full details may be found in Mackey *et al.* (2008).

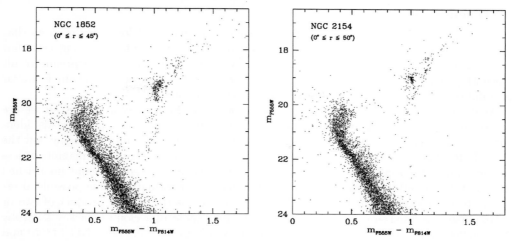

Figure 4. *HST*/ACS CMDs from our *HST* program #9891 snapshot imaging for two additional candidates for possessing peculiar turn-off morphologies. NGC 1852 potentially exhibits a bifurcated turn-off, while NGC 2154 has a turn-off spanning a large spread in colour. Both clusters suffer from relatively strong contaminating field populations; even so, Milone *et al.* (2008b) have independently confirmed the unusual nature of these two objects.

Examples for NGC 1846 are shown in the right panel of Fig. 3, aimed at reproducing the observed CMD from Fig. 2. The upper panels show synthetic CMDs constructed using a single stellar population of age 1.9 Gyr and comprised of 40% unresolved binaries with $q > 0.5$. The left-hand panel shows the synthetic CMD as generated, while the right-hand panel shows the same CMD with random photometric uncertainties added to each star. These uncertainties were generated to match those determined for detected stars in NGC 1846. These two plots clearly demonstrate that unresolved binaries add confusion and some intrinsic spread around the main-sequence turn-off, but cannot alone reproduce the peculiar observed structure for NGC 1846. Specifically, the binary sequence does not result in a distinct branch above and to the blue of the lower turn-off branch.

The lower two panels show a synthetic CMD constructed using two single stellar populations of ages 1.6 and 1.9 Gyr. As before, 40% unresolved binaries with $q > 0.5$ were included. This new CMD closely resembles the observed CMD, demonstrating that two stellar populations in NGC 1846 (and by extension NGC 1806) are necessary to explain the observed turn-off structure. NGC 1783 may require more than two turn-offs, or two more closely spaced branches.

We conclude this section by noting that NGC 1846, 1806, and 1783 are not the only intermediate-age Magellanic Cloud clusters which have been found to possess unusual main-sequence turn-off morphologies. In our *HST* #9891 data we discovered several additional candidates, the two strongest of which are shown in Fig. 4. NGC 1852 shows a potentially bifurcated turn-off, while NGC 2154 has a turn-off spanning a relatively large spread in colour. Milone *et al.* (2008b) have conducted a thorough analysis of archival *HST* data for 16 intermediate-age LMC clusters, confirming all five of our discoveries and adding an additional six. Furthermore, Glatt *et al.* (2008) have recently shown that the SMC cluster NGC 419 may well possess a peculiar turn-off morphology. To this sample of objects we should add the LMC cluster NGC 2173, which was shown several years ago to be likely to have a spread at the turn-off larger than expected from measured photometric uncertainties (Bertelli *et al.* 2003). Several younger LMC clusters, such as NGC 1868 and NGC 2011, have also been identified as possessing irregular CMDs (Santiago *et al.* 2002; Gouliermis *et al.* 2006).

3. Discussion

The rapidly-growing number of intermediate-age Magellanic Cloud star clusters that are being found to possess peculiar main-sequence turn-off morphologies suggests that such a feature may not be uncommon for this type of object. This is a surprising result since Magellanic Cloud star clusters have long been treated as prototypical single stellar populations, used to test and calibrate stellar evolution models.

As discussed by Mackey & Broby Nielsen (2007) and Mackey *et al.* (2008), if we are correct in interpreting the observed cluster CMDs in terms of internal age spreads, identifying a viable scenario for their formation is not trivial. It is difficult to imagine how the three LMC clusters discussed in this contribution, each at least an order of magnitude less massive than the peculiar Galactic globulars, might have retained sufficient unenriched gas to undergo multiple widely-separated episodes of star formation. One possible alternative to these suggestions is that each peculiar cluster is the merger product of two or more clusters formed separately within a single giant molecular cloud (Mackey & Broby Nielsen 2007). The LMC possesses numerous "double" star clusters (e.g., NGC 1850) and its low tidal field could facilitate such mergers. A second alternative, raised by several participants at this conference, is that the peculiar turn-off features may not be due to more than one stellar population, but rather a hitherto poorly-appreciated aspect of the evolution of intermediate-age stars. However, this model may have trouble explaining why a significant number of intermediate-age Magellanic Cloud clusters *do not* exhibit unusual turn-off morphologies (e.g., Milone *et al.* 2008b).

Additional observations are clearly important in constraining viable formation scenarios for these objects. In particular, it is crucial to understand in detail the chemical composition of each peculiar cluster. If these objects are not chemically homogeneous then the requirement for such large internal age spreads as are currently being assumed may well be obviated. Such measurements are under way at the time of writing, and new information should shortly be available.

References

Bertelli, G., Nasi, E., Girardi, L., Chiosi, C., Zoccali, M., & Gallart, C. 2003, *AJ*, 125, 770
Glatt, K., Grebel, E. K., Sabbi, E., *et al.* 2008, *AJ*, 136, 1703
Goudfrooij, P., Puzia, T. H., Kozhurina-Platais, V., & Chandar, R. 2008, *AJ*, submitted
Gouliermis, D. A., Lianou, S., Kontizas, M., Kontizas, E., & Dapergolas, A. 2006, *ApJ*, 652, L93
Mackey, A. D. & Broby Nielsen, P. 2007, *MNRAS*, 379, 151
Mackey, A. D., Broby Nielsen, P., Ferguson, A. M. N., & Richardson, J. C. 2008, *ApJ*, 681, L17
Marigo, P., Girardi, L., Bressan, A., Groenewegen, M. A. T., Silva, L., & Granato, G. L. 2008, *A&A*, 482, 883
Marino, A. F., Villanova, S., Piotto, G., Milone, A. P., Momany, Y., Bedin, L. R., & Medling, A. M. 2008, *A&A*, 490, 625
Milone, A. P., Bedin, L. R., Piotto, G., *et al.* 2008a, *ApJ*, 673, 241
Milone, A. P., Bedin, L. R., Piotto, G., & Anderson, J. 2008b, *A&A*, in press
Mucciarelli, A., Carretta, E., Origlia, L., & Ferraro, F. R. 2008, *AJ*, 136, 375
Pietrinferni, A., Cassisi, S., Salaris, M., & Castelli, F. 2004, *ApJ*, 612, 168
Piotto, G., Villanova, S., Bedin, L. R., *et al.* 2005, *ApJ*, 621, 777
Piotto, G., Bedin, L. R., Anderson, J., *et al.* 2007, *ApJ*, 661, L53
Piotto, G. 2008, *MemSAI*, 79, 3
Richer, H. B., Dotter, A., Hurley, J., *et al.* 2008, *AJ*, 135, 2141
Salaris, M., Cassisi, S., & Pietrinferni, A. 2008, *ApJ*, 678, L25
Santiago, B., Kerber, L., Castro, R., & de Grijs, R. 2002, *MNRAS*, 336, 139
Villanova, S., Piotto, G., King, I. R., *et al.* 2007, *ApJ*, 663, 296

The Magellanic System: Stars, Gas, and Galaxies
Proceedings IAU Symposium No. 256, 2008
Jacco Th. van Loon & Joana M. Oliveira, eds.

Star cluster evolution in the Magellanic Clouds revisited

Richard de Grijs[1,2] and Simon P. Goodwin[1]

[1]Department of Physics & Astronomy, The University of Sheffield, Hicks Building,
Hounsfield Road, Sheffield S3 7RH, UK
email: [R.deGrijs,S.Goodwin]@sheffield.ac.uk

[2]National Astronomical Observatories, Chinese Academy of Sciences, 20A Datun Road,
Chaoyang District, Beijing 100012, China

Abstract. The evolution of star clusters in the Magellanic Clouds has been the subject of significant recent controversy, particularly regarding the importance and length of the earliest, largely mass-independent disruption phase (referred to as "infant mortality"). Here, we take a fresh approach to the problem, using a large, independent, and homogeneous data set of $UBVR$ imaging observations, from which we obtain the cluster age and mass distributions in both the Large and Small Magellanic Clouds (LMC, SMC) in a self-consistent manner. We conclude that the (optically selected) SMC star cluster population has undergone at most $\sim 30\%$ (1σ) infant mortality between the age range from about $3-10$ Myr, to that of approximately $40-160$ Myr. We rule out a 90% cluster mortality rate per decade of age (for the full age range up to 10^9 yr) at a $> 6\sigma$ level. Using a simple approach, we derive a "characteristic" cluster disruption time-scale for the cluster population in the LMC that implies that we are observing the *initial* cluster mass function (CMF). Preliminary results suggest that the LMC cluster population may be affected by $< 10\%$ infant mortality.

Keywords. globular clusters: general, open clusters and associations: general, galaxies: evolution, Magellanic Clouds, galaxies: star clusters

1. Introduction

One of the most important diagnostics used to infer the formation history, and to follow the evolution of an entire star cluster population is the "cluster mass function" (CMF; i.e., the number of clusters per constant logarithmic cluster mass interval, $dN/d\log m_{cl}$). The *initial* cluster mass function (ICMF) is of particular importance. The debate regarding the shape of the ICMF, and of the CMF in general, is presently very much alive, both observationally and theoretically. This is so because it bears on the very essence of the star-forming processes, as well as on the formation, assembly history, and evolution of the clusters' host galaxies on cosmological time-scales. Yet, the observable at hand is the cluster *luminosity* function (CLF; i.e., the number of objects per unit magnitude, dN/dM_V).

The discovery of star clusters with the high luminosities and the compact sizes expected for (old) globular clusters (GCs) at young ages facilitated by the *Hubble Space Telescope (HST)* has prompted renewed interest in the evolution of the CLF (and CMF) of massive star clusters. Starting with the seminal work by Elson & Fall (1985) on the young Large Magellanic Cloud (LMC) cluster system (with ages $\lesssim 2 \times 10^9$ yr), an ever increasing body of evidence seems to imply that the CLFs of young massive clusters (YMCs) are well described by a power law of the form $dN \propto L^{1+\alpha} d\log L$, equivalent to a cluster luminosity spectrum $dN \propto L^\alpha dL$, with a spectral index $-2 \lesssim \alpha \lesssim -1.5$ (e.g., Whitmore & Schweizer 1995; Elmegreen & Efremov 1997; Miller *et al.* 1997; Whitmore *et al.* 1999,

2002; Bik *et al.* 2003; de Grijs *et al.* 2003; Hunter *et al.* 2003; Lee & Lee 2005; see also Elmegreen 2002). Since the spectral index, α, of the observed CLFs resembles the slope of the high-mass regime of the (lognormal) old GC mass spectrum ($\alpha \sim -2$; McLaughlin 1994), this observational evidence has led to the popular theoretical prediction that not only a power law, but *any* initial CLF (and CMF) will be rapidly transformed into a lognormal distribution because of (i) stellar evolutionary fading of the lowest-luminosity (and therefore lowest-mass, for a given age) objects to below the detection limit; and (ii) disruption of the low-mass clusters due to both interactions with the gravitational field of the host galaxy, and internal two-body relaxation effects leading to enhanced cluster evaporation (e.g., Elmegreen & Efremov 1997; Gnedin & Ostriker 1997; Ostriker & Gnedin 1997; Fall & Zhang 2001; Prieto & Gnedin 2008).

However, because of observational selection effects it is often impossible to probe the CLFs of YMC systems to the depth required to fully reveal any useful evolutionary signatures. As such, the young populous cluster systems in the Magellanic Clouds are as yet the best available calibrators for the canonical young CLFs that form the basis of most theoretical attempts to explain the evolution of the CLF and CMF.† It is therefore of paramount importance to understand the Magellanic Cloud cluster systems in detail.

2. Early cluster evolution in the Magellanic Clouds

The early evolution of the star cluster population in the Small Magellanic Cloud (SMC) has been the subject of considerable recent interest (e.g., Rafelski & Zaritsky 2005; Chandar, Fall & Whitmore 2006; Chiosi *et al.* 2006; Gieles, Lamers & Portegies Zwart 2007; de Grijs & Goodwin 2008). The key issue of contention is whether the SMC's star cluster system has been subject to the significant early cluster disruption processes observed in "normal", interacting and starburst galaxies, commonly referred to as "infant mortality". Chandar *et al.* (2006) argue that the SMC has been losing up to 90% of its star clusters per decade of age, at least for ages from $\sim 10^7$ up to $\sim 10^9$ yr, whereas Gieles *et al.* (2007) conclude that there is no such evidence for a rapid decline in the cluster population, and that the decreasing number of clusters with increasing age is simply caused by evolutionary fading of their stellar populations in a magnitude-limited cluster sample.

In de Grijs & Goodwin (2008) we set out to shed light on this controversy. We adopted a fresh approach to the problem, using an independent, homogeneous data set of $UBVR$ imaging observations (cf. Hunter *et al.* 2003), from which we obtained the cluster age distribution in a self-consistent manner. In Fig. 1 we present the CMFs for two subsets of our SMC cluster sample, selected based on their age distributions. In all panels of Fig. 1, we have overplotted CMFs with the canonical slope of $\alpha = -2$ (corresponding to a slope of -1 in units of $\mathrm{d}\log(M_{cl}/M_{\odot})/\mathrm{d}\log(N_{cl})$, used in these panels).

The rationale for adopting as our youngest subsample (Fig. 1a) all clusters with ages $\leqslant 10$ Myr is that at these young ages, the vast majority of the star clusters present will still be detectable, even in the presence of early gas expulsion (e.g., Goodwin & Bastian 2006) — as long as they are optically conspicuous. Fig. 1b includes our sample clusters with ages from 40 Myr to 160 Myr. While the upper age limit ensures the full inclusion of the clusters affected by the onset of the AGB stage, its exact value is rather unimportant. The lower age limit of this subsample is crucial, however. As shown by

† We point out, however, that with the latest *HST*/Advanced Camera for Surveys observations of the Antennae interacting galaxies, we are now finally getting to the point where the young (power-law) CLF shape appears to be confirmed independently down to sufficient photometric depths and for larger samples containing more massive clusters than in the Magellanic Clouds (B. Rothberg, priv. comm.).

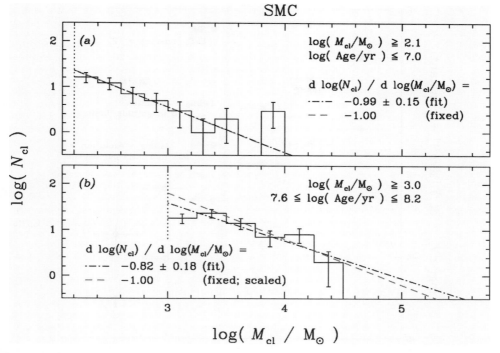

Figure 1. CMFs for statistically complete SMC cluster subsamples. Age and mass ranges are indicated in the panel legends; the vertical dotted lines indicate the lower mass (50% completeness) limits adopted. Error bars represent simple Poissonian errors, while the dash-dotted lines represent CMFs of slope $\alpha = 2$, shifted vertically as described in the text.

Goodwin & Bastian (2006), most dissolving clusters will have dispersed by an age of ~ 30 Myr, while the surviving clusters will have returned to an equilibrium state by ~ 40 Myr, when some of the early expansion will have been reversed, depending on the effective star-formation efficiency. This latter age is therefore a good lower boundary to assess the surviving star cluster population.

We explicitly exclude any star clusters aged between 10 and 40 Myr from our analysis. In this age range, it is likely that dissolving star clusters that will not survive beyond about $30-40$ Myr might still be detectable and therefore possibly contaminate our sample. In addition, this is the age range in which early gas expulsion causes rapid cluster expansion, before settling back into equilibrium at smaller radii; because of the expanded nature of at least part of the cluster sample, we might not be able to detect some of the lower-luminosity (and hence lower-mass) clusters that would again show up beyond an age of ~ 40 Myr.

In the simplest case, in which the cluster formation rate has remained roughly constant throughout the SMC's evolution (e.g., Boutloukos & Lamers 2003, their fig. 10; see also Gieles *et al.* 2007), the number of clusters would simply scale with the age range covered. In Fig. 1b we show the canonical $\alpha = -2$ CMF scaled from the best-fitting locus in Fig. 1a by the difference in (linear) age range between the panels (see de Grijs & Goodwin 2008 for details). The scaled canonical CMF in Fig. 1b is an almost perfect fit to the observational CMF (irrespective of the mass binning employed). This implies that the SMC cluster system has not been affected by any significant amount of cluster infant mortality for cluster masses greater than a few $\times 10^3$ M$_\odot$. Based on a detailed

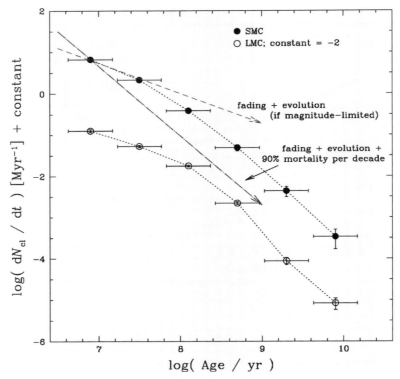

Figure 2. Age distribution of the full magnitude-limited SMC and LMC cluster samples in units of cluster numbers per Myr. The LMC sample has been shifted vertically by a constant offset for reasons of clarity. The vertical error bars are simple Poissonian errors; the horizontal error bars indicate the age range used for the generation of these data points. The dashed arrow shows the expected effects due to evolutionary fading of a magnitude-limited cluster sample; the dash-dotted arrow represents the combined effects of a fading cluster population and 90% cluster disruption per decade in $\log(\text{Age yr}^{-1})$.

assessment of the uncertainties in both the CMFs and the age range covered by our youngest subsample, we can limit the extent of infant mortality between the youngest and the intermediate age range to a maximum of $\lesssim 30\%$ (1σ). We rule out a $\sim 90\%$ (infant) mortality rate per decade of age at a $> 6\sigma$ level. This result is in excellent agreement with that of Gieles *et al.* (2007).

Moreover, Gieles *et al.* (2007) derived that for a magnitude-limited sample the decline in the number of observed clusters per unit age range, dN_{cl}/dt, as a function of age is graphically described by a slope of -0.72. In Fig. 2, we show the expected effects of evolutionary fading of a magnitude-limited cluster population as the dashed arrow. It is immediately clear that, for the SMC cluster population analysed in de Grijs & Goodwin (2008), the decline in the age distribution up to $\log(\text{Age yr}^{-1}) \simeq 7.8$ can indeed be entirely attributed to evolutionary fading. The expected effects of evolutionary fading combined with a 90% disruption rate are shown as the dash-dotted arrow in Fig. 2. The arrow clearly does not fit the observed age distribution, if we require it to pass through our youngest data point. We note, however, that the slope of this latter arrow is very similar to that of the age distribution of the full SMC sample for ages in excess of a few $\times 10^8$ yr, when secular disruption is likely to take over.

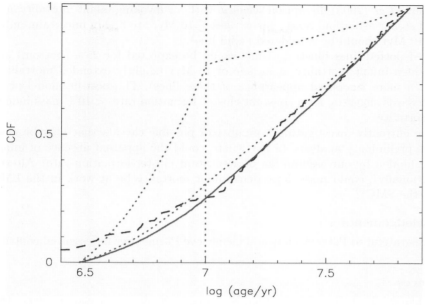

Figure 3. Cumulative distribution function (CDF) of the LMC cluster age distribution. The dashed line represents the data, smoothed by the relevant age-dependent uncertainties; the solid line is the fit for ages $> 10^7$ yr, assuming no disruption. The dotted lines are the expected CDFs for 90% (top) and 25% (bottom) mortality per decade in age.

3. Cluster disruption in the LMC

To derive the CMF of mass-limited LMC cluster subsamples, de Grijs & Anders (2006) re-analysed the *UBVR* broad-band data of Hunter *et al.* (2003). They derived that the timescale on which a 10^4 $M_\odot$ cluster is expected to disrupt is $\log(t_4^{\mathrm{dis}} \mathrm{yr}^{-1}) = 9.9 \pm 0.1$, as recently confirmed by Parmentier & de Grijs (2008) based on a detailed comparison with numerical simulations.† Such a long cluster disruption timescale results from the low-density environment of the Magellanic Clouds. It guarantees that the observed cluster mass distributions have not yet been altered significantly by secular dynamical evolution, i.e. clusters already affected by ongoing disruption have faded to below the completeness limit (see de Grijs & Anders 2006, their fig. 8). As a result, the observed mass distributions are the *initial* distributions.

We will investigate the possible effects of infant mortality among the LMC cluster population in detail in a forthcoming paper (Goodwin *et al.*, in preparation). However, a first glance at the LMC cluster population's age distribution in Fig. 2 indicates that the number of clusters populating the first $\sim 10^8$ yr can likely be fully explained by simple evolutionary fading of a magnitude-limited cluster sample, as we also concluded for the SMC, without the need to invoke infant mortality for masses $M_{\mathrm{cl}} \gtrsim 10^3$ $M_\odot$.

In Fig. 3 we display the LMC cluster population's cumulative distribution function (CDF) as a function of age (dashed line). This CDF has been smoothed by the uncertainties in the cluster ages determined for each cluster individually by de Grijs & Anders (2006). We compare the observed CDF to that of a model with a constant cluster-formation rate with and ICMF index of $\alpha = -2$. These clusters are evolved according to

---

† In fact, Parmentier & de Grijs (2008) concluded that the data do not allow us to constrain the characteristic cluster disruption timescale for a 10^4 $M_\odot$ cluster to better than $9.0 \lesssim \log(t_4^{\mathrm{dis}} \mathrm{yr}^{-1}) \lesssim 9.9$.

the Lamers *et al.* (2005) disruption scenario without invoking infant mortality, assuming standard stellar evolution. Since cluster ages $\lesssim$ 10 Myr are highly uncertain, only the fit beyond 10 Myr should be considered (solid line).

The red dotted lines illustrate what would be expected for 25% (bottom) and 90% (top) sudden infant mortality at an age of 10 Myr (slightly extended mortality would result in a more smoothed appearance of these lines). The best-fit model at 10 Myr and afterwards suggests, for a constant cluster-formation rate, < 10% mass-independent infant mortality.

We are currently investigating a number of possible caveats that may be associated with this preliminary analysis. In particular, could the apparent absence of infant mortality be hidden by our assumption of a constant cluster-formation rate? Alternatively (or additionally), could mass-dependent infant mortality be at work in the LMC (and possibly the SMC)?

Acknowledgements

We are grateful to Peter Anders and Geneviève Parmentier for essential contributions.

References

Bik, A., Lamers, H. J. G. L. M., Bastian, N., Panagia, N., & Romaniello M., 2003, *A&A*, 397, 473

Boutloukos, S. G. & Lamers, H. J. G. L. M. 2003, *MNRAS*, 338, 717

Chandar, R., Fall, S. M. & Whitmore, B. C. 2006, *ApJ*, 650, L111

Chiosi, E., Vallenari, A., Held, E. V., Rizzi, L., & Moretti, A. 2006, *A&A*, 452, 179

de Grijs, R. & Anders, P. 2006, *MNRAS*, 366, 295

de Grijs, R., Anders, P., Lynds, R., Bastian, N., Lamers, H. J. G. L. M., & O'Neill, E. J., Jr. 2003, *MNRAS*, 343, 1285

de Grijs, R. & Goodwin, S. P. 2008, *MNRAS*, 383, 1000

Elmegreen, B. G. 2002, *ApJ*, 564, 773

Elmegreen, B. G. & Efremov, Y. N. 1997, *ApJ* 480, 235

Elson, R. A. W. & Fall, S. M. 1985, *ApJ*, 299, 211

Fall, S. M. & Zhang, Q. 2001, *ApJ*, 561, 751

Gieles, M., Lamers, H. J. G. L. M. & Portegies Zwart, S. F. 2007, *ApJ*, 668, 268

Gnedin, O. Y. & Ostriker, J. P. 1997, *ApJ*, 474, 223

Goodwin, S. P. & Bastian, N. 2006, *MNRAS*, 373, 752

Hunter, D. A., Elmegreen, B. G., Dupuy, T. J., & Mortonson, M. 2003, *AJ*, 126, 1836

Lamers, H. J. G. L. M., Gieles, M., Bastian, N., Baumgardt, H., Kharchenko, N. V., & Portegies Zwart, S. F. 2005, *A&A*, 441, 117

Lee, H. J. & Lee, M. G. 2005, *JKAS*, 38, 345

McLaughlin, D. E. 1994, *PASP*, 106, 47

Miller, B. W., Whitmore, B. C., Schweizer, F., & Fall, S. M. 1997, *AJ*, 114, 2381

Ostriker, J. P. & Gnedin, O. Y. 1997, *ApJ*, 487, 667

Parmentier G. & de Grijs R. 2008, *MNRAS*, 383, 1103

Prieto, J. L. & Gnedin, O. Y. 2008, *ApJ*, 689, 919

Rafelski, M. & Zaritsky, D. 2005, *AJ*, 129, 2701

Whitmore, B. C. & Schweizer, F. 1995, *AJ*, 109, 960

Whitmore, B. C., Schweizer, F., Kundu, A., & Miller, B. W. 2002, *AJ*, 124, 147

Whitmore, B. C., Zhang, Q., Leitherer, C., Fall, S. M., Schweizer, F., & Miller, B. W. 1999, *AJ*, 118, 1551

The Magellanic System: Stars, Gas, and Galaxies
Proceedings IAU Symposium No. 256, 2008
Jacco Th. van Loon & Joana M. Oliveira, eds.

© 2009 International Astronomical Union
doi:10.1017/S1743921308028640

Gemini/GMOS detection of stellar velocity variations in the ionising cluster of 30 Dor

Guillermo Bosch[1], Elena Terlevich[2] and Roberto Terlevich[2]

[1]Facultad de Ciencias Astronómicas y Geofísicas & IALP Paseo del Bosque s/n,
1900 La Plata, Argentina
email:guille@fcaglp.unlp.edu.ar

[2]INAOE, Tonantzintla, Apdo. Postal 51, 72000 Puebla, México

Abstract. We have analysed spectra obtained with the *Gemini* Multi Object Spectrograph (GMOS) for more than 50 stars in the ionising cluster of 30 Doradus during a seven epochs observing campaign at *Gemini South*. We derive a binary candidate rate of about 50%, which is however consistent with an intrinsic 100% binary rate among massive stars. After decontaminating the sample from the stars that show binary orbital motions, we were able to calculate the "true" cluster velocity dispersion and found it to be about 8 km s^{-1}. This value implies a virial mass of about 4.5×10^5 M$_\odot$ which is consistent with previous photometric mass determinations therefore suggesting that NGC 2070 is a firm candidate for a future globular cluster.

Keywords. stars: early-type, stars: kinematics, binaries: spectroscopic, galaxies: individual (LMC), Magellanic Clouds

1. Introduction

30 Doradus in the LMC is the nearest available example of a young and massive starburst cluster. Given its proximity it is possible to perform a highly detailed study of its stellar component. The large number of massive stars present in this single cluster allows the statistical analysis of several parameters at a level of significance that is not available in the local and smaller Galactic clusters.

When analysing the stellar kinematics of the 30-Dor ionising cluster, Bosch (1999) and collaborators pointed out that the observed radial velocity dispersion of stars within the cluster, can be strongly affected by the orbital motions of massive binaries. Bosch *et al.* (2001) performed a radial velocity analysis on the stars that conform the ionising cluster of 30 Dor. They suggested that the very large value of the velocity dispersion ($\sigma \sim 35$ km s^{-1}) obtained for the OB stars within NGC 2070, the ionizing cluster in 30 Dor, could be due to the presence of an underlying binary population. From Montecarlo simulations Bosch & Meza (2001) estimated that this contribution could be as large as 36 km s^{-1}, provided all stars in the cluster belonged to a binary pair. However, it is necessary to confirm this assumption with direct evidence. The issue of binary frequency is still subject to debate, although there is a relatively high number of spectroscopic and visual binaries reported in the literature (Lada 2006).

Here we present a new set of observations of NGC 2070 obtained with the *Gemini* Multi Object Spectrograph (GMOS) at *Gemini South*. These comprise multi object optical spectroscopy of 50 early-type stars observed at least in six different epochs. The aim is to detect spectroscopic binary stars from variations in their radial velocities.

Table 1. Log of observations. Observing dates are listed in column 1 and Heliocentric Julian dates for masks I and II are shown in columns 2 and 3 respectively.

	Mask I	Mask II
07 Sep. 2005	2453620.89	
19 Dec. 2005		2453723.58
20 Dec. 2005	2453724.58	
22 Dec. 2005		2453726.68
23 Dec. 2005	2453727.65	
24 Dec. 2005		2453728.59
29 Nov. 2006	2454068.73	
30 Nov. 2006		2454069.70
24 Dec. 2006	2454093.83	
31 Dec. 2006		2454100.86
01 Jan. 2007	2454101.79	2454101.73
09 Jan. 2007		2454109.66

2. Observations and data reduction

Observations were performed at *Gemini South Observatory* (Proposals GS-2005B-Q-2 and GS-2006B-Q-21) using two multislit masks. Targets within each mask were selected from a previous imaging run with GMOS according to their spectral types as determined and compiled in Bosch *et al.* (1999).

Spectra were obtained during nights of September and December 2005 for the first run and December 2006 and January 2007 for the second run. Approximate HJD for the listed observations are shown in table 1. The instrument was set up with the B1200 grating $R \sim 3700$ centred at about 4500 Å which yields a resolution of 0.25 Å per pixel at the CCD. Although the wavelength range varies slightly in MOS spectroscopy, the region from 3900 Å to 5500 Å is covered by our spectra. The total integration time was split in three to allow for the wavelength dithering pattern needed to cover the gaps between GMOS CCDs. Overall signal to noise ratios are above 150. These ratios are measured on reduced spectra, dividing the average value of the stellar continuum by the scatter over the same spectral range.

3. Radial velocities

Radial velocities were derived measuring absorption line profiles with the aid of the `ngaussfit` task within the STSDAS/IRAF package, following a similar procedure as the one described in Bosch *et al.* (2001). This allowed us to derive individual radial velocities for each spectral line, and handle each element (or each ion in the case of He) separately. Stellar radial velocities were derived using the best set of lines available, according to the star's spectral type. Although this means we are not using the same set of lines for the whole sample, we strictly kept the same set of lines for the same star on different epochs when looking for radial velocity variations. Uncertainties introduced when fitting individual Gaussians to the absorption profiles are of the order of 5 $\mathrm{km\,s^{-1}}$.

Nebular spectra were wavelength calibrated together with the stellar spectra, to provide a strong template to check for variations of our radial velocities zero-point at different epochs. Nebular emission lines have strong narrow profiles, enabling us to cross-correlate nebular spectra obtained on different nights, using `fxcor` within IRAF, that yields accurate determinations of radial velocity differences, if present. Radial velocities derived for each night (V_{neb}) were very stable when observations on different epochs are compared and their variations $|\Delta V_{\mathrm{neb}}|$ were found to be negligible ($|\Delta V_{\mathrm{neb}}| \leqslant 1.0$ km s^{-1}, $\sigma_{\mathrm{neb}} = 3.1$ km s^{-1}) which suggests that there is no systematic shift introducing spurious variations of stellar radial velocities between epochs.

Table 2. Stellar identifications by Parker (1993) are in column 1. Columns 2 through 15 include radial velocities and uncertainties (all in km s^{-1}) for each epoch of observation. Column 16 lists the ratio between epoch-to-epoch variations and the average uncertainties.

Id	V_r	σ_I	V_r	σ_I	V_r	σ_I	V_r	σ_I	V_r	σ_I	V_r	σ_I	V_r	σ_I	$\frac{\sigma_E}{\langle\sigma_I\rangle}$
15(I)	299.1	5.5			272.7	9.0	207.5	4.9	289.3	5.7	255.2	5.6			5.9
32(II)	283.3	6.3	271.6	4.2	261.1	4.8	271.3	4.3	278.3	4.3	275.5	4.1			1.6
124(I)	266.0	4.2	239.3	3.7	209.7	6.0	268.7	3.3	243.5	3.5	261.0	5.3			5.1
171(I)	272.7	3.7	286.5	5.4	256.9	5.0	258.7	4.7	276.7	6.1	269.2	7.1			2.1
260(II)	286.0	9.1	274.7	5.9	261.0	9.4	266.6	4.2	304.5	3.8	271.5	9.1	278.4	6.1	2.3
305(II)	269.5	9.7	269.5	6.1	249.9	6.4	265.4	3.3	258.9	4.8	253.1	4.9	282.5	5.3	1.4
316(I)	286.7	5.0	366.0	6.7	302.8	3.9	214.5	4.1	349.8	8.7	302.5	5.7			9.4
485(II)	277.3	5.3	260.1	5.5	255.7	8.1	278.7	3.5	251.2	5.5	259.8	6.2	269.1	4.0	2.0
531(II)	309.3	7.1	247.6	3.7	221.2	9.3	260.9	3.7	276.9	9.0	324.6	11.3	241.7	3.5	5.3
541(I)	239.2	6.2	276.0	7.0	272.4	3.4	232.8	3.9	358.5	6.2	232.1	3.9			9.4
613(II)	246.2	4.1	275.6	4.8	332.9	9.1	257.6	3.8	264.5	4.4	245.9	7.5	236.3	5.5	5.8
649(II)	282.6	9.1	286.3	6.5	280.2	6.4	276.9	3.7	281.5	6.7	285.2	5.9	280.1	4.0	0.5
684(II)	275.7	7.9	270.4	5.4	247.8	7.8	252.7	3.4	254.8	5.0	272.5	8.7	270.7	6.4	1.9
713(I)	238.3	4.7	316.1	6.9	298.8	4.4	320.4	5.3	297.0	7.2	312.6	10.9			4.6
716(II)	274.0	3.6	273.8	7.4	266.1	4.6	271.4	3.8	275.3	4.4	279.8	4.1	277.5	3.8	1.0
747(I)	182.4	3.8	213.9	5.6	306.5	5.0	331.0	4.0	248.3	9.2	306.0	7.5			10.1
809(II)	263.8	9.4	268.7	5.5	256.0	7.9	279.3	3.8	265.2	4.0	257.6	11.7	269.1	6.1	1.2
871(II)	275.2	7.5	272.8	6.3	268.7	8.4	281.3	3.6	282.2	8.3	289.1	4.2	278.2	8.2	1.2
885(II)	276.1	5.1	289.1	4.1	286.3	7.6	234.8	3.2	270.5	8.1	267.2	5.3	283.5	4.8	3.5
905(I)	268.5	6.9	267.4	8.2	271.2	11.1	257.5	4.1	261.9	5.1	265.5	5.8			0.7
956(I)	268.4	6.6	284.5	6.4	239.1	3.7	265.7	6.5	270.3	7.1	274.3	6.8			2.5
975(II)	297.4	4.2	288.8	3.7	266.7	6.5	281.1	3.6	289.2	8.9	282.9	6.0	287.2	3.8	1.9
1022(I)	274.3	4.5	288.9	4.9	271.1	5.0	290.0	5.4	278.4	8.2	279.5	6.6			1.3
1035(II)	287.8	11.2	278.1	7.6	263.5	8.4	277.7	4.2	273.6	4.9	285.1	3.4	281.3	4.7	1.3
1063(II)	269.5	6.6	265.3	3.7	265.9	8.9	261.6	4.1	272.3	4.2	256.5	7.2	266.4	4.9	1.0
1109(I)	274.2	5.7	293.9	8.6	278.6	4.5	278.5	3.6	253.0	4.4	248.5	8.1			3.0
1139(I)	256.3	5.4	266.2	8.0	259.1	4.1	269.4	4.2	257.1	5.5	271.7	5.4			1.2
1163(I)	263.6	5.5	273.3	7.9	259.5	4.8	265.5	8.1	266.5	11.9	260.9	13.0			0.6
1218(II)	294.1	3.4	283.6	4.1	273.0	7.1	287.4	3.8	281.4	3.6	297.7	10.3	294.3	4.5	1.7
1222(I)	277.8	4.7	272.7	5.9	258.2	4.8	263.4	6.0	272.3	8.4	265.9	4.6			1.3
1247(I)	248.7	8.8	263.2	4.3	248.1	15.3	243.5	9.1	269.7	8.4	273.3	4.4			1.5
1260(I)	289.0	5.0	294.8	5.1	279.3	5.0	294.5	4.2	291.4	9.4	297.3	5.9			1.1
1339(I)	252.7	6.1	271.8	5.2	255.0	6.3	260.8	4.8	261.0	7.1	273.7	6.6			1.4
1341(I)	287.9	7.6	300.3	6.0	301.2	4.4	305.3	5.2	267.2	12.6	304.5	13.9			1.8
1350(II)	266.1	4.7	269.3	7.5	255.9	6.9	267.1	3.8	261.7	6.5	271.6	8.9	253.0	6.1	0.9
1401(I)	248.2	3.1	260.7	3.6	279.7	5.2	282.6	4.5	321.2	10.9	237.5	6.7			5.3
1468(II)	268.2	9.5	276.1	5.7	259.9	4.4	262.7	4.3	263.0	5.8	264.1	8.4	274.1	5.9	0.9
1531(I)	276.1	4.5	304.0	5.6	290.9	3.5	295.2	4.0	291.0	8.1	299.8	5.6			1.9
1553(I)	267.1	4.2	241.0	6.0	230.2	7.8	287.4	3.4	309.7	6.5	316.2	6.8			6.1
1584(II)			281.7	5.9	250.7	6.9	277.2	4.0	299.0	7.2	283.1	36.3	293.8	6.1	1.5
1604(I)	270.9	5.8	261.3	7.6	251.0	4.0	269.4	4.1	337.1	16.2	326.7	27.3			3.4
1607(II)	281.7	3.6	279.7	4.7	267.2	3.7	271.8	3.9	269.1	4.5	277.9	4.1	274.4	4.7	1.5
1614(I)	274.0	3.1	293.7	4.4	276.9	5.9	286.8	3.8	280.2	8.3	291.3	5.6			1.5
1619(I)	267.9	3.2	321.7	5.1	301.5	3.3	292.9	4.2	305.7	10.3	291.0	5.5			3.4
1729(II)	281.3	8.5	270.8	4.6	194.2	6.4	233.2	6.4	209.2	11.5	211.0	3.6	286.1	5.2	5.2
1840(II)	262.4	8.2	270.8	4.3	273.5	5.8	245.1	5.5	271.4	3.2	251.3	5.3	296.9	5.9	2.2
1969(I)	273.3	5.1	276.9	6.5	285.0	4.1	299.6	4.1	312.0	5.2	313.2	6.6			3.3
1988(I)	218.3	5.3	325.2	6.2	234.5	3.6	263.0	3.3	204.2	5.6	311.6	8.0			9.3

Table 2 lists the complete set of radial velocities determined for the sample stars. Stars are labelled following the nomenclature of Parker (1993) and data columns include average radial velocity (and its uncertainty) for each epoch. The errors listed in columns 3, 5, 7, 9, 11, 13 and 15 in Table 2 correspond to the internal error σ_I which is the quadratic sum of the standard deviation of the estimated average from the set of available lines and the minimum uncertainty (3.1 km s^{-1}) as derived from high S/N narrow nebular emission lines. Column 16 shows the ratio between the standard deviation of stellar radial velocity between different epochs and the average of σ_I.

From the different epochs' radial velocities we can check for the presence of variations with time. To quantify this, we follow the standard procedure of comparing the dispersion of the average radial velocity for each star with the average uncertainty in the determination of each radial velocity. This can be done calculating the "external" to "internal" velocity dispersion ratio (σ_E/σ_I) as defined by Abt *et al.* (1972). Radial velocity variables can then be easily flagged out as they show σ_E/σ_I above 3, which is similar to say that the variation in radial velocity is 3σ above the expected uncertainties. In addition to the results presented in Table 2, the high signal to noise ratio of our spectra has also allowed us to detect several double-lined binaries that show evident variation of their absorption line profiles from epoch to epoch.

An inspection of Table 2 shows that 17 out of 46 stars show radial velocity variations. If we add the stars that present profile variations (excluding P613 which already shows radial velocity variations in Table 2), we have our complete set of binary star candidates, which rises to 25 out of 52 stars ($\sim 48\%$).

After removing the confirmed and candidate binaries, the non-binary population of our sample decreases to 26 stars, still enough to calculate a representative value of the stellar radial velocity dispersion (σ_r). In calculating it, we must keep in mind that each radial velocity measurement has its intrinsic error (σ_{int}) which must be subtracted quadratically from the observed (σ_{obs}) velocity dispersion. The former is the average of the $\langle \sigma_I \rangle$ values over the whole sample of non radial velocity variable stars while the latter is directly derived as the standard deviation around the average radial velocity derived for the same subset of stars in the cluster and it therefore follows that $\sigma_r^2 = \sigma_{obs}^2 - \langle \sigma_{int} \rangle^2$. For our sample, $\sigma_{obs} = 10.3$ km s^{-1}, $\langle \sigma_{int} \rangle = 6.2$ km s^{-1}, which yield a value of 8.3 km s^{-1} for the actual radial velocity dispersion. This seems to confirm the suggestion by Bosch *et al.* (2001) based on simulations, that the large values derived for the stellar velocity dispersion were most probably due to the presence of binaries. As expected, if we derive the radial velocity dispersion from an individual GMOS mask observation for a single epoch, we find values as high as 30 km s^{-1}, consistent with what was previously found in Bosch *et al.* (2001) from single epoch *NTT* data.

4. Conclusions

Based on six-to-seven epoch observations with GMOS on *Gemini South*, we have presented observational evidence that shows that the binary fraction among **massive** stars in NGC 2070 may be very high. Indeed we already detect a 50% binary candidacy with only three epoch observations using medium resolution spectroscopy, which suggests that it is only a lower limit. The evidence pointing towards a high binarity fraction has an important effect on the massive end of the stellar cluster IMF, and shouldn't be overlooked.

We are aware that there are other sources for radial velocity variations among spectral features in early-type stars. However, we are confident that the variations we detected in several lines with small internal scatter are most probably due to binary nature. Regardless of the origin of the stellar radial velocity variability, the analysis of the non-variable subset provides an important result regarding kinematics of the stellar cluster itself. The radial velocity dispersion determined for the 30 Doradus ionising cluster agrees, within observational errors, with the stellar kinematics expected if the cluster is virialised and its total mass is derived from the photometric plus ionised gas masses. This suggests that the stellar cluster is far from quick disruption and stands as a candidate for surviving as a future globular cluster system.

We guess that the spread among observation epochs did not allow us to derive first order orbital parameters, which would have given us important clues about period distributions. Multiple solutions with very dissimilar periods fit our current data. More observations in further epochs should allow us to confirm the binary nature of radial velocity variables detected in this work and should allow us to derive orbital parameters and individual masses of members of the binary pairs. The success of GMOS in the MOS mode makes it a very efficient tool for discovering massive binaries at large numbers using a relatively small amount of telescope time.

References

Abt, H. A., Levy, S. G., & Gandet, T. L. 1972, *AJ*, 77, 138
Bosch, G. 1999, PhD thesis, Cambridge University
Bosch, G. & Meza, A. 2001, *RMxAAC*, 27, 11
Bosch, G., Terlevich, R., Melnick, J., & Selman, F. 1999, *A&AS*,137, 21
Bosch, G., Selman, F., Melnick, J., & Terlevich, R. 2001, *A&A*, 380, 137
Lada, C. J. 2006, *ApJ*, 640, L63
Parker, J. W. 1993, *AJ*, 106, 560

Pimms on the lawn of Wrenbury Hall.

Time for pictures!

Session VI

The Magellanic Clouds as laboratories of stellar astrophysics

The Magellanic System: Stars, Gas, and Galaxies
Proceedings IAU Symposium No. 256, 2008
Jacco Th. van Loon & Joana M. Oliveira, eds.

© 2009 International Astronomical Union
doi:10.1017/S1743921308028664

The properties of early-type stars in the Magellanic Clouds

Christopher J. Evans

UK Astronomy Technology Centre, Royal Observatory Edinburgh, Blackford Hill, Edinburgh,
EH9 3HJ, UK
email: cje@roe.ac.uk

Abstract. The past decade has witnessed impressive progress in our understanding of the physical properties of massive stars in the Magellanic Clouds, and how they compare to their cousins in the Galaxy. I summarise new results in this field, including evidence for reduced mass-loss rates and faster stellar rotational velocities in the Clouds, and their present-day compositions. I also discuss the stellar temperature scale, emphasizing its dependence on metallicity across the entire upper-part of the Hertzsprung-Russell diagram.

Keywords. stars: early-type, stars: fundamental parameters, Magellanic Clouds

1. Introduction

The prime motivation for studies of early-type stars in the Magellanic Clouds over the past decade has been to quantify the effect of metallicity (Z) on their evolution. The intense out-flowing winds in massive stars are thought to be driven by momentum transferred from the radiation field to metallic ions (principally iron) in their atmospheres; a logical consequence of this mechanism is that the wind intensities should vary with Z (Kudritzki *et al.* 1987). Monte Carlo simulations predict that, for stars with $T_{eff} > 25,000$ K, the wind mass-loss rates should scale with metallicity as $Z^{0.69}$ (Vink *et al.* 2000, 2001). This has a dramatic impact on their subsequent evolution. For example, an O-type star in the SMC should lose significantly less mass over its lifetime than a star in the Galaxy, thus retaining greater angular momentum. This could then lead to different late-phases of evolution such as the type of core-collapse supernova (SN), and offers a potential channel for long duration, gamma-ray bursts at low Z.

The Z-dependence of the *initial* rotational velocity distributions of massive stars and the importance of rotationally-induced mixing have also been active areas of research. For instance, Maeder *et al.* (1999) noted that the relative fraction of Be- to B-type stars increases with decreasing metallicity†, suggesting that this might arise from faster rotational velocities at lower Z. The recent generation of evolutionary models has explored the effects of rotational mixing (e.g., Heger & Langer 2000; Meynet & Maeder 2000), with the prediction of larger relative surface-nitrogen enhancements at faster rotation rates, and at lower Z (Maeder & Meynet 2001).

To date, we have lacked sufficient observations to explore the effects of metallicity thoroughly. Robust empirical results were needed with which to confront both the stellar wind and evolutionary models for early-type stars — here I summarise recent observations and quantitative analyses toward this objective.

---

† The sample of Maeder *et al.* comprised only one SMC cluster, NGC 330, long known to have a significant Be-fraction (e.g., Grebel *et al.* 1992) and sometimes suggested as a "pathological" case. However, new results from Martayan *et al.* (these proceedings) also find similarly large fractions for other SMC clusters.

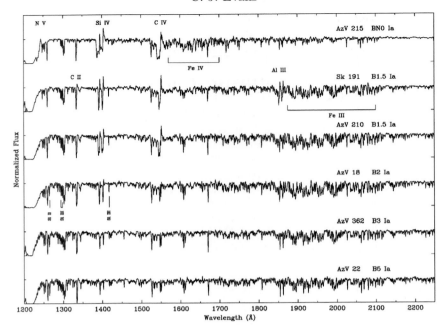

Figure 1. STIS spectra of B-type supergiants in the SMC (Evans *et al.*, 2004a). Note the distinctive (although weak) P Cygni emission in the N v, Si iv and C iv lines in AzV 215, and the iron "forests" which neatly illustrate the change in the predominant ionization stage in the early B-type domain.

2. Unique insights from ultraviolet observations

Ultraviolet (UV) spectroscopy provides an invaluable complement to optical spectroscopy in the determination of the physical parameters of early-type stars. The terminal velocity (v_∞) of the stellar wind can be measured from the saturated cores of the resonance lines, with information on the stratification and ionization of the wind provided by comparisons with model atmospheres. Unfortunately, only a handful of the brightest OB-type stars in the Clouds were within reach of high-dispersion observations with the *International Ultraviolet Explorer (IUE)*.

The *Hubble Space Telescope (HST)* and the *Far Ultraviolet Spectroscopic Explorer (FUSE)* have provided essential observations of the winds of massive stars in the Clouds. Although limited by relatively low spectral resolution, the *HST* Faint Object Spectrograph (FOS) was used in the early 1990s to observe tens of stars in the Clouds. These spectra provided morphological evidence of weaker stellar winds in the SMC when compared to Galactic standards (Walborn *et al.* 1995), as well as estimates of v_∞ (e.g., Prinja & Crowther 1998). More recently, the *HST* Space Telescope Imaging Spectrograph (STIS) and *FUSE* have both delivered spectra at resolutions of $> 10^4$, sufficient to resolve interstellar features clearly and to enable detailed comparisons with synthetic spectra.

STIS spectroscopy of 29 OB-type SMC stars was presented by Walborn *et al.* (2000) and Evans *et al.* (2004a), providing further morphological evidence for weaker wind features in SMC stars when compared to their Galactic counterparts. While the intensity of the P Cygni emission is reduced in B-type supergiants in the SMC (Fig. 1), the terminal velocities (v_∞) are not significantly slower (Evans *et al.* 2004b), consistent with the weak metallicity-dependence predicted by theory, $v_\infty \propto Z^{0.13}$ (Leitherer *et al.* 1992); in the O-type domain this relation manifests itself more clearly, e.g., Walborn *et al.* (1995).

The ratio of terminal velocity to the stellar escape velocity (v_∞/v_{esc}) for the SMC stars was used to show that the "bi-stability jump" in the behaviour of stellar winds at 21,000 K (Lamers *et al.* 1995) is a more gentle transition than previously thought (Evans *et al.* 2004b). Indeed, a more quantitative treatment of Galactic B-type supergiants found a comparable trend (Crowther *et al.* 2006), directly relating this to the distinctive change seen in the UV morphology between B0.5 and B0.7 subtypes (Walborn & Nichols-Bohlin 1987). Note that the STIS data have much broader applications as a metal-poor spectral library, useful in the context of disentangling the integrated-light observations of distant super-star-clusters (e.g., Vázquez *et al.* 2004). They have also been incorporated in the population synthesis code STARBURST99 (Leitherer *et al.* 2001).

Far-UV, high-resolution spectroscopy from *FUSE* provided access to a wealth of additional diagnostic lines (Walborn *et al.* 2002). In many cases the *FUSE* spectra have enabled the first precise measurements of v_∞ for stars previously with only *HST*-FOS or *IUE* data (e.g., Crowther *et al.* 2002; Evans *et al.* 2004c), as well as detailed studies of the wind structure and ionization (Massa *et al.* 2003). It also transpires that the P V and S IV lines in the far-UV are sensitive to the stratification of the winds, providing diagnostics of their "clumpiness" (Crowther *et al.* 2002; Evans *et al.* 2004c; Fullerton *et al.* 2006).

3. Spectral analysis with improved model atmospheres

The continued development of model atmosphere codes has also been a crucial ingredient to studies of massive stars over the past decade. The most commonly used non-LTE, line-blanketed codes that take into account spherical geometry and the effects of the stellar wind are CMFGEN (Hillier & Miller 1998; Hillier *et al.* 2003) and FASTWIND (Santolaya-Rey *et al.* 1997; Puls *et al.* 2005); both codes also include parameterizations for the effects of clumping in the wind. Compared to previous results (e.g., Vacca *et al.* 1996), these developments led to a downward revision of the temperature scale for a given spectral type (e.g., Martins *et al.* 2002; Crowther *et al.* 2002; Repolust *et al.* 2004).

In cases where the winds are less significant, plane-parallel model atmospheres from TLUSTY (Hubeny & Lanz 1995) are also used widely in the analysis of early-type stars. Even in the case of B-type supergiants, which have relatively extended atmospheres, if one excludes the features most influenced by the stellar wind (e.g., Hα and Hβ), good agreement was found in the atmospheric parameters and chemical abundances obtained from TLUSTY models compared with results obtained with FASTWIND (Dufton *et al.* 2005).

The combination of new UV data, high-resolution optical spectroscopy and improved model atmospheres led to a number of multi-wavelength analyses of individual O-type stars in the Clouds: Crowther *et al.* (2002); Hillier *et al.* (2003); Bouret *et al.* (2003); Evans *et al.* (2004c); Heap *et al.* (2006). The wind properties and chemical abundances of B-type supergiants in the SMC were also investigated (Trundle *et al.* 2004; Trundle & Lennon 2005).

While these studies began to explore some of the questions posed regarding stellar evolution in the Clouds, they lacked sufficiently large samples, in terms of the sampling of spectral types and luminosities. The FASTWIND analyses of 40 O-type stars from Massey *et al.* (2004, 2005) went some way to address the broader questions in the SMC compared to Galactic samples (see subsequent sections), but there remained a strong desire for a large, homogeneous sample which also included observations of early B-type stars to investigate the effects of rotation and rotationally-induced mixing.

4. The *VLT*-FLAMES survey of massive stars

The delivery of the FLAMES instrument to the *VLT* in 2002 was the catalyst for an *ESO* Large Programme (P.I. Smartt) to investigate the role of metallicity in the

Table 1. Present-day composition of the LMC and SMC, as traced by early B-type stars observed in the FLAMES survey. Abundances are given on the scale 12+log[X/H], with the relative fraction compared to the Solar results (Asplund *et al.* 2005) given in brackets.

Element	Solar	LMC		SMC	
C	8.39	7.73	[0.22]	7.37	[0.10]
N	7.78	6.88	[0.13]	6.50	[0.05]
O	8.66	8.35	[0.49]	7.98	[0.21]
Mg	7.53	7.06	[0.34]	6.72	[0.15]
Si	7.51	7.19	[0.48]	6.79	[0.19]
Fe	7.45	7.23	[0.51]	6.93	[0.27]

evolution of massive stars. Seven fields centred on stellar clusters were observed: NGC 3293, NGC 4755, and NGC 6611 in the Galaxy; NGC 2004 and N 11 in the LMC; NGC 330 and NGC 346 in the SMC. In total, high-resolution spectroscopy was obtained for ∼700 O- and early B-type stars (Evans *et al.* 2005, 2006). All of the observed OB-type stars have now been analysed to yield physical parameters, chemical compositions and rotational velocities, as summarised by Evans *et al.* (2008). Some of the key results from the survey are described in the rest of this review, combined with new results from other studies.

4.1. *Present-day composition of the LMC & SMC*

Determinations of chemical abundances in rapidly-rotating stars are complicated by their broadened lines—in part the reason why previous observational effort has focused mostly on narrow-lined (i.e. slowly-rotating) stars. To inform analysis of the whole FLAMES sample, the narrow-lined B-type stars ($v\sin i < 100$ km s^{-1}) were analysed first (Hunter *et al.* 2007; Trundle *et al.* 2007), yielding stellar abundances for 87 stars in the Clouds. The present-day composition of the Clouds, as traced by these B-type stars is listed in Table 1 (Mokiem *et al.* 2007b). Note that due to uncertainties in the absolute abundances, the fractions quoted for iron are relative to the Galactic results from the FLAMES survey.

Given the strong evolutionary effects on nitrogen enrichment, it is difficult to obtain the pristine value. However, the lowest abundances from the FLAMES results are in good agreement with estimates from H II regions, leading to their adoption here (see discussion by Hunter *et al.* 2007). The oxygen abundances are in excellent agreement with results from H II regions, e.g., 12+log[O/H] = 8.35 and 8.03 for the LMC and SMC, respectively, from Russell & Dopita (1992). It has been known for some time that the initial abundances of carbon and nitrogen are relatively more under-abundant than the heavier elements in the Clouds; the FLAMES results reinforce the varying fractions from element to element. Indeed, simply scaling solar abundances for quantitative work in the Clouds does not best reproduce the observed patterns.

5. Metallicity-dependent stellar winds

Analysis of O-type spectra can be a complex, time-consuming process. In addition to the usual parameters used to characterise a star (temperature, luminosity, gravity, chemical abundances), we also need to describe the velocity structure and mass-loss rate of the wind. The analysis of the FLAMES data used a semi-automated approach,

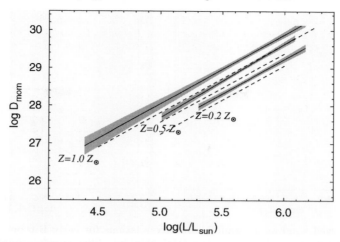

Figure 2. Comparison of the observed wind-momentum–luminosity relations (solid lines) with theoretical predictions (dashed lines) for O-type stars from Mokiem *et al.* (2007b). The upper, middle and lower relations correspond to Galactic, LMC and SMC results, respectively.

employing genetic algorithms (GA) to fit the observations with synthetic spectra from FASTWIND model atmospheres. This method was tested using a sample of Galactic stars, finding good agreement with previous results (Mokiem *et al.* 2005). The O-type spectra from the FLAMES survey were then analysed using the GA approach (Mokiem *et al.* 2006, 2007a).

The effect of metallicity on wind intensities in O-type stars was investigated by Massey *et al.* (2005). From comparisons of the wind-momenta (D_{mom}, a function of the mass-loss rate, terminal velocity and stellar radius) for a sample of 22 stars in the Clouds, Massey *et al.* found evidence for an offset with Z. This result is seen more clearly in the FLAMES results, as shown in Fig. 2, providing compelling evidence for reduced intensities at decreased metallicities. Fig. 2 also shows the theoretical predictions using the prescription from Vink *et al.* (2001). The relative separations are in good agreement, with the FLAMES results finding a scaling of $Z^{0.72-0.83}$ (with the exponent depending on assumptions regarding clumping in the winds), as compared to $Z^{0.69\pm0.10}$ from theory (Mokiem *et al.* 2007b).

There is also quantitative evidence for weaker winds in early B-type supergiants in the SMC compared to their Galactic counterparts—somewhat reassuring given that they are the direct descendants of massive O-type stars! Fig. 3 shows the SMC results from Trundle *et al.* (2004) and Trundle & Lennon (2005), compared to Galactic results from Crowther *et al.* (2006).

These observational tests are important for a number of areas—including considerations of the feedback from massive stars to the local interstellar medium in the context of star formation (see review by Oliveira, these proceedings). The reduced mass-loss rates at lower Z mean that less angular momentum is lost, i.e. an evolved star in the SMC would be expected to retain a larger fraction of its initial rotational velocity compared to a similar star in the Galaxy. Indeed, the rotational velocity distribution for the unevolved (i.e. luminosity class IV or V) SMC stars, appears to have preferentially faster velocities when compared to Galactic results (albeit limited in terms of its statistical significance; Mokiem *et al.* 2006). This could offer a potential channel for long duration, gamma-ray bursts at low metallicity (e.g., Yoon *et al.* 2006).

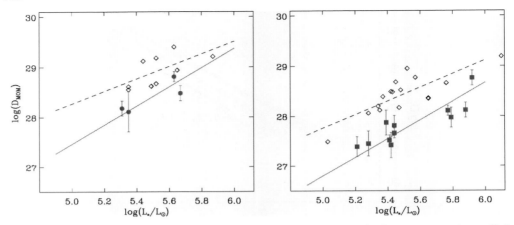

Figure 3. Observed wind-momentum–luminosity relations for early B-type supergiants (*left:* $T_{eff} > 23.5$ kK; *right:* $T_{eff} < 23.5$ kK) in the SMC (solid line, filled circles/squares) from Trundle *et al.* (2004) and Trundle & Lennon (2005), compared to Galactic results (dashed line, open diamonds) from Crowther *et al.* (2006). Figure from Trundle *et al.* (2006).

Reduced mass-loss also impacts on the end products of massive O-type stars. As summarised by Crowther (2007), in the "Conti Scenario" stars in the Milky Way can expect to pass through the following phases, depending on their initial masses (M_i):

- $M_i > 75$ $M_\odot$: O → WN(H-rich) → LBV → WN(H-poor) → WC → SNIc;
- $M_i = 40$–75 $M_\odot$: O → LBV → WN(H-poor) → WC → SNIc;
- $M_i = 25$–40 $M_\odot$: O → LBV/RSG → WN(H-poor) → SNIb.

With reduced mass-loss, the threshold to reach the WC phase in the SMC will move upwards from $40\,M_\odot$. This is reflected in the small relative number of WC to WN stars at low metallicity, as compared to the ratio seen in, e.g., the solar neighbourhood and M 31 (see Fig. 8 from Crowther 2007).

Moreover, a clearer picture is emerging of the progenitors of core-collapse supernovae, with increasing evidence that type II-P supernovae are descendant from red supergiants (see discussion by Smartt *et al.* 2008, and references therein). As such, one would expect that the ratio of type II to type Ib/c supernovae also varies with Z, with initial evidence of this recently reported by Prieto *et al.* (2008).

6. The stellar temperature scale as a function of metallicity

6.1. *O- and early B-type stars*

Stellar temperatures have long been known to depend on luminosity class—supergiants, with their lower gravities and extended atmospheres, are found to be cooler than dwarfs of the same spectral type. The inclusion of line-blanketing effects in the calculation of model atmospheres led to an overall downward revision of the stellar temperature scale (see Section 3). However, this effect is less dramatic at lower metallicities because of the diminished cumulative opacity from the metal lines—there is less "back warming" by trapped radiation, and so a hotter model is required to reproduce the observed spectral line-ratios (ionization balance). This effect is clearly seen in the temperatures obtained for O-type stars in the SMC, which are hotter than those found for Galactic stars with the same spectral type (Massey *et al.* 2005; Mokiem *et al.* 2006; Fig. 2). The temperatures for stars in the LMC are seen to fall neatly between the SMC and Galactic results (Mokiem

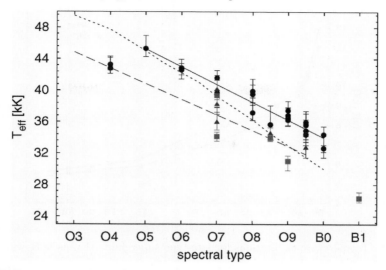

Figure 4. Effective temperatures for the O-type SMC stars analysed by Mokiem *et al.* (2006). The solid line is the fit to the SMC dwarfs, compared to the calibration for Galactic dwarfs from Martins *et al.* (2005). The dotted line is the SMC scale from Massey *et al.* (2005) for luminosity class V and III stars.

et al. 2007a). A similar *Z*-dependence has also been seen for the first time in the early B-type stars observed with FLAMES (Trundle *et al.* 2007).

6.2. *Late-type supergiants*

The classification of O- and early B-type stars is based, primarily, on the relative line-ratios of different ionization stages of the same element (helium and also silicon). The *Z*-dependence of the temperature scale is a consequence of the abundance effects in the model atmosphere calculations.

For evolved, luminous supergiants there is a more direct effect of the metal abundances on the stellar temperature scale. For instance, the primary classification criterion in the A-type domain is the ratio of the Ca K line to the blend of the Ca H and the Hϵ line. Thus, a cooler temperature is needed to reproduce the criterion for a given spectral type at SMC metallicity than in a Galactic star. This effect is the origin of the cooler temperatures reported by Venn (1999), which led her to reclassify those stars. However, the classifications should be employed on purely morphological grounds, i.e. T$_{\text{eff}}$ can be *f(Z)* at a given spectral type, but the spectral type should be independent of environment. Evans & Howarth (2003) investigated the scale of this effect, finding lower temperatures in the SMC by up to 10% (Fig. 5). This effect, although less significant, is also seen in F- and G-type supergiants.

Temperatures determined recently for M-type supergiants in the Galaxy and the Clouds also find a *Z*-dependence, again with the SMC stars cooler at a given type (Levesque *et al.* 2005, 2006, 2007). The primary classification criterion in this domain is the intensity of the TiO bands, so cooler temperatures are also required to yield the intensity necessary for a given spectral type—an effect first noted by Humphreys (1979), while attempting to explain the later types seen in M-type supergiants in the Milky Way and LMC compared to the SMC. A similar trend in temperatures is also seen between the LMC and SMC results for K-type supergiants although, somewhat intriguingly, the Galactic stars appear slightly cooler at the early K types (Fig. 6, Levesque *et al.* 2007).

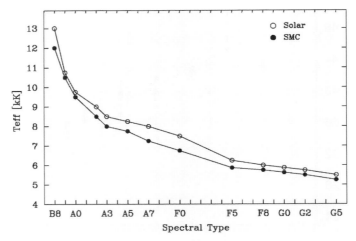

Figure 5. Temperature calibrations for late-type supergiants (Evans & Howarth 2003).

In summary, effective temperatures (for a given spectral type) are a function of metallicity over the whole of the upper Hertzsprung-Russell diagram—from the most massive O-type stars, through to the coolest M-type supergiants. The typical variations between solar and SMC metallicity are relatively small ($\sim 5-10\%$), but should be considered when adopting temperature estimates on the basis of a known spectral type.

7. Evidence for faster rotation at low Z

The first large study of the Z-dependence of stellar rotational velocities was the work by Keller (2004), who observed $\sim$100 B-type stars in the LMC and compared their rotation rates with published Galactic results. Keller found that the cluster members had mean rotational velocities that were larger than field stars (at the corresponding metallicity), and that the LMC cluster members were rotating more quickly than those in Galactic clusters (at just under a 2σ significance). Since then, a number of surveys have used FLAMES to investigate these effects further in early B-type stars (in which the effects of mass-loss are relatively small, thereby removing a further complication to the evolution of $v\sin i$).

From FLAMES observations of B- and Be-type stars in fields centred on NGC 330 and NGC 2004, Martayan et al. (2007) found evidence for faster rotational velocities at the lower metallicity of the SMC. Similar conclusions were found from the analysis of Hunter et al. (2008a; Fig. 5), in which the SMC stars are found to rotate more quickly than those in the Galaxy, with a significance at the 3σ level. Given the overlapping target fields, an independent check on the methods was possible between the results of Martayan et al. and Hunter et al., finding good agreement in the velocity distributions of normal B-type stars in the NGC 330 field.

To complicate this picture slightly, new results from additional FLAMES observations (Royer et al., this meeting) find velocity distributions in the LMC and SMC that are consistent with being drawn from the same parent population—this raises the question of whether local metallicity variations could account for the differences? Given the large global offset between the LMC and SMC, this seems relatively unlikely, but it appears that something is still eluding us at the current time!

8. Reconciling the 'Hunter diagram' with evolutionary predictions

To investigate the impact of rotation on surface nitrogen abundances in the FLAMES sample, new evolutionary models were calculated using the chemical compositions from

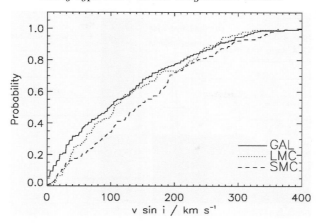

Figure 6. Cumulative distribution functions for the rotational velocities of Galactic field stars, compared with LMC and SMC results from FLAMES (Hunter et al., 2008a).

Table 1. Fig. 6 shows the nitrogen abundances, as a function of $v\sin i$ for the B-type stars in the LMC from the FLAMES survey (Hunter *et al.* 2008b). Typical uncertainties in the abundances are $\sim 0.2-0.3$ dex, so the scatter of the results indicates genuine differences in the N-enrichment in both the core-hydrogen-burning stars (dwarfs and giants, left-hand panel) and the supergiants (right-hand panel).

Two groups ("Groups 1 and 2") appear inconsistent with the predicted abundances. The dark grey (blue) points in Group 1 comprise rapidly-rotating stars that appear to have undergone little chemical mixing, and yet they have surface gravities that indicate they are near the end of core hydrogen burning. These results are at odds with the evolutionary models that predict nitrogen abundances some ~ 0.5 dex greater for the more massive stars (in which mixing is expected to be most efficient). Note that there was no evidence for binarity in the spectra of many of these stars.

The 14 (apparently single) core-hydrogen-burning stars in Group 2 are equally puzzling as they are rotating very slowly ($v\sin i < 50$ km s^{-1}), but show significant N-enrichment. For a random orientation, we could expect about two of these to be rapidly-rotating stars viewed pole-on. This seems an unlikely explanation for all 14, and Hunter *et al.* concluded that the majority are intrinsically slow rotators. Recent studies of Galactic β-Cepheid stars have found a correlation between nitrogen enrichment and magnetic fields (Morel *et al.* 2006); perhaps the enrichments in Group 2 are related to magnetic fields.

The results for the supergiants can be considered as two groups—Group 3, with relatively normal levels of N-enrichment, and Group 4, with much larger abundances. Simplistically these could be pre-RSG (Group 3) and post-RSG (Group 4) stars. However, while the abundances of the Group 4 stars are consistent with the predicted level, the models cannot reproduce their effective temperature in the Hertzsprung-Russell diagram. Some of these show evidence of binarity, so mass transfer may also be important.

So, rotationally-induced mixing appears to play a key role in the enrichment of surface nitrogen in massive stars, but it appears that there are also other processes at work, particularly at low rotational velocities—presenting new challenges to the evolutionary models.

9. Massive binaries

The effects of binarity/multiplicity on the formation and subsequent evolution of high-mass stars are a vibrant area of research. One of the key ingredients missing from

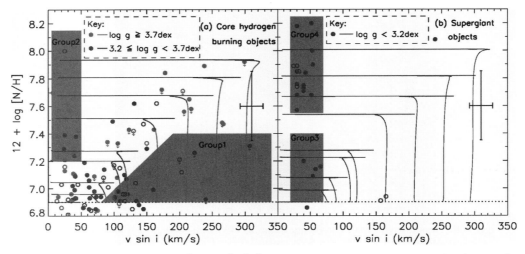

Figure 7. Nitrogen abundances (12+log[N/H]) compared to projected rotational velocities for core-hydrogen-burning (left-hand panel) and supergiant (right-hand panel) B-type stars in the LMC (Hunter *et al.* 2008b). The solid lines are new evolutionary tracks, open circles are radial velocity variables, downward arrows are upper limits, and the dotted horizontal line is the LMC baseline nitrogen abundance.

current theories of both star formation and cluster evolution is a robust binary fraction of high-mass stars (and the distribution of the relative mass-ratios in these systems). Observational effort in this area has been somewhat piecemeal to date (e.g., Sana *et al.* 2008), with Zinnecker & Yorke (2007) highlighting the need for multi-epoch radial velocity surveys of stellar clusters to provide better constraints to theoretical models of star formation.

One of the serendipitous aspects of the FLAMES survey was the large number of spectroscopic binaries discovered. The time sampling of the service-mode observations did a reasonable (but not thorough) job of binary detection in three of the fields in the Clouds, with lower limits to the binary fraction of 25–35% (Evans *et al.* 2006). As illustrated by new multi-epoch observations in 30 Doradus with GMOS (Bosch *et al.*, these proceedings), the true binary fraction in young clusters could be much larger.

10. Closing remarks

There has been huge observational and theoretical progress in our understanding of massive star evolution in the Clouds in the past decade. The intensity of stellar winds, the effective temperature scale, and the rotational velocities of OB-type stars are all found to be dependent on the metal content of their local environment.

Multi-object spectrographs (e.g., FLAMES, 2dF and GMOS) have truly opened-up our knowledge of the stellar content of the Clouds over the past decade. We are beginning to discover many new examples of the peculiar members of the 'OB Zoo' (e.g., Vz, f?p, nfp, B[e], etc.) and with future surveys we will be able to explore the evolutionary connection of these rare sub-types to the morphologically normal population—are they critical, short-lived phases that every massive star experiences?

Moreover, key questions remain regarding the binary fraction of massive stars, binary evolution, and the efficiency of rotationally-induced mixing in O-type stars. The *VLT*-FLAMES Tarantula Survey is an *ESO* Large Programme (P.I. Evans) approved in July

2008 to address these issues, via multi-epoch observations of ~1,000 stars in 30 Doradus in the LMC.

Lastly, it is worth noting that *IUE* left a large spectral archive of Galactic OB-type stars, which has continued to yield new results. While *FUSE* has observed a large number of early-type stars (giving us a valuable window on the wind parameters revealed in the far-UV), it is remarkable that high-resolution spectroscopy from *HST* exists for only ~40 stars in the Clouds. The 1200–1900 Å region is uniquely important in the study of massive stars and, given the absence of future UV missions at the advanced stages, new *HST* observations after the servicing mission would provide immense legacy value.

Acknowledgements

Grateful thanks to Nolan Walborn for careful reading of this manuscript.

References

Asplund, M., Grevesse, N. & Sauval, A. J. 2005, in: T. G. Barnes III & F. Bash (eds.), *Cosmic Abundances as Records of Stellar Evolution and Nucleosynthesis*, ASP Conf. Series, Vol. 336, p. 25

Bouret, J.-C., Lanz, T., Hillier, D. J., Heap, S. R., Hubeny, I., Lennon, D. J., Smith, L. J., & Evans, C. J. 2003, *ApJ*, 595, 1182

Crowther, P. A., Hillier, D. J., Evans, C. J., Fullerton, A. W., De Marco, O., & Willis, A. J. 2002, *ApJ*, 579, 774

Crowther, P. A., Lennon, D. J., & Walborn, N.R. 2006, *A&A*, 446, 279

Crowther, P. A. 2007, *ARAA*, 45, 177

Dufton, P. L., Ryans, R. S. I., Trundle, C., Lennon, D. J., Hubeny, I., Lanz, T., & Allende Prieto, C. 2005, *A&A*, 434, 1125

Evans, C. J. & Howarth, I. D. 2003, *MNRAS*, 345, 1223

Evans, C. J., Lennon, D. J., Walborn, N. R., Trundle, C., & Rix, S. A. 2004a, *PASP*, 116, 909

Evans, C. J., Lennon, D. J., Trundle, C., Heap, S. R., & Lindler, D. J. 2004b, *ApJ*, 607, 451

Evans, C. J., Crowther, P. A., Fullerton, A. W., & Hillier, D. J. 2004c, *ApJ*, 610, 1021

Evans, C. J., Smartt, S. J., Lee, J.-K, et al. 2005, *A&A*, 437, 467

Evans, C. J., Lennon, D. J., Smartt, S. J., & Trundle, C. 2006, *A&A*, 456, 623

Evans, C. J, Hunter, I., Smartt, S., et al. 2008, *Msngr*, 131, 25

Fullerton, A. W., Massa, D. L., & Prinja, R. K. 2006, *ApJ*, 637, 1025

Grebel, E. K., Richtler, T., & de Boer, K. S. 1992, *A&A*, 254, L5

Heap, S. R., Lanz, T., & Hubeny, I. 2006, *ApJ*, 638, 409

Heger, A. & Langer, N. 2000, *ApJ*, 544, 1016

Hillier, D. J. & Miller, D. L. 1998, *ApJ*, 496, 407

Hillier, D. J., Lanz, T., Heap, S. R., Hubeny, I., Smith, L. J., Evans, C. J., Lennon, D. J., & Bouret, J. C. 2003, *ApJ*, 588, 1039

Hubeny, I. & Lanz, T. 1995, *ApJ*, 439, 875

Humphreys, R. M. 1979, *A&A*, 231, 384

Hunter, I., Dufton, P. L., Smartt, S. J., et al. 2007, *A&A*, 466, 277

Hunter, I., Lennon, D. J., Dufton, P. L., Trundle, C., Simn-Daz, S., Smartt, S. J., Ryans, R. S. I., & Evans, C. J. 2008a, *A&A*, 479, 541

Hunter, I., Brott, I., Lennon, D. J., et al. 2008b, *A&A*, 676, L29

Keller, S. C. 2004, *PASA*, 21, 310

Kudritzki, R.-P., Pauldrach, A., & Puls, J. 1987, *A&A*, 173, 293

Lamers, H. J. G. L. M., Snow, T. P., & Lindholm, D. M. 1995, *ApJ*, 455, 269

Leitherer, C., Robert, C., & Drissen, L. 1992, *ApJ*, 401, 596

Leitherer, C., Leão, J. R. S., Heckman, T. M., Lennon, D. J., Pettini, M., & Robert, C. 2001, *ApJ*, 550, 724

Levesque, E. M., Massey, P., Olsen, K. A. G., Plez, B., Josselin, E., Maeder, A., & Meynet, G. 2005, *ApJ*, 628, 973

Levesque, E. M., Massey, P., Olsen, K. A. G., Plez, B., Meynet, G., & Maeder, A. 2006, *ApJ*, 645, 1102

Levesque, E. M., Massey, P., Olsen, K. A. G., & Plez, B. 2007, *ApJ*, 667, 202

Maeder, A., Grebel, E. K., & Mermilliod, J.-C. 1999, *A&A*, 346, 459

Maeder, A. & Meynet, G. 2001, *A&A*, 373, 555

Martayan, C., Frémat, Y., Hubert, A.-M., Floquet, M., Zorec, J., & Neiner, C. 2007, *A&A*, 462, 683

Martins, F., Schaerer, D., & Hillier, D.J. 2002, *A&A*, 382, 999

Martins, F., Schaerer, D., & Hillier, D. J. 2005, *A&A*, 436, 1049

Massa, D., Fullerton, A. W., Sonneborn, G., & Hutchings, J. B. 2003, *ApJ*, 586, 996

Massey, P., Bresolin, F., Kudritzki, R. P., Puls, J., & Pauldrach, A. W. A. 2004, *ApJ*, 608, 1001

Massey, P., Puls, J., Pauldrach, A. W. A., Bresolin, F., Kudritzki, R. P., & Simon, T. 2005, *ApJ*, 627, 477

Meynet, G. & Maeder, A. 2000, *A&A*, 361, 101

Mokiem, M. R., de Koter, A., Puls, J., Herrero, A., Najarro, F., & Villamariz, M. R. 2005, *A&A*, 441, 711

Mokiem, M. R., de Koter, A., Evans, C. J., *et al.* 2006, *A&A*, 456, 1131

Mokiem, M. R., de Koter, A., Evans, C. J., *et al.* 2007a, *A&A*, 465, 1003

Mokiem, M. R., de Koter, A., Vink, J. S., *et al.* 2007b, *A&A*, 473, 603

Morel, T., Butler, K., Aerts, C., Neiner, C., & Briquet, M. 2006, *A&A*, 457, 651

Prieto, J. L., Stanek, K. Z., & Beacom, J. F. 2008, *ApJ*, 673, 999

Prinja, R. K. & Crowther, P. A. 1998, *MNRAS*, 300, 828

Puls, J., Urbaneja, M. A., Venero, R., Repolust, T., Springmann, U., Jokuthy, A., & Mokiem, M. R. 2005, *A&A*, 435, 669

Repolust, T., Puls, J., & Herrero, A. 2004, *A&A*, 415, 349

Russell, S. C. & Doptia, M. A. 1992, *ApJ*, 384, 508

Sana, H., Gosset, E., Nazé, Y., Rauw, G., & Linder, N. 2008, *MNRAS*, 386, 447

Santolaya-Rey, A. E., Puls, J., & Herrero, A. 1997, *A&A*, 323, 488

Smartt, S. J., Eldridge, J. J., Crockett, R. M., & Maund, J. R. 2008, *MNRAS*, submitted, arXiv:0809.0403

Trundle, C., Lennon, D. J., Puls, J., & Dufton, P. L. 2004, *A&A*, 417, 217

Trundle, C. & Lennon, D. J. 2005, *A&A*, 434, 677

Trundle, C., Lennon, D. J., Puls, J., Dufton, P. L., & Evans, C. J. 2006, in: H. J. G. L. M. Lamers, N. Langer, T. Nugis, & K. Annuk (eds.), *Stellar Evolution at Low Metallicity: Mass Loss, Explosions, Cosmology*, ASP Conf. Series, Vol. 353, p. 127

Trundle, C., Dufton, P. L., Hunter, I., Evans, C. J., Lennon, D. J., Smartt, S. J., & Ryans, R.S.I. 2007, *A&A*, 471, 625

Vacca, W. D., Garmany, C. D., & Shull, J. M. 1996, *ApJ*, 460, 914

Vázquez, G. A., Leitherer, C., Heckman, T. M., Lennon, D. J., de Mello, D. F., Meurer, G. R., & Martin, C. L. 2004, *ApJ*, 600, 162

Venn, K. A. 1999, *ApJ*, 518, 405

Vink, J. S., de Koter, A., & Lamers, H. J. G. L. M. 2000, *A&A*, 362, 295

Vink, J. S., de Koter, A., & Lamers, H. J. G. L. M. 2001, *A&A*, 369, 574

Walborn, N. R. & Nichols-Bohlin, J. 1987, *PASP*, 99, 40

Walborn, N. R., Lennon, D. J., Haser, S. M., Kudritzki, R.-P., & Voels, S. A. 1995, *PASP*, 107, 104

Walborn, N. R., Lennon, D. J., Heap, S. R., Lindler, D. J., Smith, L. J., Evans, C. J., & Parker, J. Wm. 2000, *PASP*, 112, 1243

Walborn, N. R., Fullerton, A. W., Crowther, P. A., Bianchi, L., Hutchings, J. B., Pellerin, A., Sonneborn, G., & Willis, A.J. 2002, *ApJS*, 141, 443

Yoon, S.-C., Langer, N., & Norman, C. 2006, *A&A*, 460, 199

Zinnecker, H. & Yorke, H. W. 2008, *ARAA*, 45, 481

The Magellanic System: Stars, Gas, and Galaxies
Proceedings IAU Symposium No. 256, 2008
Jacco Th. van Loon & Joana M. Oliveira, eds.

Stellar evolution models at the Magellanic Cloud metallicities

Raphael Hirschi[1,2], Sylvia Ekström[3], Cyril Georgy[3], Georges Meynet[3] and André Maeder[3]

[1]Astrophysics group, Keele University, Lennard-Jones Lab., Keele, ST5 5BG, UK
email: r.hirschi@epsam.keele.ac.uk

[2]IPMU, University of Tokyo, Kashiwa, Chiba 277-8582, Japan

[3]Observatoire Astronomique de l'Université de Genève, CH-1290, Sauverny, Switzerland

Abstract. The Magellanic Clouds are great laboratories to study the evolution of stars at two metallicities lower than solar. They provide excellent testbeds for stellar evolution theory and in particular for the impact of metallicity on stellar evolution. It is important to test stellar evolution models at metallicities lower than solar in order to use the models to predict the evolution and properties of the first stars. In these proceedings, after recalling the effects of metallicity, we present stellar evolution models including the effects of rotation at the Magellanic Clouds metallicities. We then compare the models to various observations (ratios of sub-groups of massive stars and supernovae, nitrogen surface enrichment and gamma-ray bursts) and show that the models including the effects of rotation reproduce most of the observational constraints.

Keywords. stars: mass loss, stars: rotation, supernovae: general, stars: Wolf-Rayet, galaxies: evolution, Magellanic Clouds, gamma rays: bursts

1. Introduction

Massive stars play a key role in the evolution of galaxies via radiative, kinetic and chemical feedback. A current hot topic is the evolution and properties of the first massive stars since these stars took part in the assembly of the first structures in the Universe. The first massive stars died a long time ago and only via simulations can we predict their properties. The simulations can be constrained by observing the chemical signature of these stars in halo low-mass extremely metal-poor stars (Beers & Christlieb 2005) and by comparing stellar evolution models with observations of massive stars at metallicities lower than solar. The Magellanic Clouds (MCs) are the best place to constrain massive star models since they contain numerous young stars with initial metallicities as low as one fifth solar and they are close enough to observe stars individually. In Sect. 2, we recall the main effects of metallicity. In Sect. 3, we compare models with observations at MC metallicities and conclude in Sect. 4.

2. The impact of metallicity

The metallicity (Z) has several effects on the properties of massive stars (see for example Heger *et al.* 2003; Chieffi & Limongi 2004; Meynet *et al.* 1994; Mowlavi *et al.* 1998). First, low-Z massive stars are more compact due to lower opacity. Second, mass loss is generally weaker at low Z. The metallicity (Z) dependence of mass-loss rates is usually described using the formula:

$$\dot{M}(Z) = \dot{M}(Z_\odot)(Z/Z_\odot)^\alpha \qquad (2.1)$$

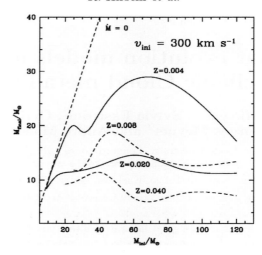

Figure 1. Final mass versus initial mass of models at different metallicities. As the metallicity decreases, mass loss decreases and therefore the final mass increases. At the high mass end, mass loss remains important even at $Z_{\rm SMC}$ ($\simeq 0.004$) and the final mass is much smaller than when mass loss is ignored (straight dashed line).

The exponent α varies between $0.5-0.6$ (Kudritzki & Puls 2000, Kudritzki 2002) and $0.7-0.86$ (Vink *et al.* 2001; Vink & de Koter 2005) for O-type and WR stars respectively (See Mokiem *et al.* 2007 for a recent comparison between mass-loss prescriptions and observed mass-loss rates). As a result of the weak mass loss at low Z, the final mass increases at low Z as shown in Fig. 1.

Until very recently, most models use at best the total metal content present at the surface of the star to determine the mass-loss rate. However, the surface chemical composition becomes very different from the solar mixture, due either to mass loss in the WR stage or by internal mixing (convection and rotation) after the main sequence. It is therefore important to know the contribution from each chemical species to opacity and mass loss. Recent studies (Vink *et al.* 2000; Vink & de Koter 2005) show that iron is the dominant element concerning radiation line-driven mass loss for O-type and WR stars. In the case of WR stars, there is however a plateau at low metallicity due to the contributions from light elements like carbon, nitrogen and oxygen (CNO). In between the hot and cool parts of the HR-diagram, mass loss is not well understood. Observations of the LBV stage indicate that extremely high mass loss may take place during eruptions (see, e.g., Smith *et al.* 2003) and there is no indication of a metallicity dependence (see for example Pustilnik *et al.* 2008). In the red supergiant (RSG) stage, the rates generally used are still those of Nieuwenhuijzen & de Jager (1990). More recent observations indicate that there is a very weak dependence of dust-driven mass loss on metallicity. Van Loon *et al.* (2005) provide recent mass-loss-rate prescriptions in the RSG stage. From a theoretical point of view, for dust formation, nucleation seed components like silicon and titanium are dominant (van Loon 2000; van Loon 2006; Ferrarotti & Gail 2006). The ratio of carbon to oxygen is important to determine which kind of molecules and dust form. If the ratio of carbon to oxygen is larger than one, then carbon-rich dust would form, and more likely drive a wind since they are more opaque than oxygen-rich dust at low metallicity (Höfner & Andersen 2007).

How do rotation induced processes vary with metallicity? The surface layers of massive stars usually accelerate due to internal transport of angular momentum from the core to the envelope. Since at low Z, stellar winds are weak, this angular momentum dredged up by meridional circulation remains in the star, and the star more easily reaches critical rotation. At the critical limit, matter can easily be launched into a keplerian disk, which probably dissipates under the action of the strong radiation pressure of the star. The efficiency of meridional circulation (dominating the transport of angular momentum) decreases towards lower Z because the Gratton-Öpik term of the vertical velocity of the outer cell is proportional to $1/\rho$. On the other hand, shear mixing (dominating the mixing of chemical elements) is more efficient at low Z. Indeed, the star is more compact and therefore the gradients of angular velocity are larger and the mixing timescale (proportional to the square of the radius) is shorter. This leads to stronger internal mixing of chemical elements at low Z (Meynet & Maeder 2002).

3. Comparison between models and observations

There are many observational constraints that stellar evolution models need to successfully reproduce. In this section, we present models including the effects of rotation and show how they can reproduce several of these constraints.

3.1. *Sub-types of massive stars and supernovae*

The first test is the ratio of subtypes of massive stars and supernovae. Meynet & Maeder (2005) show that rotating models well reproduce the ratio of WR to O-type stars, which tests the lifetime of the different stages of stellar evolution. The ratio of type Ib or Ic SNe to type II provides constraints on the initial mass ranges of stars finishing their life giving rise to these different supernova types. Prieto *et al.* (2008) present new observational values for the variation with the metallicity of the number ratio (SN Ib+SN Ic)/SN II to which theoretical predictions can be compared. We discuss here the predictions of single star models for the type Ib/Ic supernovae frequency. Since these supernovae do not show any H-lines in their spectrum, they should have as progenitors stars having removed *at least* their H-rich envelope by stellar winds, i.e. their progenitors should be WR stars of the WNE type (stars with no H at their surface and presenting He and N lines) or of the WC/WO type. Considering that all models ending their lifetime as a WNE or WC/WO phase will explode as a type Ibc supernova, it is possible to compute the variation with the metallicity of the number ratio of type Ibc to type II supernovae.

The result is shown in Fig. 2. One sees that this ratio increases with the metallicity. This is due to the fact that the minimum initial mass of stars ending their life as WNE, or WC/WO stars decreases with increasing metallicity. Single rotating star models can reasonably well reproduce the observed trend with the metallicity. They however give slightly too small values with respect to the observations, and the difference may be explained by the binary stars contribution to the formation of SN Ibc. Models accounting for single and binary channel (but without rotation) are shown as a dotted line (Eldridge *et al.* 2008). They provide a good fit to the observations. But in that case, most of the supernovae originate from the binary channel, leaving little place for the single star scenario. These models would also predict that most of the WR stars are the outcome of close binary evolution, which is not confirmed by observations (see Foellmi *et al.* 2003a,b). Most likely, both the single and binary channels contribute. We also computed (SN Ib + SN Ic)/ SN II ratios with the assumption that all models massive enough to form a black hole (BH) do not produce a SN. Comparing the theoretical prediction with this assumption (dashed line in Fig. 2) with the observed rates, we see that in the case no supernova event occurs when a BH is formed, single star models might still account for a significant fraction of the type Ibc supernovae for $Z > 0.02$, whereas at low Z, no

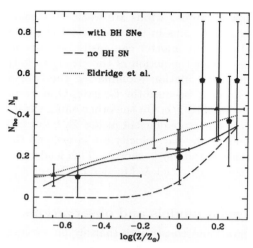

Figure 2. Rate of SN Ibc / SN II if all models produce a SN (solid line) or if models producing a black hole do not explode in a SN (dashed line). Pentagons are observational data from Prieto *et al.* (2008), and triangles are data from Prantzos & Boissier (2003). The dotted line represents the binary models of Eldridge *et al.* (2008).

contribution is expected from single stars. It is however likely that some stars forming black holes produce a SN.

3.2. *Nitrogen surface enrichment*

Surface enrichment in helium and nitrogen are good probes of internal mixing in main sequence stars. While non-rotating models predict no such enrichment before the first dredge-up occurring in the red supergiant stage, rotating models predict enrichment already during the main sequence. This is shown in Fig. 3, where the nitrogen enrichment is plotted along the z axis and the HR diagram in the x-y plane.

The recent FLAMES survey of massive stars (see contribution by Evans) has provided a wealth of data on massive stars (mass loss, initial composition, surface rotational velocities and chemical enrichment, etc.). Hunter *et al.* (2008) study the correlation between surface nitrogen enrichment and rotation velocities and find that a majority of stars in their sample are consistent with models of single rotating stars. They also find two groups of stars that apparently disagree with the models: 1) a group of fast rotating stars showing little nitrogen enrichment and 2) a group of slowly rotating stars with high nitrogen enrichment. They mention binarity and magnetic fields as possible sources of discrepancies between the observations and the models. The surface enrichment is not only a function of the rotational velocity, it also depends on the mass, the chemical composition, the evolutionary stage and the multiplicity. These observations have recently been re-analysed by Maeder *et al.* (2008) who show that when the variation of the other intervening parameters is limited, a good agreement between models of single rotating stars and observations is found, and the two discrepant groups disappear or are significantly reduced.

3.3. *Gamma-Ray Bursts progenitors and Oe/Be stars*

Gamma-Ray Bursts have now been firmly connected with type Ic SNe (see Woosley & Bloom 2006 for a review). Models including the effects of rotation are able to reproduce the frequency and the upper metallicity limit of observed GRBs assuming that only type Ic SNe are able to produce GRBs (Hirschi *et al.* 2005). However, these models (not

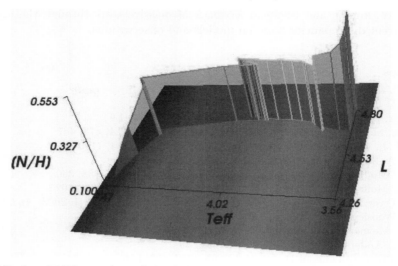

Figure 3. Surface [N/H] ratio (z axis) divided by a factor ten along the evolutionary track in the HR diagram (x-y plane) for a rotating 15 M$_\odot$ model at solar metallicity. Rotating models predict surface enrichment in nitrogen and helium already on the main sequence whereas non-rotating models predict no enrichment until the first dredge-up.

including the effects of magnetic fields) overestimate the initial rotation rate of pulsars. The inclusion of the effects of magnetic fields according to Spruit (2002) allows a better reproduction of the initial pulsar periods (Heger *et al.* 2005). Along with gravity waves, magnetic fields are one possible cause for the flat rotation profile of the Sun (Eggenberger *et al.* 2005). Although it becomes much harder for the core of massive stars to retain enough angular momentum until the core collapse, there is still an evolutionary scenario, the so-called chemically homogeneous evolution, leading to the production of fast rotating cores at the pre-SN stage and therefore enabling MHD explosions and GRBs (see Yoon *et al.* 2006; Woosley & Heger 2006). The theoretical GRB event rates obtained by Yoon *et al.* (2006) are in good agreement with observations apart from the upper metallicity limit, which is lower than the observed one (Modjaz *et al.* 2008). Although the upper Z limit is still an open issue, it is very likely that critically rotating massive stars like Oe and Be stars are the progenitors of GRBs. Ekström *et al.* (2008) study in detail the evolution of stars towards the critical limit and theoretical results compare well with observations of Oe/Be stars (see Martayan *et al.* 2007 and contribution by Martayan in this volume).

4. Conclusion and outlook

We have shown that models of massive stars including the effects of rotation are able to reproduce many observable constraints at the Magellanic Cloud metallicities. These models are therefore apt to simulate the evolution of the first massive stars and predict their impact on the early evolution of the Universe. There are still open issues like the upper Z limit for GRBs or the importance of binarity and magnetic fields. Large surveys like the FLAMES survey of massive stars are very useful to quantitatively constrain stellar evolution models and we will certainly learn a lot in the near future. New grids of models are underway at metallicities ranging from solar down to zero in order to predict the radiative, kinetic and chemical feedback of massive stars through the ages and to resolve current discrepancies between models and observations.

References

Beers, T. C. & Christlieb, N. 2005, *ARAA*, 43, 531

Chieffi, A. & Limongi, M. 2004, *ApJ*, 608, 405

Eggenberger, P., Maeder, A., & Meynet, G. 2005, *A&A*, 440, L9

Ekström, S., Meynet, G., Maeder, A., & Barblan, F. 2008, *A&A*, 478, 467

Eldridge, J. J., Izzard, R. G., & Tout, C. A. 2008, *MNRAS*, 384, 1109

Ferrarotti, A. S. & Gail, H.-P. 2006, *A&A*, 447, 553

Foellmi, C., Moffat, A. F. J., & Guerrero, M. A. 2003a, *MNRAS*, 338, 360

Foellmi, C., Moffat, A. F. J., & Guerrero, M. A. 2003b, *MNRAS*, 338, 1025

Heger, A., Fryer, C. L., Woosley, S. E., Langer, N., & Hartmann, D. H. 2003, *ApJ*, 591, 288

Heger, A., Woosley, S. E., & Spruit, H. C. 2005, *ApJ*, 626, 350

Hirschi, R., Meynet, G., & Maeder, A. 2005, *A&A*, 443, 581

Höfner, S. & Andersen, A. C. 2007, *A&A*, 465, L39

Hunter, I., Brott, I., Lennon, D. J., *et al.* 2008, *ApJ*, 676, L29

Kudritzki, R.-P. 2002, *ApJ*, 577, 389

Kudritzki, R.-P. & Puls, J. 2000, *ARAA*, 38, 613

Maeder, A., Meynet, G., Ekström, S., & Georgy, C. 2008, in Comm. in Asteroseismology, Contribution to the Proceedings of the 38th LIAC, in press (arXiv.0810.0657)

Martayan, C., Floquet, M., Hubert, A. M., *et al.* 2007, *A&A*, 472, 577

Meynet, G. & Maeder, A. 2002, *A&A*, 390, 561

Meynet, G. & Maeder, A. 2005, *A&A*, 429, 581

Meynet, G., Maeder, A., Schaller, G., Schaerer, D., & Charbonnel, C. 1994, *A&AS*, 103, 97

Modjaz, M., Kewley, L., Kirshner, R. P., *et al.* 2008, *AJ*, 135, 1136

Mokiem, M. R., de Koter, A., Vink, J. S., *et al.* 2007, *A&A*, 473, 603

Mowlavi, N., Meynet, G., Maeder, A., Schaerer, D., & Charbonnel, C. 1998, *A&A*, 335, 573

Nieuwenhuijzen, H. & de Jager, C. 1990, *A&A*, 231, 134

Prieto, J. L., Stanek, K. Z., & Beacom, J. F. 2008, *ApJ*, 673, 999

Prantzos, N. & Boissier, S. 2003, *A&A*, 406, 259

Pustilnik, S. A., Tepliakova, A. L., Kniazev, A. Y., & Burenkov, A. N. 2008, *MNRAS*, 388, L24

Smith, N., Gehrz, R. D., Hinz, P. M., *et al.* 2003, *AJ*, 125, 1458

Spruit, H. C. 2002, *A&A*, 381, 923

van Loon, J. Th. 2000, *A&A*, 354, 125

van Loon, J. Th. 2006, in H. J. G. L. M. Lamers, N. Langer, T. Nugis, & K. Annuk (eds.), *Stellar Evolution at Low Metallicity: Mass Loss, Explosions, Cosmology*, ASP Conf. Ser. 353, p. 211

van Loon, J. Th., Cioni, M. -R. L., Zijlstra, A. A., & Loup, C. 2005, *A&A*, 438, 273

Vink, J. S. & de Koter, A. 2005, *A&A*, 442, 587

Vink, J. S., de Koter, A., & Lamers, H. J. G. L. M. 2000, *A&A*, 362, 295

Vink, J. S., de Koter, A., & Lamers, H. J. G. L. M. 2001, *A&A*, 369, 574

Woosley, S. E. & Bloom, J. S. 2006, *ARAA*, 44, 507

Woosley, S. E. & Heger, A. 2006, *ApJ*, 637, 914

Yoon, S. -C., Langer, N., & Norman, C. 2006, *A&A*, 460, 199

The Magellanic System: Stars, Gas, and Galaxies
Proceedings IAU Symposium No. 256, 2008
Jacco Th. van Loon & Joana M. Oliveira, eds.

A survey of the most massive stars
in the Magellanic Clouds

Alceste Z. Bonanos

Carnegie Institution of Washington, 5241 Broad Branch Road, Washington, DC 20015, USA
email: bonanos@dtm.ciw.edu

Abstract. Despite the large impact very massive stars ($> 30\,M_\odot$) have in astrophysics, their fundamental parameters remain uncertain. I present results of a survey aiming to characterize the most massive stars in the Magellanic Clouds. The survey targets the brightest, blue, eclipsing binaries discovered by the OGLE microlensing survey, for which masses and radii are measured to 5%. Such precise data are rare and provide constraints for theories of massive star formation and evolution at low metallicities.

Keywords. binaries: eclipsing, binaries: spectroscopic, stars: early-type, stars: fundamental parameters, stars: individual (LMC-SC1-105), Magellanic Clouds

1. Introduction

The fundamental parameters of very massive stars ($\geqslant 30\,M_\odot$) remain uncertain, despite the large impact massive stars have in astrophysics, both individually and collectively (see review by Massey 2003). The equations of stellar structure allow for stars with arbitrarily large masses, however the mechanisms to form massive stars (accretion and mergers; e.g., Bally & Zinnecker 2005) and the associated instabilities (see Elmegreen 2000; Zinnecker & Yorke 2007 and references therein) are not well understood, hindering theoretical predictions on the existence of an upper limit on the stellar mass. The "mass discrepancy" problem, i.e. the disagreement between masses derived from parameters determined by fitting stellar atmosphere models to spectra and from evolutionary tracks (see e.g., Massey *et al.* 2000; Repolust *et al.* 2004, for a comparison), still affects studies of single massive stars, even though significant progress has been made in both stellar atmosphere (see review by Herrero 2008) and stellar evolution models (e.g., Meynet & Maeder 2003). The parameters of single stars also suffer from suspected multiplicity, which in many cases cannot be determined.

The only model-independent way to obtain accurate fundamental parameters of distant massive stars and to resolve the "mass discrepancy" problem is to use eclipsing binaries (see review by Andersen 1991). In particular, double-lined spectroscopic binary systems exhibiting eclipses in their light curves are extremely powerful tools for measuring masses and radii of stars. The most massive stars measured in eclipsing binaries are galactic Wolf-Rayet stars of WN6ha spectral type: NGC 3603-A1 ($M_1 = 116\pm31\,M_\odot$, $M_2 = 89\pm16\,M_\odot$; Schnurr *et al.* 2008), and WR 20a ($M_1 = 83.0 \pm 5.0\,M_\odot$ and $M_2 = 82.0 \pm 5.0\,M_\odot$; Rauw *et al.* 2004; Bonanos *et al.* 2004) in Westerlund 2, presenting a challenge for both stellar evolution and massive star formation models (Yungelson et al. 2008; Zinnecker & Yorke 2007) and raising the issue of the frequency and origin of "binary twins" (Pinsonneault & Stanek 2006; Krumholz & Thompson 2007). Such systems are of particular interest, since massive binaries might be progenitors of gamma-ray bursts (e.g., Fryer *et al.* 2007), especially in the case of Population III, metal-free stars (see Bromm & Loeb 2006).

Analogs of these heavyweight champions, if not more massive binaries, are bound to exist in the young massive clusters at the center of the Galaxy (Center, Arches, Quintuplet), in nearby super star clusters (e.g., Westerlund 1, R 136), in Local Group galaxies (e.g., LMC, SMC, M 31, M 33) and beyond (e.g., M 81, M 83, NGC 2403). A systematic wide-ranging survey of these clusters and galaxies is currently underway. The goal is to provide data with which to test star formation theories, stellar atmosphere and stellar evolution models for both single and binary stars as a function of metallicity, and the theoretical predictions on the upper limit of the stellar mass. The adopted strategy involves two steps: a variability survey to discover eclipsing binaries in these massive clusters and nearby galaxies, which is followed by spectroscopy to derive parameters of the brightest — thus most luminous and massive — blue systems. Figure 1 illustrates the extent of our knowledge of precise fundamental parameters of massive stars. It presents published mass-radius measurements from eclipsing binaries, accurate to better than 10% for the more massive component. This Table consists of only 14 very massive stars with better than 10% mass-radius measurements, located in 3 galaxies. Of these, WR 20a and

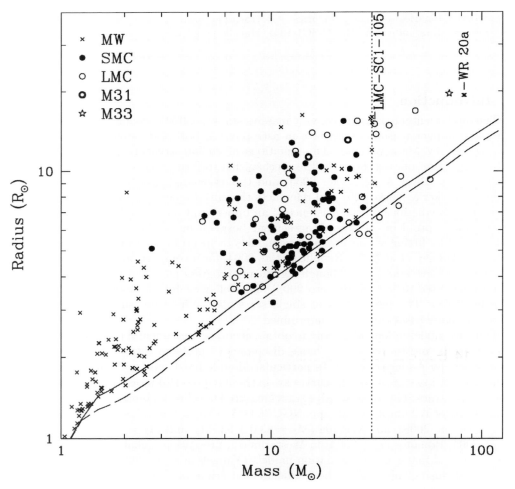

Figure 1. Mass and radius determinations of stars in eclipsing binaries, accurate to $\leqslant 10\%$ and complete $\geqslant 30\,\mathrm{M_\odot}$ from the literature. The solid line is the $Z = 0.02$ ZAMS from Schaller *et al.* (1992); the dashed line is the $Z = 0.008$ ZAMS from Schaerer *et al.* (1993).

M 33 X-7 (Orosz *et al.* 2007) are the most massive and noteworthy. M 33 X-7 contains a very massive 70.0 ± 6.9 M$_\odot$ O-type giant and a record-breaking 15.65 M$_\odot$ black hole, challenging current evolutionary models, which fail to explain such a large black hole mass. Without accurate measurements for a large sample of massive stars, theoretical models will remain unconstrained.

2. LMC-SC1-105

A survey to determine accurate parameters for several massive eclipsing binaries in the low metallicity ($Z = 0.008$) LMC was undertaken, with the purpose of increasing the sample and improving our understanding of these rare systems. Several candidates were selected from the OGLE-II catalog of eclipsing binaries in the LMC (Wyrzykowski *et al.* 2003) as the brightest systems with $B - V < 0$ mag. LMC-SC1-105, or OGLE 053448.26−694236.4, has $I_{max} = 13.04$ mag, $V_{max} = 12.97$ mag, $B_{max} = 12.81$ mag and a preliminary semi-detached classification (Figure 2). A total of 9 spectra of LMC-SC1-105 near

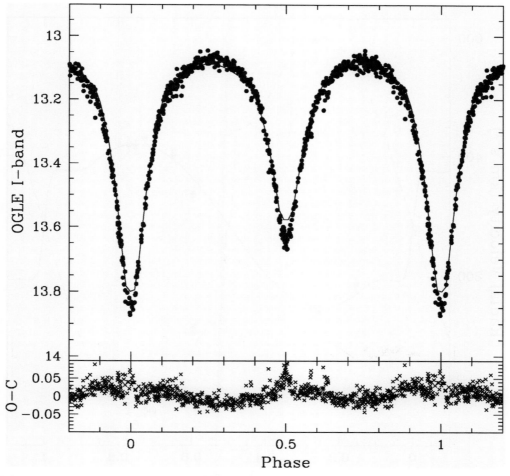

Figure 2. Phased OGLE I−band light curve of LMC-SC1-105. The best fit model from PHOEBE (solid curve) assumes a semi-detached configuration with the secondary filling its Roche lobe. The residuals suggest the presence of an accretion stream and hot spots (not modeled), arising from mass-transfer onto the primary.

quadrature phases were acquired over 4 runs on 2 telescopes at *Las Campanas Observatory*, Chile. Inspection of the quadrature spectra reveals that LMC-SC1-105 exhibits the *"Struve-Sahade effect"*. This term refers to the variable strength of the spectral lines of the secondary star (or primary star in some cases) in a double-lined spectroscopic binary (see Howarth *et al.* 1997 and references therein). Following the criteria of Walborn & Fitzpatrick(1990), the spectral types of the primary and secondary are O8 V and O8 III–V at phase 0.75, respectively. The Struve-Sahade effect causes the spectral types of both stars to change: at phase 0.25 the stars appear to have types O7V and O8.5 III-V.

Radial velocities were measured from the spectra via two dimensional cross correlation (Zucker & Mazeh 1994) and PHOEBE (Prša & Zwitter 2005) was used to derive the light curve parameters (Figure 3). The final values for the masses and radii are $M_1 = 30.9 \pm 1.0\,M_\odot$ and $R_1 = 15.1 \pm 0.2\,R_\odot$ for the primary, and $M_2 = 13.0 \pm 0.7\,M_\odot$ and $R_2 = 11.9 \pm 0.2\,R_\odot$ for the secondary. The semi-detached configuration of LMC-SC1-105 with the less massive star filling its Roche lobe, along with the main sequence classification of the primary and possible (sub)giant classification of the secondary, point to the system being

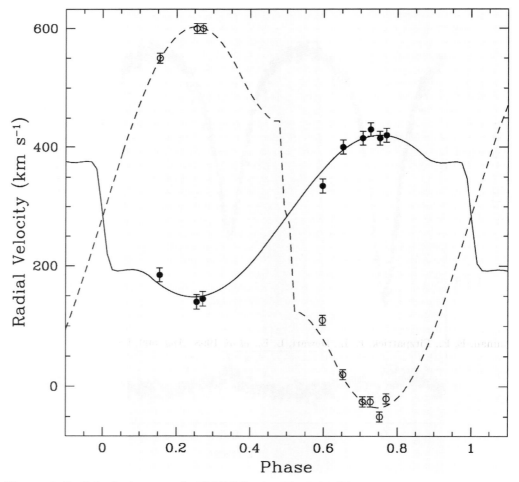

Figure 3. Radial velocity curve for LMC-SC1-105. The TODCOR measurements are shown as filled circles for the primary and open circles for the secondary; overplotted is the best fit model from PHOEBE, denoted by a solid line for the primary and a dashed line for the secondary.

in a slow-mass-transfer stage of case A binary evolution (see Bonanos 2009 for further details).

3. Conclusions

The parameters of LMC-SC1-105 were determined from the light curves available from the OGLE and MACHO surveys and newly acquired high resolution spectroscopy that targeted quadrature phases. The system was found to contain a very massive main sequence primary (30.9 ± 1.0 M$_\odot$) and a possibly evolved Roche lobe-filling secondary. The spectra display the Struve-Sahade effect, which is present in all the He I lines, causing the spectral classification to change with phase, and could be related to the mass transfer occurring in the system. LMC-SC1-105 could further be used as a distance indicator to the LMC. However, in addition to accurate radii, accurate flux (i.e. effective temperatures) and extinction estimates are necessary for accurate distances. Eclipsing binaries have been used to derive accurate and independent distances to the LMC (e.g., Guinan *et al.* 1998), the SMC (Harries *et al.* 2003; Hilditch *et al.* 2005), M 31 (Ribas *et al.* 2005) and most recently to M 33 (Bonanos *et al.* 2006).

The accurate parameters determined for LMC-SC1-105 contribute valuable data on very massive stars, increasing the current sample of 14 very massive stars with accurate parameters to 15, which despite their importance remain poorly studied. Such data serve as an external check to resolve the "mass discrepancy" problem, as Burkholder *et al.* (1997) have shown, and to constrain stellar atmosphere, evolution and formation models. Further systematic studies of massive binaries in nearby galaxies are needed to extend the sample of 50 SMC eclipsing binaries (Harries *et al.* 2003; Hilditch *et al.* 2005) to higher masses and metallicities and populate the sparsely sampled parameter space (mass, metallicity, evolutionary state) with accurate measurements of their masses and radii. The method of targeting very massive stars in bright blue eclipsing binaries can therefore be employed towards this goal.

References

Andersen, J. 1991, *AAPR*, 3, 91
Bally, J. & Zinnecker, H. 2005, *AJ*, 129, 2281
Bonanos, A. Z., Stanek, K. Z., Udalski, A., *et al.* 2004, *ApJ*, 611, L33
Bonanos, A. Z., Stanek, K. Z., Kudritzki, R. P., *et al.* 2006, *ApJ*, 652, 313
Bonanos, A. Z. 2009, *ApJ*, 691, 407
Bromm, V. & Loeb, A. 2006, *ApJ*, 642, 382
Burkholder, V., Massey, P., & Morrell, N. 1997, *ApJ*, 490, 28
Elmegreen, B. G. 2000, *ApJ*, 539, 342
Fryer, C. L., Mazzali, P. A., Prochaska, J., *et al.* 2007, *PASP*, 119, 1211
Guinan, E. F., Fitzpatrick, E. L., Dewarf, L. E., *et al.* 1998, *ApJ*, 509, L21
Harries, T. J., Hilditch, R. W., & Howarth, I. D. 2003, *MNRAS*, 339, 157
Herrero, A. 2008, *RMxAC*, 33, 15
Hilditch, R. W., Howarth, I. D., & Harries, T. J. 2005, *MNRAS*, 357, 304
Howarth, I. D., Siebert, K. W., Hussain, G. A. J., *et al.* 1997, *MNRAS*, 284, 265
Krumholz, M. R. & Thompson, T. A. 2007, *ApJ*, 661, 1034
Massey, P. 2003, *ARAA*, 41, 15
Massey, P., Waterhouse, E., & DeGioia-Eastwood, K. 2000, *AJ*, 119, 2214
Meynet, G. & Maeder, A. 2003, *A&A*, 404, 975
Orosz, J. A., McClintock, J. E., Narayan, R., *et al.* 2007, *Nature*, 449, 872
Pinsonneault, M. H. & Stanek, K. Z. 2006, *ApJ*, 639, L67
Prša, A. & Zwitter, T. 2005, *ApJ*, 628, 426
Rauw, G., De Becker, M., Nazé, Y., *et al.* 2004, *A&A*, 420, L9

Repolust, T., Puls, J., & Herrero, A. 2004, *A&A*, 415, 349

Ribas, I., Jordi, C., Vilardell, F., *et al.* 2005, *ApJ*, 635, L37

Schaerer, D., Charbonnel, C., Meynet, G., *et al.* 1993, *AAPS*, 102, 339

Schaller, G., Schaerer, D., Meynet, G., & Maeder, A. 1992, *AAPS*, 96, 269

Schnurr, O., Casoli, J., Chené, A. -N., Moffat, A. F. J., & St-Louis, N. 2008, *MNRAS*, 89, 38

Walborn, N. R. & Fitzpatrick, E. L. 1990, *PASP*, 102, 379

Wyrzykowski, Ł., Udalski, A., Kubiak, M., *et al.* 2003, *Acta Astronomica*, 53, 1

Yungelson, L. R., van den Heuvel, E. P. J., Vink, J. S., *et al.* 2008, *A&A*, 477, 223

Zinnecker, H. & Yorke, H. W. 2007, *ARAA*, 45, 481

Zucker, S. & Mazeh, T. 1994, *ApJ*, 420, 806

Some more Pimms. . .

The Magellanic System: Stars, Gas, and Galaxies
Proceedings IAU Symposium No. 256, 2008
Jacco Th. van Loon & Joana M. Oliveira, eds.

The WFI Hα spectroscopic survey of the Magellanic Clouds: Be stars in SMC open clusters

Christophe Martayan[1,2], Dietrich Baade[3] and Juan Fabregat[4]

[1] Royal Observatory of Belgium, 3 avenue circulaire 1180 Brussels, Belgium
email: martayan@oma.be

[2] GEPI, Observatoire de Paris, CNRS, Université Paris Diderot; 5 place Jules Janssen 92195 Meudon Cedex, France

[3] ESO – European Organisation for Astronomical Research in the Southern Hemisphere, Karl-Schwarzschild-Str. 2, D-85748 Garching b. Muenchen, Germany
email: dbaade@eso.org

[4] Observatorio Astronómico de Valencia, edifici Instituts d'investigació, Poligon la Coma, 46980 Paterna Valencia, Spain, email: Juan.Fabregat@uv.es

Abstract. At low metallicity, B-type stars show lower loss of mass and, therefore, angular momentum so that it is expected that there are more Be stars in the Magellanic Clouds than in the Milky Way. However, till now, searches for Be stars were only performed in a very small number of open clusters in the Magellanic Clouds. Using the *ESO*/WFI in its slitless spectroscopic mode, we performed a Hα survey of the Large and Small Magellanic Cloud. Eight million low-resolution spectra centered on Hα were obtained. For their automatic analysis, we developed the ALBUM code. Here, we present the observations, the method to exploit the data and first results for 84 open clusters in the SMC. In particular, cross-correlating our catalogs with OGLE positional and photometric data, we classified more than 4000 stars and were able to find the B and Be stars in them. We show the evolution of the rates of Be stars as functions of area density, metallicity, spectral type, and age.

Keywords. techniques: spectroscopic, surveys, stars: emission-line, Be, galaxies: individual (SMC), Magellanic Clouds, galaxies: star clusters

1. Introduction

Emission line stars (ELS) range from young to evolved stars (T Tauri, Herbig Ae/Be, WR, Planetary Nebulae, etc.), from hot to cool stars (Classical Be star, Oe, Supergiants star, Mira Ceti, Flare stars, etc.). Among the ELS, here, we focus on the classical Be stars. They are non-supergiant B type stars, which have displayed at least once emission lines in their spectra, mainly in the Balmer series of hydrogen. The emission lines come from a circumstellar decretion disk formed by episodic matter ejection of the central star. It appears that the Be phenomenon is related to fast rotation and probably additional properties such as non-radial pulsation or magnetic fields. For a comprehensive review of Be stars in the Milky Way we refer the reader to Porter & Rivinius (2003). It seems also that low metallicity plays a role (Kudritzki *et al.* 1987): at low metallicity, typical of the Small Magellanic Cloud (SMC), the stellar radiatively driven winds are less efficient than at high metallicity (typical of the Milky Way, MW), thus the mass loss is lower and the stars keep more angular momentum. As a consequence, B-type stars rotate faster in the SMC/LMC than in the MW (Martayan *et al.* 2007).

It is then expected that metallicity has also an effect on the Be-phenomenon itself as reported by Maeder *et al.* (1999) or Wisniewski & Bjorkman (2006), while the evolutionary phase could also play a role (Fabregat & Torrejón 2000). In most cases, the study of open clusters was done by using photometric observations, in the MW. The work by McSwain & Gies (2005) is a typical example. To test these issues and improve the comparisons between SMC and MW, spectroscopic observations of stars in open clusters have to be done. In the SMC, a survey of emission line objects was performed by Meyssonnier & Azzopardi (1993) using photographic plates. However, they were not able to study the stars in open clusters but mainly in the field. This paper deals with our slitless spectroscopic survey of ELS and Oe/Be/Ae stars in SMC open clusters, while in the whole SMC 3 million spectra were obtained.

2. Observations, data-reduction

To increase the number of open clusters studied in the SMC (1 to 6 in the previous photometric studies of Maeder *et al.* 1999 or Wisniewski & Bjorkman 2006), to improve the statistics, and to quantify the evolution of the rates of Be stars to B stars with decreasing metallicity, we performed a slitless spectroscopic survey of the SMC. The observations were obtained on September 25, 2002 with the *ESO* Wide Field Imager (WFI, see Baade *et al.* 1999) at the 2.2-m *MPG Telescope* located at La Silla in Chile. We used its slitless spectroscopic mode with the R50 grism and a narrow filter centered on Hα to reduce the crowding of observed areas. We recall that this kind of instrumentation is not sensitive to the diffuse ambient nebulosities and does not allow weak emission lines to be found. With 10-minutes exposures with WFI, it is possible to detect emission lines with EW ⩾ 10 Å or with a relative intensity to the continuum equal to 2 down to V magnitudes around 17.5 mag. For fainter stars, due to the noise in the spectra, only stronger emission lines are within reach. For an example of this kind of observations, see the study by Martayan *et al.* (2008a) in the MW NGC 6611 open cluster and the Eagle Nebula.

The CCD image treatment was performed using IRAF tasks. The spectra extraction was done using the SExtractor code (Bertin & Arnouts 1996). The treatment of spectra and search for stars with emission were done by using the ALBUM code (see Martayan *et al.* 2008b).

Details about the samples used in this study are given in Table 1.

3. Results

After data-reduction, the objects were sorted by categories (definite or candidate ELS, absorption stars), and the astrometry of the stars was performed with an accuracy of 0.5″ with the ASTROM package of Wallace & Gray (2003).

Classification of stars

We then cross-matched our catalogues with OGLE photometric catalogues in SMC open clusters from Udalski *et al.* (1998), Pietrzyński & Udalski (1999), and Udalski (2000), in order to obtain the B, V, I magnitudes of the stars, and the reddening and age of open clusters. With this information, the absolute magnitudes and the dereddened colour indices of the stars were derived.

Several open clusters are found with high ratios of Be to B stars, occasionally even higher than in NGC 330, an open cluster already known for its rich Be content. In order to abstract from the large variations from one open cluster to another in their frequency

Table 1. Details and comparison of samples used in this study.

	SMC (this study)	MW from McSwain & Gies (2005)
Number open clusters	84	54
Definite Be stars	109	52
Candidate Be stars	54	116
Total calssical Be stars	163	168
Ae stars	7	57
Oe stars	6	3
Other ELS (not MS)	90	
Unclassified ELS[1]	49	
NGC 346[2]	54	
B stars	1384	1741
other stars	2683	508

Notes:
[1] No OGLE photometry available for these stars.
[2] NGC 346 is a complex young SMC open cluster, which contains both classical Be stars and Herbig Be/Ae stars but also other kinds of ELS (WR, T Tauri). Because of the risk of possible confusions, the classical Be stars from this cluster are not included in the statistics of Be stars.

of Be stars, the stars were grouped in a global sample. We classified the 4300 stars in our sample by using the calibration provided by Lang (1992). Fig. 1 shows the corresponding global HR colour/magnitude diagrams. From these results it seems that the SMC sample is complete for the ELS till spectral types B3–B4, while for absorption stars it is complete till B5.

Ratios of Be to B stars vs. metallicity
To highlight and quantify a potential trend of the rates of Be stars to B stars with decreasing metallicity, we compared these ratios by spectral-type categories with the study of McSwain & Gies (2005) in the MW. The same calibration for the classification of the stars was used in the MW. The comparison of the rates is shown in Fig. 2. Down to the completness limit of our study of ELS, one can see that the ratios are several times higher in the SMC than in the MW. Thus, there is an impact of the metallicity on the number of Be stars probably corresponding to the increase of the rotational velocities in the SMC in comparison with the MW.

Distribution of Be stars with spectral types.
The distribution of Be stars by spectral type categories in the SMC and MW was also studied. In the MW, the results from McSwain & Gies (2005) in open clusters, and from Zorec & Frémat (2005) in the field, were used. The result is shown in Fig. 3. It appears that the maximum of Be stars is reached at the spectral type B2 in the SMC and MW. There is also another peak at B5–B6, which can be seen in the MW studies (complete for the whole B-main sequence). The first peak corresponds to the maximum of the emission intensity, the second one to the combined effects of the decrease of the emission intensity and the increase of the initial mass function of stars towards late type stars.

Distribution of open clusters with Be stars vs. age
From preliminary results it appears that predominantly young open clusters (age < 100 Myr) are found to contain classical Be (CBe) stars, but depending on the types of the stars, certain CBe stars could have reached the terminal-age main sequence (TAMS). For example in the case of B0e stars, the TAMS is reached in ~ 10 Myr. Some other old

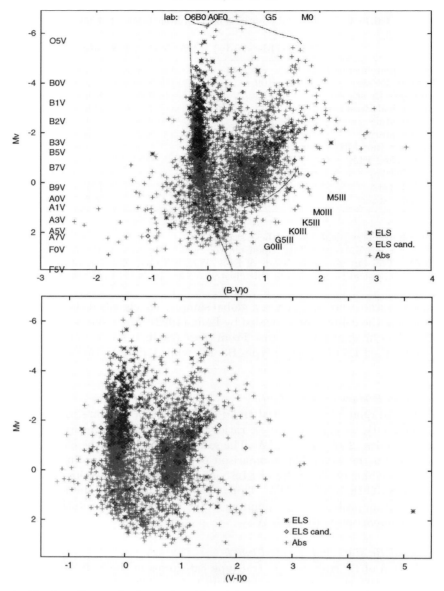

Figure 1. Absolute V magnitude vs. dereddened $(B-V)$ or $(V-I)$ color (respectively top and bottom) for SMC stars of our sample. Blue crosses correspond to definite ELS, blue diamonds correspond to candidate ELS, and red "+" to absorption stars.

open clusters (age > 100 Myr) host CBe stars. From these results, it seems that certain CBe could be born as CBe stars and some others appear during their evolution. This point needs more investigations star by star.

4. Conclusions

Preliminary studies have shown a trend of the increase of the fraction of Be stars with decreasing metallicity. However, up to now, the studies were only performed on a very limited sample of open clusters (less than 10 in the SMC) and with photometric data.

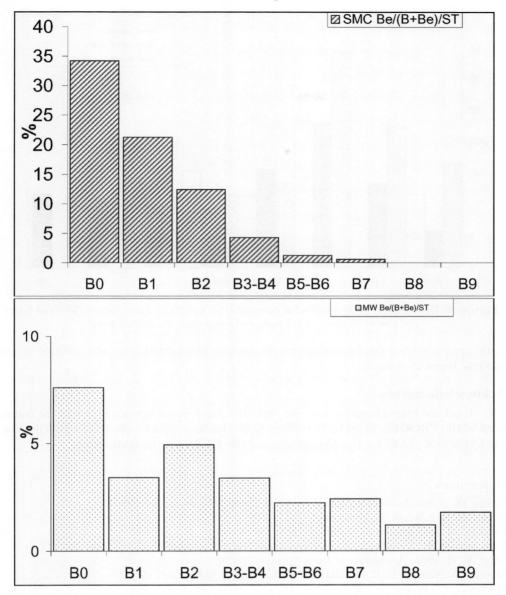

Figure 2. Ratios of definite Be stars to B stars as a function of spectral type in the SMC (top), and in the MW (bottom, from McSwain & Gies 2005).

Using the *ESO*/WFI in its slitless spectroscopic mode, we observed 84 open clusters in the SMC. Thanks to different codes and OGLE data, we were able to find and classify the emission line stars and absorption stars (~ 4300 stars). The ratios of Be stars to B stars in the SMC and MW were studied. The comparison allows to quantify the increase of the number of Be stars with decreasing metallicity. Be stars in the SMC are 2 to 4 times more abundant than in the MW depending on the spectral types. It seems also that early Be stars follow the same distribution in the SMC and MW with a maximum at the spectral type B2. About the stellar phases at which Be stars appear, from our preliminary results, it seems that certain Be stars could be born as Be stars, while others

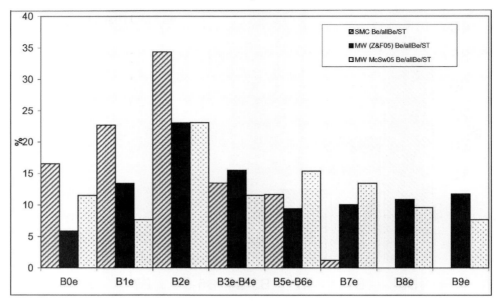

Figure 3. Distribution of Be stars with spectral type in SMC (open clusters, hatched bars)
and the MW (field, filled bars and open clusters, dotted bars).

could appear during the main sequence depending also probably on the metallicity and
spectral types of stars.

Acknowledgements

C.M. acknowledges funding from the ESA/Belgian Federal Science Policy in the framework of the PRODEX program (C90290). C.M. thanks support from ESO's DGDF 2006, the IAUS SOC/LOC for the IAU grant, and the FNRS for the travel grant.

References

Baade, D., Meisenheimer, K., Iwert, O., *et al.* 1999, *The Messenger*, 95, 15
Bertin, E. & Arnouts, S. 1996, *A&AS*, 117, 393
Fabregat, J. & Torrejón, J. M. 2000, *A&A*, 357, 451
Kudritzki, R. P., Pauldrach, A., & Puls, J. 1987, *A&A*, 173, 293
Lang, K. R. 1992, *Astrophysical Data*, Springer Verlag
Maeder, A., Grebel, E. K., & Mermilliod, J. -C. 1999, *A&A*, 346, 459
Martayan, C., Frémat, Y., Hubert, A. -M., *et al.* 2007, *A&A*, 462, 683
Martayan, C., Floquet, M., Hubert, A.-M., *et al.* 2008a, *A&A*, 489, 469
Martayan, C., Baade, D., Hubert, A. -M., *et al.* 2008b, *The 2007 ESO instrument calibration workshop*, p. 595
McSwain, M. V. & Gies, D. R. 2005, *ApJS*, 161, 118
Meyssonnier, N. & Azzopardi, M. 1993, *A&AS*, 102, 451
Pietrzyński, G. & Udalski, A. 1999, *AcA*, 49, 157
Porter, J. M. & Rivinius, T. 2003, *A&A*, 115, 1153
Udalski, A., Szymański, M., Kubiak, M., *et al.* 1998, *AcA*, 48, 147
Udalski, A. 2000, *AcA*, 50, 279
Wallace, P. T. & Gray, N. 2003, *User guide of ASTROM*
Wisniewski, J. P. & Bjorkman, K. S. 2006, *ApJ*, 652, 458
Zorec, J. & Frémat, Y. 2005, *SF2A-2005: Semaine de l'Astrophysique Française*, p. 361

The Magellanic System: Stars, Gas, and Galaxies
Proceedings IAU Symposium No. 256, 2008
Jacco Th. van Loon & Joana M. Oliveira, eds.

© 2009 International Astronomical Union
doi:10.1017/S1743921308028706

A comprehensive study of the link between star-formation history and X-ray source populations in the SMC

Vallia Antoniou[1,2], Andreas Zezas[1] and Despina Hatzidimitriou[2]

[1]Harvard-Smithsonian Center for Astrophysics, 60 Garden St., Cambridge, MA 02138, USA
email: vantoniou@cfa.harvard.edu

[2]Physics Department, University of Crete, P.O. Box 2208, GR-710 03, Heraklion, Crete, Greece

Abstract. Using *Chandra*, *XMM-Newton* and optical photometric catalogs we study the young X-ray binary (XRB) populations of the Small Magellanic Cloud (SMC). We find that the Be/X-ray binaries (Be-XRBs) are observed in regions with star-formation (SF) rate bursts $\sim$ 30–70 Myr ago, which coincides with the age of maximum Be-star formation, while regions with strong but more recent SF (e.g., the Wing) are deficient in Be-XRBs. Using the 2dF spectrograph of the *Anglo-Australian Telescope* (*AAT*) we have obtained optical spectra of 20 High-Mass X-ray Binaries (HMXBs) in the SMC. All of these sources were proved to be Be-XRBs. Similar spectral-type distributions of Be-XRBs and Be field stars in the SMC have been found. On the other hand, the Be-XRBs in the Galaxy follow a different distribution than the isolated Be stars in the Galaxy, in agreement with previous studies.

Keywords. stars: emission-line, Be, stars: formation, galaxies: individual (SMC), Magellanic Clouds, X-rays: binaries

1. Introduction: why observe the SMC?

The SMC is the best target to study a, as complete as possible, XRB population. Similar studies in the Galaxy are hampered by extinction and distance uncertainties. The Large Magellanic Cloud (LMC) is much more extended than the SMC, requiring large area coverage to obtain sufficient numbers of XRBs, while other Local Group galaxies are too far to reach the quiescent population of HMXBs (typical $L_X \sim 10^{32-34}$ erg s^{-1}; van Paradijs & McClintock 1995). This way we are able to construct the luminosity function of the HMXBs in the central region of the SMC (see Zezas *et al.*, these proceedings), and compare these luminosity functions with state of the art XRB synthesis models (e.g., Belczyński *et al.* 2008) and luminosity functions in other star-forming galaxies. In addition, the SMC hosts a large number of HMXBs (e.g., Haberl & Pietsch 2004; McBride *et al.* 2008; Antoniou *et al.* 2008a).

2. X-ray study of the SMC

Using the ACIS-I detector on board *Chandra* we observed 5 fields (P.I. A. Zezas) in the central part of the SMC (the so called SMC "bar"), with typical exposure times of 8–12 ks. These observations yielded a total of 158 sources, down to a limiting luminosity of $\sim 4 \times 10^{33}$ erg s^{-1} (in the 0.7–10 keV band), reaching the luminosity range of quiescent HMXBs. The analysis of the data, the source-list and their X-ray luminosity functions (XLFs) are presented in Zezas *et al.* (in preparation), while their optical counterparts and resulting classification are presented in Antoniou *et al.* (2008a).

Our *XMM-Newton* survey (P.I. A. Zezas) consists of 5 observations in the outer parts of the SMC, performed with the 3 EPIC (MOS1, MOS2 and PN) detectors in the full frame mode. The observed fields were selected to sample stellar populations in a range of ages ($\sim$ 10–500 Myr; based on the SF history of Harris & Zaritsky 2004). One of these fields was affected by high background flares and it is not included in the current study. We detected 186 sources down to a limiting luminosity of $\sim 3.5 \times 10^{33}$ erg s^{-1} in the 0.2–12 keV band. More details on the data analysis and the final source-list (including the XLFs) will be presented in Antoniou *et al.* (in preparation).

3. Candidate Be-XRBs

In order to identify the Be-XRBs that lie in our fields we first study the X-ray properties of the sources. Be-XRBs show pulsations and have hard 1–10 keV spectra (i.e. with a power-law energy index of $\Gamma < 1.6$; e.g., Yokogawa *et al.* 2003), which are signatures of accretion onto strongly magnetized neutron stars. This information is derived from X-ray spectral fits. However, for sources with small number of counts we use X-ray color–color diagrams (e.g., Prestwich *et al.* 2003).

The next step is the identification of an early (O or B) type star as the optical counterpart of these selected hard X-ray sources. We cross-correlate their position with optical photometric catalogs within a search radius calculated from the combination, in quadrature, of the astrometric uncertainty of the corresponding optical catalog, and the positional uncertainty for each X-ray source. In the present work, we have used two optical catalogs: the OGLE-II (Udalski *et al.* 1998) and the Magellanic Clouds Photometric Survey (MCPS; Zaritsky *et al.* 2002). In crowded fields, actual photometric (as well astrometric) uncertainties can be larger, due to source confusion, which we consider to be more severe in the MCPS catalog, due to the larger pixel size, and worse overall seeing. However, because of the incomplete coverage of the *Chandra* and *XMM-Newton* fields by the OGLE-II survey ($\sim$ 70% and < 40%, respectively), we supplemented the optical data with the MCPS catalog, which fully covers the observed fields.

In order to identify an early-type counterpart, we use the locus of early (O and B) type stars in the V, $B - V$ color-magnitude diagram. We define this locus by using data from the 2dF spectroscopic survey of SMC stars (the most extended such catalog available; Evans *et al.* 2004). In addition, we perform Monte-Carlo simulations and we estimate the chance-coincidence probability for the O or B type stars to be $\leqslant$ 19% for the *Chandra* fields and $\sim$ 8% for the *XMM-Newton* fields.

Following the above approach, we find 9 and 7 new candidate Be-XRBs within the *Chandra* and *XMM-Newton* fields, respectively. Moreover, our results are consistent with previous classifications in all cases of overlap (18 for the *Chandra* and 1 for the *XMM-Newton* in total; all Be-XRBs). If we add to the above numbers the confirmed and candidate Be-XRBs that lie in our fields but have not been detected in our surveys (from the latest census of Magellanic Clouds HMXBs of Liu *et al.* 2005), we have a total of 29 and 9 Be-XRBs in the *Chandra* and *XMM-Newton* fields, respectively. We note that because of the transient nature of these systems, their numbers can be considered only as lower limits, but are nonetheless representative of the relative populations in the observed fields.

4. The "overabundance" of SMC Be-XRBs with respect to the Galaxy

It is widely accepted that the SMC hosts an unusual large number of HMXBs and Be-XRBs when compared to the Galaxy (see contribution by Coe, this volume). In order to investigate this, we have to minimize the age effects or variations due to SF rate differences for populations of different ages. This is feasible by studying the Be-XRBs with respect to their related stellar populations, i.e. the ratio of Be-XRBs to OB stars within an area. For the Be-XRBs in the *Chandra* SMC fields and the Galaxy we used sources with an X-ray luminosity limit of $\sim 10^{34}$ erg s^{-1}, while for the Galaxy we only kept sources within 10 kpc of the Sun. For the Galactic HMXBs we used the compilation of Liu *et al.* (2006), while for the SMC we supplemented the catalog of Liu *et al.* (2005) with our candidate SMC Be-XRBs (see §3). The OB stars for the SMC fields are derived from the MCPS catalog (based on the V magnitude and $B - V$ color), while for the Galaxy we used the compilation of Reed (2001).

We find that Be-XRBs are ~ 2 times more common in the SMC when compared to the Galaxy, thus *there is still a residual excess that cannot be accounted for by the difference in the SF rate*. However, this residual excess can be attributed to the lower metallicity of the SMC (~ 0.2 Z$_\odot$). Population synthesis models predict a factor of ~ 3 higher numbers of HMXBs in galaxies with metallicities similar to that of the SMC, when compared to the Galaxy (Dray 2006). In addition there is observational evidence for a trend of higher proportion of Be stars in lower metallicity environments (at least in the case of younger systems — Wisniewski & Bjorkman 2006; Martayan *et al.* 2007). We thus conclude that on its own, neither the lower SMC metallicity nor the enhanced SF rate at the age of ~ 40 Myr ago, can produce the observed 'overabundance' of SMC Be-XRBs (see also Antoniou *et al.* 2008a and McBride *et al.* 2008).

5. SMC X-ray source populations as a function of age

The SMC "bar" hosts young stellar populations (typically < 100 Myr; e.g., Harris & Zaritsky 2004), and the vast majority of SMC pulsars (Galache *et al.* 2008). Shtykovskiy & Gilfanov (2007) found that the age distribution of the HMXBs peaks at ~ 20–50 Myr after the SF event, while McSwain & Gies (2005) observed a strong evolution in the fraction of Be stars with age up to 100 Myr, with a maximum at $7.4 < \log(\mathrm{age}) < 7.8$. These results motivated us to investigate the connection of the SF history of our *Chandra* and *XMM-Newton* fields with their population of Be-XRBs.

Using data from Harris & Zaritsky (2004) we derive the recent SF history in our *Chandra* and *XMM-Newton* fields, by calculating the average SF history of the MCPS regions ($\sim 12' \times 12'$) encompassed by them. We find that:
(*i*) For the *Chandra* fields, the most recent major burst peaked ~ 42 Myr ago, and it had a duration of ~ 40 Myr. In addition, there were older SF episodes (~ 0.4 Gyr ago) with lower intensity but longer duration, as well as a more recent episode (~ 11 Myr) observed only in one of the *Chandra* fields.
(*ii*) For one of our *XMM-Newton* fields, the most recent major burst occurred ~ 67 Myr ago. We also observed two fields with very young populations (most recent major burst at ~ 11 and ~ 17 Myr ago, respectively). Both these fields have additional less intense bursts ~ 67 Myr ago.

5.1. *Link between SF and the XRB populations*

In order to investigate the link between SF and the XRB populations, we calculate the average SF history for the MCPS regions ($\sim 12' \times 12'$; Harris & Zaritsky 2004) that host one or more Be-XRBs (candidate and confirmed) detected in our *Chandra* and *XMM-Newton* surveys. The SF history at each region is weighted by the encompassed number of Be-XRBs. We repeat this exercise for the MCPS regions that do not have any known Be-XRB in our surveys. The two SF histories are presented in Fig.1 (upper panel).

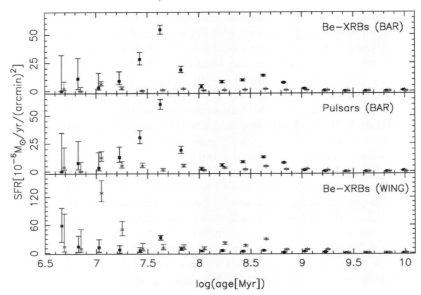

Figure 1. (*upper panel*) The average SF history for the MCPS regions (using data from Harris & Zaritsky 2004) which overlap with our *Chandra* and *XMM-Newton* fields and host one or more (shown in black) or none (shown in gray) detected Be-XRBs (candidate and confirmed). (*middle panel*) The same plot as above but for MCPS regions with and without X-ray pulsars, and (*bottom panel*) with and without Be-XRBs from the *Chandra* Wing survey (P.I. Coe, AO6). For clarity a small offset of log(age[Myr]) ~ 0.025 has been applied in the distributions of areas without Be-XRBs and/or pulsars.

Following the above comparison, we construct (middle panel of Fig.1) the SF history for the MCPS regions (overlapping with any of our fields) that host one or more known X-ray pulsars† (shown in black), and for those that do not host such sources (shown in gray). A large fraction of these pulsars also appears in the Be-XRBs sample, since most of the companions of the SMC pulsars are Be stars. However, for completeness we present both (upper and middle panel, respectively), since the pulsars are X-ray selected while the Be-XRBs used above are selected based on the optical properties of the companion stars. In total, in our *Chandra* fields lie 19 X-ray pulsars, while in the *XMM-Newton* fields only 3. As expected, the pattern in their SF history is very similar to that of Be-XRBs. In Fig.1 (bottom panel) we also present the average SF history for the MCPS regions with any (shown in black) and without (shown in grayq) Be-XRBs detected in the *Chandra* Wing survey (P.I. M. Coe, AO6). This survey covered 20 fields, however 3 of those were not used here because they do not overlap with any MCPS region, while all 4 Be-XRBs are also X-ray pulsars (Schurch *et al.* 2007).

From the above analysis we find that the number of Be-XRBs peaks for stellar populations of ages ~ 30 − 70 Myr. This is consistent with the study of McSwain & Gies (2005), who find that Be stars develop their decretion disks at ages of ~ 25–80 Myr, with a peak at ~ 35 Myr. OB stars formed during this episode are expected to reach the maximum rate of decretion disk formation at the current epoch. We also find that the other two peaks (~ 11 and ~ 422 Myr) observed in the SF history of regions with Be-XRBs do not give any Be-XRBs as expected. The first one (at ~ 11 Myr) is too early to give any pulsar

† Using the on-line census of Malcolm Coe (http://www.astro.soton.ac.uk/~mjc/ as of 06/05/2007).

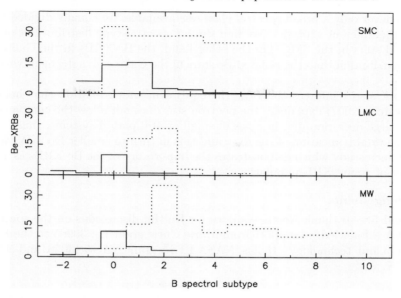

Figure 2. Comparison of the B spectral subtype distributions of Be-XRBs (solid histograms) to isolated Be stars (dashed histograms) in the SMC (top panel), the LMC (middle panel), and the Milky Way (bottom panel). Negative spectral subtypes correspond to O-type stars.

Be-XRB, while the second SF rate peak (at ~ 422 Myr) cannot result in Be-XRB formation, since by that time all OB stars have become supernovae. The peak at ~ 422 Myr ago is also observed in the global SF history of the SMC, and it temporally coincides with past perigalactic passages of the SMC with the Galaxy (Harris & Zaritsky 2004). A similar study by Shtykovskiy & Gilfanov (2007), reached the same conclusions, however these authors did not attribute the large number of SMC Be-XRBs to the timescale of maximum Be star formation.

Furthermore, the lack of a large number of Be-XRBs in the SMC Wing is consistent with the present study. As it is shown in Fig. 1 (bottom panel), the Wing has a weaker than the "bar" SF burst at the age of enhanced formation of Be stars (i.e. at ~ 42 Myr), while its most recent intense SF burst occurred only ~ 11 Myr ago. Thus we do not expect a significant number of SMC Wing Be-XRBs, at least comparable to that in the SMC 'bar'. Since in the present study we used the number of Be-XRBs from a single observation of each field and for fields covering both the SMC "bar" and the Wing, we were able to minimize the effects of the transient Be-XRB nature.

6. Optical spectroscopy of 20 SMC Be-XRBs

Using the 2dF spectrograph of the AAT we observed 20 HMXBs (Antoniou *et al.* 2008b) detected with *Chandra* (Zezas *et al.*, in prep.) and *XMM-Newton* (Haberl & Pietsch 2004). All of these sources were proved to be Be-XRBs. The spectral classification of 6 previously classified Be-XRBs have been revisited, while we estimate that our spectral types are accurate to better than ± 1 subclass in most cases, especially for earlier than B2 spectral types.

The distribution of spectral types of our Be-XRB sample shows a peak at B1.5. In Figure 2 we present the B spectral subtype distributions of Be-XRBs (solid histograms) and of isolated Be stars (dashed histograms) in the SMC (top panel), the LMC (middle panel), and the Galaxy (bottom panel). Negative spectral subtypes correspond to O-type stars. Whenever only a broad spectral class was available, we equally divided the contribution in the different subtypes. We find similar spectral-type distributions of Be-XRBs and Be field stars in the SMC. On the other hand, the Be-XRBs in the Galaxy follow a different distribution than the isolated Be stars in the Galaxy, in agreement with previous studies.

This work also reinforces the P_{orb}–Hα equivalent-width relation that holds for Be-XRBs. As Reig (2007) explained, the neutron star does not allow the companion star to develop a large decretion disc in cases of small orbital period systems, thus its presence leads to the tidal truncation of the disc, and this in turn to smaller Hα EW values. This is the first such study which demonstrates the importance of the Be-XRBs as a dominant component of young XRB populations.

Acknowledgements

We would like to thank Nolan Walborn for fruitful discussions on the spectral classification. VA acknowledges support from Marie Curie grant no. 39965 to the Foundation for Research and Technology - Hellas, NASA LTSA grant NAG5-13056, and NASA grant GO2-3117X.

References

Antoniou, V., Zezas, A., Hatzidimitriou, D., & McDowell, J. 2008a, *ApJ*, submitted

Antoniou, V., Hatzidimitriou, D., Zezas, A., & Reig, P. 2008b, *ApJ*, submitted

Belczyński, K., Kalogera, V., Rasio, F. A., Taam, R. E., Zezas, A., Bulik, T., Maccarone, T. J., & Ivanova, N. 2008, *ApJS*, 174, 223

Dray, L. M. 2006, *MNRAS*, 370, 2079

Evans, C. J., Howarth, I. D., Irwin, M. J., Burnley, A. W., & Harries, T. J. 2004, *MNRAS*, 353, 601

Galache, J. L., Corbet, R. H. D., Coe, M. J., Laycock, S., Schurch, M. P. E., Markwardt, C., Marshall, F. E., & Lochner, J. 2008, *ApJS*, 177, 189

Haberl, F. & Pietsch, W. 2004, *A&A*, 414, 667

Harris, J. & Zaritsky, D. 2004, *AJ*, 127, 1531

Liu, Q. Z., van Paradijs, J., & van den Heuvel, E. P. J. 2005, *A&A*, 442, 1135

Liu, Q. Z., van Paradijs, J., & van den Heuvel, E. P. J. 2006, *A&A*, 455, 1165

Martayan, C., Frémat, Y., Hubert, A. -M., Floquet, M., Zorec, J., & Neiner, C. 2007, *A&A*, 462, 683

McBride, V. A., Coe, M. J., Negueruela, I., Schurch, M. P. E., & McGowan, K. E. 2008, *MNRAS*, 388, 1198

McSwain, M. V. & Gies, D. R. 2005, *ApJS*, 161, 118

Prestwich, A. H., Irwin, J. A., Kilgard, R. E., Krauss, M. I., Zezas, A., Primini, F., Kaaret, P., & Boroson, B. 2003, *ApJ*, 595, 719

Reed, B. C. 2001, *PASP*, 113, 537

Reig, P. 2007, *MNRAS*, 377, 867

Schurch, M. P. E., Coe, M. J., McGowan, K. E., *et al.* 2007, *MNRAS*, 381, 1561

Shtykovskiy, P. E. & Gilfanov, M. R. 2007, *Astron. Lett.*, 33, 437

Udalski, A., Szymański, M., Kubiak, M., Pietrzyński, G., Wozniak, P., & Żebruń, K. 1998, *AcA*, 48, 147

van Paradijs, J. & McClintock, J. E. 1995, in W. H. G. Lewin, J. van Paradijs, & E. P. J. van den Heuvel (eds.), *X-ray Binaries* (Cambridge: CUP), p. 58

Wisniewski, J. P. & Bjorkman, K. S. 2006, *ApJ*, 652, 458

Yokogawa, J., Imanishi, K., Tsujimoto, M., Koyama, K., & Nishiuchi, M. 2003, *PASJ*, 55, 161

Zaritsky, D., Harris, J., Thompson, I. B., Grebel, E. K., & Massey, P. 2002, *AJ*, 123, 855

The Magellanic System: Stars, Gas, and Galaxies
Proceedings IAU Symposium No. 256, 2008
Jacco Th. van Loon & Joana M. Oliveira, eds.

Properties of X-ray binaries in the Magellanic Clouds from *RXTE* and *Chandra* observations

R. H. D. Corbet[1,2], M. J. Coe[3], K. E. McGowan[3], M. P. E. Schurch[3], L. J. Townsend[3], J. L. Galache[4] and F. E. Marshall[2]

[1] University of Maryland, Baltimore County/CRESST, 1000 Hilltop Circle, Baltimore, MD 21250, USA

[2] NASA Goddard Space Flight Center, Greenbelt, MD 20771, USA

[3] School of Physics and Astronomy, University of Southampton, SO17 1BJ, UK

[4] Harvard-Smithsonian Center for Astrophysics, 60 Garden Street, Cambridge, MA 02138, USA

Abstract. The X-ray binary population of the SMC is very different from that of the Milky Way consisting, with one exception, entirely of transient pulsating Be/neutron star binaries. We have now been monitoring these SMC X-ray pulsars for over 10 years using the *Rossi X-ray Timing Explorer* with observations typically every week. The *RXTE* observations have been complemented with surveys made using the *Chandra* observatory. The *RXTE* observations are non-imaging but enable detailed studies of pulsing sources. In contrast, *Chandra* observations can provide precise source locations and detections of sources at lower flux levels, but do not provide the same timing information or the extended duration light curves that *RXTE* observations do. We summarize the results of these monitoring programs which provide insights into both the differences between the SMC and the Milky Way, and the details of the accretion processes in X-ray pulsars.

Keywords. stars: emission-line, Be, stars: neutron, pulsars: general, Magellanic Clouds, X-rays: binaries

1. Introduction

Mass transfer in high-mass X-ray binaries (HMXBs) may occur in 3 different ways from the OB star component. (i) The mass-donor primary star may fill its Roche lobe. These systems are very luminous ($\sim 10^{38}$ erg s^{-1}) but are very rare. (ii) If the system contains a supergiant primary with an extensive stellar wind then accretion from the wind may take place. These systems have modest luminosity ($\sim 10^{36-37}$ erg s^{-1}) but are rather more common. (iii) For systems containing a Be star accretion takes place from the circumstellar envelope. These have a wide range of luminosities (10^{34-39} erg s^{-1}) and are very common, but are transient.

In most HMXBs the accreting object is a highly magnetized neutron star. Accretion is funneled onto the magnetic poles of the neutron star and we see pulsations at the neutron star spin period. If the pulse periods of HMXBs are plotted against their corresponding orbital periods then it is seen that sources divide into three groups in this diagram which correspond to the three modes of mass transfer (Corbet 1986). In particular there is strong correlation between pulse period and orbital period for the Be star systems. The positions of sources in this diagram is thought to depend on the accretion torques experienced by the neutron stars and hence on the circumstellar environments around the primary stars. These classes of HMXB are well-studied in the Galaxy and we wish to

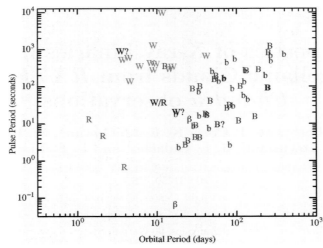

Figure 1. Pulse period vs. orbital period for HMXBs. "R" = Roche lobe overflow, "W" = wind accretion, "B" = Galactic Be star source, "b" = SMC Be star source, "β" = LMC Be star source.

know how the HXMB populations compare in other galaxies. Because of their proximity, the SMC and LMC make them the easiest external galaxies to investigate.

Initial estimates of the HMXB population of the SMC were based on the mass of the SMC. The SMC mass is a few percent of the mass of the Galaxy and about 65 Galactic X-ray pulsars are known. Therefore, 1 or 2 X-ray pulsars would be expected in the SMC. The larger fraction of Be stars in the SMC increased the estimate to ~ 3. The first X-ray pulsar discovered in the SMC was SMC X-1 in 1970s. Its luminosity can reach $\sim 10^{39}$ erg s^{-1} and it has a 0.71 s pulse period and a 3.89 day orbital period. The mass-donating companion is a Roche-lobe filling B0I star. In 1978 two transients, SMC X-2 and SMC X-3, were found (Clark *et al.* 1978). The three pulsars then known agreed with the simple prediction, although all three were surprisingly bright.

2. *RXTE* observations of the SMC

RXTE was launched in 1995 and its primary instrument is the Proportional Counter Array (PCA). The *RXTE* PCA has a 2° FWZI, 1° FWHM field of view. The PCA is non-imaging, but it has a large collecting area of up to 7,000 cm^2. The *RXTE* observing program is extremely flexible and almost all observations are time constrained. These include monitoring, phase constrained, and target of opportunity observations as well as observations coordinated with other observatories both in space and ground-based.

Serendipitous *RXTE* PCA slew observations in 1997 showed a possible outburst from SMC X-3 (Marshall *et al.* 1997). A follow-up pointed *RXTE* observation showed a complicated power spectrum with several harmonic, almost-harmonic, and non-harmonic peaks. Imaging *ASCA* observations were then made of this region and they showed the presence of two separate pulsars. However, neither of these pulsars coincided with the position of SMC X-3. A revised look at the *RXTE* power spectrum revealed three pulsars simultaneously active with periods of 46.6, 91.1, and 74.8 s (Corbet *et al.* 1998).

Since 1997 we have monitored one or more positions weekly using the *RXTE* PCA. The flexible observing program of *RXTE* has enabled us to carry out a regular monitoring

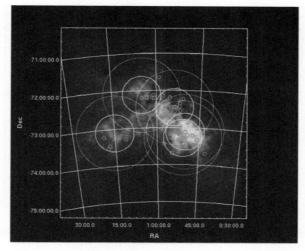

Figure 2. H I Image of the SMC. Large circles = PCA FOV (FWHM and FWZI) at different monitoring positions. Small circles show locations of X-ray pulsars.

program that would not have been possible with other satellites. The typical observation duration has been about 10,000 seconds. We use power spectra of the light curves to extract pulsed flux from any X-ray pulsars in the FOV. The sensitivity to pulsed flux is $\sim 10^{36}$ erg s^{-1} at the distance of the SMC. From this program we have detected many transient sources and all identified optical counterparts have been found to be Be stars. The SMC HMXB pulsar population has now been found by ourselves and other investigators to be much larger than originally thought. Our naming convention for SMC pulsars is SXPx, where "x" is the pulse period, for *SMC X-ray Pulsar*. This convention is particularly useful for X-ray pulsars discovered with *RXTE* for which a precise position is not yet available. For detailed light curves and analyses see Laycock *et al.* (2005) and Galache *et al.* (2008). In addition, we have recently been able to measure orbital parameters from Doppler modulation of the pulse period of SXP 18.3 (Schurch *et al.* 2008).

The Be pulsar spin period/orbital period correlation is believed to be related to the structure of the extended envelopes of Be stars. SMC and Milky Way Be stars have differences, for example, the SMC metallicity is far lower and the Be phenomenon is more common in the SMC. Is this reflected in the P_s/P_{orb} relation? That is, are there significant differences between Be star envelopes in the SMC and the Galaxy? For a linear fit (to the log—log diagram) the intercept is related to Be star mass loss rates and the gradient is related to the radial structure of Be star envelopes.

Currently 23 SMC Be X-ray pulsars now have measured orbital periods. The periods have been measured by several techniques. These include: X-ray flux monitoring with *RXTE*, pulse timing with *RXTE* (one system) and optical observations from MACHO and OGLE. In comparison, 24 Galactic Be X-ray pulsars now have measured orbital periods. We find that for the SMC and Galactic systems the intercepts are the same, the gradients are the same, and the scatter about the fits are the same. Thus, the metallicity difference between the two galaxies gives no measurable effect on the spin period/orbital period relationship and the Be star envelopes in SMC and Galaxy are apparently similar.

Galache (2006) proposes that the frequency of outbursts per orbit (X-ray "outburst density" or X_{od}) depends on the orbital period. Long period systems are more likely to show an outburst at periastron. The reason for this correlation is not yet clear.

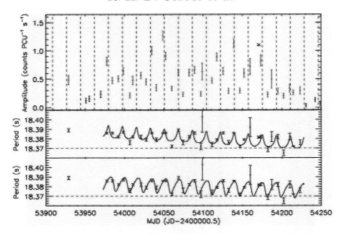

Figure 3. The extended outburst from SXP 18.3. The top panel shows the amplitude of the pulsed flux. The two lower panels show two possible timing solutions. The middle panel shows the preferred solution with the orbital period fixed at the photometric period. (Schurch *et al.* 2009).

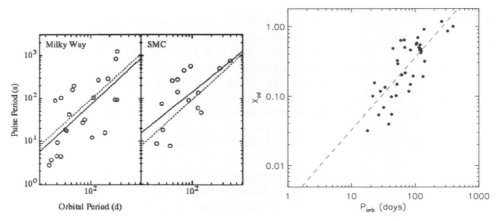

Figure 4. *Left:* A comparison of the P_s/P_{orb} relationships for the SMC and the Galaxy. *Right:* The relationship between outburst density and orbital period proposed by Galache (2006).

3. *Chandra* SMC Wing survey

A possible connection between hydrogen column density (N_H) and HMXB location was proposed by Coe *et al.* (2005). To investigate this we undertook a survey of the SMC wing using Chandra. We observed 20 fields with ~10 ks observation time per field. 523 sources were detected (McGowan *et al.* 2008a), but only ~5 of these were HMXBs (McGowan *et al.* 2007) and the majority of sources are probably background AGNs. There thus appear to be fewer X-ray pulsars in the wing than the bar. This is despite that fact that the most luminous SMC HMXB, SMC X-1 is located in the wing.

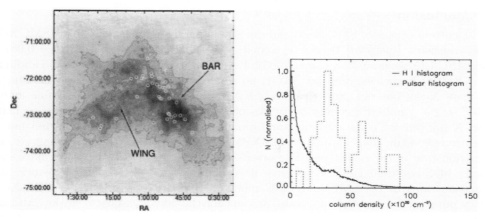

Figure 5. *Left:* The location of SMC pulsars superimposed on an H I contour map. *Right:* Histogram of SMC H I distributions and corresponding histogram of H I columns at the location of the X-ray pulsars (Coe *et al.* 2005).

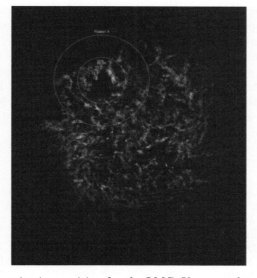

Figure 6. The PCA monitoring position for the LMC. Known pulsar positions are marked.

4. *RXTE* monitoring of the LMC

The SMC appears to be very abundant in Be X-ray pulsars. This was only known after regular observations of the SMC started. The known LMC X-ray pulsar population is more modest. There is one Roche lobe overflow source, and a few Be systems. To investigate the LMC population in more detail we undertook an *RXTE* monitoring program similar to the one used for the SMC. However, the angular size of the LMC is larger so we restricted the program to monitoring one position that was already know to contain several X-ray sources. We analyzed data from our one year monitoring program, together with archival data from other programs (Townsend *et al.*, in preparation). In the monitoring region 4 of the 5 known X-ray pulsars were detected. However, no new X-ray pulsars were discovered. This implies that the X-ray pulsar content of the LMC is more like that of the Galaxy than the SMC.

5. Conclusion

The current census of SMC X-ray pulsars is: 1 supergiant Roche lobe filler (SMC X-1); $\sim$ 50 transients (likely all Be star systems); 1 possible Crab-like pulsar ($P = 0.087$ s) from *ASCA* (Yokogawa & Koyama 2000); 1 Anomalous X-ray Pulsar (AXP) candidate ($P = 8.02$ s) from *Chandra* and *XMM*; no supergiant wind accretion systems and no low-mass X-ray binaries. Supergiant wind systems should easily be detectable at our $\sim 10^{36}$ erg s^{-1} pulsed flux sensitivity. An obvious question is: why are there so many SMC X-ray pulsars? The current star formation rate in the SMC is reported not to be extremely high. The lifetime of HMXBs is short which implies an enhanced star formation rate in the recent past. However, supergiant wind systems, which have even shorter lifetimes than Be star systems, have not been found. Models of historic star formation rates in the SMC and LMC must be compatible with the observed X-ray binary populations, and they most also account for the differences between the SMC and LMC.

There are also similarities between the SMC and Galactic pulsar populations. The SMC and Galactic Be star systems have identical (within errors) P_s/P_{orb} relationships. The LMC X-ray pulsar population also appears to be more similar to that of the Galaxy. The large and growing SMC X-ray pulsar database has considerable potential for understanding the astrophysics of accretion processes. It facilitates comparative studies, such as pulse profile morphology, as a function of luminosity. Or, luminosity effects can be removed and we can examine the effects of other parameters such as magnetic field strength. The SMC is nearby and optical counterparts can be observed with modest size telescopes. In particular, MACHO and OGLE lightcurves exist for many counterparts (e.g., Coe *et al.* this volume; McGowan *et al.* 2008b). The overall X-ray pulsar properties can tell us about the evolutionary similarities and differences of a very nearby galaxy compared to our own.

References

Clark, G., Doxsey, R., Li, F., Jernigan, J. G., & van Paradijs, J. 1978, *ApJ*, 221, L37

Coe, M. J., Edge, W. R. T., Galache, J. L., & McBride, V. A. 2005, *MNRAS*, 356, 502

Corbet, R. H. D. 1986, *MNRAS*, 220, 1047

Corbet, R. H. D., Marshall, F. E., Lochner, J. C., Ozaki, M., & Ueda, Y. 1998, *IAUC,* 6803

Galache, J. L. 2006, PhD thesis, University of Southampton

Galache, J. L., Corbet, R. H. D., Coe, M. J., Laycock, S., Schurch, M. P. E., Markwardt, C., Marshall, F. E., & Lochner, J. 2008, *ApJS*, 177, 189

Laycock, S., Corbet, R. H. D., Coe, M. J., Marshall, F. E., Markwardt, C., & Lochner, J. 2005, *ApJS*, 161, 96

Marshall, F. E., Lochner, J. C., & Takeshima, T. 1997, *IAUC*, 6777

McGowan, K. E., Coe, M. J., Schurch, M. P. E., *et al.* 2007, *MNRAS*, 376, 759

McGowan, K. E., Coe, M. J., Schurch, M. P. E., *et al.* 2008a, *MNRAS*, 383, 330

McGowan, K. E., Coe, M. J., Schurch, M. P. E., Corbet, R. H. D., Galache, J. L., & Udalski, A. 2008b, *MNRAS*, 384, 821

Schurch, M. P. E., Coe, M. J., Galache, J. L., *et al.* 2009, *MNRAS*, 392, 361

Yokogawa, J. & Koyama, K. 2000, in D. W. E. Green (ed.) *IAUC*, 7361

The Magellanic System: Stars, Gas, and Galaxies
Proceedings IAU Symposium No. 256, 2008
Jacco Th. van Loon & Joana M. Oliveira, eds.
© 2009 International Astronomical Union
doi:10.1017/S174392130802872X

Optical properties of High-Mass X-ray Binaries (HMXBs) in the Small Magellanic Cloud

M. J. Coe[1], R. H. D. Corbet[2], K. E. McGowan[1], V. A. McBride[1], M. P. E. Schurch[1], L. J. Townsend[1], J. L. Galache[3], I. Negueruela[4] and D. Buckley[5]

[1]School of Physics & Astronomy, University of Southampton, SO17 1BJ, UK

[2]University of Maryland, Baltimore County, Mail Code 662, NASA Goddard Space Flight Center, Greenbelt, MD 20771, USA

[3]Harvard-Smithsonian Center for Astrophysics, 60 Garden Street, Cambridge, MA 02138, USA

[4]Departamento de Física, Ingeniería de Sistemas y Teoría de la Señal, Universidad de Alicante, Apdo. 99, 03080 Alicante, Spain

[5]South African Astronomical Observatory, Observatory, 7935, Cape Town, South Africa

Abstract. The SMC represents an exciting opportunity to observe the direct results of tidal interactions on star birth. One of the best indicators of recent star birth activity is the presence of significant numbers of High-Mass X-ray Binaries (HMXBs) — and the SMC has them in abundance! We present results from nearly 10 years of monitoring these systems plus a wealth of other ground-based optical data. Together they permit us to build a picture of a galaxy with a mass of only a few percent of the Milky Way but with a more extensive HMXB population. However, as often happens, new discoveries lead to some challenging puzzles — where are the other X-ray binaries (e.g., black hole systems) in the SMC? And why do virtually all the SMC HMXBs have Be star companions? The evidence arising from these extensive optical observations for this apparently unusual stellar evolution are discussed.

Keywords. stars: emission-line, Be, stars: evolution, pulsars: general, galaxies: individual (SMC), Magellanic Clouds, X-rays: binaries

1. Introduction

The Be/X-ray systems represent the largest sub-class of massive X-ray binaries. A survey of the literature reveals that of the 115 identified massive X-ray binary pulsar systems (identified here means exhibiting a coherent X-ray pulse period), most of the systems fall within this Be counterpart class of binary. The orbit of the Be star and the compact object, presumably a neutron star, is generally wide and eccentric. X-ray outbursts are normally associated with the passage of the neutron star close to the circumstellar disk (Okazaki & Negueruela 2001). A detailed review of the X-ray properties of such systems may be found in Sasaki *et al.* (2003) and a review of the optical properties can be found in Coe *et al.* (2005).

Fig. 1 shows the current numbers for the different types of X-ray binary populations that are found in the Milky Way and the SMC. Since the number of LMXBs is thought to scale linearly with the mass of hydrogen in the galaxy, then the ratio of ~ 100 in masses between the two objects explains the lack of LMXBs known in the SMC. But where are the supergiant and Black Hole systems in the SMC?

We currently know of ~ 40 optically identified systems. This represents by far the largest homogeneous population of X-ray binaries in any galaxy including the Milky

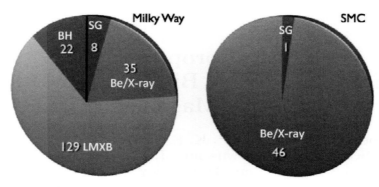

Figure 1. Relative X-ray binary populations in the SMC and the Milky Way.

Way. For all these systems OGLE & MACHO lightcurves exist for $\geqslant 10$ years enabling the confirmation of counterparts, the identification of binary periods, seeking correlated optical/X-ray flaring etc. These data are supported by follow-up spectroscopy (*SALT*, *AAT* & *ESO*) establishing spectral classes, circumstellar disk status and links into binary evolution.

2. Optical Binary modulation

Some 10–15 of the systems in the SMC have been observed to show a strong optical modulation at the binary period. Some of them also show evidence for quasi-periodic behaviour probably associated with Non Radial Pulsations (see, for example, Schmidtke & Cowley 2006). SXP 327 is an exceptional member of the SMC X-ray binary pulsar systems in that it shows a very strong optical modulation at the binary period (Coe *et al.* 2008, see Fig. 2). Another source, SXP 46.6, has also recently been shown by McGowan *et al.* (2008) to exhibit optical flaring at the same phase as X-ray outbursts, but not in the same strong and consistent manner as SXP 327. Those authors discuss the probable cause of this phenomenon as lying in the periodic disturbance of the Be stars circumstellar disk. At the time of periastron passage Okazaki & Negueruela (2001) have shown that the disk can be perturbed from its stable, resonant state with a resulting increase in surface area and, hence, optical brightness. What is very unusual about this system, SXP 327, is that there is not one, but at least two outbursts every binary cycle at phases 0.0 and 0.25 (i.e. separated by about 11 d). In addition, the average profile seems to also show a third peak at phase 0.55 — which could be close to apastron if the main peak represents periastron. Fig. 2 shows that the colours of the system reflect the optical brightness. It is obvious from this figure that the correlation between colour and flux occurs throughout the binary cycle even though it is most prominent at the time of the outbursts. The direction of the correlation is to make the system bluer when brighter — perhaps an indication of X-ray heating contributing to the colour changes.

On a longer timescale, the average optical modulation varies from year to year (see Fig. 3), probably indicative of major changes in the disk structure on the same timescales as the well-known V/R ratio changes.

3. Population evolution

There is considerable interest in the evolutionary path of High-Mass X-ray binary systems (HMXBs), and, in particular, the proper motion of these systems arising from

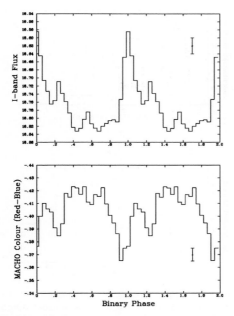

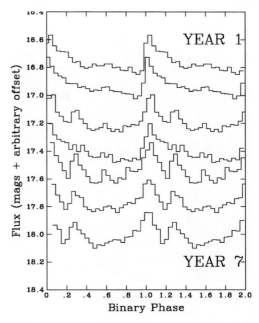

Figure 2. Strong correlated colour effects in SXP 327 when folded at binary period of 46 d. N.B. the double peaked structure.

Figure 3. Annual variation in the binary profile of the optical photometry of SXP 327 obtained from folding the OGLE III data.

the kick velocity imparted when the neutron star was created. Portegies Zwart (1995) and van Bever & Vanbeveren (1997) investigate the evolutionary paths such systems might take and invoke kick velocities of the order 100–400 km s^{-1}. In an investigation into bow shocks around galactic HMXBs Huthoff & Kaper (2002) used *Hipparcos* proper motion data to derive associated space velocities for Be/X-ray and supergiant systems. From their results, an average value of 48 km s^{-1} is found for the 7 systems that they were able to fully determine the three dimensional motion. Since this is rather lower than the theoretical values it is important to seek other empirical determinations of this motion.

Coe (2005) used a sample of 17 SMC Be/X-ray binaries to address what may be learnt about kick velocities by looking at the possible association of HMXBs in the SMC with the nearby young star clusters from which they may have emerged as runaway systems (see Fig. 4). Here we extend this work to 37 systems.

In order to determine whether the SXP sources may have originated from a nearby stellar cluster, the coordinates of the 37 SXP objects were compared to those of the RZ clusters (Rafelski & Zaritsky (2005). For every SXP its position was compared to the location of all of the RZ clusters and the identification of the nearest cluster neighbour obtained. The average distance between the pairs of objects was found to be 3.54 arcminutes. The histogram of the distances between each SXP source and the nearest RZ cluster is shown in the upper panel of Fig. 5. Obviously it is important to ensure that the SXP-RZ cluster distances are significantly closer than a sample of randomly distributed points. One way to determine this is simply to just use the RZ cluster data and find the average cluster-cluster separation. This gives a value of 6.13 arcminutes. Alternatively, the minimum distance between 100,000 random points and, in each case, the nearest RZ cluster was found. The average value was found to be 5.30 arcminutes and the corresponding histogram is shown in the lower panel of Fig. 5. From comparing the two histograms it is clear that there does exist a much closer connection between

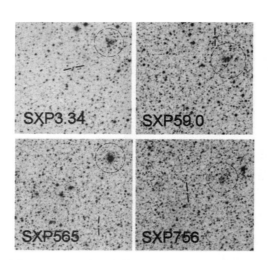

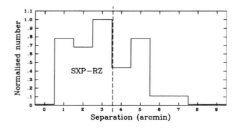

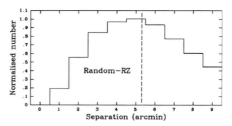

Figure 4. The fields around 4 Be/X-ray binary systems as a clue to HMXB evolution. In each case we note the presence of a nearby cluster catalogued by Rafelski & Zaritsky (2004).

Figure 5. Upper panel: histogram of the distances between each SXP and its nearest neighbour RZ cluster. Lower panel: histogram of the minimum distance between 100,000 random points and RZ clusters. In each case the dotted vertical line shows the position of the mean of the distribution.

SXP sources and RZ clusters than expected randomly. A paired student t-test of the two distributions gives a probability of only 7% that the two distributions could have been drawn from the same parent population.

Using a value of 60 kpc for the distance to the SMC, then 3.54 arcminutes corresponds to 60 pc. Savonije & van den Heuvel (1977) estimate the maximum possible lifetime of the companion Be star after the creation of the neutron star to be ~ 5 million years. So 60 pc indicates the minimum average transverse velocity of the SXP systems is 16 km s^{-1}. van den Heuvel *et al.* (2000) interpreted the Hipparcos results for galactic HMXBs in terms of models for kick velocities, and obtained values around 15 km s^{-1}.

4. Spectral classification

With the advent of arcsecond resolution X-ray telescopes the number of optically identified Be/X-ray binaries (all but one of the HMXBs in the SMC are Be/X-ray binaries) in the SMC has risen dramatically over the last few years. As there are clear differences in the numbers of HMXBs between the Milky Way and SMC, which can be ascribed to metallicity and star formation, there may be other notable differences in the populations. Most fundamentally, how do the metallicity and star formation rate reflect on the spectral distribution of the optical counterparts to the Be/X-ray binary population of the SMC?

Negueruela (1998) showed that the spectral distribution of Be stars occurring in Be/X-ray binary systems is significantly different from that of isolated Be stars in the Milky Way. Whereas isolated Galactic Be stars show a distribution beginning at the early

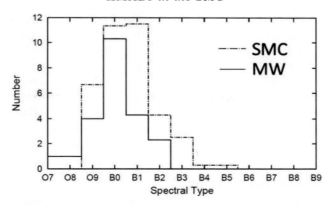

Figure 6. Spectral distribution, determined from blue spectra of ~ 40 Be/X-ray binaries in the SMC, as compared the distribution of Be/X-ray binaries in the Galaxy (McBride *et al* 2008).

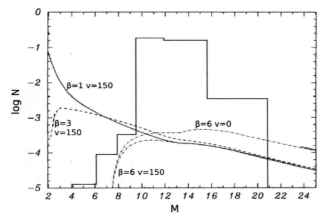

Figure 7. The absolute normalised number distribution of Be stars with a neutron star companion for four evolutionary scenarios. The solid histogram represents the spectral distribution of SMC Be/X-ray binaries from McBride *et al* (2008). v represents the supernova kick velocities (in km s^{-1}), while β represents the amount of angular momentum lost per unit mass loss from the system during evolution. Original figure from Portegies Zwart (1995).

B-types and continuing through until A0, the Be star companions of X-ray binaries show a clear cutoff near spectral type B2.

McBride *et al.* (2008) carried out detailed blue spectral observations and used these data to classify each counterpart. The spectral distribution of SMC Be/X-ray binaries is shown in Fig. 6. The distribution shows a similarity to the spectral distributions of the Galactic (Negueruela 1998) and LMC (Negueruela & Coe 2002) Be/X-ray binaries. The spectral distribution of Be/X-ray binary counterparts in the SMC peaks at spectral type B1, compared to the LMC and Galaxy distributions, which peak at B0. The Galactic and LMC distributions show a sharp cutoff at B2, whereas there are 5 SMC objects with possible spectral types beyond B2. But the exact spectral type cannot be determined with certainty in these cases. A Kolmogorov-Smirnov test of the difference between the SMC and Galactic distributions gives a K-S statistic $D = 0.22$, indicating that the null hypothesis (which is that the two distributions are the same) cannot be rejected even at significances as low as 90%. Hence, it is likely that both Galactic and SMC Be/X-ray binary counterparts are drawn from the same population.

Fig. 7 shows the arbitrarily scaled spectral distribution of SMC Be/X-ray binaries superimposed on the predicted spectral distribution of Be/X-ray binaries (from Portegies Zwart 1995). As with the Milky Way and LMC distributions, the SMC distribution cuts off around spectral type B2 (~ 8 M$_\odot$), indicating that there may be significant angular momentum losses in the binary system prior to the Be/X-ray binary evolutionary phase. A possible interpretation of the fact that there is no significant metallicity dependence of the spectral distributions of Be/X-ray binaries is that the angular momentum is lost through mechanisms other than the stellar winds of early-type components of these systems.

5. Future work and conclusions

This next year should provide us with a wealth of new high-energy and optical data of the XRB population in the SMC:

• Weekly X-ray (*RXTE*) monitoring campaign of the SMC Bar will continue for as long as possible.

• For the period July–Sep 2008 we will obtain *VLT* high resolution spectral data on 21 systems every week. We will use the detailed line profiles to study circumstellar disk structures, correlating with X-ray outbursts, as well as RV measurements to identify/confirm binary periods.

• From November 2008 till July 2009 *ESA*'s *INTEGRAL* observatory will carry out a detailed study of the whole of the SMC to a total depth of 2 Msec (equivalent to ~ 70 nights of telescope time).

So, in summary, we have an excellent homogeneous population of High-Mass X-ray Binaries. The combined optical & X-ray data are proving to be a superb laboratory for exploring X-ray binary evolutionary and accretion processes. In addition, the population as a whole has crucial differences with their partners in the Milky Way that need explaining. Finally, the High-Mass X-ray Binaries are providing us with invaluable insights into the recent history of star formation in the SMC.

References

Coe, M. J. 2005, *MNRAS*, 358, 1379
Coe, M. J., Edge, W. R. T., Galache, J. L., & McBride, V. A. 2005, *MNRAS*, 356, 502
Coe, M. J., Schurch, M., Corbet, R. H. D, Galache, J., McBride, V. A., Townsend, L. J., & Udalski, A. 2008, *MNRAS*, 387, 724
Huthoff, F. & Kaper, L. 2002, *A&A*, 383, 999
McBride, V. A., Coe, M. J., Negueruela, I., Schurch, M. P. E., & McGowan, K. E. 2008, *MNRAS*, 388, 1198
McGowan, K. E., Coe, M. J., Schurch, M. P. E., Corbet, R. H. D., Galache, J. L., & Udalski, A. 2008, *MNRAS*, 384, 821
Negueruela, I. 1998, *A&A*, 338, 505
Negueruela, I. & Coe, M. J. 2002, *A&A*, 385, 517
Okazaki, A. & Negueruela, I. 2001, *A&A*, 377, 161
Portegies Zwart, S. F. 1995, *A&A*, 296, 691
Rafelski, M. & Zaritsky, D. 2005, *AJ*, 129, 270
Sasaki, M., Pietsch, W., & Haberl, F. 2003, *A&A*, 403, 901
Savonije, G. J. & van den Heuvel, E. P. L. 1977, *ApJ*, 214, L19
Schmidtke, P. & Cowley, A. 2006, *AJ*, 132, 919
van Bever, J. & Vanbeveren, D. 1997, *A&A*, 322, 116
van den Heuvel, E. P. L., Portegies Zwart, S. F., Bhattacharya, D., & Kaper, L. 2000, *A&A*, 364, 563

The Magellanic System: Stars, Gas, and Galaxies
Proceedings IAU Symposium No. 256, 2008
Jacco Th. van Loon & Joana M. Oliveira, eds.

© 2009 International Astronomical Union
doi:10.1017/S1743921308028731

Pulsating variable stars in the Magellanic Clouds

Gisella Clementini

INAF Osservatorio Astronomico di Bologna, Via Ranzani n. 1, 40127 Bologna, Italy
email: `gisella.clementini@oabo.inaf.it`

Abstract. Pulsating variable stars can be powerful tools to study the structure, formation and evolution of galaxies. I discuss the role that the Magellanic Clouds' pulsating variables play in our understanding of the whole Magellanic System, in light of results on pulsating variables produced by extensive observing campaigns like the MACHO and OGLE microlensing surveys. In this context, I also briefly outline the promise of new surveys and astrometric missions which will target the Clouds in the near future.

Keywords. stars: oscillations, Cepheids, delta Scuti, stars: variables: other, Magellanic Clouds, cosmology: distance scale

1. Introduction

The Large Magellanic Cloud (LMC) and the Small Magellanic Cloud (SMC) represent the nearest templates where we can study the stellar populations and galaxy interactions in detail, and they are where we set up and verify the astronomical distance scale. The pulsating variable stars can play a fundamental role in this context, offering several advantages with respect to normal stars. The light variation caused by the periodic expansion/contraction of the surface layers makes the pulsating stars easier to recognize than normal stars, even when stellar crowding is severe. Their main parameter, the pulsation period, is measured with great precision, is unaffected by distance and reddening, and is directly related to intrinsic stellar quantities such as the star mass, radius, and luminosity. Among pulsating variables, the Classical Cepheids (CCs) are the brightest stellar standard candles after Supernovae. The Period–Luminosity relation (PL), for which we celebrate this year the 100th anniversary of discovery by Henrietta Leavitt, makes them primary distance indicators in establishing the cosmic distance scale. On the other hand, since pulsating variables of different types are in different evolutionary phases, they can be used to identify stellar components of different ages in the host system: the RR Lyrae stars and the Population II Cepheids (T2Cs) tracing the oldest stars ($t > 10$ Gyr); the Anomalous Cepheids (ACs) tracing the intermediate-age component (~ 4–8 Gyr); and the CCs tracing the young stellar populations (50–200 Myr). The role of pulsating stars becomes increasingly important in stellar systems like the Magellanic Clouds (MCs) where stars of different age and metal abundance share the same region of the color–magnitude diagram (CMD). The RR Lyrae stars, in particular, belonging to the oldest generation of stars, eyewitnessed the first epochs of their galaxy's formation and thus can provide hints on the early formation and assembling of the MCs system.

Our knowledge of the pulsating variable stars and the census of the MCs variables have made dramatic steps forward thanks to the microlensing surveys which, as a by-product, revealed and measured magnitudes and periods for thousands of variables in both Clouds. The overwhelming amount of information which these surveys have produced, not fully exploited yet, allowed for the first time to study the properties of primary distance

Table 1. Main properties of different types of pulsating variable stars[1].

Class	Pulsation Period (days)	M_V (mag)	Population	Evolutionary Phase
δ Cephei (CCs)[2]	$1 \div 100$[3]	$-7 \div -2$	I	Blue Loop
δ Scuti stars (δ Sc)	< 0.5	$2 \div 3$	I	MS–PMS
β Cephei	< 0.3	$-4.5 \div -3.5$	I	MS
RV Tauri	$30 \div 100$	$-2 \div -1$	I, II	post-AGB
Miras[4]	$100 \div 1000$	$-2 \div 1$	I, II	AGB
Semiregulars (SRs)[4]	> 50	$-3 \div 1$	I, II	AGB
RR Lyrae (RRL)	$0.3 \div 1$	$0.0 \div 1$	II	HB
W Virginis (T2Cs)[5]	$10 \div 50$	$-3 \div 1$	II	post-HB
BL Herculis (T2Cs)[5]	< 10	$-1 \div 0$	II	post-HB
SX Phoenicis (SX Phe)	< 0.1	$2 \div 3$	II	MS
Anomalous Cepheids (ACs)	$0.3 \div 2.5$	$-2 \div 0$	?	HB turnover

Notes:
[1] Adapted from Marconi (2001).
[2] δ Cephei variables are more commonly known as Classical Cepheids (CCs).
[3] A few CCs with periods longer than 100 days are known in both Clouds, in NGC 6822, NGC 55, NGC 300 and in I Zw 18 (Bird *et al.* 2008, and references therein). Unfortunately, CCs with $P > 100$ days are generally saturated in the OGLE and MACHO photometry. They are now being observed with smaller telescopes in order to extend the PL relation to longer periods (W. Gieren, private communication).
[4] Miras and SRs often are jointly refereed to as red variables or long period variables (LPVs).
[5] W Virginis and BL Herculis variables are often referred to as Population II or Type II Cepheids (T2Cs).

indicators such as the CCs and the RR Lyrae stars based on statistically significant numbers, as well as to reveal unknown features and new types of variables. These topics are the subject of the present review.

2. Position on the HR diagram and main properties

Table 1 presents an overview of the currently known major types of pulsating variables, along with their main characteristics (typical period, absolute visual magnitude, parent stellar population and evolutionary phase). Fig. 1 shows the loci occupied by different types of pulsating variables in the HR diagram, along with lines of constant radius and a comparison with stellar evolutionary tracks for masses in the range from 1 to 30 $M_\odot$. The two long-dashed lines running almost vertically through the diagram mark the boundaries of the so-called "classical instability strip". Going from low to high luminosities in its domain we find: δ Scuti variables (luminosity: $\log L/L_\odot \sim 1$, Spectral Type: A0–F0, mass: $M \sim 2\ M_\odot$), the RR Lyrae stars ($\log L/L_\odot \sim 1.7$, Spectral Type: A2–F2, $M < 1\ M_\odot$), and the Cepheids (ACs: $\log L/L_\odot \sim 2$, Spectral Type: F2–G6, $M \sim 1$–$2\ M_\odot$; T2Cs: $\log L/L_\odot \sim 2$, Spectral Type: F2–G6, $M \sim 0.5\ M_\odot$; and CCs: $\log L/L_\odot \sim 3 - 5$, Spectral Type: F6–K2, $M \sim 3$–$13\ M_\odot$). Once extrapolated beyond the main sequence the instability strip crosses the region of the pulsating hydrogen-rich DA white dwarfs. Red

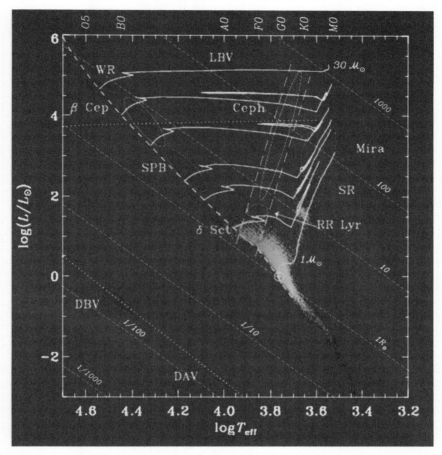

Figure 1. Position of different types of pulsating stars in the HR diagram. The heavy dashed line stretching from the upper left to the lower right is the main sequence of stars with solar abundances. Lines of constant radius (from 1/1000 to 1000 $R_\odot$) are shown, as well as tracks corresponding to masses in the range from 1 to 30 $M_\odot$. The two long-dashed lines indicate the position of the classical instability strip. Different acronyms mean: WR: Wolf-Rayet stars; LBV: luminous blue variables; SPB: slowly pulsating B stars; SR: semiregular variables; DVB, DAV: pulsating He white dwarfs (DB), and hydrogen-rich pulsating white dwarfs (DA). Adapted from Gautschy & Saio (1995).

variables (Miras and SRs) are situated instead below the red edge of the instability strip, at temperatures corresponding to spectral types K–M and luminosities $\log L/L_\odot \sim 2-4$.

Given their complex stellar populations the MCs host samples of all the various types of pulsating stars shown in Fig. 1, although in varying proportions. As a combination of evolutionary/stellar population effects, but also due to the magnitude limit and time resolution of the currently available variability surveys, up to now RR Lyrae stars, Cepheids and red variables are by far the most frequent and best studied variables in the MCs. This is shown in Fig. 2, which displays the color magnitude diagram of a region close to the LMC bar, from the variability study of Clementini *et al.* (2003), with the different types of pulsating variables (Cepheids, RR Lyrae and δ Scuti stars) marked by different symbols, and the locus of red pulsating variables outlined by a large grey box.

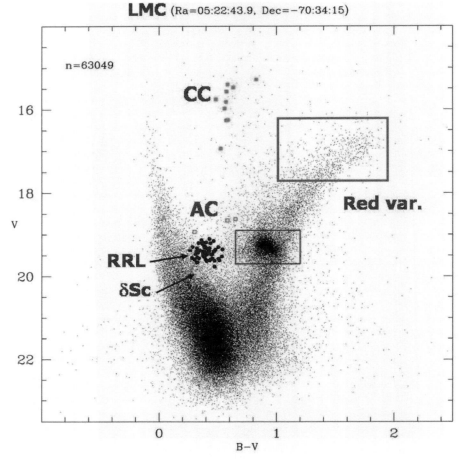

Figure 2. Position of major types of pulsating stars in the CMD of a region close to the LMC bar. Adapted from Clementini *et al.* (2003).

3. The MCs pulsating variables in numbers

Until the last decade of the twentieth century our knowledge of the MCs variables relied mainly upon photographic and photoelectric data and small and inhomogeneous samples. The situation drastically changed at the beginning of the nineties when the MACHO (http://wwwmacho.mcmaster.ca/) and EROS (http://eros.in2p3.fr/) microlensing experiments followed by OGLE (http://ogle.astrouw.edu.pl/) in 1997 began to regularly monitor the MCs for microlensing events and led to the discovery of thousands of pulsating stars. Since then, an increasing number of photometric surveys spanning the whole wavelength spectrum have taken a census of the MCs pulsating variables and allowed to study in detail their pulsation properties on the basis of multiband light curves with hundreds of phase points spanning several years observations. Some of these surveys are summarized in Table 2. Also listed in the table are the relatively few spectroscopic studies available so far for the MCs pulsating variables and, in the last part of the table, the new photometric, spectroscopic and astrometric surveys which are planned for the next decade.

In recent years the photometry of pulsating variables was progressively extended to the infrared region of the wavelength spectrum because the light variation of pulsating stars is smoother, reddening effects become negligible, and calibrating relations such as

Table 2. Surveys of MCs variable stars.

Visual	Infrared	Spectroscopy	Astrometry
EROS[1]	*ISO*	Luck & Lambert (1992)	
MACHO[2]	DENIS	Luck *et al.* (1998)	
SUPERMACHO[3]	SIRIUS	Romaniello *et al.* (2005, 2008)	
OGLE II–III[4]	2MASS	Gratton *et al.* (2004)	
MOA[5]	SAGE[6]	Borissova *et al.* (2004, 2006)	
STEP@*VST*[7]	VMC [8]	STEP@*VLT*	
Gaia[9]		Gaia	*Gaia*
LSST[10]			

Notes:
[1] http://eros.in2p3.fr/
[2] http://wwwmacho.mcmaster.ca/
[3] http://www.ctio.noao.edu/~supermacho/
[4] http://ogle.astrouw.edu.pl/~ogle/; http://bulge.astro.princeton.edu/~ogle/
[5] http://www.phys.canterbury.ac.nz/moa/
[6] http://sage.stsci.edu/index.php
[7] http://vstportal.oacn.inaf.it/node/40/
[8] http://star.herts.ac.uk/~mcioni/vmc/
[9] http://gaia.esa.int
[10] http://www.lsst.org/lsst_home.shtml

the *PL* relations become tighter when moving to the red region of the spectrum (see Figs. 4 and 6 of Madore & Freedman 1991; Fouqué *et al.* 2003; Ngeow & Kanbur 2008; Freedman *et al.* 2008). The MCs variables make no exception, and several of the past and future surveys listed in Table 2 (e.g., *ISO*, DENIS, VMC) cover in fact the infrared spectral range extending to the mid-infrared domain (3.6–8.0 μm) with the SAGE survey on the *Spitzer* satellite.

The first results from the microlensing surveys of the MCs were published at the end of the nineties. Alcock *et al.* (1996), announced the discovery of about 8000 RR Lyrae stars as a result of the MACHO microlensing survey of the LMC, and Alcock *et al.* (1998, 1999a,b) reported on the identification of about 2000 LMC Cepheids. At same time, large numbers of Cepheids were found in the SMC by the EROS survey (Sasselov *et al.* 1997; Bauer *et al.* 1999), while Wood *et al.* (1999) discovered about 1400 variable stars (Miras, semi-regulars, contact and semi-detached binaries) defining five distinct parallel period-luminosity sequences on the red and asymptotic giant branches of the LMC. OGLE observations of the Clouds started only in 1997. However, the OGLE II and III surveys represent the largest by area-coverage and the deepest and more complete census of the MCs variables. First results from the OGLE III survey of the LMC were recently published by Udalski *et al.* (2008). The new survey extends over about 40 square degrees and is about 1–1.5 magnitudes deeper than the OGLE II survey. Preliminary results on the LMC variables discovered by OGLE III have been presented during this conference by I. Soszynski. They are summarized in Table 3 and compared with results from the OGLE II survey. The number of LMC CCs has been almost doubled by OGLE III, and significant numbers of ACs and δ Scuti stars were also discovered (see Soszyński's talk

Table 3. Number of pulsating stars in the OGLE surveys of the MCs.

Class	LMC	SMC	Survey	Reference
CCs	1416	2144	OGLE II	Udalski *et al.* (1999a,b,c)
				Soszyński *et al.* (2000)
	3361		OGLE III	Soszyński (this conference)
T2Cs	14		OGLE II	Kubiak & Udalski (2003)
ACs	Yes		OGLE III	Soszyński (this conference)
RR Lyrae stars	7612	571	OGLE II	Soszyński *et al.* (2002, 2003)
Miras & SRs	3221		OGLE II	Soszyński *et al.* (2005)
Small Amplitude Red Giants (SARGs)	15400	3000	OGLE II–III	Soszyński *et al.* (2004, 2005)
δ Scuti stars	Yes		OGLE III	Soszyński (these proceedings)

in this conference). The increased sample of CCs traces very nicely the bar and gas-rich regions of the LMC. The RR Lyrae stars, instead, are evenly distributed on the whole field observed by OGLE III and outline the more spheroidal distribution of the LMC's oldest stellar component. The OGLE III Wesenheit ($W_{\mathrm{I}} = I - 1.55(V - I)$) PL diagram spans a total range of about 14 magnitudes, reaching about 1–1.5 magnitudes fainter than the OGLE II PL diagram (see Soszyński 2006). On this diagram the LMC δ Scuti stars locate on the extension to fainter magnitudes of the CCs PL, as suggested by McNamara *et al.* (2007), based on the analysis of the very few δ Scuti stars which were known in the LMC.

In the following, I will specifically address some major results which the surveys listed in Tables 2 and 3 have produced in the study of Cepheids, RR Lyrae stars and red variables and, in turn, in our knowledge of the MCs system.

3.1. *The PL relation of Classical Cepheids*

The CCs are primary standard candles that allow to link the local distance scale to the cosmological distances needed to determine the Hubble constant, H_0.

The CCs PL relation, discovered by H. Leavitt at the beginning of the twentieth century as she was picking up variables on photographic plates of the MCs, is unquestionably one of the most powerful tools at our disposal for determining the extragalactic distance scale. The extraordinary large number of CCs discovered in the MCs by the MACHO and OGLE surveys, allowed to derive the PL relations on unprecedented statistically significant and homogeneous samples of CCs. Udalski *et al.* (1999a) used fundamental-mode (FU) CCs in the LMC and SMC to derive the following PL relations:

$$V_0(\mathrm{LMC}) = -2.760 \log P - 17.042, \quad \sigma = 0.159 \text{ mag}, \quad (649 \quad \mathrm{FU} \quad \mathrm{CCs})$$

$$V_0(\mathrm{SMC}) = -2.760 \log P - 17.611, \quad \sigma = 0.258 \text{ mag}, \quad (466 \quad \mathrm{FU} \quad \mathrm{CCs})$$

The slope of these relations is in very good agreement with the slope of theoretical PL relations computed by Caputo *et al.* (2000) from nonlinear convective pulsation models of CCs ($M_{\mathrm{V}} = -2.75 \log P - 1.37, \quad \sigma = 0.18$ mag).

In order to use the PL's of the MCs CCs to measure distances outside the Clouds the zero point of the PL relation is generally fixed by using Galactic CCs whose absolute magnitudes are known from parallax measurements and/or Baade-Wesselink analyses or, alternatively, by assuming a value for the distance to the LMC. In the latter case, the zero-point problem thus shifts to the problem of having a robust distance determination for the LMC. The *HST* key program (Freedman *et al.* 2001) used the slope of the CCs *PL* relations by Udalski *et al.* (1999a) and a zero-point consistent with an assumed true distance modulus for the LMC of $\mu_{\rm LMC}$=18.5 mag to measure distances to 31 galaxies with distances from 700 kpc to 20 Mpc. These then served to calibrate other, more far-reaching secondary distance indicators to determine the Hubble constant in a region of constant Hubble flow (see Freedman *et al.* 2001, but also Saha *et al.* 2001, and Tammann *et al.* 2008, for different conclusions on the value of H_0).

In spite of the success in measuring distances up to 20 Mpc, a number of basic questions concerning the CCs *PL* relation still need an answer (see Fouqué *et al.* 2003, for a nice review on this topic). Is the CCs *PL* relation universal, as suggested, for instance, by Fouqué *et al.* (2007), so that we are allowed to apply the LMC *PL* to CCs in other galaxies? Does it depend on metal abundance, as also suggested by nonlinear pulsation models (see, e.g., Marconi *et al.* 2005; Bono *et al.* 2008, and references therein)? Is it linear or does it break at periods around 10 days, as a number of studies (Tammann & Reindl 2002; Tammann *et al.* 2002; Kanbur & Ngeow 2004; Sandage *et al.* 2004; Ngeow *et al.* 2005, 2008) are now suggesting? And, how reliably do we know the distance to the LMC and the distance modulus of $\mu_{\rm LMC}$=18.5 mag adopted by the *HST* key program? I will try to address this latter question in Section 4. Romaniello *et al.* (2008) provide a summary of the rather controversial results on the metallicity sensitivity of the Cepheid distances, in the literature of the past twenty years. These authors use direct spectroscopic measurements of iron abundance for Galactic and MC Cepheids to study the metallicity sensitivity of the CCs *PL* and conclude that the V-band *PL* is metallicity dependent, while no firm conclusions can be reached for the K-band *PL*. However, in their recent paper based on OGLE II and SAGE observations of the LMC CCs, Neilson *et al.* (2008) find that the infrared *PL* relations as well have additional uncertainty due to a metallicity dependence. Clearly, elemental abundance estimates for larger numbers of CCs spanning broad metallicity ranges, increased samples of photometric data in the infrared domain, and a fine tuning of all the parameters involved in the definition of the CCs *PL* relations are needed to quantitatively assess the metallicity dependence of both zero-point and slope of the CCs *PL* relations. Hopefully, most of the questions still pending on the CCs *PL* relation will find more definite answers from the new surveys of the MCs variables planned for the next decade (see Section 5).

3.2. *The MCs RR Lyrae stars*

The RR Lyrae stars are the primary distance indicators for stellar systems mainly composed by an old stellar component. They follow an absolute magnitude-metallicity relation in the visual band: $M_V - [\rm Fe/H]$ (Sandage 1981a,b) and a tight ($\sigma \sim 0.05$ mag) Period–Luminosity–Metallicity relation in the K-band: $PL_K Z$, (Bono *et al.* 2003; Catelan *et al.* 2004; Sollima *et al.* 2008, and references therein). Some observational and theoretical studies (see, e.g., Bono *et al.* 2003; Di Criscienzo *et al.* 2004, and references therein) have suggested that the $M_V - [\rm Fe/H]$ relation is not linear, becoming steeper when moving to larger metal content.

RR Lyrae stars have been found in all Local Group (LG) galaxies irrespective of morphological type and, although much fainter than the CCs, have been observed and measured as far as in the Andromeda galaxy. Alcock *et al.* (1996) discovered about 8000 RR Lyrae stars in the LMC, among which a fairly large number of double-mode pulsators (Alcock *et al.* 1997). Results from the study of the light curves (Alcock *et al.* 2000, 2003, 2004) showed that the average periods of the LMC RR Lyrae stars differ from what is observed for the Milky Way (MW) variables. Alcock *et al.*'s results were later confirmed and strengthened by the OGLE II studies of the LMC and SMC RR Lyrae stars (Soszyński et al. 2002, 2003). These findings suggest differences in the star formation history and rule out both MCs as possible contributors to the assembling of the MW halo.

Spectroscopic data of about 250 LMC RR Lyrae were obtained by Gratton et al. (2004), and Borissova et al. (2004, 2006). They were used to estimate the metal abundance, radial velocity and radial velocity dispersion of the LMC RR Lyrae population. The LMC RR Lyrae stars are metal-poor, with average metal abundance of $\langle[Fe/H]\rangle = -1.48/1.54$ dex and spread of about $0.2-0.3$ dex. The radial velocity dispersion, $\sigma_{v_r} = 50$ km s^{-1}, does not vary significantly with increasing distance from the LMC center (Borissova et al. 2006 and references therein), and is higher than the velocity dispersion of any other LMC population previously measured, thus providing empirical evidence for a kinematically hot, metal-poor halo in the LMC. Gratton et al. (2004) combined their spectroscopic measurements with high accuracy V magnitudes for about a hundred RR Lyrae stars by Clementini et al. (2003) to derive the luminosity–metallicity relation of the LMC variables, for which they found the following linear relation: $M_V = 0.214$ ([Fe/H] + 1.5) + 19.064. The slope of this relation agrees very well with slopes derived for the luminosity-metallicity relation of the MW RR Lyrae stars (Fernley et al. 1998) and horizontal branch stars in the globular clusters of M 31 (Rich et al. 2005), thus supporting the idea that the luminosity-metallicity relation of the RR Lyrae stars is, in first approximation, linear and universal.

Both the $M_V - $ [Fe/H] and the $PL_K Z$ relations were extensively used to measure distances to the LMC field and globular cluster's stars (see, e.g., Clementini et al. 2003; Dall'Ora et al. 2004; Szewczyk et al. 2008). Results from these studies are summarized in Table 4.

3.3. *The red variables*

The red variables are, typically, highly evolved stars in the Asymptotic Giant Branch (AGB) phase. Their atmospheres pulsate with typical periods in the range several tens to several hundreds days, and amplitudes ranging from 0.1 up to 6 magnitudes. The class includes the first ever recorded pulsating star: Mira, the prototype of variables with the largest visual amplitudes of any class of pulsating stars: the Miras. The light curves of these variables are often semiregular and multiperiodic, with short brightness outbursts observed sometimes on top of the periodic light change. Mass loss and dust emission, typical of the AGB evolutionary phase, further complicate the scenario, and even the mode of radial pulsation of these stars has long remained a matter of debate. Although the study of these variables most benefited from the long-term photometric monitoring of the Clouds by the microlensing surveys, and then from the combination of the visual data with infrared photometry, still they remain perhaps the least understood of all variable stars.

Wood et al. (1999) found that about 1400 red and asymptotic giant branch stars observed in the LMC by the MACHO survey were long period variables. They identified 5 distinct parallel I-band PL sequences, (labeled from "A" to "E" in their Fig. 1), and derived a first tentative classification of the red variables. By combining the MACHO photometry with infrared J and K data Wood (2000) further refined this classification, and definitely identified the Miras as fundamental mode pulsators falling on a single PL_K relation corresponding to sequence "C" of Wood et al. (1999). The SR variables are instead first to third overtone, or even fundamental mode, pulsators falling on sequence "B", the small amplitude red variables are on sequence "A", the long secondary period variables on sequence "D", and, finally, the contact binaries are on sequence "E". This classification was confirmed by various authors (e.g., Lebzelter et al. 2002). Since the Miras are bright, large amplitude variables, their PL relation is an important distance indicator for old and intermediate age populations. A new calibration of the Miras PL_K

relation was recently derived by Whitelock *et al.* (2008) using 53 LMC Miras with periods less than 420 days.

The number of red variables identified around the tip of the MCs red and asymptotic giant branches has massively increased in the last years. Fraser *et al.* (2005) detected about 22 000 red variables by combining the 8 year light-curve database from the MACHO survey of the LMC, with 2MASS infrared J, H, K photometry. The OGLE II and III surveys detected more than 3000 Miras and SR variables in the LMC (Soszyński *et al.* 2005), and about 15 400 and 3000 small amplitude red giant variables (SARGs) respectively in the LMC and in the SMC. These variables appear to be a mixture of AGB and red giant branch pulsators (Soszyński *et al.* 2004). Ita *et al.* (2004a,b), combining results from the OGLE II and the SIRIUS near-infrared JHK surveys, found that variable red giants in the SMC form parallel sequences on the PL_K plane, just like those found by Wood in the LMC. Moreover, Wood's original sequences were found to split into several separate subsequences above and below the tip of the LMC and SMC red giant branches. Slightly different relations were also found for carbon- and oxygen-rich variables. The number of PL sequences identified in the red variable domain was brought to fourteen by Soszyński *et al.* (2007), who also found that the slopes of the PL relation for Miras and SR variables seem to be the same in the LMC and SMC. The number of PLs is expected to increase further once the analysis of the red variables in the OGLE III database will be completed.

4. The distance to the LMC

Because the LMC is the first step of the extragalactic distance ladder, the knowledge of its distance has a tremendous impact on the entire astronomical distance scale. Benedict *et al.* (2002) published an historical summary of distances to the LMC from different indicators. Their Fig. 8 provides an impressive overview of the dispersion in the LMC distance moduli (μ_{LMC}) published during the ten-year span from 1992 to 2001. The last decade has seen dramatic progress in the calibration of the different distance indicators. The dispersion in μ_{LMC} has definitely shrunk, and values at the extremes of Benedict *et al.*'s distribution (18.1 and 18.8 mag, respectively) are not seen very often in the recent literature.

Table 4 summarizes some recent determinations of LMC distances based on pulsating variables. Far from being exhaustive, this table is only meant to highlight some recent advances in the distance determinations based on major types of pulsating variables found in the LMC. Although systematic differences still exist and need to be worked upon, (for instance the metallicity-corrected PL relation based on revised *Hipparcos* parallaxes for CCs, van Leeuwen *et al.* (2007), gives a somewhat shorter modulus, as does also the PL based on new values for the *p*-factor used to transform radial velocities into pulsational velocities in the Baade-Wesselink analyses of CCs), the controversy between the so-called "short" and "long" distances to the LMC seems to have largely vanished, and there is now a substantial convergence of the most reliable standard candles on a distance modulus for the LMC around 18.5 mag (Clementini *et al.* 2003; Walker 2003; Alves 2004; Romaniello *et al.* 2008).

5. The new surveys

Among the new photometric surveys, STEP and VMC, expected to start at the end of 2009, will repeatedly observe the Clouds allowing to study variable stars.

STEP (The SMC in Time: Evolution of a Prototype interacting dwarf galaxy, see the poster contribution to this conference by Ripepi *et al.*), will use the *VLT Survey Telescope*

Table 4. Distances to the LMC from pulsating variables.

Method	Distance modulus	Reference
PL, LMC δ Scuti stars	18.50 ± 0.22	McNamara *et al.* (2007)
Model fitting, δ Scuti stars	18.48 ± 0.15	McNamara *et al.* (2007)
Model fitting, Bump Cepheids	$18.48 \div 18.58$ 18.55 ± 0.02 18.54 ± 0.018	Bono *et al.* (2002) Keller & Wood (2002) Keller & Wood (2006)
Model fitting, field RR Lyrae stars	18.54 ± 0.02	Marconi & Clementini (2005)
$M_V - $ [Fe/H], field RR Lyrae stars	$18.46 \pm 0.07^{(1)}$	Clementini *et al.* (2003)
$PL_K Z$, field RR Lyrae stars	18.48 ± 0.08	Borissova *et al.* (2004)
$PL_K Z$, RR Lyrae stars in Reticulum	$18.52 \pm 0.01 \pm 0.12$	Dall'Ora *et al.* (2004)
$PL_K Z$, field RR Lyrae stars	$18.58 \pm 0.03 \pm 0.11$	Szewczyk *et al.* (2008)
PL_K, CCs $PLC_{J,K}$, CCs PL_W, CCs PL_W, CCs	18.47 ± 0.03 18.45 ± 0.04 18.52 ± 0.03 18.39 ± 0.03	van Leeuwen *et al.* (2007) metallicity-corrected

Notes:
[1] This distance modulus was derived using values from Clementini *et al.* (2003) for $\langle V(RR) \rangle$ and the reddening, and the assumption of $M_V = 0.59 \pm 0.03$ mag for the absolute visual magnitude of RR Lyrae stars at metal abundance [Fe/H] $= -1.5$ (Cacciari & Clementini 2003).

(VST) to obtain V, B and i' single-epoch photometry of the SMC, reaching below the galaxy main sequence turn-off; as well as shallow time-series photometry of the Wing and Bridge toward the LMC, for which no previous variability survey exists yet, reaching variables as faint as the RR Lyrae stars.

VMC (*VISTA* near-infrared survey of the Magellanic Clouds, see the poster paper by Cioni *et al.* in these proceedings) is instead an *ESO* public survey, which will obtain near-infrared YJK_s photometry of the whole Magellanic System (LMC, SMC and Bridge) with the subset of K_s exposures taken in time-series fashion to study variable stars.

These two surveys together will provide new multiband data to study the spatially resolved star formation history of the MCs, and will allow to reconstruct the 3-dimensional structure of the whole Magellanic System using various types of pulsating stars (Classical, Type II and Anomalous Cepheids, RR Lyrae, δ Scuti, and Miras).

The astrometric satellite *Gaia*, planned for launch in 2011, is one of the *European Space Agency* (*ESA*) cornerstone missions. During its lifetime of nominally 5 years, *Gaia* will scan the entire sky repeatedly, with an average frequency of about 80 measurements per object over the five-year time span, and will provide astrometry, 2-color photometry, and slit-less spectroscopy in the Ca triplet domain (847–874 nm) for all sources brighter than $V \sim 20$ mag (about 1.3×10^9 stars in total). Expected errors of the Gaia measurements are: $\sigma_\pi = 10$–25 μarcsec at $V \sim 15$ mag for parallaxes, and $\sigma = 15$ km s^{-1} at $V < 6$–17 mag for radial velocities. This is the domain of the bright pulsating variables in the MCs, which, if the satellite performs as expected, will then have their parallax, magnitude, radial velocity and metal abundance directly measured by *Gaia*. The direct measurements via trigonometric parallaxes of distances for Magellanic Cloud CCs and Miras will thus allow the calibration with unprecedented precision of these most important primary standard candles.

6. Acknowledgments

It is a pleasure to thank Marcella Marconi for comments and suggestions on a preliminary version of this review, and Thomas Lebzelter for useful directions on the literature of the red variables. A special thanks goes to the editors, Jacco van Loon & Joana Oliveira for patiently waiting for my Conference Proceedings.

References

Alcock, C., Allsman, R. A., Axelrod, T. S., *et al.* 1996, *AJ*, 111, 1146
Alcock, C., Allsman, R. A., Alves, D., *et al.* 1997, *ApJ*, 482, 89
Alcock, C., Allsman, R. A., Alves, D. R., *et al.* 1998, *AJ*, 115, 1921
Alcock, C., Allsman, R. A., Alves, D. R., *et al.* 1999a, *ApJ*, 511, 185
Alcock, C., Allsman, R. A., Alves, D. R., *et al.* 1999b, *AJ*, 117, 920
Alcock, C., *et al.* 2000, *ApJ*, 542, 257
Alcock, C., Allsman, R. A., Alves, D. R., *et al.* 2003, *ApJ*, 598, 597
Alcock, C., Alves, D. R., Axelrod, T. S., *et al.* 2004, *AJ*, 127, 334
Alves, D. 2004, *NewAR*, 48, 659
Bauer, F., Afonso, C., Albert, J. N., *et al.* 1999, *A&A*, 348, 175
Benedict, G., McArthur, B. E., Fredrick, L. W., *et al.* 2002, *AJ*, 123, 473
Bird, J. C., Stanek, K. Z., & Prieto, J. L. 2008, *ApJ*, accepted, (arXiv:0807.4933)
Bono, G., Castellani, V., & Marconi, M. 2002, *ApJ*, 565, L83
Bono, G., Caputo, F., Castellani, V., Marconi, M., Storm, J., & Degl'Innocenti, S. 2003, *MNRAS*, 344, 1097
Bono, G., Caputo, F., Fiorentino, G., Marconi, M., & Musella, I. 2008, *ApJ*, 684, 102
Borissova, J., Minniti, D., Rejkuba, M., & Alves, D., 2006, *A&A*, 460, 459
Borissova, J., Minniti, D., Rejkuba, M., Alves, D., Cook, K. H., & Freeman, K. C. 2004, *A&A*, 423, 97
Cacciari, C. & Clementini, G. 2003, in D. Alloin & W. Gieren (eds.), *Lecture Notes in Physics*, 635, 105
Caputo, F., Marconi, M., & Musella, I. 2000, *A&A*, 354, 610
Catelan, M., Pritzl, B. J., & Smith, H. A. 2004, *ApJS*, 154, 633
Clementini, G., Gratton, R. G., Bragaglia, A., Carretta, E., Di Fabrizio, L., & Maio, M. 2003, *AJ*, 125, 1309
Dall'Ora, M., Storm, J., Bono, G., *et al.* 2004, *ApJ*, 610, 269
Di Criscienzo, M., Marconi, M., & Caputo, F. 2004, *ApJ*, 612, 1092
Fernley, J., Skillen, I., Carney, B. W., Cacciari, C., & Janes, K. 1998, *MNRAS*, 293, L61
Fouqué, P., Storm, J., & Gieren, W. 2003, in D. Alloin & W. Gieren (eds.), *Lecture Notes in Physics*, 635, 21
Fouqué, P., Arriagada, P., Storm, J., *et al.* 2007, *A&A*, 476, 73
Fraser, O. J., Hawley, S. L., Cook, K. H., & Keller, S. C. 2005, *AJ*, 129, 768
Freedman, W. L., Madore, B. F., Gibson, B. K, *et al.* 2001, *ApJ*, 553, 47
Freedman, W. L., Madore, B. F., Rigby, J., Persson, S. E., & Sturch, L. 2008, *ApJ*, 679, 71
Gautschy, A. & Saio, H. 1995, *ARAA*, 34,551
Gratton, R. G., Bragaglia, A., Clementini, G., Carretta, E., Di Fabrizio, L., Maio, M., & Taribello, E. 2004, *A&A*, 421, 937
Ita, Y., Tanabé, T., Matsunaga, N., *et al.* 2004a, *MNRAS*, 347, 720
Ita, Y., Tanabé, T., Matsunaga, N., *et al.* 2004b, *MNRAS*, 353, 705
Kanbur, S. & Ngeow, C. 2004, *MNRAS*, 350, 962
Keller, S. C. & Wood, P. R 2002, *ApJ*, 578, 144
Keller, S. C. & Wood, P. R 2006, *ApJ*, 642, 841

Kubiak, M. & Udalski, A. 2003, *Acta Astronomica*, 53, 117

Lebzelter, T., Schultheis, M., & Melchior, A. L. 2002, *A&A*, 393, 573

Luck, R. E., Moffett, T. J., Barnes, T. G., III, & Gieren, W. P. 1998, *AJ*, 115, 605

Luck, R. E. & Lambert, D. L. 1992, *ApJS*, 79, 303

Madore, B. F. & Freedman, W. L. 1991, *PASP*, 103, 933

Marconi, M. 2001, available at http://www.oa-teramo.inaf.it/scuola2001/ under Contributi Docenti

Marconi, M. & Clementini, G. 2005, *AJ*, 129, 2257

Marconi, M., Musella, I., & Fiorentino, G. 2005, *ApJ*, 632, 590

McNamara, D. H., Clementini, G., & Marconi, M. 2007, *AJ*, 133, 2752

Neilson, H. R., Ngeow, C. -C., Kanbur, S. M., & Lester, J. B. 2008, *ApJ*, in press, arXiv:0810.3001

Ngeow, C. -C., Kanbur, S. M., Nikolaev, S., Buonaccorsi, J., Cook, K. H., & Welch, D. L. 2005, *MNRAS*, 363, 831

Ngeow, C. -C., Kanbur, S. M., & Nanthakumar, A. 2008, *A&A*, 447, 621

Ngeow, C. -C., & Kanbur, S. M. 2008, *A&A*, 679, 76

Rich, M. R., Corsi, C. E., Cacciari, C., Federici, L., Fusi Pecci, F., Djorgovski, S. G., & Freedman, W. L. 2005 *AJ*, 129, 2670

Romaniello, M., Primas, F., Mottini, M., Groenewegen, M. A. T., Bono, G., & François, P. 2005, *A&A*, 429, L37

Romaniello, M., Primas, F., Mottini, M., *et al.* 2008, *A&A*, 488, 731

Saha, A., Sandage, A., Tammann, G. A., Dolphin, A. E., Christensen, J., Panagia, N., & Macchetto, F. D. 2001, *ApJ*, 562, 314

Sandage, A. 1981a, *ApJ*, 244, L23

Sandage, A. 1981b, *ApJ*, 248, 161

Sandage, A., Tammann, G. A., & Reindl, B. 2004, *A&A*,424, 43

Sasselov, D. D., Beaulieu, J. P., Renault, C., *et al.* 1997, *A&A*, 324, 471

Sollima, A., Cacciari, C., Arkharov, A. A. H., Larionov, V. M., Gorshanov, D. L., Efimova, N. V., & Piersimoni, A. 2008, *MNRAS*, 384, 1583

Soszyński, I., Udalski, A., Szymański, M., Kubiak, M., Pietrzyński, G., Woźniak, P. R., & Żebruń, K. 2000, *Acta Astronomica*, 50, 451

Soszyński, I., Udalski, A., Szymański, M., *et al.* 2002, *Acta Astronomica*, 52, 369

Soszyński, I., Udalski, A., Szymański, M., *et al.* 2003, *Acta Astronomica*, 53, 93

Soszyński, I., Udalski, A., Kubiak, M., Szymański, M., Pietrzyński, G., Żebruń, K., Szewczyk, O., & Wyrzykowski, L. 2004, *Acta Astronomica*, 54, 129

Soszyński, I., Udalski, A., Kubiak, M., *et al.* 2005, *Acta Astronomica*, 55, 331

Soszyński, I. 2006, *Mem. S.A.It.*, 77, 265

Soszyński, I., Dziembowski, W. A., Udalski, A., *et al.* 2007, *Acta Astronomica*, 27, 201

Szewczyk, O., Pietrzyński, G., Gieren, W., *et al.* 2008, *AJ*, 136, 272

Tammann, G. A. & Reindl, B. 2002, *ApSS*, 280, 165

Tammann, G. A., Reindl, B., Thim, F., Saha, A., & Sandage, A. 2002, in N. Metcalfe & T. Shanks (eds.), *A New Era in Cosmology*, ASPC 283 (San Francisco: ASP), p. 258

Tammann, G. A., Sandage, A., & Reindl, B. 2008, *A&AR*, 15, 289

Udalski, A., Szymański, M., Kubiak, M., Pietrzyński, G., Soszyński, I., Woźniak, P. R., & Żebrun, K. 1999a, *Acta Astronomica*, 49, 201

Udalski, A., Soszyński, I., Szymański, M., Kubiak, M., Pietrzyński, G., Woźniak, P. R., & Żebruń, K. 1999b, *Acta Astronomica*, 49, 223

Udalski, A., Soszyński, I., Szymański, M., Kubiak, M., Pietrzyński, G., Woźniak, P. R., & Żebruń, K. 1999c, *Acta Astronomica*, 49, 437

Udalski, A., Soszyński, I., Szymański, M., *et al.* 2008, *Acta Astronomica*, 58, 89

van Leeuwen, F., Feast, M. W., Whitelock, P. A., & Laney, C. D, 2007, *MNRAS*, 379, 723

Walker, A. 2003, in D. Alloin & W. Gieren (eds.), *Lecture Notes in Physics*, 635, 625

Whitelock, P. A., Feast, M. W., & van Leeuwen, F. 2008, *MNRAS*, 386, 313

Wood, P. R., Alcock, C., Allsman, R. A., *et al.* 1999, in T. Le Bertre, A. Lèbre, & C. Waelkens (eds.), *Asymptotic Giant Branch Stars*, IAUS 191, p. 151

Wood, P. R. 2000, *PASA*, 17, 18

The Magellanic System: Stars, Gas, and Galaxies
Proceedings IAU Symposium No. 256, 2008
Jacco Th. van Loon & Joana M. Oliveira, eds.

© 2009 International Astronomical Union
doi:10.1017/S1743921308028743

Thermally-pulsing asymptotic giant branch stars in the Magellanic Clouds

Paola Marigo[1], Léo Girardi[2], Alessandro Bressan[1], Martin A. T. Groenewegen[3], Bernhard Aringer[4], Laura Silva[5] and Gian Luigi Granato[5]

[1] Dipartimento di Astronomia, Università di Padova, Vicolo dell'Osservatorio 2, I-35122 Padova, Italy — email: paola.marigo@unipd.it

[2] Osservatorio Astronomico di Padova – INAF, Vicolo dell'Osservatorio 5, I-35122 Padova, Italy

[3] Royal Observatory of Belgium, Ringlaan 3, B-1180 Brussels, Belgium

[4] Department of Astronomy, University of Vienna, Tuerkenschanzstr. 17, A1180 Wien, Austria

[5] Osservatorio Astronomico di Trieste – INAF, Via Tiepolo 11, I-34131 Trieste, Italy

Abstract. We present the latest results of a theoretical project aimed at investigating the properties of thermally-pulsing asymptotic giant branch (TP-AGB) stars in different host systems. For this purpose, we have recently calculated calibrated synthetic TP-AGB tracks — covering a wide range of metallicities ($0.0001 \leqslant Z \leqslant 0.03$) up to the complete ejection of the envelope by stellar winds (Marigo & Girardi 2007) — and used them to generate new sets of stellar isochrones (Marigo *et al.* 2008). The latter are converted to about 25 different photometric systems, including the mid-infrared filters of *Spitzer* and *AKARI* as the effect of circumstellar dust from AGB stars is taken into account. First comparisons with AGB data in the MC field and stellar clusters are discussed.

Keywords. stars: AGB and post-AGB, stars: carbon, stars: evolution, globular clusters: individual (47 Tuc), Magellanic Clouds, infrared: stars

1. Introduction

Owing to their intrinsic brightness and distinctive spectral features, TP-AGB stars play an important role in many properties of their host systems. For instance, the TP-AGB contribution to the total luminosity of single-burst stellar populations reaches a maximum of about 40% at ages from 1 to 3 Gyr (Frogel *et al.* 1990), and account for most of the bright-infrared objects in resolved galaxies, as clearly demonstrated by DENIS, 2MASS, SAGE, S^3MC, and *AKARI* IRC data (Cioni *et al.* 1999; Nikolaev & Weinberg 2000; Blum *et al.* 2006; Bolatto *et al.* 2007; Ita *et al.* 2008) for the Magellanic Clouds (MC). Despite its large relevance, the description of the TP-AGB phase in evolutionary models of galaxies has always been far from detailed and physically accurate.

Our goal is to provide complete sets of TP-AGB models useful for the evolutionary population synthesis of galaxies — both resolved and unresolved into stars — taking into consideration all the processes and physical inputs which are known to be critical in determining the evolution along this phase, and reproducing its basic observables. The Magellanic Clouds play a key role in our project, as we can rely on rich and complete samples of well-observed TP-AGB stars, to be used in the calibration of the basic model parameters.

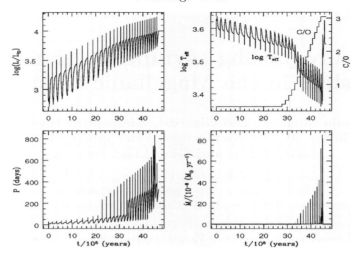

Figure 1. Predicted evolution of luminosity, effective temperature, surface C/O ratio, pulsation period, and mass-loss rate during the TP-AGB evolution of a 1.8 $M_\odot$, $Z = 0.008$ star from Marigo & Girardi (2007). Notice the sudden cooling of the track as the star becomes of C-type (C/O > 1), which is due to the abrupt change in the main sources of molecular opacity.

2. Synthetic TP-AGB tracks and their calibration

The TP-AGB evolutionary tracks used here have been described in Marigo & Girardi (2007) and Marigo *et al.* (2008). They include a series of crucial aspects such as: the complex luminosity evolution over the pulse-cycle, the third-dredge-up efficiency as a function of stellar mass and metallicity, the derivation of the effective temperatures from the integration of stellar envelope models, hot-bottom burning at the base of convective envelopes, the dramatic changes in low-temperature opacities when the stellar envelopes become C-rich (Marigo 2002), different mass loss descriptions for M- and C-type stars, the dependence of mass loss on pulsation period, the transition from the first overtone pulsation mode to the fundamental one, etc.

The evolutionary tracks contain a few free parameters (mainly related to the third dredge-up) which were calibrated so as to reproduce two basic observables of the LMC and SMC: the carbon star luminosity functions (Groenewegen 2002), and the C- and M-type lifetimes as a function of turn-off mass as derived from MC star clusters (Girardi & Marigo 2007a). One example of our TP-AGB evolutionary tracks is presented in Fig. 1.

3. From stellar tracks to synthetic samples of AGB stars

The TP-AGB evolutionary tracks were attached to the Girardi et al. (2000) tracks for the previous evolutionary phases, and used to build the Marigo *et al.* (2008) isochrones. They consider the reprocessing of radiation by dust in the circumstellar envelopes of mass-losing stars. The web interface in http://stev.oapd.inaf.it/cmd provides these isochrones — and their derivatives, such as luminosity functions and integrated magnitudes — for more than 25 optical-to-far-infrared photometric systems, including those more useful to the study of MC AGB stars (e.g., OGLE, DENIS, 2MASS, *Spitzer* IRAC+MIPS, and *AKARI*).

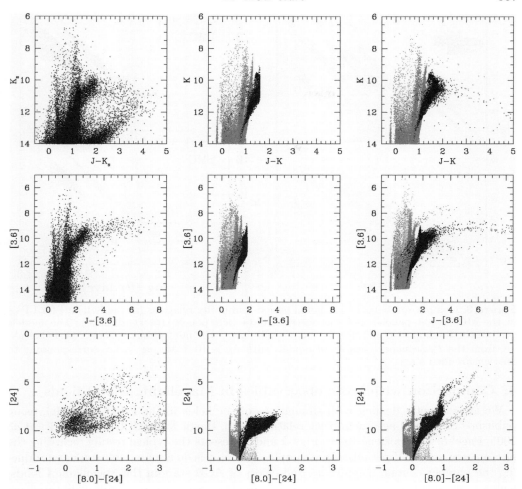

Figure 2. Comparison between 2MASS and *Spitzer* near- and mid-infrared data for the LMC (left panels), and the simulated photometry without (middle panels) and with (right panels) the effect of AGB circumstellar dust. Predicted carbon stars are marked with black points.

In the process of building the isochrones, we have used the sequences of synthetic C star spectra from Loidl *et al.* (2001). These cover well the $T_{\rm eff}$ range of C stars in the Magellanic Clouds and in the Milky Way galaxy, but not the wide intervals of surface gravities and C/O ratios expected for such stars. Moreover, they have been computed for C stars of initial solar metallicity. We are now extending the database of C star spectra, using the COMARCS code (Lebzelter *et al.* 2008), so as to cover the complete parameter space expected from the evolutionary tracks (Aringer *et al.*, in preparation).

This set of theoretical models have been included in the TRILEGAL code (Girardi *et al.* 2005) for simulating the photometry of resolved stellar populations. The code has been adapted (Girardi & Marigo 2007b) so as to deal with the luminosity and temperature variations driven by thermal pulse cycles, and to consider additional variables such as the pulsation period of LPVs, the mass loss rate, the optical depth of circumstellar dust, etc. This allows us to simulate complete samples of AGB stars together with their main optical-to-infrared properties, for any history of star formation and chemical enrichment.

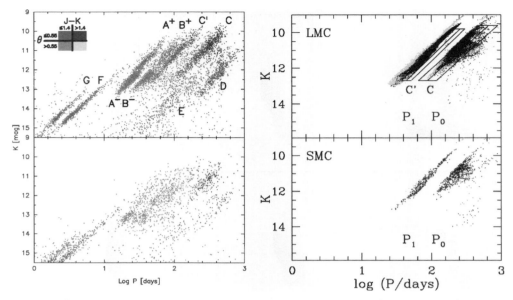

Figure 3. Comparison between predicted period-luminosity relations (in the K-band) for LPVs in the MCs (right panels) and the observed data (left panels; Ita *et al.* 2004). The models contain the fundamental and first overtone modes only. The faint plume at longer periods leaking from the P_0 sequence consists of models with $\dot{M} > 10^{-6}$ M$_\odot$ yr^{-1}, i.e. corresponding to dust-enshrouded stars.

4. Comparison with basic observables in Magellanic Cloud fields

We are presently dealing with simulations of the fields in the LMC and SMC, using published SFHs and age–metallicity relations (e.g., Harris & Zaritsky 2004; Javiel *et al.* 2005; Pagel & Tautvaišienė 1998). Figs. 2 and 3 presents the typical results regarding the photometry and LPV periods of the sample. The models do a reasonable job in reproducing the number counts, magnitudes and colors of AGB stars in the Magellanic Clouds, which is no surprise since the evolutionary models have been originally calibrated using MC data. Looking at the quantitative details, however, we are accumulating several hints on the aspects of the models which *need* to be improved. One of the most obvious deficiencies is in the description of the pulsation periods for C stars (Fig. 3), which requires more reliable pulsation models (e.g., computed with the right molecular opacities, as in Lebzelter & Wood 2007) than we have used so far. Moreover, although our models describe well the sequences of mass-losing stars in mid-infrared CMDs (Fig. 2), they fail to describe the detailed color distributions. This likely means that we have to adjust the mass loss prescriptions, which, in turn, would imply the recalculation and recalibration of the TP-AGB tracks. We are implementing such iterative process, which is however not simple and quite demanding in terms of CPU time. No doubt the final target is well worth of the effort.

5. Comparison with basic observables in 47 Tuc

The old globular cluster 47 Tuc is of particular interest to us, thanks to the detailed information available for its AGB stars, and to its metallicity which is comparable to those found in young and intermediate-age populations of the MCs. Figure 4 presents the period-luminosity diagram of 47 Tuc LPVs (Lebzelter & Wood 2005). They

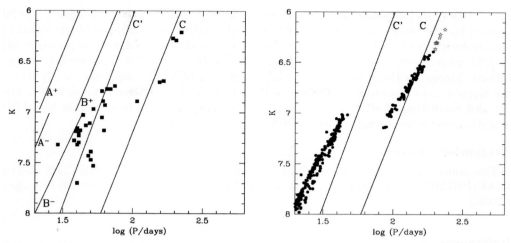

Figure 4. Comparison between predicted period–luminosity relations (right panel) for LPVs in 47 Tuc (actually simulated for a much more massive cluster), and the Lebzelter & Wood (2005) data (left panel). Starred symbols correspond to TP-AGB models with $\dot{M} > 10^{-7}$ $M_\odot$ yr^{-1}.

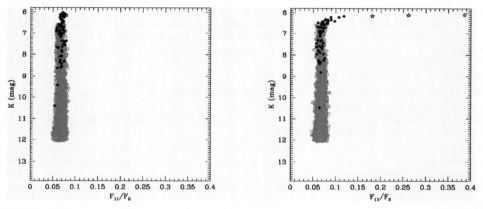

Figure 5. Theoretical CMDs for 47 Tuc simulated in the *AKARI* filters, without (left panel) and with (right panel) the effect of AGB circumstellar dust. TP-AGB stars are plotted with black dots, and those with $\dot{M} > 5\ 10^{-7}$ $M_\odot$ yr^{-1} are marked with starred symbols.

clearly distribute on two sequences, with the switch from the first overtone to the fundamental mode occurring at $K \sim 6.7$ mag. Our models show the same transition, but at $K \sim 7$ mag.

Moreover, our simulations (Fig. 5) reproduce the behavior of the mid-IR colors that has been recently measured by *AKARI* (Ita *et al.* 2007): the only stars that present a prominent mid-IR excess are those at the very tip of the AGB (see also van Loon *et al.* 2006; Lebzelter *et al.* 2006). This reinforces the idea that the bulk of dust formation occurs at the end of the AGB, with the role of RGB stars being quite marginal, if not negligible (see also Boyer *et al.* 2008, for ω Cen).

6. Other perspectives

In addition to the detailed work allowed by the MCs and by 47 Tuc, we are involved in different projects aiming at the calibration of the TP-AGB tracks at lower and higher metallicities. In particular, it has become clear that dwarf galaxies provide a useful testbed for the models at small metallicities, using either near-infrared observations of nearby dwarfs (such as Leo II and Leo I; Gullieuszik *et al.* 2008, and work in preparation) or *HST* photometry of more distant dwarfs (Dalcanton *et al.* 2008, and work in preparation). Moreover, the Spitzer mid-infrared spectrum of Virgo ellipticals clearly shows the signatures of mass-losing AGB stars (Bressan *et al.* 2006; Clemens *et al.* 2009). The detailed modelling of all these galaxies will eventually help us to constrain the AGB evolution over a wide range of metallicities.

Acknowledgements

This study was partially supported by the University of Padova (CPDA052212) and by INAF/PRIN07 (CRA 1.06.10.03). B.A. acknowledges funding by Austrian FWF project P19503.

References

Blum, R. D., Mould, J. R., Olsen, K. A., *et al.* 2006, *AJ*, 132, 2034
Bolatto, A. D., Simon, J. D., Stanimirović, S., *et al.* 2007, *ApJ*, 655, 212
Boyer, M. L., McDonald, I., van Loon, J. Th., *et al.* 2008, *AJ*, 135, 1395
Bressan, A., Panuzzo, P., Buson, L., *et al.* 2006, *ApJ*, 639, L55
Cioni, M. R., Habing, H. J., Loup, C., *et al.* 1999, in P. Whitelock, & R. Cannon (eds.), *The Stellar Content of Local Group Galaxies*, IAUS 192 (San Francisco: ASP), p. 65
Clemens, M. S., Bressan, A., Panuzzo, P., Rampazzo, R., Silva, L., Buson, L., & Granato, G. L. 2009, *MNRAS*, 392, 982
Dalcanton, J. J., *et al.* 2008, *ApJS*, submitted
Frogel, J. A., Mould, J., & Blanco, V. M. 1990, *ApJ*, 352, 96
Girardi, L., Bressan, A., Bertelli, G., *et al.* 2000, *A&AS*, 141, 371
Girardi, L., Groenewegen, M. A. T., Hatziminaoglou, E., & da Costa L. 2005, *A&A*, 436, 895
Girardi, L. & Marigo, P. 2007a, *A&A*, 462, 237
Girardi, L. & Marigo, P. 2007b, in F. Kerschbaum, C. Charbonnel, & R. F. Wing (eds.), *Why Galaxies Care About AGB Stars: Their Importance as Actors and Probes*, ASP Conference Series 378 (San Francisco: ASP), p. 20
Groenewegen, M. A. T. 2002, arXiv:astro-ph/0208449
Gullieuszik, M., Held, E. V., Rizzi, L., *et al.* 2008, *MNRAS*, 388, 1185
Harris, J. & Zaritsky, D. 2004, *AJ*, 127, 1531
Ita, Y., Tanabé, T., Matsunaga, N., *et al.* 2004, *MNRAS*, 347, 720
Ita, Y., Tanabé, T., Matsunaga, N., *et al.* 2007, *PASJ*, 59, 437
Ita, Y., Onaka, T., Kato, D., *et al.* 2008, *PASJ*, 60, S435
Javiel, S. C., Santiago, B. X., & Kerber, L. O. 2005, *A&A*, 431, 73
Lebzelter, T. & Wood, P. R. 2005, *A&A*, 441, 1117
Lebzelter, T., Posch, T., Hinkle, K., *et al.* 2006, *ApJ*, 653, L145
Lebzelter, T. & Wood, P. R. 2007, *A&A*, 474, 643
Lebzelter, T., Lederer, M. T. , Cristallo, S. *et al.* 2008, *A&A*, 486, 511
Loidl, R., Lançon, A., & Jørgensen, U. G. 2001, *A&A*, 371, 1065
Marigo, P. 2002, *A&A*, 387, 507
Marigo, P. & Girardi, L. 2007, *A&A*, 469, 239
Marigo, P., Girardi, L., Bressan, A., *et al.* 2008, *A&A*, 482, 883
Nikolaev, S. & Weinberg, M. D. 2000, *ApJ*, 542, 804
Pagel, B. E. J. & Tautvaišienė, G. 1998, *MNRAS*, 299, 535
van Loon, J. Th., McDonald, I., Oliveira, J. M., *et al.* 2006, *A&A*, 450, 339

The Magellanic System: Stars, Gas, and Galaxies
Proceedings IAU Symposium No. 256, 2008
Jacco Th. van Loon & Joana M. Oliveira, eds.

© 2009 International Astronomical Union
doi:10.1017/S1743921308028755

On the self-consistent physical parameters of LMC intermediate-age clusters

Leandro O. Kerber[1,2] **and Basílio X. Santiago**[3]

[1]IAG/USP, São Paulo, Brazil
email: kerber@astro.iag.usp.br

[2]INAF-OAPd, Padova, Italy

[3]IF/UFRGS, Porto Alegre, Brazil

Abstract. The LMC clusters with similar ages to the Milky Way open clusters are in general more metal-poor and more populous than the latter, being located close enough to allow their stellar content to be well resolved. Therefore, they are unique templates of simple stellar population (SSP), being crucial to calibrate models describing the integral light as well as to test the stellar evolution theory. With this in mind we analyzed *HST*/WFPC2 (V, $B - V$) colour-magnitude diagrams (CMDs) of 15 populous LMC clusters with ages between ~ 0.3 Gyr and ~ 4 Gyr using different stellar evolutionary models. Following the approach described by Kerber, Santiago & Brocato (2007), we determined accurate and self-consistent physical parameters (age, metallicity, distance modulus and reddening) for each cluster by comparing the observed CMDs with synthetic ones generated using isochrones from the PEL and BaSTI libraries. These determinations were made by means of simultaneous statistical comparison of the main-sequence fiducial line and the red clump position, offering objective and robust criteria to select the best models. We compared these results with the ones obtained by Kerber, Santiago & Brocato (2007) using the Padova isochrones. This revealed that there are significant trends in the physical parameters due to the choice of stellar evolutionary model and treatment of convective core overshooting. In general, models that incorporate overshooting presented more reliable results than those that do not. Furthermore, the Padova models fitted better the data than the PEL and BaSTI models. Comparisons with the results found in the literature demonstrated that our derived metallicities are in good agreement with the ones from the spectroscopy of red giants. We also confirmed that, independent of the adopted stellar evolutionary library, the recovered 3D distribution for these clusters is consistent with a thick disk roughly aligned with the LMC disk as defined by field stars. Finally, we also provide new estimates of distance modulus to the LMC center, that are marginally consistent with the canonical value of 18.50 mag.

Keywords. Hertzsprung-Russell diagram, galaxies: individual (LMC), Magellanic Clouds, galaxies: star clusters

1. Introduction

The LMC contains a rich system of stellar clusters, with more than three thousand cataloged objects (Bica *et al.* 2008) and covering ages from few Myr to about 13 Gyr. There are about one hundred that can be considered as populous ones ($> 10^5$ stars), which offer the opportunity to recover the age-metallicity relation for the LMC by means of accurate age (from CMD analysis) and metallicity (from spectroscopy analysis) determinations. Furthermore, the objects with ages between ~ 0.3 and 4 Gyr — the intermediate-age LMC clusters (IACs) — are more metal poor than the open clusters in the Milky Way, being therefore fundamental pieces in the local universe to calibrate integrated light models as well as to test the evolutionary models in the sub-solar metallicity regime.

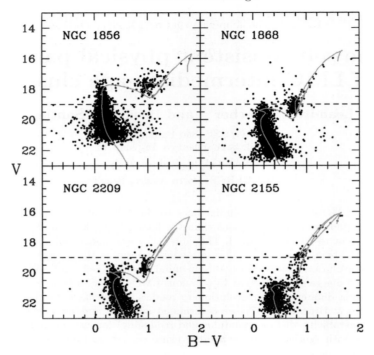

Figure 1. Examples of observed *HST*/WFPC2 (*V*, *B − V*) CMDs for 4 stellar clusters in our sample. These clusters are in a sequence of age, from ∼ 0.3 Gyr (NGC 1856) to ∼ 3.5 Gyr (NGC 2155). The best solutions found by us using the Padova isochrones (KSB07) are also shown in the figure.

Taking the advantage of the superior photometric quality of the *HST* and the intrinsic large stellar statistics for the populous IACs we have applied a statistical method to recover accurate physical parameters — not only age, but also metallicity, distance and reddening — for these objects in a self-consistent way. By self-consistency we mean the ability to simultaneously infer these parameters from the same data-set without any prior assumptions about any of them. The method, presented by Kerber, Santiago & Brocato (2007) (hereafter KSB07) using the Padova isochrones (Girardi *et al.* 2002), joins CMD modelling and statistical analysis to objectively determine which are the synthetic CMDs that best reproduce the observed ones. In the present work we expanded this analysis to two other stellar evolutionary libraries: PEL (or Pisa, Castellani *et al.* 2003) and BaSTI (or Teramo, Pietrinferni *et al.* 2004). This allowed us to quantify how the recovered physical parameters depend on the adoption of different stellar evolutionary libraries, including the treatment for the convective core overshooting. Furthermore, since we are determining the individual distance to each cluster, we could also probe the three dimensional distribution of these clusters, which seems to be roughly aligned with LMC disk (KSB07; Grocholski *et al.* 2007), and to obtain new determinations of distance modulus to the LMC centre.

2. CMDs: data vs. model

HST/WFPC2 data. We analysed *HST*/WFPC2 (*V*, *B − V*) CMDs for a sample of 15 IACs in the LMC covering ages from ∼ 0.3 to ∼ 4 Gyr. These data come from Brocato *et al.* (2001) and are photometrically quite homogeneous, reaching typically *V* ∼ 22 mag.

Table 1. The adopted stellar evolutionary libraries.

	Padova	PEL (Pisa)	BaSTI (Teramo)
Convective core:			
Classical ($\Lambda_{OV}/H_p = 0$) ?	no	yes	yes
Overshooting $\Lambda_{OV}/H_p \neq 0$	~ 0.25	0.25	0.25
Z (0.001$-Z_\odot$)	0.001, 0.002, 0.004, 0.006 0.008, 0.012, 0.015, 0.019 (8 values)	0.001, 0.004 0.008 (3 values)	0.001, 0.002, 0.004 0.008, 0.010, 0.0198 (6 values)
Age (0.10$-$4 Gyr)	33 values	24 values	22 values

As can be seen in Fig. 1, these CMDs cover at least 2 magnitudes along the main sequence (MS) and clearly display the core helium burning stars in the red clump (RC).

CMD modelling. We are modelling these CMDs as SSPs, where the basic steps to generate a synthetic CMD are the following: 1) we choose the stellar evolutionary model; 2) we choose an age and metallicity. These two steps are equivalent to picking up an isochrone from the adopted stellar evolutionary library; 3) we apply our choice of distance modulus and reddening to the isochrone; 4) we distribute the synthetic stars following a Salpeter IMF and a fraction of binaries of 30%; 5) as a last step we introduce the photometric errors and completeness as determined from the data.

The stellar evolutionary libraries. In Table 1 we can see the basic differences in the three stellar evolutionary libraries that were used. All of them offer models with treatment for convective core overshooting with similar values for Λ_{OV}. However, only PEL and BaSTI also offer models without overshooting — the classical or canonical models. Another important difference between the libraries is their discreteness or the number of steps in age or metallicity. In this respect Padova offers a significant larger number of possibilities than the other libraries, specially PEL, where there are only three values of metallicity available.

Statistical comparisons. To determine the best models we applied a statistical tool to compare simultaneously the observed and synthetic MS and RC. For the MS we compared the differences in colour along the observed and theoretical fiducial lines of each cluster using a χ^2 statistics to assess their similarity. For the evolved stars we computed a CMD distance, δ_{RC}, between the median positions of real and artificial RC stars. So, the best models are the ones that have the minimum values in both statistics.

3. Self-consistent physical parameters

Table 2 summarises the results. The first line lists the typical formal random uncertainties in each parameter. For more details on how these uncertainties are estimated we refer to KSB07. In the remaining lines, the first (second) number in each entry is the systematic (rms) difference in the specific comparison being made. We compared our previous results based on Padova with the literature, as well as the distinct evolutionary models among themselves. We detail these comparisons below.

Padova vs. literature. Figure 2 shows the comparisons between our results using the Padova isochrones (KSB07) with the ones found in the literature. We identify an underestimation in age in the previous works, which were based on ground-based observations (e.g., Elson & Fall 1988). The amplitude of the effect is ~ 0.30 in log(age) for clusters

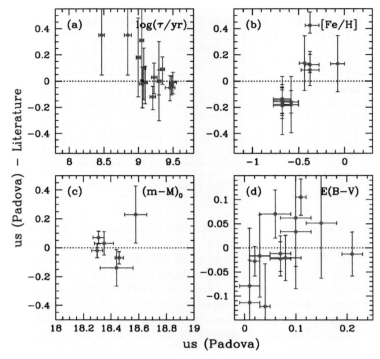

Figure 2. Differences in the physical parameters determined by us (using the Padova isochrones) in relation to the ones found in the literature (see text for details).

younger than 1 Gyr. On the other hand, for the clusters older than this age limit, all of them analysed with high quality data (*HST* or *VLT*) by other authors (e.g., Rich *et al.* 2001), this difference seems disappear.

A very interesting comparison can be done based on spectroscopy of red giants, in particular using the analysis of Calcium triplet (Olszewski *et al.* 1991; Grocholski *et al.* 2006). These comparisons reveal that, despite the scatter of 0.20, our CMD analysis predicts metallicities without any systematic effect. So, this scatter can be considered as a realistic random uncertainty in our method based on a pure CMD analysis.

Concerning the distance modulus, another interesting comparison can be done with the results based on the RC in the K band as a distance indicator (Grocholski *et al.* 2007). For the six stellar clusters in common, despite the significant scatter found, the systematic differences are very small. The same can be said in relation to the reddening, where in this case the literature results come from analysis of integrated optical colours (McLaughlin & van der Marel 2005).

Classical vs. overshooting. As expected (Gallart, Zoccali & Aparicio 2005), inside the same library, models with treatment for the convective core overshooting recovered older ages in relation to the ones obtained by the classical models. Despite the systematic differences in log(age) of 0.08 for the PEL models and 0.13 for the BaSTI models, there is no clear trend in the other parameters.

PEL or BaSTI vs. Padova. The comparisons involving different libraries with convective core overshooting reveal that the adoption of a specific stellar evolutionary model can produce a significant bias in the physical parameters. Taking the Padova library as a reference, the PEL library systematically overestimates distance modulus by 0.12 (or $\sim$ 2.8 kpc at 50 kpc), whereas the BaSTI models systematically overestimate ages by

Table 2. Self-consistent physical parameters. First line: typical formal uncertainties. Other lines: systematic and *rms* differences between the results from different stellar evolutionary models.

	log(age/yr)	[Fe/H]	$(m - M)_0$	$E(B - V)$
typical formal uncertainties	0.05	0.10	0.08	0.02
Padova – literature	+0.30 / 0.10^1 +0.02 / 0.11^2	+0.02 / 0.20	−0.02 / 0.13	−0.01 / 0.07
PEL overshooting – classical	+0.08 / 0.09	+0.06 / 0.11	+0.01 / 0.10	−0.02 / 0.03
BaSTI overshooting – classical	+0.13 / 0.09	+0.02 / 0.20	+0.04 / 0.14	0.00 / 0.04
PEL (over) – Padova	+0.00 / 0.06	+0.04 / 0.17	+0.12 / 0.11	0.00 / 0.04
BaSTI (over) – Padova	+0.12 / 0.05	+0.06 / 0.12	+0.03 / 0.05	−0.01 / 0.03

Notes: [1] age < 1 Gyr ; [2] age > 1 Gyr.

0.12 in log(age) (~ 0.30 Gyr for an age of 1 Gyr). Apart from these, no other trend was observed in the other parameters.

4. The quality of the fit

To determine the quality of the fit we took simultaneously the MS and RC fit into account. We define a parameter called n for each model as

$$n = \sqrt{(\chi^2/\sigma_\chi)^2 + (\delta_{RC}/\sigma_\delta)^2} \;,$$

where σ means the standard deviation in each statistics as determined by control experiments. So, the best model is the one that minimizes this parameter ($n_{\min}$).

As can be seen in Fig. 3, there is no clear age dependence in the quality of the fit. On the other hand, on average, models with overshooting fit better the data than the classical ones. Furthermore, the quality of the fit is similar for the PEL and BaSTI libraries, but significant higher for the Padova models.

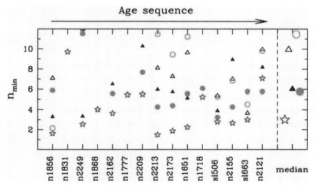

Figure 3. The quality of the fit for all clusters as determined by the $n_{\min}$. The clusters are ordered according to age, with the youngest on the left. The symbols code the different choices of stellar evolutionary models: Padova (stars), PEL with (solid triangles) and without (open triangles) overshooting, BaSTI with (solid circles) and without (open circles) overshooting. The median values for all clusters are also shown in the figure.

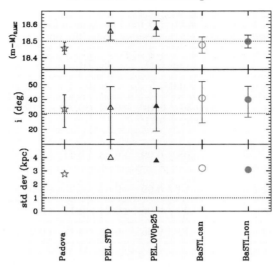

Figure 4. LMC distance (upper panel), disk inclination (middle panel) and standard deviation (lower panel) in the fitted 3D distribution for our sample of 15 stellar clusters.

5. 3D distribution and LMC distance

Since we determined the individual distances to each cluster we were able to probe the 3D distribution of these objects and also to provide new determinations of the distance to the LMC centre. These results are summarized in Fig. 4, which reveals that: 1) independent of the adopted stellar evolutionary library, the IACs have a spatial distribution roughly aligned (disk inclination $\sim 35°$) with the LMC disk as defined by field stars (in this case Cepheid stars, Nikolaev *et al.* 2004); 2) the IACs seem to belong to a thick disk, and the spatial scatter around the best-fit geometry is slightly smaller for models with overshooting; 3) the determined distance moduli to the LMC center are marginally consistent with the canonical value of 18.50 mag (Clementini *et al.* 2003); the PEL models recovered systematically larger values than the ones obtained using the other libraries.

References

Bica, E., Bonatto, C., Dutra, C. M., & Santos, J. F. C. 2008, *MNRAS*, 389, 678
Brocato, E., Di Carlo, E., & Menna, G. 2001, *A&A*, 374, 523
Castellani, V., Degl'Innocenti, S., Marconi, M., Prada Moroni, P. G., & Sestito, P. 2003, *A&A*, 404, 465
Clementini, G., Gratton, R., Bragaglia, A., *et al.* 2003, *AJ*, 125, 1309
Elson, R. A. & Fall, S.M. 1988, *AJ*, 96, 1383
Gallart, C., Zoccali, M., & Aparicio, A. 2005, *ARAA*, 43, 387
Girardi, L., Bertelli, G., Bressan, A., *et al.* 2002, *A&A*, 391, 195
Grocholski, A. J., Cole, A. A., Sarajedini, A., Geisler, D., & Smith, V. V. 2006, *AJ*, 132, 1630
Grocholski, A. J., Sarajedini, A., Olsen, K., Tiede, G., & Mancone, C. 2007, *AJ*, 134, 680
Kerber, L. O., Santiago, B. X., & Brocato, E. 2007, *A&A*, 462, 139 (KSB07)
McLaughlin, D. E., & van der Marel, R. P. 2005, *AJ*, 161, 304
Nikolaev, S., Drake, A. J., Keller, S. C., *et al.* 2004, *ApJ*, 601, 260
Olszewski, E. W., Schommer, R. A., Suntzeff, B., & Harris, H. 1991, *AJ*, 101, 515
Pietrinferni, A., Cassisi, S., Salaris, M., & Castelli, F. 2004, *ApJ*, 612, 168
Rich, R. M., Shara, M. M., & Zurek, D. 2001, *AJ*, 122, 842

The Magellanic System: Stars, Gas, and Galaxies
Proceedings IAU Symposium No. 256, 2008
Jacco Th. van Loon & Joana M. Oliveira, eds.

A study of AGB stars in LMC clusters

Thomas Lebzelter[1], Michael T. Lederer[1], Sergio Cristallo[2], Oscar Straniero[2] and Kenneth H. Hinkle[3]

[1]Department of Astronomy, University of Vienna, Türkenschanzstraße 17,
A-1180 Vienna, Austria
email: lebzelter@astro.univie.ac.at

[2]INAF, Teramo, Italy
email: straniero@oa-teramo.inaf.it

[3]NOAO, Tucson, USA
email: hinkle@noao.edu

Abstract. LMC clusters offer an outstanding opportunity to investigate the late stages of stellar evolution of stars in the mass range between 1.5 and 2 $M_\odot$. In this presentation we will focus on our results on mixing events during the evolution along the Asymptotic Giant Branch (AGB). Surface abundances have been determined for a number of cluster AGB stars from high resolution near infrared spectra. We show for the first time the evolution of C/O and $^{12}C/^{13}C$ ratios along a cluster AGB. The change of both quantities due to dredge up events is compared with model predictions. Our results indicate the late occurrence of a moderate extra-mixing in some cases.

Keywords. stars: abundances, stars: AGB and post-AGB, stars: evolution, galaxies: individual (LMC), Magellanic Clouds, galaxies: star clusters

1. Introduction

Stars of low and intermediate mass contribute significantly to the cosmic cycle of matter. The evolutionary phase during which they actually enrich the interstellar medium with freshly produced elements, however, is a quite short one, namely the Asymptotic Giant Branch (AGB) phase. The enrichment process itself can be roughly separated into three (of course connected) steps: the production of the elements in the interior, the dredge up of the material to the surface, and the mass loss (e.g., Busso *et al.* 1999)

In this paper the focus is set to the dredge up process. The dredge up events on the AGB are called the third dredge up, with the first dredge up occurring at the bottom of the RGB and the second dredge up acting in intermediate mass stars after the ignition of core He burning. Stellar evolution models (e.g., Herwig 2000; Straniero *et al.* 2006; Karakas & Lattanzio 2007) agree qualitatively on the basic characteristics of the third dredge up, but details on onset and efficiency are still awaiting constraints from observational tests.

With our study of AGB stars in stellar clusters we want to contribute to our understanding of the mixing processes during the AGB phase. Stellar clusters, generally consisting of single stellar populations, provide the required sample of AGB stars homogeneous in mass and metallicity. The intermediate age clusters in the Magellanic Clouds provide samples of stars that allow to measure the change of surface abundances due to these mixing events. The probes we use in our study are the C/O ratio and the carbon isotopic ratio $^{12}C/^{13}C$. Both quantities are affected by the dredge up of ^{12}C produced in the stellar interior.

2. Target selection

Our project aims to obtain the above mentioned abundance ratios in a variety of clusters in the age range 1.4 to 1.9 Gyr corresponding to turn-off masses between roughly 1.4 and 1.8 $M_\odot$. The lower mass limit is close to the typical model predicted minimum mass for third dredge up. The upper limit is given by the need for a sufficient number of stars currently on the AGB. Among the LMC clusters fulfilling this requirement we started our study with NGC 1846 and NGC 1978. In both clusters a number of AGB stars have been identified already before by Lloyd Evans (1983) and Frogel *et al.* (1990). Furthermore, luminous carbon stars were detected in both of them (Frogel *et al.* 1990) indicating that third dredge up occurred in these clusters.

Both clusters have a similar metallicity around [Fe/H] $= -0.4$ (Grocholski *et al.* 2006; Mucciarelli *et al.* 2007). The AGB stars of NGC 1846 have a mass of about 1.8 $M_\odot$ (Lebzelter & Wood 2007), while the colour—magnitude diagram of NGC 1978 indicates a somewhat lower mass around 1.4–1.5 $M_\odot$ (Mucciarelli *et al.* 2007). NGC 1978 is elongated, but no indication for a second population (resulting from a merging of two clusters) could be found (Mucciarelli *et al.* 2007). On the opposite Mackey & Broby Nielsen (2007) claim the detection of two populations in NGC 1846 separated in age by about 300 Myr. Both clusters were found to be very homogeneous in metallicity.

3. Observations and abundance determination

Spectra of a large fraction of stars along the whole AGB were observed in both clusters. We used the Phoenix Spectrograph at *Gemini South* to obtain spectra around the CO 3–0 band head and strong OH lines in the H-band. In addition we observed the same stars also in a small region in the K-band including lines of ^{12}CO and ^{13}CO.

Abundance and isotopic ratios were determined by fitting of the observations with synthetic spectra. These spectra are based on model atmospheres that were calculated with the COMARCS code, a modified version of the MARCS code. Starting values for T_eff and L were derived from broad-band, near-infrared photometry. The metallicity of the cluster was taken as fixed. While spectra of both oxygen and carbon rich stars were obtained, difficulties in modelling the C-stars lead to higher uncertainties for the abundances in these objects. A detailed description on data analysis, synthetic spectra calculation and abundance determination can be found in Lebzelter *et al.* (2008).

4. Results

4.1. *NGC 1846*

We present the results for the two clusters separately. More details on the findings for NGC 1846 have been published already in Lebzelter *et al.* (2008). For the O-rich stars in NGC 1846 we determined C/O ratios between 0.2 and 0.65 with an uncertainty from ± 0.05 to ± 0.1 dex. Carbon isotopic ratios between 12 and 60 were found for these stars. Both dredge up indicators show obvious variations from star to star, but nicely correlated with each other as expected.

Indeed the most luminous O-rich star in NGC 1846 has the highest C/O and ^{12}C/^{13}C and for the least luminous star in our sample we found the lowest values. However, there is no simple relation in between. This may be explainable by the luminosity changes due to the thermal pulse cycle which can significantly affect a star's location in the colour-magnitude-diagram. It is also possible that NGC 1846 indeed harbours two populations and the observed AGB consists of a mixture of stars of different age. There may also

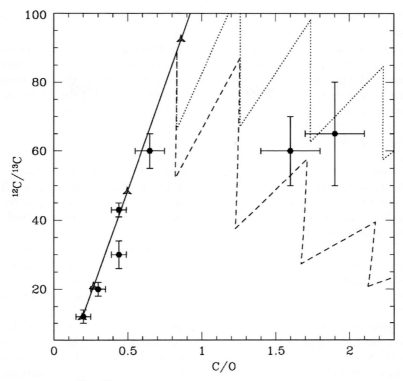

Figure 1. C/O versus ^{12}C/^{13}C for all the sampled stars. Predicted values from models with and without extra-mixing are shown. Filled circles with error bars are our measurements in NGC 1846. Solid line refers to our reference model, while the dotted and dashed ones refer to models with $T_{\mathrm{lim}} = 35 \times 10^6$ K and $T_{\mathrm{lim}} = 40 \times 10^6$ K, respectively. The symbols along the solid model line mark the values predicted after each dredge up event.

be a correlation with the pulsation mode — see Lebzelter *et al.* (2008) for a detailed discussion.

Only for two C-stars we could derive a C/O and ^{12}C/^{13}C ratio with reasonable error bars. While the C/O ratio increases as expected to values around 1.8, the isotopic ratio was found to stay close to 60, the value found for the most luminous O-rich star.

We also made a comparison of our findings for that cluster with predictions from the FRANEC stellar evolution code (Chieffi *et al.* 1998). Figure 1 illustrates the outcome of this comparison. The most likely model based on the global cluster parameters is indicated by a solid line. The model predictions for the O-rich stars are nicely reproduced by the observations if we assume an oxygen overabundance of about +0.2 dex (leading to a C/O ratio of 0.2 at the beginning of the AGB phase). On the other hand, the ^{12}C/^{13}C ratio found in the C-stars cannot be explained with standard models for the third dredge up. The most obvious explanation for this observational pattern is the occurrence of a mixing process able to bridge the radiative gap between the cool bottom of the convective envelope and the hot H-burning zone. A similar process seems to explain some abundance anomalies in red giants stars (see, e.g., Charbonnel 1995). It is usually referred to as extra or deep mixing. The values found indicate that the envelope material of the C-stars could have been exposed to a temperature on the order of 30–40 $\times$ 10^6 K. At that temperature and on an AGB time scale, ^{12}C is partially converted into ^{13}C, but only marginally into

[14]N. The two dotted lines in Fig. 1 show the predicted values if extra-mixing is included (see Lebzelter *et al.* 2008 for details).

Beside the C/O ratio we also made an attempt to measure the change of the fluorine abundance along the AGB. Abundances were derived from only one blended HF line, thus especially the absolute values have to be taken with some caution. A clear increase in the F abundance with luminosity was seen in the O-rich NGC 1846 AGB stars. Models predict also an increase in the F abundance, but less than observed.

4.2. *NGC 1978*

This cluster is an interesting candidate for comparison with our findings for NGC 1846. While having a very similar metallicity it is clearly older and thus the stars on the AGB have a smaller mass. As mentioned above, this mass is close to the predicted minimum mass for the occurrence of third dredge up.

Five oxygen rich AGB stars were analysed in this cluster. Opposite to NGC 1846 they all show very similar values of C/O and $^{12}C/^{13}C$ close to 0.2 and 12, respectively. Thus we see no indication for an ongoing dredge up in these stars, although they cover some range in luminosity. Again, only 2 C-stars could be analysed giving C/O ratios of 1.3 and 1.35, respectively. Also different from NGC 1846, the C-stars in that cluster show very high values of the carbon isotopic ratio close to the value predicted by the standard model. According to these preliminary results no extra-mixing has to be included to reproduce the observations in this cluster. Compared with the findings for field stars (Smith & Lambert 1990) the isotopic ratio measured in these objects is untypically high. We note that it is even above the solar value (89).

In Fig. 2 the resulting abundance ratios for NGC 1978's AGB stars are plotted. The big gap between the O-rich stars and the C-stars is intriguing and at the moment lacks a clear explanation. We list here several possible reasons for this gap:

- The cluster may consist of two stellar populations with different age. The O-rich stars and the C-rich stars would then have different masses and ages. The existence of two populations has been suggested by Alcaino *et al.* (1999), but a more recent detailed study by Mucciarelli *et al.* (2007) could not confirm this.

- We may assume an abundance scatter, especially in the α-elements, within the cluster. In that case the starting composition of the AGB stars would show some scatter which affected the dredge up efficiency.

- Models predict combinations of stellar parameters, where the dredge up is so efficient that the star turns from O-rich to C-rich chemistry during one dredge up event (e.g., Herwig *et al.* 2000). However, the predicted metallicity where such an event should occur is much lower than the value found for NGC 1978.

- As the derived mass of the AGB stars in this cluster is very close to the mass limit for third dredge up predicted by stellar evolution models the findings are possibly explained by the following scenario: the O-rich stars found at the bottom of the AGB have a slightly lower mass than the C-stars at the top of the AGB. The limit for third dredge up may be exactly between the mass of the C-rich and O-rich stars. Then the O-rich stars would have not enough mass to encounter a third dredge up. However, dredge up models predict a "mild" dredge up (not producing a C-star) at the lower mass end which is not observed.

At this point none of the scenarios gives a satisfying explanation for the observations. A more detailed comparison with models will be given in a forthcoming paper (Lederer *et al.* in preparation).

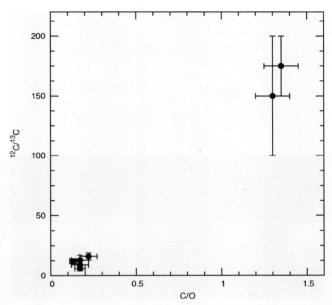

Figure 2. C/O versus ^{12}C/^{13}C for the sampled stars in NGC 1978. The lack of dredge up in the O-rich stars is obvious. The large error bars in the C-rich case result from both uncertainties in the determination of the stellar parameters and the insufficient accuracy of the line lists. For the other C-stars only a lower C/O limit could be derived which is approximately 1.5.

5. Conclusions

Our observations clearly illustrate that AGB stars in globular clusters are well suited to test and constrain the description of the AGB phase in stellar evolution models. The need for an inclusion of extra-mixing has been demonstrated. The exact conditions for the occurrence of this mixing require further investigation. We will continue our study by extending the observational material towards other sets of stellar parameters and by intensifying comparison with stellar models. Finally, we emphasize the need for further improvements in the area of molecular line lists and dynamical model atmospheres to fully explore the wide variety of stellar clusters in the Magellanic Clouds.

Acknowledgements

This work was supported by the Austrian FWF under project number P18171−N02 and P20046−N16. MTL has been supported by the Austrian Academy of Sciences (DOC programme). Based on observations obtained at the Gemini Observatory. The spectra were obtained as part of program GS-2005B-C-7.

References

Alcaino, C., Liller, W., Alvarado, F., *et al.* 1999, *A&AS*, 135, 103
Busso, M., Gallino, R., & Wasserburg, J. 1999, *ARAA*, 37, 239
Charbonnel, C. 1995, *ApJ*, 453, 41
Chieffi, A., Limongi, M., & Straniero, O. 1998, *ApJ*, 502, 737
Frogel, J. A., Mould, J., & Blanco, V. M., 1990, *ApJ*, 352, 96
Grocholski, A. J., Cole, A. A. Sarajedini, A., *et al.* 2006, *AJ*, 132, 1630
Herwig, F. 2000, *A&A*, 360, 952
Herwig, F., Blöcker, T., & Driebe, T. 2000, *MemSAI*, 71, 745

Karakas, A. & Lattanzio, J. 2007, *PASA*, 24, 103

Lebzelter, T. & Wood, P. R. 2007, *A&A*, 475, 643

Lebzelter, T., Lederer, M. T., Cristallo, S., *et al.* 2008, *A&A*, 486, 511

Lloyd Evans, T. 1983, *MNRAS*, 204, 985

Mackey, A. D. & Broby Nielsen, P. 2007, *MNRAS*, 379, 151

Mucciarelli, A., Ferraro, F. R., Origlia, L., & Fusi Pecci, F. 2007, *AJ*, 133, 2053

Smith, V. V. & Lambert, D. L. 1990, *ApJS*, 72, 387

Straniero, O., Gallino, R., & Cristallo, S. 2006, *N.Phys.A*, 777, 311

At the conference dinner inside Wrenbury Hall.

Session VII

The final stages of stellar evolution and feedback

The Magellanic System: Stars, Gas, and Galaxies
Proceedings IAU Symposium No. 256, 2008
Jacco Th. van Loon & Joana M. Oliveira, eds.

The production of dust in the Magellanic Clouds

G. C. Sloan[1]

[1]Department of Astronomy, Cornell University
108 Space Sciences, Ithaca, NY 14853-6801, USA
email: sloan@isc.astro.cornell.edu

Abstract. The sensitivity of the Infrared Spectrograph on the *Spitzer Space Telescope* has enabled detailed surveys of mass-losing stars in the Large and Small Magellanic Clouds. Comparisons of samples from these galaxies and the Milky Way reveal how the dust produced by evolved stars depends on the metallicity of the host environment. Oxygen-rich stars show several trends with metallicity. In more metal-poor environments, fewer of them show dust excesses, the circumstellar SiO absorption grows weaker, the quantity of silicate dust decreases, and alumina dust grows rare. As carbon stars grow more metal-poor, the amount of circumstellar acetylene gas increases, while the amount of trace dust elements like SiC and MgS decreases. However, there is little dependence on metallicity in the amount of amorphous carbon dust produced by carbon stars, because they produce the carbon needed to make dust themselves. As galaxies grow more metal-poor, the composition of the dust they produce should grow more carbon rich.

Keywords. stars: AGB and post-AGB, stars: atmospheres, stars: carbon, stars: mass loss, stars: supergiants, Magellanic Couds, infrared: stars

1. Introduction

The sensitivity of the Infrared Spectrograph (IRS) (Houck *et al.* 2004) on the *Spitzer Space Telescope* (Werner *et al.* 2004) makes possible spectroscopy of individual supergiants and stars on the asymptotic giant branch (AGB) in nearby galaxies to distances of over 100 kpc. As a result, multiple observing programs have investigated mass loss and dust production in several Local Group galaxies. This paper focuses on the Large Magellanic Cloud (LMC) and Small Magellanic Cloud (SMC). Both Clouds have a complex structure and history (as this conference has demonstrated so well). To assume that each system can be described with a single distance and metallicity is somewhat of an oversimplification. Nonetheless, this assumption is the first step in using Local Group galaxies as probes of how mass loss and dust production vary with metallicity. Making this leap reveals trends relevant not just to the Local Group but also to more distant galaxies where *Spitzer* can detect dust but cannot resolve the separate components which produce it.

Five *Spitzer* programs have spectroscopically surveyed evolved stars in the Magellanic Clouds (see the review by Sloan *et al.* 2008b). The majority of the initial papers have focused on the carbon stars in the samples, in both the LMC (Zijlstra *et al.* 2006; Leisenring *et al.* 2008) and the SMC (Sloan *et al.* 2006; Lagadec *et al.* 2007). These papers have led to the discovery of several dependencies of the gas and dust properties around carbon stars on metallicity. As the samples grow more metal poor, the amount of SiC and MgS decreases relative to the dominant component, amorphous carbon, while the absorption from acetylene (C_2H_2) actually grows stronger. In the Milky Way, carbon stars show absorption bands from both acetylene and HCN, but in the Magellanic samples, HCN absorption is generally absent (Matsuura *et al.* 2006).

The MC_DUST program is a guaranteed time program by the IRS team to examine evolved stars in both the LMC and SMC. Sloan *et al.* (2008a) recently presented a detailed analysis of the full spectroscopic sample, giving the first in-depth look at the evolved oxygen-rich stars in both Magellanic Clouds. A comparison of this sample with the carbon stars published by the other groups reveals some fundamental differences between the metallicity dependencies of carbon-rich and oxygen-rich dust.

2. Naked stars

Table 1. Fraction of naked oxygen-rich stars.

Sample	Period (days)		
	$\leqslant 250$	250–700	> 700
Galaxy	2 of 106	5 of 269	0 of 11
LMC	2 of 3	2 of 9	0 of 8
SMC	1 of 1	6 of 7	0 of 1

Naked stars (i.e. stars with no obvious dust emission in their infrared spectra) account for more than one quarter of the total MC_DUST sample. Table 1 compares the fraction of naked stars in the oxgyen-rich MC_DUST sample to the Galactic sample defined by Sloan and Price (1995, 1998) from observations by the Low-Resolution Spectrometer (LRS) on the *Infrared Astronomical Satellite*. Moving to progressively lower metallicities increases the percentage of naked stars, as is most obvious by examining the sources with periods between 250 and 700 days.

Table 2. SiO band strengths.

Sample	W (μm)
Galactic SWS	0.35±0.11
Galactic IRS	0.28±0.06
LMC	0.26±0.11
SMC	0.16±0.08

Most late-type oxygen-rich giants have an SiO molecular absorption band at 8 μm. Table 2 compares the equivalent width of this band in two Galactic samples and the Magellanic Clouds. Heras *et al.* (2002) defined the Galactic SWS sample from observations by the Short Wavelength Spectrometer (SWS) aboard the *Infrared Space Observatory*. The Galactic IRS sample comes from the IRS program of Sloan *et al.* (in preparation). As the sample grows more metal poor, the strength of the SiO absorption weakens. SiO is the building block of silicate dust, and its growing weakness indicates that silicates are more difficult to form in more metal-poor stars.

3. Oxygen-rich dust

The MC_DUST sample contains 17 sources with clearly identifiable silicate dust emission, 14 of which show no evidence of self-absorption at 10 μm. To analyze these spectra, Sloan *et al.* (2008a) applied the technique originally developed for the LRS database by Sloan & Price (1995, 1998). This technique classifies the spectrum by first fitting and

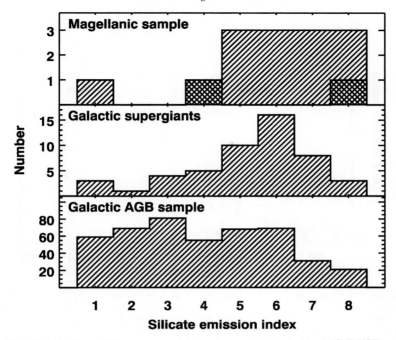

Figure 1. The distribution of the oxygen-rich dust chemistries in the MC_DUST sample, compared to Galactic supergiants and AGB stars. In the top panel, the two SMC sources are cross-hatched; the remainder are LMC sources. SE1–3 spectra arise from alumina-rich dust and are largely absent in the Magellanic sources considered here. (Adapted from Fig. 27 of Sloan *et al.* 2008a.)

subtracting an estimated stellar continuum and then quantifying the shape of the 10 μm feature with the remaining flux at 10, 11, and 12 μm. Spectra are classified as SE (for silicate emission) 1 through 8. Classifications of SE1–3 correspond to dust emission dominated by amorphous alumina grains (Al_2O_3), while the other end of the sequence (SE6–8) is dominated by amorphous silicates (Egan & Sloan 2001).

Figure 1 shows that amorphous alumina is largely absent from the Magellanic sample, while it is common in the Galactic AGB sample. The distribution of Magellanic SE indices looks much like the Galactic supergiant sample, even though the majority of the Magellanic sources are confirmed to be on the AGB.

Sloan & Price (1995) defined the dust emission contrast (DEC) in their oxygen-rich sample as the ratio of the dust excess to the stellar contribution integrated from 7.7 to 14.0 μm. This measure quickly assesses the amount of dust in the spectrum. Figure 2 plots the DEC for those oxygen-rich variables with known periods in the MC_DUST and Galactic samples. The amount of dust generally increases with increasing pulsation period in the Galactic sample, but the Magellanic samples lag behind. For periods < 500 days, most of the LMC and SMC sample are naked. From this point up to periods of 700 days, most LMC sources show some dust, but less than the Galactic sample, while the SMC sources are still naked. For periods < 700 days, the amount of dust clearly decreases with lower metallicity.

At longer periods, the situation is complicated by the presence of younger, more massive stars which may not reflect the average metallicity of their host galaxy. In Figure 2, most of the longer-period variables in the LMC show strong dust emission. Of the four

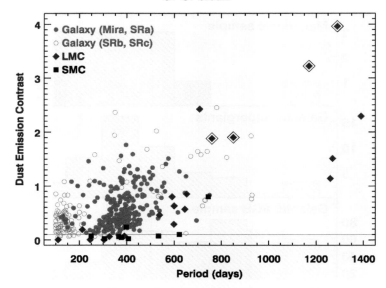

Figure 2. Dust emission contrast as a function of pulsation period for the MC_DUST sample of Magellanic stars and a Galactic sample observed by the LRS. The horizontal line separates naked from dusty stars. The four outlined LMC sources are known or suspected supergiants. (Based on Fig. 26 of Sloan *et al.* 2008a.)

outlined in Figure 2, the two with periods close to 800 days are certain supergiants, and the two with DEC > 3 are probably super-AGB sources.

4. Carbon stars

The key measure used in the study of Magellanic carbon stars is the [6.4]−[9.3] color, which is defined in the Manchester method to measure the amount of amorphous carbon in the spectra. Amorphous carbon shows no obvious spectral structure in the infrared; instead, its opacity drops steadily with wavelength in the infrared, falling as λ^{-2}. The [6.4]−[9.3] color measures the contribution of amorphous carbon in two narrow wavelength ranges which are mostly free of molecular absorption bands or dust emission features. Groenewegen *et al.* (2007) applied radiative transfer models to all of the Magellanic carbon stars in the samples published through 2007 and showed that the [6.4]−[9.3] color is closely correlated with the mass-loss rate. More significantly, they noted that there was little evidence that the mass-loss rate from carbon stars changed with metallicity.

Sloan *et al.* (2008a) examined the dust properties of the carbon stars in the Galactic and Magellanic samples as a function of pulsation period. Figure 3 shows that there is no apparent difference between these samples, All trace the same relation of increasing [6.4]−[9.3] color, and thus mass-loss rate, with increasing period, no matter the initial metallicity of the sample.

5. Discussion

Infrared spectroscopy from *Spitzer* reveals an important difference in how the production of oxygen-rich and carbon-rich dust depends on metallicity in nearby Local Group galaxies. While the production of oxygen-rich dust declines in more metal-poor environments, no change in carbon-rich dust is detected. Observations of the Magellanic Clouds

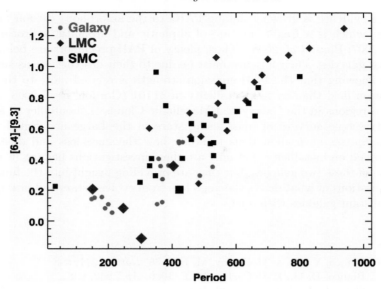

Figure 3. The [6.4]–[9.3] color as a function of pulsation period for Magellanic and Galactic carbon stars. The [6.4]–[9.3] color traces mass-loss rate. No difference in mass-loss rate as a function of period is apparent in the samples. (Based on Fig. 29 of Sloan *et al.* 2008a).

show that the fraction of AGB stars which are carbon rich increases at lower metallicities (Blanco *et al.* 1978, 1980). This increase results from a drop in the lower mass limit for carbon stars with lower initial metallicities (Renzini & Voli 1981). These trends lead to the conclusion that more metal-poor galaxies will produce less oxygen-rich dust and more carbon-rich dust.

The observed trends are relevant to the question of how to power the mass loss from oxygen-rich AGB stars. Woitke (2006) modeled the process in both oxygen-rich and carbon-rich stars, and he found that while the opacity of carbon-rich dust can drive the mass loss, oxygen-rich dust is too transparent to transfer sufficient radiation pressure outward to the gas. Lagadec & Zijlstra (2008) have recently examined observations of evolved stars in the Magellanic Clouds and other Local Group galaxies. They find that the final phase of high mass loss on the AGB, the superwind, has different triggers in the carbon-rich and oxygen-rich environments. For carbon stars, the superwind begins when the C/O ratio crosses a metallicity-dependent threshhold, while for oxygen-rich sources, the superwind begins at a critical luminosity which grows higher as the metallicity decreases. This conclusion still does not answer the question of what drives the mass loss from the oxygen-rich stars. In metal-poor stars, the lack of Si should limit the production of silicate grains, making a solution even more difficult than in the Milky Way. Perhaps some combination of pulsation velocity, molecular opacity, and grain opacity is sufficient to drive the gas over the escape velocity.

The mass-loss rate from carbon stars does not depend on metallicity in any obvious way. This observation helps address the question of why emission from polycyclic aromatic hydrocarbons (PAHs) is weak in metal-poor galaxies (e.g., Engelbracht *et al.* 2005; Wu *et al.* 2006). Either the PAHS are destroyed more efficiently by the harsher interstellar radiation in these systems (Galliano *et al.* 2005; Madden *et al.* 2006) or the precursors to PAHs are not forming (Galliano *et al.* 2008). If anything, the results from Magellanic carbon stars suggest the injection of *more* carbon-rich grains into the interstellar medium, not less. The evidence is growing linking PAHs to the amorphous carbon produced by carbon stars, which is a fragile mixture of aliphatic and aromatic hydrocarbons (e.g., Sloan *et al.* 2007; Pino *et al.* 2008). Thus, plenty of PAH precursors are being produced in metal-poor galaxies. Their absence must be due to their destruction, as supported by recent work showing that the PAH emission strength is related more to the harshness of the radiation field than to the metallicity in M 101 (Gordon *et al.* 2008) and within extended H II regions in the Galaxy and Magellanic Clouds (Lebouteiller *et al.* 2008).

The spectroscopic surveys of mass-losing stars in the Large and Small Magellanic Clouds with *Spitzer* have allowed us to study how the mass-loss and dust formation processes depend on metallicity. Not only are these investigations helping us understand the evolution of these two galaxies, but they are providing insight into the fundamentally important questions of what drives the mass loss from evolved stars and how this process seeds more distant galaxies with dust.

References

Blanco, B. M., Blanco, V. M., & McCarthy, M. F. 1978, *Nature*, 271, 638
Blanco, V. M., Blanco, B. M., & McCarthy, M. F. 1980, *ApJ*, 242, 938
Egan, M. P. & Sloan, G. C. 2001, *ApJ*, 558, 165
Engelbracht, C. W., Gordon, K. D., Rieke, G. H., Werner, M. W., Dale, D. A., & Latter, W. B. 2005, *ApJ*, 628, 29
Galliano, F., Madden, S. C., Jones, A. P., Wilson, C. D., & Bernard, J. -P. 2005, *A&A*, 434, 867
Galliano, F., Dwek, E., & Chanial, P. 2008, *ApJ*, 672, 214
Gordon, K. & Clayton, G. C. 1998, *ApJ*, 500, 816
Gordon, K. D., Engelbracht, C. W., Rieke, G. H., Misselt, K. A., Smith, J. -D. T., & Kennicutt, R. C., Jr. 2008, *ApJ*, 682, 336
Groenewegen, M. A. T., Wood, P. R., Sloan, G. C., *et al.* 2007, *MNRAS*, 376, 313
Heras, A. M., Shipman, R. F., Price, S. D., *et al.* 2002, *A&A*, 394, 539
Houck, J. R., Roellig, T. L., van Cleve, J., *et al.* 2004, *ApJS*, 154, 18
Lagadec, E., Zijlstra, A. A., Sloan, G. C., *et al.* 2007, *MNRAS*, 376, 1270
Lagadec, E. & Zijlstra, A. A. 2008, *MNRAS*, 390, L59
Lebouteiller, V., Bernard-Salas, J., Brandl, B., Whelan, D. G., Wu, Y., Charmandaris, V., Devost, D., & Houck, J. R. 2008, *ApJ*, 680, 398
Leisenring, J. M., Kemper, F., & Sloan, G. C. 2008, *ApJ*, 681, 1557
Madden, S. C., Galliano, F., Jones, A. P., & Sauvage, M. 2006, *A&A*, 446, 877
Matsuura, M., Wood, P. R., Sloan, G. C., *et al.* 2006, *MNRAS*, 371, 415
Pino, T., Dartois, E., Cao, A. -T., *et al.* 2008, *A&A*, 490, 665
Renzini, A. & Voli, M. 1981, *A&A*, 94, 175
Sloan, G. C. & Price, S. D. 1995, *ApJ*, 451, 758
Sloan, G. C. & Price, S. D. 1998, *ApJS*, 119, 141
Sloan, G. C., Kraemer, K. E., Matsuura, M., Wood, P. R., Price, S. D., & Egan, M. P. 2006, *ApJ*, 645, 1118
Sloan, G. C., Jura, M., Duley, W. W., *et al.* 2007, *ApJ*, 664, 1144
Sloan, G. C., Kraemer, K. E., Wood, P. R., Zijlstra, A. A., Bernard-Salas, J., Devost, D., & Houck, J. R. 2008a, *ApJ*, 686, 1056
Sloan, G. C., *et al.* 2008b, in D. G. Luttermoser, B. J. Smith, & R. E. Stencel (eds.), *The Biggest, Baddest, Coolest Stars*, in press
Werner, M., Roellig, T. L., Low, F. J., *et al.* 2004, *ApJS*, 154, 1
Woitke, P. 2006, *A&A*, 452, 537
Wu, Y., Charmandaris, V., Hao, L., Brandl, B. R., Bernard-Salas, J., Spoon, H. W. W., & Houck, J. R. 2006, *ApJ*, 639, 157
Zijlstra, A. A., Matsuura, M., Wood, P. R., *et al.* 2006, *MNRAS*, 370, 1961

The Magellanic System: Stars, Gas, and Galaxies
Proceedings IAU Symposium No. 256, 2008
Jacco Th. van Loon & Joana M. Oliveira, eds.

Dust around red supergiants in the Magellanic Clouds

Geoffrey C. Clayton[1], **W. Freeman**[1], **S. Bright**[1], **P. Massey**[2],
K. D. Gordon[3], **E. Levesque**[4], **B. Plez**[5], **K. Olsen**[6] and **J. Nordhaus**[7]

[1] Department of Physics & Astronomy, Louisiana State University, Baton Rouge, LA 70803,
USA
email: gclayton@fenway.phys.lsu.edu

[2] Lowell Observatory, 1400 W Mars Hill Rd., Flagstaff, AZ 86001, USA
email: phil.massey@lowel.edu

[3] Space Telescope Science Institute, 3700 San Martin Drive Baltimore, MD 21218
email: kgordon@stsci.edu

[4] Institute for Astronomy, University of Hawaii, 2680 Woodlawn Drive, Honolulu, HI 96822,
USA
email: emsque@ifa.hawaii.edu

[5] GRAAL, Universite de Montpellier II, CNRS, 34095 Montpellier, France
email: bertrand.plez@graal.univ-montp2.fr

[6] Cerro Tololo Inter-American Observatory, NOAO, Casilla 603, La Serena, Chile
email: kolsen@noao.edu

[7] Department of Astrophysical Sciences, Princeton University, Princeton, NJ 08544, USA
email: nordhaus@astro.princeton.edu

Abstract. It is both surprising and exciting to find that young galaxies at high redshift contain large dust masses. For galaxies at $z > 5$, after only 1 Gyr, there has not been time for low-mass stars to have evolved to the AGB phase and produce dust. In such galaxies, Type II SNe and red supergiants (RSGs) may even dominate the dust production rate. It has long been known that RSG atmospheres produce dust, but little is known about it. We are pursuing three parallel studies to better understand RSG dust. First, we are using optical spectra and JHK photometry to characterize the optical and near-IR extinction curves of the RSGs. Second, we are using the optical spectra combined with 2MASS, IRAC and MIPS photometry to estimate the dust mass loss rates from Local Group RSGs. In addition, we will use our Monte Carlo radiative transfer models to analyze the emission from dust in the circumstellar shells. Third, the final piece of the puzzle is being provided by obtaining new IRS spectra of LMC and SMC RSGs. We plan to use the IRS to make a systematic study of the dust properties in RSG shells in the LMC and SMC so that we can probe how they may vary with a large range of galactic metallicities. The derived stellar SEDs and extinction curves will be combined with *Spitzer* IRAC and MIPS photometry and IRS spectra for use as inputs to our Monte Carlo codes which will be used to study the composition, size distributions and clumpiness of the dust.

Keywords. circumstellar matter, supergiants, dust, extinction, Magellanic Clouds

1. Introduction

Red Supergiants (RSGs) are the evolved, He-burning descendants of moderately massive ($\leqslant 40$ M$_\odot$) O and B stars. In reviews of the origin of cosmic dust, the role of RSGs is often ignored, with the primary sources of dust given as SNe and low-mass AGB stars. However, for primordial galaxies, RSG dust will play a more important role due the absence of old low-mass AGB stars. So, in extreme environments (such as primordial

411

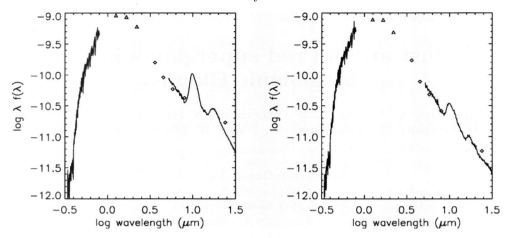

Figure 1. Two of our target RSGs, LMC 119219 (left) and LMC 116895 (right). The plotted spectrum to the blue of 1 μm is the visible spectrum. The triangles are the 2MASS JHK, and the diamonds are the IRAC 3.6, 4.5, 5.8 & 8.0 μm, and MIPS 24 μm photometry. The solid line in the IR is the *Spitzer*/IRS spectrum.

galaxies, lacking in AGBs), we expect that RSGs may be a primary source of dust along with the Type II SNe (Massey *et al.* 2005).

It has long been known that RSG atmospheres are "smokey" and produce dust at distances of 5–10 stellar radii from the star (Danchi *et al.* 1994), but it has only just been realized that a significant fraction of RSGs in Galactic OB associations and clusters show several magnitudes of excess visual extinction compared to OB stars in the same regions. Massey *et al.* (2005) demonstrate that this is in fact just what we *should* expect, given the amount of dust produced by RSGs (Josselin *et al.* 2000). Furthermore, the stars with the highest amount of extra visual extinction also show significant near-UV (NUV) excesses compared to MARCs stellar atmosphere modes (Plez *et al.* 1992) reddened by the standard reddening law. The discrepancy in the NUV is striking. The size of the NUV discrepancy is well correlated with the amount of extra extinction, the dust production rate $\dot{M}_{\rm d}$ as measured from the 12-μm excess, and the bolometric luminosity of the star computed from the K-band (Massey *et al.* 2005).

2. Infrared emission from dust around red supergiants

We have been amassing a unique dataset which can be used to investigate the amount and nature of dust produced by RSGs in very different environments. To this end, optical spectra are being obtained for RSGs in various galaxies in the Local Group. Moderate-resolution, high S/N, spectrophotometry has been obtained of 36 RSGs in the LMC and 37 RSGs in the SMC (Levesque *et al.* 2006). These spectra were taken using the *CTIO Blanco* 4 m telescope and cover the range 3500–9000 Å (Figure 1). Similar high-quality data were recently obtained for a sample of M 31 RSGs using the *MMT* 6.5 m. It is planned to expand the sample to RSGs in M 33 and NGC 6822 in the near future. The stars in our sample have been definitively identified as RSGs by relying on radial velocities to distinguish foreground dwarfs from RSGs in the LMC, SMC and M 31.

In order to investigate dust around the RSGs, we need to measure their emission in the infrared. High energy photons from the central star warms dust lying in a surrounding

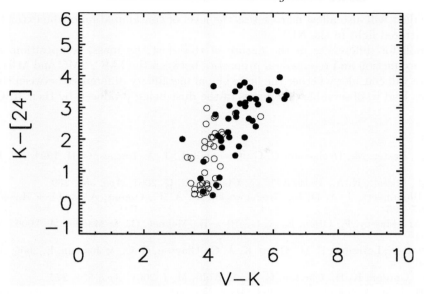

Figure 2. Colour–colour plot of LMC (filled circles) and SMC (open circles) RSGs. The $K - [12]$ color has been used to estimate dust mass-loss rates (Josselin *et al.* 2000). We plan to calibrate the $K - [24]$ color using the MIPS 24 μm photometry.

shell. The warm dust re-emits this energy in the IR. Near-IR photometry in the JHK bands is available from the 2MASS archive, as well as new JHK photometry for the M 31 stars we obtained using the *KPNO* 4 m telescope. In addition, we have a *Spitzer* archival program to get IRAC and MIPS photometry of RSGs in Local Group galaxies. We will use the derived stellar SEDs and IR photometry as inputs to the radiative transfer codes described below. By measuring the amount of dust produced by RSGs in different environments (low vs. high metallicity), we can quantify their role in galactic dust-content evolution, which will have an impact on our understanding of massive star evolution and dust production. In particular, we will quantify the circumstellar infrared excess, and the relationship of mass loss to stellar properties, using the $K - [24]$ colour index as a measure of mass loss similar to what has been done previously with the $K - [12]$ colours (Josselin *et al.* 2000). See Figure 2.

3. Modeling

The size, shape, and composition of the dust grains are inputs to the radiative transfer models. We will apply the DIRTY Monte Carlo code (Misselt *et al.* 2001; Gordon *et al.* 2001) to model radiative transfer in the circumstellar shells around the RSGs. We expect that many of our sample RSGs in the LMC, SMC and M 31 will have significant detections in the IRAC bandpasses, as well as at 24 μm with MIPS. Figure 1 shows the IR excess from the shells around two of the LMC stars in our sample. The stellar SEDs and extinction curves, derived from the optical and near-IR data, will be combined with the *Spitzer* IR photometry to use as inputs to the radiative transfer codes. We will model the dust shells using amorphous carbon and silicate dust grains as well as Polycyclic aromatic hydrocarbons molecules (PAHs) appropriate to these shells. This modeling will allow us to infer the total circumstellar dust mass around the sample RSGs. The estimated optical depth of the dust will also allow us to assess the level of contamination of the extinction curves by scattered light in the NUV.

We will look for differences in the nature of this dust, its mass, composition, size distribution, extinction and re-emission properties between the LMC, SMC and M 31. In particular, we will consider whether the large global metallicity differences between these three galaxies lead to observable differences in the dust being produced by their RSGs.

References

Danchi, W. C., Bester, M., Degiacomi, C. G., Greenhill, L. J., & Townes, C. H. 1994, *AJ*, 107, 1469

Gordon, K. D., Misselt, K. A., Witt, A. N., & Clayton, G. C. 2001, *ApJ*, 551, 269

Josselin, E., Blommaert, J. A. D. L., Groenewegen, M. A. T., Omont, A., & Li, F. L. 2000, *A&A*, 357, 225

Levesque, E. M., Massey, P., Olsen, K. A. G., Plez, B., Meynet, G., & Maeder, A. 2006, *ApJ*, 645, 1102

Massey, P., Plez, B., Levesque, E. M., Olsen, K. A. G., Clayton, G. C., & Josselin, E. 2005, *ApJ*, 634, 1286

Misselt, K. A., Gordon, K. D., Clayton, G. C., & Wolff, M. J. 2001, *ApJ*, 551, 277

Plez, B., Brett, J. M., & Nordlund, Å. 1992, *A&A*, 256, 551

At the conference dinner inside Wrenbury Hall.

The Magellanic System: Stars, Gas, and Galaxies
Proceedings IAU Symposium No. 256, 2008
Jacco Th. van Loon & Joana M. Oliveira, eds.

© 2009 International Astronomical Union
doi:10.1017/S1743921308028809

The optically bright post-AGB population of the LMC

Els van Aarle[1], Hans van Winckel[1], Tom Lloyd Evans[2] and Peter R. Wood[3]

[1] Instituut voor Sterrenkunde, Celestijnenlaan 200D BUS 2401, 3001 Leuven, Belgium
email: els.vanaarle@ster.kuleuven.be, hans.vanwinckel@ster.kuleuven.be

[2] School of Physics & Astronomy, University of St Andrews,
North Haugh, St Andrews KY16 9SS, Scotland, UK
email: thhle@st-andrews.ac.uk

[3] Mount Stromlo Observatory, Cotter Road, Weston Creek, ACT 2611, Australia
email: wood@mso.anu.edu.au

Abstract. The detected variety in chemistry and circumstellar shell morphology of the limited sample of Galactic post-AGB stars is so large, that there is no consensus yet on how individual objects are linked by evolutionary channels. The evaluation is complicated by the fact that the distances and hence luminosities of these objects are poorly known. In this contribution we report on our project to overcome this problem by focusing on a significant sample of post-AGB stars with known distances: those in the LMC. Via cross-correlation of the infrared SAGE-SPITZER catalogue with optical catalogues we selected a sample of 322 LMC post-AGB candidates based on their position in the various colour-colour diagrams. We determined the fundamental properties of 82 of them, using low resolution optical spectra that we obtained at *Siding Spring* and *SAAO*. We selected a subsample to be studied at high spectral resolution in order to obtain accurate abundances of a wide range of species. This will allow us to connect the theoretical predictions with the obtained surface chemistry at a given luminosity and metallicity. By this, we want to constrain important structure parameters of the evolutionary models. Preliminary results of the selection process are presented.

Keywords. stars: AGB and post-AGB, circumstellar matter, stars: evolution, Hertzsprung-Russell diagram, galaxies: individual (LMC), Magellanic Clouds

1. Introduction

Post-AGB stars are stars of low and intermediate mass that evolve rapidly from the Asymptotic Giant Branch (AGB) towards the Planetary Nebula (PN) phase before cooling down as a white dwarf. The evolutionary stage of these objects is still badly understood. In the Galactic sample, the post-AGB stars display for instance a much broader chemical diversity than can be predicted based on the theoretical evolutionary models: stars with similar global properties can show completely different photospheric abundance patterns (see Van Winckel 2003 and references therein). The connection between the initial metallicity and the occurrence of a third dredge-up also remains unclear. Furthermore, the very different circumstellar geometries and kinematics of post-AGB outflows can not be explained yet (see Balick & Frank (2002); Sahai *et al.* (2007), and references therein). The fact that all these questions remain unresolved, is a consequence of the lack of fundamental insight in how the internal structural evolution leads to the chemical enrichment of the stellar surface as well as of the poor understanding of the physical processes that are responsible for the high, often collimated, mass-loss rates that we observe. In addition to gaining insight in their own evolutionary phase, studying post-AGB stars can also help us to better understand the AGB evolution as well as the PNe physics (e.g., Van Winckel 2003; Hrivnak 2003; García Lario 2006). Thanks to the broad spectral range of their electromagnetic spectrum, a simultaneous study of the stellar photosphere and the circumstellar environnement is possible: the central star is responsible for the UV- and optical emission while the cool circumstellar envelope radiates in the infrared.

Resolving the mysteries of post-AGB stars is, however, rather difficult since this evolutionary phase is relatively short. Hence, not many post-AGB stars are known and detailed studies of individual objects prevail in the literature (for a list of post-AGB stars in the Galaxy, see Szczerba *et al.* 2007). Moreover, the poorly constrained distances of the Galactic sample prevent good theoretical diagnostics. The main goal of our project is therefore to study a significant sample of post-AGB stars with a known distance and hence luminosity: those in the LMC. We first discuss the selection of our sample and go into detail on some preliminary results. We end with conclusions and a short description of the future plans.

2. Selection of the sample

The post-AGB stars we observe in the Galaxy can be roughly divided into two different classes according to the shape of their SED. The most well known of these types has a double peaked SED in which the peak at longer wavelengths corresponds to the emission of a detached shell of expanding and cooling dust. If the optical depth in the line of sight is not too large, the central star is visible at shorter wavelengths. The second type of SED shows a broad infrared excess, which indicates that there is still hot dust in the system. These SED's are indicative of a disc rather than an expanding envelope (De Ruyter *et al.* 2006) and recent interferometric results show that the dusty discs are very compact and likely very stable (e.g., Deroo *et al.* 2006; Gielen *et al.* 2008).

For the selection of our sample, we made use of the magnitudes provided in the point source catalogue of the SAGE-SPITZER LMC survey (Meixner *et al.* 2006). Our initial selection criteria to look for optically bright post-AGB stars included a flux limit of $F(24) < 1\,\mathrm{Jy}$, with $F(24)$ the flux at 24 μm, to avoid the majority of the supergiants. To avoid the bulk of the young stellar objects, we imposed that $2\,\mathrm{mJy} < F(24)$. The existence of the two types of SED's is reflected in the other criteria we chose. To detect post-AGB stars with a freely expanding, detached dust shell, we focused on the presence of cool dust in the system and therefore imposed that $F(24) > F(8)$. The selection criterion for disc sources was inspired by the Galactic sample for which we folded the SEDs with the filter transmission curves of *Spitzer*. We also looked for the presence of cool dust, but we made the criterion less strong: $F(24) > 0.5 \times F(8)$. Furthermore we imposed a $J - K < 1.0$ mag as an additional criterion. This has the advantage that it also helps to avoid the bulk of the M-type stars.

On their poster for this conference, Toshiya Ueta and collaborators (see Ueta *et al.* these proceedings) used similar criteria based on SAGE colours, to select post-AGB stars in the LMC. The main difference between our method and theirs is that we focus on optically bright post-AGB stars while they include also the obscured ones in their criteria. This is justified by the different goals we aim at: they plan to make a list of all post-AGB stars of the LMC while we are selecting a range of stars to study in more detail.

430 objects fulfill our colour and magnitude conditions. We then correlated the positions with three optical catalogues: Massey's catalogue of the UBVR CCD survey of the

Magellanic clouds (Massey 2002) which gave 68 matches, the Guide Star Catalog, Version 2.3.2 (2006) which resulted in 313 matches and the LMC stellar catalog of Zaritsky *et al.* (2004) which gave 275 matches. The total sample at this point contained 381 stars. A search on SIMBAD allowed to remove the stars that are known not to be post-AGB stars and this resulted in a final sample of 322 stars.

To characterise the sample more, we obtained low-resolution optical spectra ($\sim$ 3.7 Å) at the *Siding Spring Observatory* in Australia or at *SAAO* for 82 of the 322 stars so far. The spectra were reduced using IRAF and following the standard recipies for long-slit spectral reduction. The main goal of this, is to determine the spectral type of the different objects and to examine their membership of the LMC. We detected all spectral types from O to M with number density peaks around B and M. About one third of the post-AGB candidates show emission lines, which might indicate the presence of hot circumstellar gas (and dust).

3. SEDs

To determine the luminosities of our sample, we developed a method based on a Monte Carlo simulation to minimise simultaneously for the total reddening as wel as for the effective temperature of the star. First, we determined separately the best atmosphere model to be used for each star. We allowed a small range in temperatures, constrained by the spectral type which we deduced from the low-resolution optical spectra. The effective temperatures of all models used in the routine range from 3000 up to 10 000 K in steps of 250 K, from 10 000 up to 13 000 K in steps of 500 K, from 13 000 to 35 000 K in steps of 1000 K and from 35 000 up to 50 000 K in steps of 2500 K. Log g is not a free parameter since only one model was used for each temperature. For atmosphere models with effective temperatures higher than or equal to 3500 K, we used Kurucz models with a metallicity of -0.3. For the range of T_{eff} from 3000 to 3250 K we used MARCS-models with metallicity -1.0. A lower limit of 0.04 was imposed for the value of $E(B-V)$ to correct for the interstellar extinction. By use of the Monte Carlo Method with 150 steps and a Gaussian distribution of the errors on the photometric points, the best model for each star is determined minimising the χ^2 between the rescaled atmosphere model and the deredded photometric data used only up to 25 000 Å. The spectra were left out of this process since the uncertainties on the calculated flux remain too large. If other photometric data are available, the magnitudes from the Guide Star Catalogue were not involved in the procedure since they differ significantly from those in the other two catalogues.

The model that was most frequently found as a solution is used to determine the values and errors of the other free parameters. In this second step of our minimisation procedure we computed these different values, again by rescaling the atmosphere model and dereddening the photometry up to 25 000 Å with a lower limit of 0.04 for the value of $E(B-V)$. The final solution is calculated by the Monte Carlo Method with 150 steps and the error deduced is the standard deviation of all physically sound solutions: if a reasonable amount (2/5) of other solutions existed, the ones where $E(B-V)$ equalled 0.04 were deleted. The luminosity was calculated then by integration under the final atmosphere model.

This minimisation procedure gives very decent results (see Fig. 1 for some examples), except for spectral types O and B, since there is a lack of photometry at wavelengths lower than 3600 Å, where still a significant part of the luminosity is radiated. The calculated luminosities remain, however, very sensitive to changes of the atmosphere model, the

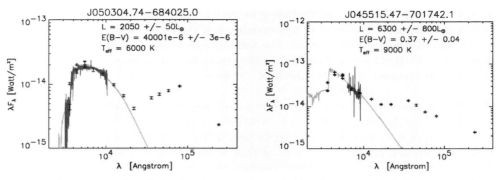

Figure 1. Some examples of the SED's we computed. The figure on the left corresponds to a post-AGB candidate with a freely expanding, detached shell, while the one on the right has a disc.

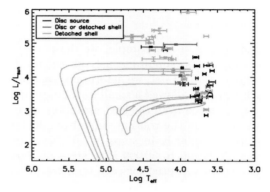

Figure 2. The position in the HR diagram of the post-AGB candidates for which we already obtained a low resolution optical spectrum. The evolutionary tracks are from Blöcker (1995) and correspond to stars with an initial mass of 1, 3, 4, 5 and 7 $M_\odot$.

value of $E(B - V)$ and variability of the photometry. Based on the info in the SED, the low resolution optical spectra and our colour selection criteria, we found that 60 of the 82 observed stars remain good post-AGB candidates. 30 have a freely expanding, detached shell and 25 are disc sources. The remaining five could be either.

In Fig. 2 we show the position in the HR diagram of all post-AGB candidates with a low resolution optical spectrum. The different candidates show a large spread in luminosities and represent therefore a wide range of different initial masses. This is very interesting in the light of the final goal of our project. By allowing a broader range in temperatures in our minimisation procedure, we could calculate an approximation to the luminosity of the objects without low resolution optical spectrum. The HR diagram of these objects is shown in Fig. 3. We notice that many sources, mainly the ones with a freely expanding, detached shell, have a low luminosity. Disc candidates show, on average, a lower effective temperature than the other objects. As less massive stars evolve slower, the expected bias towards lower mass objects is also observed.

4. Conclusions and future plans

Since other types of stars like YSOs or supergiants show similar colours as post-AGB stars, the low resolution optical spectra are needed for a full characterisation of the

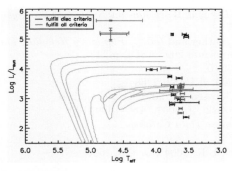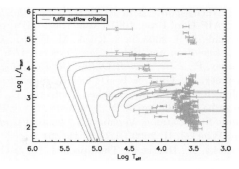

Figure 3. HR diagram containing the 260 post-AGB candidates for which we have not obtained low resolution optical spectra yet. The evolutionary tracks are from Blöcker (1995) and correspond to stars with an initial mass of 1, 3, 4, 5 and 7 $M_\odot$.

objects. This is why only 75% of the stars of our initial sample could still be considered to be post-AGB candidates, based on the information from the spectra. This rather high percentage does emphasize the efficiency of our selection criteria.

So far, we have a good sample of post-AGB candidates with a freely expanding, detached shell for which we already have low resolution optical spectra. Since this sample shows a rather large spread in luminosities, it will be very interesting to analyse it further in the light of the chemical evolution of post-AGB stars. Our initial sample also contains an extensive number of post-AGB candidates with expanding dust with a rather low luminosity. Obtaining low resolution optical spectra of these sources will form an interesting extension to the existing group of objects and give us an even broader look on post-AGB evolution. Also in the LMC, like in our Galaxy, there is a significant number of post-AGB candidates which seem to harbor a disc.

The uncertainty on the luminosities we computed based on these SEDs remains, however, large. The values are sensitive to changes of the atmosphere model, the value of $E(B - V)$ and variability of the photometry.

During next LMC observing season, we want to obtain low-resolution optical spectra for all objects of the extensive sample of post-AGB stars with a freely expanding, detached shell and a rather low luminosity. These objects can form an interesting extension of the existing sample. Furthermore, we plan to investigate the sample for which we already have low resolution spectra in more detail. For this, we were awared telescope time to obtain high-quality, high-resolution optical spectra of a small sample of representative candidate post-AGB stars. These spectra will allow us to study the chemical abundance patterns in detail. The ultimate goal is to confront the atmospheric enrichment distributions, with evolutionary model predictions. This will allow us to investigate the occurence of a third dredge up as a function of the stellar mass and metallicity.

References

Balick, B. & Frank, A. 2002, *ARAA*, 40, 439

Blöcker, T. 1995, *A&A*, 299, 755

Deroo, P., Van Winckel, H., Min, M., *et al.* 2006, *A&A*, 438, 987

De Ruyter, S., Van Winckel, H., Maas, T., Lloyd Evans, T., Waters, L. B. F. M., & Dejonghe, H. 2006, *A&A*, 448, 641

García-Lario, P. 2006, in M. J. Barlow & R. H. Méndez (eds.), *Planetary Nebulae in our Galaxy and Beyond* IAU Symposium 234 (Cambridge: CUP), p. 63

Gielen, C., Van Winckel, H., Waters, L. B. F. M., Min, M., & Dominik, C. 2008, in R. Guandalini, S. Palmerini, & M. Busso (eds.), *The IXth Torino Workshop on Evolution and*

Nucleosynthesis in AGB Stars and the IInd Perugia Workshop on Nuclear Astrophysics, AIP Conf.Proc. 1001 (New York: AIP), p. 357

Hrivnak, B. J. 2003, in S. Kwok, M. Dopita, & R. Sutherland (eds.), *Planetary Nebulae: Their Evolution nad Role in the Universe*, IAU Symposium 209 (ASP), p. 113

Massey, P., 2002, *VizieR Online Data Catalog*, 2236, 0

Meixner, M., Gordon, K. D., Indebetouw, *et al.* 2006, *AJ*, 132, 2268

Sahai, R., Morris, M., Sánchez Contreras, C., & Claussen, M. 2007, *AJ*, 134, 2200

Szczerba, R., Siódmiak, N., Stasińska, G., & Borkowski, J. 2007, *A&A*, 469, 799

Guide Star Catalog, Version 2.3.2 2006, *VizieR Online Data Catalog*

Van Winckel, H. 2003, *ARAA*, 41, 391

Zaritsky, D., Harris, J., Thompson, I. B., & Grebel, E.K. 2004 *AJ*, 128, 1606

The Magellanic System: Stars, Gas, and Galaxies
Proceedings IAU Symposium No. 256, 2008
Jacco Th. van Loon & Joana M. Oliveira, eds.

The population of Magellanic Cloud planetary nebulae

Letizia Stanghellini

National Optical Astronomy Observatory, 950 N. Cherry Ave, Tucson AZ 85719, USA

Abstract. In this review we address the progress that has been made toward the understanding of Magellanic Cloud planetary nebulae (PNe) and their evolution since the last Magellanic Cloud Symposium. Planetary nebulae in the Magellanic Clouds are the key probes of stellar and circumstellar evolution, both for their known distances and relative vicinity, and for their broad metallicity range A selection of recent results is presented, including the *HST* study of PNe and their central stars, the study of the population of Magellanic Cloud PNe based on abundance analysis, the recent *Spitzer* analysis of their dust contents, and the use of Magellanic Cloud PNe to constrain the distance scale of Galactic PNe.

Keywords. stars: AGB and post-AGB, stars: evolution, stars: winds, outflows, planetary nebulae: general, Magellanic Clouds

1. Introduction

Planetary nebulae (PNe) are direct probes of the evolution of low- and intermediate-mass stars (LIMS, $\sim$1–8 $M_\odot$). LIMS constitute a major component of stellar mass in all types of galaxies and in the intra-cluster medium; they go through the asymptotic giant branch (AGB) phase, which is characterized by very high IR luminosities, high mass-loss rates, and the production of carbon and nitrogen. Through PNe, this important phase of stellar evolution is made observable across different galaxy types, and as far as $\sim$30 Mpc. The progeny of LIMS in galaxies are especially important since LIMS are the major producers of nitrogen in the universe, and they supply as much carbon as massive stars. The knowledge of AGB and PN evolution in different environments, and, in particular, at different metallicities, is essential to soundly constrain the models of stellar and Galactic evolution.

Planetary nebulae in the Magellanic Clouds have always been of great interests for the metallicity baseline that the Clouds offer, $Z \sim 0.2$–1 $Z_\odot$ (Russell & Bessell 1989). The population of Magellanic Cloud PNe is thus the benchmark for AGB studies at moderately low metallicity, essential for the understanding of the integrated light in unresolved galaxies (Maraston 2005). While the Magellanic Cloud PNe are typically 50 times farther away than their Galactic counterparts, their distance uncertainties are very low, $\sim$ 5% compared to $\sim$ 50% or more for Galactic PNe (Stanghellini *et al.* 2008), making the former the best absolute probes of LIMS evolution. Nonetheless, the Magellanic Cloud PNe are close enough to be studied in much detail, both spectroscopically and via imaging. Furthermore the low selective reddening toward the Clouds represent a further advantage with respect to the Galactic PN population.

This paper is an overview of recent results in the field of Magellanic Cloud PNe, with particular regard to space-based observations. In §2 we explore the advances in this field since last Magellanic Cloud Symposium. In §3 we discuss PN morphology in the Clouds, and how this physical property of PNe is correlated to stellar evolution. Section 4 deals with PN abundances and the study of their progenitor environment through the α-elements, i.e., those elements whose yields do not change during LIMS evolution. Section 5 looks at the central stars (CS) of Magellanic Cloud PNe, who constitute the only CS population of any significant size in the universe whose physical parameters, such as luminosity and mass, are known in absolute terms. Section 6 gives an overview on recent studies on dust in Magellanic Cloud PNe, based on *Spitzer* observations. Section 7 shows how Magellanic Cloud PNe can forward the field of Galactic PNe also, by framing the calibration of the Galactic PN distance scale. Finally, in §8, we present a summary and future endeavors in this field.

2. Progress since the last IAU Symposium

In the last decade the field of Magellanic Cloud PNe has advanced greatly in several directions. The use of systematic ground-based surveys has allowed the discovery and spectroscopic confirmation of many more Magellanic Cloud PNe, more than doubling their known population size. About 230 PNe were known in the LMC (Leisy *et al.* 1997) at the time of the last Magellanic Clouds IAU Symposium (Dopita 1999), while recently Reid & Parker (2006) have identified $\sim$700 PN candidates in the LMC, $\sim$300 of which have been spectroscopically confirmed (Reid & Parker, this volume). In the SMC, Jacoby & De Marco (2002) analyzed the $\sim$60 PNe that define a complete sample 6 magnitudes down the planetary nebula luminosity function (PNLF) cutoff for the central 2.8 deg^2 of the SMC. This makes the SMC the first galaxy where PNe are resolved, and the PNLF features can be studied and interpreted based on a magnitude-limited, complete population. Ground-based spectroscopy is essential not only to confirm a target as a bona fide PN, but also to analyze the plasma and determine the elemental abundances. In the last decade, thanks especially to the work of Leisy & Dennefeld (2006), and Costa *et al.* (2000), and Idiart *et al.* (2007) also, a large database of Magellanic Cloud PN abundances has become available.

Space astronomy has been essential to forward the field of Magellanic Cloud PNe. The extensive use of the *HST*, which has the capability of resolving them spatially, allowed morphological studies of extragalactic PNe, and made their central stars directly observable. Central stars of Magellanic Cloud PNe have been studied in detail by Villaver *et al.* (see §5), and, given the known distances of Magellanic Cloud PNe, it had been possible to locate quite accurately these stars on the *HST* diagram for direct comparison with the stellar evolutionary tracks. Ultraviolet spectroscopy has allowed the detection of carbon emission lines in Magellanic Cloud PNe (Stanghellini *et al.* 2005), and the derivation of their carbon abundances.

As *Spitzer* has become available, the IR spectra of Magellanic Cloud PNe became observable, affording dust studies both in imaging (Hora *et al.* 2008) and spectroscopy (Stanghellini *et al.* 2007); furthermore, *Spitzer* spectroscopy reaches out to many atomic transitions that are elusive in the optical, allowing precise abundance calculations (Bernard-Salas *et al.* 2004, 2008). Another important aspect that has advanced in the field of Magellanic Cloud PNe is the publication of sets of stellar models, both synthetic (Marigo 2001, and this volume) and evolutionary (e.g., Karakas & Lattanzio 2007), based on initial conditions that reflect those of the Magellanic Cloud populations. These models give the yields of the element of stellar evolution in relation to the few final thermal pulses, when the PN is ejected, thus are readily comparable with the PN observations.

The wealth of new data and models make the Magellanic Cloud PNe the ideal laboratory to study LIMS populations and evolution at various metallicities; these have been used to explore, and trying to answer, some open questions in the field on PNe, in

Table 1. Magellanic Cloud PN morphology.

	LMC	SMC
round	29%	35%
elliptical	17%	29%
bipolar	34%	6%
bipolar core	17%	24%
point-symmetric	3%	6%

particular, the focus has been on the different PN morphologies, and how these originate and evolve; the chemistry of PNe as they evolve in different metallicity environments, and how do they contribute to cosmic recycling; the evolution of the PN central stars (CS); the PNLF in the Magellanic Clouds, and how can it be used to constrain and scale the extragalactic distance scale; the role of dust in PNe in PN ejection, evolution, and morphology, and the nature of dust in PNe of different metallicities; finally the use of Magellanic Cloud PNe as calibrator of the Galactic PN distance scale. In the next few sections we will explore several of these topics.

3. Magellanic Cloud PN morphology and the evolutionary connection

Planetary nebula morphology reveals the evolutionary history of the final phases of the LIMS life. Magellanic Cloud PNe are on average 0.5 arcsec across, thus they are spatially resolvable only with observations from space. In the last decade many samples of Magellanic Cloud PN images have been acquired with the *HST*, forming a large database to study the morphological evolutionary connection (Shaw *et al.* 2001; Stanghellini *et al.* 2003; Shaw *et al.* 2006). These studies include 114 LMC and 35 SMC PNe, representing approximately 2/3 of all Magellanic Cloud PNe known at the time of these surveys, and populate the 5 bright magnitude bins of the PNLF (in terms of $m_{\lambda 5007}$). Shapes of Magellanic Cloud PNe can be grouped into four major classes, round, elliptical, bipolar, and bipolar core (or ring) PNe, just as their Galactic counterparts. In Table 1 we show the statistics of the morphological types of Magellanic Cloud PNe, where we enlarge the sample described above by including the PNe already observed with the *HST* by Dopita, Vassiliadis, and collaborators (see Dopita 1999) We note that asymmetric PNe (bipolar PNe, and those ones showing other asymmetries such as bipolar cores) are much more common in the LMC than the SMC, and, in particular, the number of bipolar PNe in the SMC is very low.

The origin of PN morphology can be ascribed to the mechanism of mass ejection at the tip of the AGB and to the condition of the circumstellar medium at that evolutionary phase: The round and most elliptical shapes can be formed, according to hydrodynamic models, via ballistic expansion. On the other hand, the evolution into bipolar shape of the AGB ejecta needs the presence of an equatorial enhancement, at the time of the envelope ejection. What creates the conditions for the enhancement is still controversial. Most of the models for highly asymmetric PN evolution involve either rotation and magnetic fields (García-Segura 1997), or common-envelope (CE) processes (Morris 1981; Soker 1998). The former set of models agree with a progenitor-mass dependency of morphological evolution, while the latter set of models, those involving the CE evolution, are mass independent, since the chance of close binary evolution unlikely depends on the progenitor mass.

The Magellanic Cloud data have helped to clarify that the process forming bipolar PNe is necessarily mass dependent for the majority of observed bipolar PNe. It had been shown that asymmetry is related to higher LIMS mass. From stellar evolution we know that the third dredge-up, and the hot bottom burning process, occur only in the most massive AGB stars ($M/M_\odot > 3$–4). These processes have the net effect to reduce carbon and enhance nitrogen (and N/O) both in the Magellanic Cloud and the solar-metallicity models (Marigo 2001; Karakas & Lattanzio 2007). In Table 2 we show the average abundances of the key evolutionary elements for those Magellanic Cloud PNe whose morphology have been classified via the *HST* images. The abundances are from Leisy & Dennefeld (2006) excluding uncertain values. Carbon abundances are also from Stanghellini *et al.* (2005) and from new ACS prism spectra (Stanghellini *et al.*, in preparation). The averages are given for whole samples, and for the morphological groups of symmetric (round and elliptical), and asymmetric (bipolar core and bipolar), PNe. For the LMC, the sample of bipolar PNe is large enough to have a separate entry for this specific type. It is evident that PN chemistry closely correlates to their morphology. A look at the LMC averages in Table 2 shows that carbon is depleted in asymmetric PNe, whereas nitrogen is strongly enriched.

Figure 1 shows the averages and ranges of the N/H ratios in SMC ($Z = 0.004$), LMC ($Z = 0.008$), and Galactic ($Z = 0.016$) PNe, plotted against the galaxy metallicity. The Galactic data are from Stanghellini *et al.* (2006). There is no doubt that, whatever makes the asymmetric PNe acquire their shape, it has to be closely correlated with the nitrogen yield, and this, in turn, is correlated with the progenitor mass, for all metallicities. There are few bipolar PNe with low nitrogen (and/or high carbon) abundance. In these cases, the mechanism forming the bipolarity is mass-independent, and could be ascribed to CE evolution, following close binary interaction. Interestingly, binary evolutionary models show that close binary evolution does not enhance nitrogen nor deplete carbon (Izzard

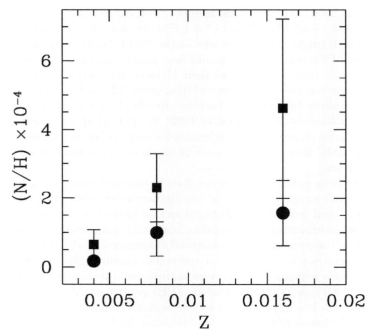

Figure 1. Average nitrogen abundances in the populations of SMC, LMC, and Galactic PNe with known morphology, plotted against the mean galaxy metallicity. Averages are plotted with filled circles and squares for symmetric (round and elliptical) and asymmetric (bipolar core and bipolar) PNe respectively. Bars represent data ranges.

Table 2. Evolutionary connection.

	LMC	SMC
$<C/H> \times 10^4$		
whole sample	2.49±2.18	3.71±3.66
round, elliptical	3.96±2.00	4.55±3.86
bipolar core, bipolar	2.10±2.13	2.00
bipolar	0.47±0.45	...
$<N/H> \times 10^4$		
whole sample	1.48±1.75	0.29±0.33
round, elliptical	1.00±1.35	0.18±0.68
bipolar core, bipolar	2.30±1.99	0.65±0.87
bipolar	2.68±2.11	...
$<N/O>$		
round, elliptical	0.57±0.89	0.12±0.09
bipolar core, bipolar	1.31±1.45	...
bipolar	1.54±1.61	...

et al. 2006), thus can not describe the formation of the majority of SMC, LMC, and Galactic bipolar PNe.

4. Abundances in Magellanic Cloud PNe, and metallicity gradients

The abundance of oxygen, neon, and other α-elements in Magellanic Cloud PNe probes the chemical environment at the time of PN progenitor formation, since α-elements can be considered *primordial*, as they are mostly produced in massive stars in primary nucleosynthesis, and not in LIMS. From the literature selection described in the previous section we looked at the primordial element distribution in Magellanic Cloud PNe. In Figure 2 we show the distribution of oxygen and neon in the LMC (open symbols) and the SMC (filled symbols) PNe, where morphological types have been also coded. The correlation is well defined, as expected from elements in lockstep evolution. The average Ne/O ratio is $<Ne/O> = 0.17 \pm 0.09$ both for the LMC and the SMC, while it is 0.27 in Galactic PNe (Stanghellini *et al.* 2006). The lower ratio at low metallicity shows that oxygen and neon abundances do not always scale with metallicity in lockstep. We have checked the PN abundance distribution of α-elements across the face of the Clouds, looking for a metallicity gradient. We have not found a clear metallicity gradient in either Cloud, nor a relation between α-element abundance and the location of the star forming regions. The morphological type distribution across the LMC has also been studied (Stanghellini *et al.* 2002) to show no particular morphological segregation, in agreement with a short crossing time of these galaxies compared to the timeframe of LIMS evolution.

5. Central stars of Magellanic Cloud PNe

The direct imaging of central stars (CS) of Magellanic Cloud PNe is only possible through space imaging. Villaver *et al.* (2003, 2004, 2007) have analyzed ∼50 LMC and SMC CS images acquired with STIS and WFC2 on the *HST*, and, by measuring their luminosity and temperature (with the aid of ground-based spectroscopy), estimated the masses of ∼20 CS. Villaver *et al.* (2007) determined that $<M/M_\odot> = 0.65\pm0.1$ both for the LMC and the SMC PNe, which is slightly higher than what has been estimated for Galactic CS. The possible reason why Magellanic Cloud CS are more massive than their Galactic counterparts is that, at lower metallicity, the mass loss at the TP-AGB is less efficient (Villaver *et al.* 2003, 2004). Interestingly, Villaver *et al.* found no clear cut relation between stellar mass and PN morphology, indicating that by the time the superwind is over there is no longer memory of the initial stellar mass.

The CS of the Magellanic Cloud PNe are among the few CS with known distances. As such, they offer the opportunity to test the PNLF, and, in particular, to see which type of stars illuminates the PNe that populate the high luminosity cutoff of the PNLF, which is used to calibrate the extragalactic distance scale. In Figure 3 we show the relation between the measured PN radii and the CS luminosities in the Clouds. Stellar and nebular data are from the papers quoted above. We see that the brightest CS are those hosted by compact PNe in both the LMC and the SMC, with $R_{\rm phot} < 0.5$ pc. This is expected, as the evolution of central stars (on the HR diagram) follows a luminosity plateau right after PN ejection, and then evolves toward the WD cooling line. Nonetheless, this effect has never been shown empirically before, since the distance of Galactic PNe are too uncertain for these types of studies. Central stars have masses between ∼0.55 and 1.4 $M_\odot$. We found that the stars populating the bright end of the PNLF are those with $M \sim 0.65$–0.7 $M_\odot$, rather than the most massive ones.

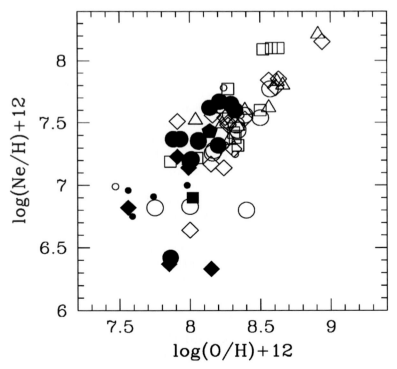

Figure 2. Neon vs. oxygen abundances, in log scale, for LMC (open symbols) and SMC (filled symbols) PNe. Shape of symbol indicates morphological type; Circle: round, diamonds: elliptical, triangles: bipolar core, squares: bipolar, and small circles: unknown morphology PNe.

6. Dust in Magellanic Cloud PNe

The study of dust in Magellanic Cloud PNe had become possible with the launch of the *Spitzer* Space Telescope; only a few bright Magellanic Cloud PNe were within observing reach with earlier technology. The IRS/*Spitzer* spectra in the 1–4 μm range had proven essential to study both the circumstellar dust and gas (Bernard-Salas *et al.* 2004, 2008; Stanghellini *et al.* 2007) at a distance of $\sim$50 kpc. The importance of studying dust in PNe resides in the fact that mass loss occurring toward the end of the AGB phase is still not completely understood. Theoretical studies indicate that the pressure on the dust grains produces the mass loss at the AGB tip (Willson 2000), and that mass-loss efficiency is directly proportional to metallicity, through the dependence of the absorption coefficient. The data available to date seem to agree with this correlation: there are fewer obscured AGB stars in the Magellanic Clouds than in the Galaxy (Groenewegen *et al.* 2000); the C-rich to O-rich ratio of AGB stars is higher at lower metallicity (Cioni & Habing 2003); and, as discussed in §3, there are fewer aspheric PNe in the SMC than in the LMC, showing that asymmetry is rarer in lower metallicity environments.

On this basis, Stanghellini *et al.* (2007) examined a homogeneous sample of IRS spectra to determine the IR/dust properties of the Magellanic Cloud PNe whose morphology had been previously determined through the *HST* images. Half of the analyzed spectra are featureless, except for the nebular emission lines and a weak dust continuum; the other half shows dust features in the form of solid state emission superimposed on the dust continuum. In most cases the solid state features are recognized as carbon-rich dust emission such as SiC and PAH, while oxygen-rich dust signatures were observed only in three PNe.

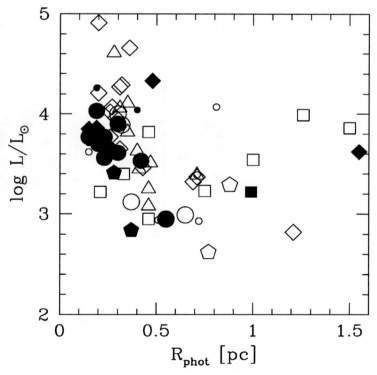

Figure 3. CS luminosities vs. PN physical radii, for LMC (open symbols) and SMC (filled symbols) PNe. Shape of symbol indicates morphological type as in Fig. 2.

In Figure 4 we show how the dust features in the IRS spectra correlate with the chemistry of nebular gas, and with morphology. Both panels show the IR luminosity (derived from a black-body fit of the dust continua, after subtracting the solid state features and the nebular emission lines) versus the carbon abundance of the PNe. Symbols in the left panel characterize the dust properties, while on the right panel they indicate the different PN morphologies. Symbols in the right panel are the same as in Figures 2 and 3, while in the left panel triangles indicate featureless, diamonds carbon-rich dust, and squares oxygen-rich dust PNe from the IRS spectra. In both panels, open symbols indicate LMC, and filled symbols SMC, PNe. From the left panel we infer that most featureless IRS spectra PNe (triangles) are in the lower luminosity part of the diagram, possibly indicating the rather short time span in which solid state features can be observed after PN ejection. This is also confirmed by the dependency of the features on the nebular radii (Stanghellini et al. 2007). It is evident from the Figure that the carbon-rich dust PNe (diamonds) correspond to higher carbon abundances, and oxygen-rich dust PNe (squares) are those with low carbon. The right panel shows that all round and elliptical PNe (circles and diamonds) correspond to carbon-rich dust, or featureless, IRS spectra, and none of the oxygen-rich dust PNe are round or elliptical. The differences in gas and morphology between LMC and SMC PNe is clear in Figure 4, where most of the SMC PNe whose carbon abundances and IRS spectra are available are either round or elliptical, with carbon-rich dust.

Comparison of the IRS data with the literature disclosed a sharp difference between dust features in the Magellanic Cloud and Galactic PN populations, in that all observed Galactic PNe show solid state features (García-Lario et al., in preparation). The ratio of carbon-rich dust to oxygen-rich dust PNe is ~11 in the SMC, ~4.5 in the LMC, while it is estimated to be close to unity in the Galaxy (García-Lario et al., in preparation), indicating that the population metallicity has enormous impact on AGB dust formation.

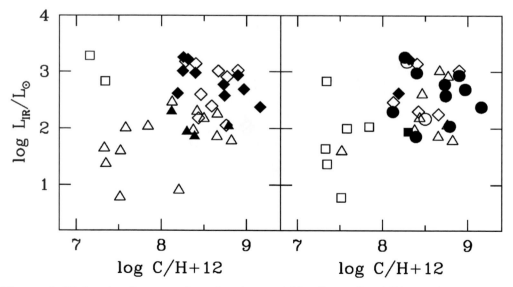

Figure 4. IR Luminosity vs. carbon abundance of Magellanic Cloud PNe. Left panel: dust properties (triangles: no dust features; diamonds: carbon-rich dust; and squares: oxygen-rich dust PNe). Right panel: morphology (circles: round; diamonds: elliptical; squares: bipolar, triangles: bipolar core PNe). In both panels, open symbols are for LMC, and filled symbols for SMC, PNe (adapted from Stanghellini et al. 2007).

7. Magellanic Cloud PNe as distance calibrators

There are ~1800 PNe in the Galaxy (Acker *et al.* 1992), but their distances are uncertain with the exception of a few PNe where a cluster or binary membership could be established. To get the distances of most Galactic PNe it is customary to use relations between physical parameters, and calibrate these relations with the parameters of the few PNe with known distances. The most recent calibrations of the surface-brightness to ionized-mass scale (Daub 1982; Cahn *et al.* 1992) constitute the most used catalogs of PN distances. Nonetheless, even the best calibrations are based on very few data points, as model-independent Galactic PNe distances are few. On the other hand, there are more than a hundred PNe in the Magellanic Clouds whose ionized-mass and surface-brightness are available through *HST* observations, and they can be used to calibrate the Galactic scale. Stanghellini *et al.* (2008) have shown that the re-calibration is reliable, and found distances for hundreds of Galactic PNe with the new scale. The resulting distances are in excellent agreement with the Galactic parallaxes and cluster membership distances.

8. Future endeavors

The past decade has proven incredibly productive in the field of Magellanic Cloud PNe, especially thanks to space and ground-based surveys that have become available. Large databases have allowed the exploration of the PN abundances, yields, central stars, evolution and tracing of the different populations of PNe. Systematic studies of LMC and SMC PNe disclose the tight connection between dust, gas, and morphology. Dust production in AGB and post-AGB stars is directly related to metallicity. It seems that, for the most part, symmetric PNe are the progeny of the lower end of the LIMS mass range, while bipolarity is associated with the higher-mass progenitors, thus the mechanism that produces bipolarity must depend on progenitor mass, in most observed cases.

An initial set of abundances have become available in this decade, and hopefully will be expanded soon to include all known Magellanic Cloud PNe. The new abundance databases will include the IR emission lines, thus be less dependent on the ionization correction factor modeling that is typically used to obtain elemental concentrations. The use of a refurbished *HST* will be very important to extent the morphological approach. There are many other aspects of the Magellanic Cloud PN field that could not have been explored in this paper, but in particular the comparison between the PN properties, and those of their progenitors (AGB stars) and progeny (WD) in a complete evolutionary scheme is the most interesting avenue for future discoveries. Models of the stars and nebulae together, including dust, will eventually encompass observations of the whole life and death of the LIMS at different metallicities, to directly understand the contribution of LIMS to galaxies in terms of their mass, metallicity, and observing wavelength.

Acknowledgements

Thanks to the Organizers for a very interesting Symposium, and to my collaborators for their input in the work presented here.

References

Acker, A., Marcout, J., Ochsenbein, F., Stenholm, B., & Tylenda, R. 1992, *Strasbourg – ESO catalogue of galactic planetary nebulae* (Garching: European Southern Observatory)

Bernard-Salas, J., Houck, J. R., Morris, P. W., Sloan, G. C., Pottasch, S. R., & Barry, D. J. 2004, *ApJS*, 154, 271

Bernard-Salas, J., Pottasch, S. R., Gutenkunst, S., Morris, P. W., & Houck, J. R. 2008, *ApJ*, 672, 274

Cahn, J. H., Kaler, J. B., & Stanghellini, L. 1992, *A&AS*, 94, 399

Cioni, M. -R. L. & Habing, H. J. 2003, *A&A*, 402, 133

Costa, R. D. D., de Freitas Pacheco, J. A., & Idiart, T. P. 2000, *A&AS*, 145, 467

Daub, C. T. 1982, *ApJ*, 260, 612

Dopita, M. A. 1999, in Y. -H. Chu, N. Suntzeff, J. Hesser, & D. Bohlender (eds.), *New Views of the Magellanic Clouds*, IAUS 190, p. 332

García-Segura, G. 1997, *ApJ*, 489, L189

Groenewegen, M. A. T., van Loon, J. T., Whitelock, P. A., Wood, P. R., & Zijlstra, A. A. 2000, in R. F. Wing (ed.), *The Carbon Star Phenomenon*, IAUS 177 (Dordrecht: Kluwer), p. 385

Hora, J. L., Cohen, M., Ellis, R. G., *et al.* 2008, *AJ*, 135, 726

Idiart, T. P., Maciel, W. J., & Costa, R. D. D. 2007, *A&A*, 472, 101

Izzard, R. G., Dray, L. M., Karakas, A. I., Lugaro, M., & Tout, C. A. 2006, *A&A*, 460, 565

Jacoby, G. H. & De Marco, O. 2002, *AJ*, 123, 269

Karakas, A. & Lattanzio, J. C. 2007, *PASA*, 24, 103

Leisy, P. & Dennefeld, M. 2006, *A&A*, 456, 451

Leisy, P., Dennefeld, M., Alard, C., & Guibert, J. 1997, *A&AS*, 121, 407

Maraston, C. 2005, *MNRAS*, 362, 799

Marigo, P. 2001, *A&A*, 370, 194

Morris, M. 1981, *ApJ*, 249, 572

Reid, W. A. & Parker, Q. A. 2006, *MNRAS*, 373, 521

Russell, S. C. & Bessell, M. S. 1989, *ApJS*, 70, 865

Shaw, R. A., Stanghellini, L., Mutchler, M., Balick, B., & Blades, J. C. 2001, *ApJ*, 548, 727

Shaw, R. A., Stanghellini, L., Villaver, E., & Mutchler, M. 2006, *ApJS*, 167, 201

Soker, N. 1998, *ApJ*, 496, 833

Stanghellini, L., Shaw, R. A., Mutchler, M., Palen, S., Balick, B., & Blades, J. C. 2002, *ApJ*, 575, 178

Stanghellini, L., Shaw, R. A., Balick, B., Mutchler, M., Blades, J. C., & Villaver, E. 2003, *ApJ*, 596, 997

Stanghellini, L., Shaw, R. A., & Gilmore, D. 2005, *ApJ*, 622, 294

Stanghellini, L., Guerrero, M. A., Cunha, K., Manchado, A., & Villaver, E. 2006, *ApJ*, 651, 898

Stanghellini, L., García-Lario, P., García-Hernández, D. A., Perea-Calderón, J. V., Davies, J. E., Manchado, A., Villaver, E., & Shaw, R. A. 2007, *ApJ*, 671, 1669

Stanghellini, L., Shaw, R. A., & Villaver, E. 2008, *ApJ*, 689, 194

Villaver, E., Stanghellini, L., & Shaw, R. A. 2003, *ApJ*, 597, 298

Villaver, E., Stanghellini, L., & Shaw, R. A. 2004, *ApJ*, 614, 716

Villaver, E., Stanghellini, L., & Shaw, R. A. 2007, *ApJ*, 656, 831

Willson, L. A. 2000, *ARAA*, 38, 573

The Magellanic System: Stars, Gas, and Galaxies
Proceedings IAU Symposium No. 256, 2008
Jacco Th. van Loon & Joana M. Oliveira, eds.

© 2009 International Astronomical Union
doi:10.1017/S1743921308028822

On the huge mass loss of B[e] supergiants in the Magellanic Clouds

M. Kraus[1], M. Borges Fernandes[2] and F. X. de Araújo[3]

[1]Astronomický ústav, Akademie věd České republiky, Fričova 298, 251 65 Ondřejov,
Czech Republic
email: kraus@sunstel.asu.cas.cz

[2]UMR 6525 H. Fizeau, Univ. Nice Sophia Antipolis, CNRS,
Observatoire de la Côte d'Azur, Av. Copernic, 06130 Grasse, France
email: Marcelo.Borges@obs-azur.fr

[3]Observatório Nacional, Rua General José Cristino 77, 20921-400 São Cristovão,
Rio de Janeiro, Brazil
email: araujo@on.br

Abstract. B[e] supergiants are known to possess circumstellar disks in which molecules and dust can form. The formation mechanism and the resulting structure of these disks is, however, still controversial. Nevertheless, to protect the disk material from the dissociating stellar radiation and to allow for dust formation in the vicinity of a luminous supergiant star, the amount of mass comprised within these disks must be huge. We study the amount of hydrogen neutral material by means of an analysis of the strong [O I] emission lines in our optical high-resolution FEROS spectra of two B[e] supergiants, the edge-on system S 65 in the SMC, and the pole-on system R 126 in the LMC. In addition, we study the possible disk dynamics of S 65, based on a simultaneous line-profile modeling. We find that the [O I] emission lines in S 65 must originate either from an outflowing disk, in which the outflow velocity is slowly decreasing outwards, or from a Keplerian rotating ring, resulting from an ejection event.

Keywords. circumstellar matter, stars: individual (R 126, S 65), stars: mass loss, supergiants, stars: winds, outflows, Magellanic Clouds

1. Introduction

B[e] supergiants in the Magellanic Clouds, even though studied in great detail, are still far from being understood. Their non-spherically symmetric wind is proven by, e.g., polarimetric observations (Magalhães 1992; Magalhães *et al.* 2006; Melgarejo *et al.* 2001), and the presence of a dusty circumstellar disk is confirmed by their strong infrared excess emission (e.g., Zickgraf *et al.* 1986) and the strong CO band emission (McGregor *et al.* 1988a,b, 1989; Morris *et al.* 1996). Based on the analysis of the [O I] lines, it has recently even been suggested that the disks around B[e] supergiants are neutral in hydrogen right from the stellar surface, and that molecules and dust are forming (and can exist) already close to the stellar surface (Kraus *et al.* 2007). In fact, for the LMC B[e] supergiant R 126, Kastner *et al.* (2006) found that the inner edge of the dusty disk is located at about $360\,R_*$, which is 3 times closer to the star than the value of $\sim 1000\,R_*$ suggested by Zickgraf *et al.* (1985).

To guarantee that the disk material is neutral so that molecules and dust can exist, a huge disk mass must be postulated. Since the disks around supergiants cannot be pre-main sequence in origin, the disk formation must be linked to non-spherical, i.e., predominantly equatorial, high mass loss of the central star, either by continuous steady

Table 1. Parameters of the two B[e] supergiants.

Star	T_{eff} [K]	$R_*[R_\odot]$	$\log L_*/L_\odot$	$v\sin i$[km/s]	inclination	$M_*[M_\odot]$	Reference
S 65	17 000	81	5.7	150	$\pm$ edge-on	~ 35	Zickgraf *et al.* (1986)
R 126	22 500	72	6.1	?	$\pm$ pole-on	> 60	Zickgraf *et al.* (1985)

material outflow, or by some violent, ring-shaped mass ejection event(s). In the former scenario, the [O I] emission can be modeled with the so-called outflowing disk scenario, in which the wind material in the equatorial region is streaming out (more or less) radially, forming a disk with constant opening angle. In the ejection scenario, the ring material is disconnected from the star, freely expanding and probably revolving the star on Kepler orbits.

Here, we study the [O I] line emission from two B[e] supergiants, the SMC star S 65 and the LMC star R 126, whose stellar parameters are summarised in Table 1. For both stars we obtained high-resolution spectra in 1999 with FEROS attached to the *ESO* 1.52-m telescope (agreement *ESO*/ON-MCT) in La Silla (Chile). In addition, in order to flux calibrate the line emission, low-resolution spectra were obtained with the Boller & Chivens spectrograph, also at the *ESO* 1.52-m telescope (agreement *ESO*/ON-MCT).

2. Modeling of the [O I] lines

Oxygen has about the same ionization potential as hydrogen. The detection of emission lines from O I thus means, that these lines must be generated within a region, in which hydrogen is predominantly neutral as well. Since free electrons are (besides the less efficient, but nevertheless important, neutral hydrogen atoms) the main collision partners to excite the lowest energy levels in O I, from which the forbidden emission lines arise, we introduce the parameter q_e as the ionization fraction, defined by $n_e = q_e n_H$, with $q_e = q_{\mathrm{metals}} + q_{H^+}$. Assuming that all elements with ionization potential lower than that of hydrogen and oxygen are fully ionized, delivers an upper limit of $q_{\mathrm{metals}} < 5 \times 10^{-5}$ for SMC metallicity, and $q_{\mathrm{metals}} < 1.6 \times 10^{-4}$ for LMC metallicity. This means, that if hydrogen is ionized by only 1%, it will still deliver the dominant amount of free electrons. Since we do not know the amounts of ionized hydrogen and ionized metals within the disk, q_e is a free parameter at first. The hydrogen density distribution in the disk, $n_H(r)$, follows from the equation of mass continuity, i.e. $n_H(r) \sim F_{\mathrm{m,Disk}}/(r^2 v_{\mathrm{out}})$, where the parameters $F_{\mathrm{m,Disk}}$ and v_{out} represent the disk mass flux and outflow velocity, respectively.

The SMC B[e] supergiant S 65. We start our investigation with the star S 65, which is assumed to be oriented more or less edge-on (see Table 1). The disk opening angle is set to $\sim 20°$, and we first adopt the outflowing disk scenario.

Our optical spectra display 3 [O I] lines, of which two originate from the same upper level (see left panel of Fig. 1). This means that their line ratio is determined by pure quantum mechanics. Only line ratios with the $\lambda 5577$ Å line are thus sensitive to temperature and density. To constrain these two parameters, we use in the following the $\lambda 6300/\lambda 5577$ line ratio. The observed value of this ratio is plotted as the dotted line in the top right panel of Fig. 1. For the computation of the line ratios, we use different (but constant) electron temperatures and calculate for each input value of q_e the [O I] line luminosities. For this, the density parameter, $F_{\mathrm{m,Disk}}/v_{\mathrm{out}}$ is varied until the $\lambda 6300$ line luminosity can be fitted, delivering a certain value for the line ratio with the $\lambda 5577$ line.

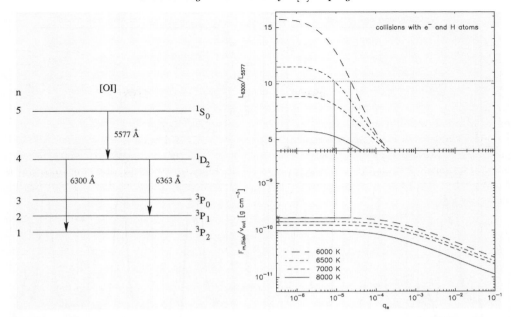

Figure 1. Left: The 5-level atom with the observed transitions indicated. Right: Determination of the possible ranges in disk electron temperature and ionization fraction, q_e, based on the observed [O I] line ratio (top panel), and the disk density parameter (bottom panel).

These line ratio values for different temperatures and as a function of q_e are included in the top right panel of Fig. 1.

Comparison with the observed value shows, that the ranges within the possible parameter space are quite narrow, restricting the temperature to $T_e \simeq 6000$–6500 K, and the ionization fraction to $q_e \simeq 9 \times 10^{-6} - 2.5 \times 10^{-5}$. This low value of q_e confirms indeed that the disk must be predominantly neutral ($q_{H^+} < 2 \times 10^{-5}$). In addition, the small ranges in q_e and T_e confine the density parameter, $F_{m,\text{Disk}}/v_{\text{out}}$, to a value of $\sim 1.8 \times 10^{-10}$ g cm^{-3} (bottom right panel of Fig. 1).

The increase in line luminosity with distance from the star is shown in the top left panel of Fig. 2 for all three [O I] lines. Saturation of the lines happens at a distance of roughly $500\,R_*$, with the $\lambda\,5577$ line saturating even much closer to the star. The arrows to the right indicate the observed line luminosity values.

Inspection of the observed [O I] line profiles (lower panels of Fig. 2) shows that the lines are double-peaked, with the $\lambda\,5577$ line showing the widest peak separation. With the knowledge that S 65 is seen edge-on, a double-peak structure alone does not give any hint on the underlying kinematics, because both, an outflowing disk with constant outflow velocity, as well a narrow Keplerian rotating ring, result in the identical, double-peaked line profile (left panel of Fig. 3). To test the outflowing disk scenario, we calculate the line profiles for a constant outflow velocity of $v_{\text{out}} \simeq 22$ km s^{-1} as can be derived from the peak separation of the $\lambda\,5577$ line. In addition, to account for the broader wings of the line, some Gaussian shaped turbulent velocity of $v_{\text{turb}} \simeq 8$ km s^{-1} needs to be included. The resulting fit to the $\lambda\,5577$ line is shown by the dashed line in the bottom left panel of Fig. 2. Using the same model parameters, delivers identical line profiles for the other two [O I] lines (dotted lines), which do, however, not agree with the observed ones. Instead, to fit these two lines, which are formed further away from the star, we need to assume that the outflow velocity has slowed down to $v_{\text{out}} \simeq 16$ km s^{-1}, while the turbulence has

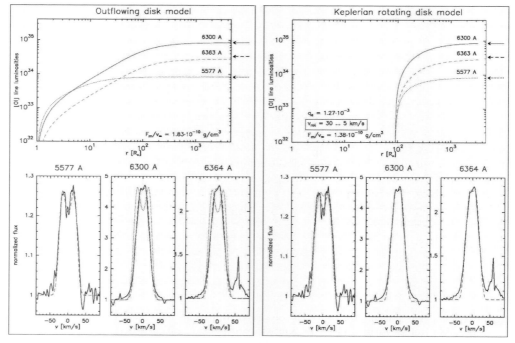

Figure 2. Results for the [O I] line luminosities (top panels) and line profile fits (bottom panels) for the two competing scenarios. For details see text.

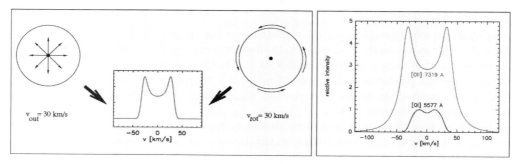

Figure 3. Left: Both scenarios, an edge-on seen outflowing disk or ring, and a narrow Keplerian rotating ring, result in the identical, double-peaked line profile. Right: Expected strength and shape of the [O II] $\lambda 7319$ line in the Keplerian rotating disk scenario.

slightly increased to $v_{\text{turb}} \simeq 11$ km s^{-1} (dashed lines). Such a scenario is, however, quite reasonable and does not contradict or discard the outflowing disk scenario.

Nevertheless, the $\lambda 6300$ and $\lambda 6364$ lines, which are produced at larger distances from the star, show a much narrower peak-separation than the $\lambda 5577$ line, which is produced much closer to the star. This might be a clear indication for Keplerian rotation. In a second attempt we, therefore, calculate the [O I] line luminosities and the profiles with a Keplerian rotating disk model. The peak separation of the $\lambda 6300$ and $\lambda 6364$ lines delivers a minimum rotation velocity of about 5 km s^{-1}, which sets the outer edge of the [O I] line emission region to about 3000 R_*. The width of the $\lambda 5577$ line defines a maximum rotation velocity of about 30 km s^{-1}, placing the inner edge of the disk to about 90 R_*.

The line luminosity calculations (top right panel of Fig. 2) indicate that in this case, the disk ionization fraction, q_e, must be higher by about a factor of 100. This trend is clear, since the line formation is now forced to happen at much larger distances from the star, where the O I density is generally much lower. Therefore, a higher number of collision partners (i.e., free electrons) is necessary to produce the same luminosity. Nevertheless, even with this much higher ionization fraction, the disk can still be considered as neutral in hydrogen ($q_{H^+} < 1.3 \times 10^{-3}$).

Calculating the shapes of the line profiles, it turns out that the $\lambda\,6300$ and $\lambda\,6364$ lines can be excellently fitted (dashed lines in the bottom right panel of Fig. 2), while the same model fails to reproduce the $\lambda\,5577$ line profile (dotted line). Instead, to fit the $\lambda\,5577$ line profile we need to request that at the inner edge of the Keplerian rotating disk the material is still slowly expanding, with $v_{out} \simeq 9$ km s^{-1}. Also this scenario seems to be reasonable, since the disk material originates from some mass loss or material ejection.

If the [O I] lines result from a Keplerian rotating disk, the question arises whether this disk extends down to the stellar surface, or whether the material is detached from the stellar surface. In the case that the disk extends down to (or close to) the stellar surface, the disk parts within $90\,R_*$, which do not contribute to the [O I] emission, must consequently be ionized. Since O II is also known to have forbidden emission lines in the optical spectral range, we calculate the line luminosity and profile of the [O II] $\lambda\,7319$ line from the inner disk parts, extending from the stellar surface to $90\,R_*$. The line profile and intensity is then compared to the weakest of our [O I] lines, i.e., the $\lambda\,5577$ line (right panel of Fig. 3). The [O II] line turns out to be much broader (due to the higher rotation velocities closer to the star) and about 5 times more intense than the $\lambda\,5577$ line. It should, therefore, be clearly detectable. However, our spectra do not show indications for any [O II] line. This might indeed indicate that, if the Keplerian rotating disk scenario is the correct one, the material must be detached from the star, speaking in favour of a (single or multiple) mass ejection event rather than a steady outflow. The best-fit parameters of both scenarios for S 65 are summarised in Table 2. Also included are the resulting amounts of gas mass enclosed within the [O I] line forming regions.

Table 2. Best fit parameters of the two competing scenarios for S 65.

	Outflowing Disk	Detached Keplerian Ring
Extend of line forming region	$1\,R_* - 500\,R_*$	$90\,R_* - 3000\,R_*$
v_{out}	$22\dots16$ km/s	$9\dots0$ km/s
v_{rot}	—	$30\dots5$ km/s
v_{turb}	$8\dots11$ km/s	—
Total M_{gas} in [O I] forming regions	$\sim 3.3 \times 10^{-3}$ M$_\odot$	$\sim 1.5 \times 10^{-2}$ M$_\odot$

The LMC B[e] supergiant R 126. The modeling of the [OI] lines from R 126 has been discussed in detail by Kraus *et al.* (2007). Therefore, we give here only a short summary of the main results and conclusions.

The pole-on orientation of R 126 (see Table 1) does not allow for a detailed line profile calculation as in the case of S 65. For the line-luminosity calculations, we followed the same procedure as described above, assuming an outflowing disk scenario that is predominantly neutral in hydrogen right from the stellar surface. With this model, we derive a disk electron temperature of $T_e \simeq 8000$ K, an ionization fraction of $q_e \simeq 4 \times 10^{-4}$, and a density parameter of $F_{\mathrm{m,Disk}}/v_{\mathrm{out}} \simeq 2.2 \times 10^{-11}$ g cm^{-3}. As in the case of S 65, the [O I] line luminosities clearly saturate within about 500 R_*.

Interestingly, observations with the *Spitzer Space Telescope* performed by Kastner *et al.* (2006) revealed that the inner edge of the dusty disk must be located at about 360 R_*, which is even closer to the stellar surface than the [O I] saturation region found from our analysis. In addition, for dust to exist, the disk temperatures must have dropped below the dust sublimation temperature of about 1500 K. This means that our model assumptions were even too conservative, and that the disk mass loss rate must be even (much) higher than our assumed value, confirming the high-mass character of the B[e] supergiant stars' disks.

3. Conclusions

Based on the modeling of both, the observed line luminosities of the [O I] lines and their line profiles, we found that the two B[e] supergiants studied must have high-density disks, which are predominantly neutral in hydrogen. The total amount of ionized hydrogen is found to be less than $\sim 0.1\%$. While for the LMC star R 126 an outflowing disk scenario seems to be the most plausible one, our analysis of the SMC star S 65 showed that for this star both models, i.e., either an outflowing disk, or a detached Keplerian rotating ring, delivered reasonably good fits to the line profiles. For this star, further investigations, which help to distinguish between the two scenarios, are thus necessary.

Acknowledgements

M.K. acknowledges financial support from GA AV ČR grant number KJB300030701, and M.B.F. from the Centre National de la Recherche Scientifique (CNRS). M.K. and M.B.F. are further grateful to the IAU and the organizers for their financial support to attend this IAU Symposium.

References

Kastner, J. H., Buchanan, C. L., Sargent, B., & Forrest, W. J. 2006, *ApJ*, 638, L 29

Kraus, M., Borges Fernandes, M., & de Araújo, F. X. 2007, *A&A*, 463, 627

Magalhães, A. M. 1992, *ApJ*, 398, 286

Magalhães, A. M., Melgarejo, R., Pereyra, A., & Carciofi, A. C. 2006, in M. Kraus & A. S. Miroshnichenko (eds.), *Stars with the B[e] Phenomenon* (San Francisco: ASP), p. 147

McGregor, P. J., Hillier, D. J., & Hyland, A. R. 1988a, *ApJ*, 334, 639

McGregor, P. J., Hyland, A. R., & Hillier, D. J. 1988b, *ApJ*, 324, 1071

McGregor, P. J., Hyland, A. R., & McGinn, M. T. 1989, *A&A*, 223, 237

Melgarejo, R., Magalhães, A. M., Carciofi, A. C., & Rodrigues, C. V. 2001, *A&A*, 377, 581

Morris, P. W., Eenens, P. R. J., Hanson, M. M., Conti, P. S., & Blum, R. D. 1996, *ApJ*, 470, 597

Zickgraf, F. -J., Wolf, B., Stahl, O., Leitherer, C., & Klare, G. 1985, *A&A*, 143, 421

Zickgraf, F. -J., Wolf, B., Stahl, O., Leitherer, C., & Appenzeller, I. 1986, *A&A*, 163, 119

The Magellanic System: Stars, Gas, and Galaxies
Proceedings IAU Symposium No. 256, 2008
Jacco Th. van Loon & Joana M. Oliveira, eds.

High-exitation nebulae around Magellanic Wolf-Rayet stars

Manfred W. Pakull[1]

[1] CNRS, Observatoire Astronomique 11, rue de l'Université, F67000 Strasbourg, France
email: pakull@astro.u-strasbg.fr

Abstract. The SMC harbours a class of hot nitrogen-sequence Wolf-Rayet stars (WNE) that display only relatively weak broad He II λ4686 emission indicative of their low mass-loss rates and which are therefore hard to detect. However, such stars are possible emitters of strong He^+ Lyman continua which in turn could ionize observable He III regions, i.e. highly excited H II regions emitting nebular He II λ4686 emission. We here report the discovery of a second He III region in the SMC within OB association NGC 249 within which the weak-lined WN star SMC WR10 is embedded. SMC WR10 is of special importance since it is a single star showing the presence of atmospheric hydrogen. While analysing the spectrum in the framework of two popular, independent WR atmosphere models we found strongly discrepant predictions (by 1 dex) of the He^+ continuum for the same input parameters. A second interesting aspect of the work reported here concerns the beautiful MCELS images which clearly reveal a class of strongly O III λ5007 emitting (blue-coded) nebulae. Not unexpectedly, most of the "blue" nebulae are known Wolf-Rayet bubbles, but new bubbles around a few WRs are also detected. Moreover, we report the existence of blue nebulae without associated known WRs and discuss the possibility that they reveal weak-wind WR stars with very faint stellar He II λ4686 emission. Alternatively, such nebulae might hint at the hitherto missing population of relatively low-mass, hot He stars predicted by massive binary evolution calculations. Such a binary system is probably responsible for the ionization of the unique He II λ4686-emitting nebula N 44C.

Keywords. shock waves, stars: winds, outflows, stars: Wolf-Rayet, ISM: bubbles, HII regions, Magellanic Clouds

1. Wolf-Rayet stars and their He^+ ionizing radiation

My interest for high-excitation H II regions dates back to the late 1980s. At that time I tried to understand the formation of the extended He II λ4686 nebular recombination line in the He III region around the massive black-hole candidate LMC X-1 (Pakull & Angebault 1986) in terms of photoionization and to draw conclusions about the otherwise unobservable EUV/soft X-ray ionizing continuum. X-ray ionization has been a well-understood process, but hitherto mainly applied to power-law ionizing continua in AGN. It turned out that the only known examples of nebular λ4686 emission in H II regions concerned two Wolf-Rayet ring nebulae, one around the presumably very hot Galactic WO star WR102 and the other one surrounding a similar object in the Local Group galaxy IC 1613 (see, e.g., Pakull & Motch 1989b and references therein). Another prominent He III region is located in the LMC, namely the small region N44C (= NGC 1936) within the larger H II complex N44. Here, the ionizing star is not a WR, but appears to be a normal O7 star. This led Pakull & Motch (1989a) to propose an interpretation in terms of a fossil X-ray ionized nebula, possibly due to the transient source LMC X-5. As a historical remark, I'd like to mention that the presence of highly ionized He III regions around some WO stars led Terlevich & Melnick (1985) to suggest that a population of very hot ($T_{rad} > 80$–100 kK) luminous stars might be responsible for the high ionization observed in AGN, thus challenging the general understanding in terms of massive accreting black holes. For such stars they coined the fitting term *Warmers*.

In 1991, at the Bali IAU Symposium 143 on Wolf-Rayet stars two groups (Niemela *et al.* 1991; Pakull 1991; Pakull & Bianchi 1991) independently announced the discovery of He III regions around a few Magellanic WR stars. They were SMC WR7 (located in N 76) in the SMC, and in the LMC Brey 2 (located in N 79) and Brey 40a (within an anonymous H II region). The spectral types of these Magellanic "Wolf-Rayet Warmers" (hereafter WRW) came as a surprise: these stars are early WN types rather than the more advanced WO types as one might have naively expected from the previously recognized objects. Moreover, it was shown that the two known Magellanic WOs do not excite observable He III regions (Pakull 1991), i.e., a WO is not necessarily a WRW! Since then, and although several additional Magellanic WR stars have been discovered in the meantime, no new WRW turned up in the MCs.

Since the pioneering work by Schmutz, Leitherer & Gruenwald (1992) we know that the ionizing radiation from WR stars does not only depend on the atmospheric "core" temperature of the star (i.e., at a radius where the wind is still subsonic), but also on the mass-loss rate. In short, a strong wind from a WN star (consisting mainly of helium) will absorb all He$^+$ Lyman continuum photons (i.e., $h\nu > 54$ eV), not unlike an ionization-bound Strömgren sphere. If however the wind is sufficiently weak, a large fraction of the hard ionizing photons will escape and create observable He III regions in the surrounding ISM. Grids of the most recent Wolf-Rayet star models which among others also take into account line-blankening are described in Smith *et al.* (2002) (CMFGEN models) and in Hamann & Gräfener (2004) (PoWR models). Note that the CMFGEN grid is implemented in the popular Starburst99 software package. As will be shown below, and contrary to what is often claimed or assumed, the two models strongly disagree on the predicted power of the emitted He$^+$ Lyman continuum emission.

The study of Wolf-Rayet (WR) stars currently witnesses a renaissance since it was realised that these stars — or rather a not-yet-specified rare subclass thereof — appear to be the direct progenitors of (long) gamma ray bursts (cf., Yoon & Langer 2005). One important ingredient of currently favoured models is very high rotation of the core at the time of collapse/jet formation which however is not easily realized due to expected angular momentum loss suffered during their evolution due to stellar wind.

At low metallicity, WR winds are now known (cf., Vink & de Koter 2005 and references therein) to be weaker than those of their more metal-rich Galactic counterparts, and it is possible that under these circumstances (low Z, very rapid rotation) massive stellar evolution proceeds quasi-chemically homogeneously (Yoon & Langer 2005). In this scenario, the stars remain in the blue part of the HR diagram and evolve from the main sequence directly "bluewards" to high effective temperatures ($T_{eff} > 70$–100 kK) even if their outer layers still contain hydrogen. It is however an open question whether homogeneous evolution is indeed realized in nature.

2. SMC WR10: A second Warmer in the Small Cloud

Shortly after the somewhat unexpected discovery of several new WR stars in the SMC (Massey & Duffy 2001) we obtained long-slit spectra of the newly identified WR10 which is located within the H II region NGC 249 (Fig. 1). Although this object has subsequently also been observed with various similar instrumental set-ups no other observer seems to have noticed (or have cared to look at) the relatively strong *extended* He II $\lambda4686$ emission that is depicted in Fig. 2.

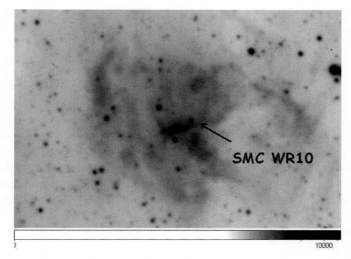

Figure 1. *ESO NTT*/EMMI Hα image of the SMC H II region NGC 249 in which the Wolf-Rayet star SMC WR10 (WN3ha) is embedded. North is up and East is to the left. The image covers $3' \times 2'$.

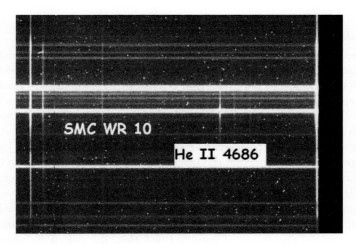

Figure 2. *ESO NTT*/EMMI Long-slit spectrum of SMC WR10 and the surrounding H II region NGC 249. The strong vertical lines to the left and to the right are nebular Hγ and Hβ, respectively. The He II λ4686 emitting He III region extends $45''$ corresponding to some 10 pc in the SMC.

The particular importance of SMC WR10 for our understanding of the physics and evolution of WR stars comes from the fact that this WN3ha star is most likely single (Foellmi, Moffat & Guerro 2003) and therefore probably not a result of massive binary evolution. Moreover, the star is faint ($V \sim 16.0$ mag), displays a rather weak ($EW = 25$ Å) and narrow ($FWHM = 30$ Å) stellar He II λ4686 emission line (Crowther & Hadfield 2006), and it contains substantial amounts of hydrogen in its atmosphere.

An important diagnostic for WRWs is the flux ratio, R, between the narrow nebular and the broad stellar He II λ4686 emission. R directly measures the fraction of He$^+$-ionizing radiation that escapes the Wolf-Rayet wind. For SMC WR10 one finds $R \sim 5$

which is similar to the R values found for WRWs Brey 2 and SMC WR7. Fitting this set of observations with the Potsdam WR model WNL grid (cf., Hamann & Gräfener 2004) one readily derives the following stellar parameters:

$T_\star = 100 \pm 10$ kK, $R_\star = 2.4 \pm 0.3$ R$_\odot$, $L = 10^{5.7 \pm 0.2}$ L$_\odot$, $\dot{M} = 2\,10^{-6}$ M$_\odot$.

Note in particular the small derived mass-loss rate as compared to more typical Wolf-Rayet winds that are 10 to 100 times more powerful. Interestingly, the Smith *et al.* (2002) CMFGEN models suggest a significantly smaller temperature $T_\star = 80$ kK for the same He$^+$/H^0 Lyman continuum ratio. In other words, for the same core temperature, CMFGEN models predict roughly $10\times$ more He$^+$-ionizing photons than the corresponding PoWR models! On my request, Wolf-Rainer Hamann kindly computed weak-wind PoWR models with exactly the same input parameters as the hot low metallicity (0.2 Z$_\odot$) stars listed in Smith *et al.* (2002). He found the same discrepancy and suggested that possibly different assumed gravities of the underlying cores might play a role here.

As far as I am aware no standard evolutionary track leads to objects like SMC WR10. However, as mentioned earlier, for very rapid rotation evolution of massive stars is expected (Yoon & Langer 2005) to proceed directly towards high effective temperatures into the region of the Hertzsprung-Russell diagram where SMC WR10 is located. Quite possibly, this is the first time that quasi-chemically homogeneous evolution has been substantiated.

3. Highly excited "blue" nebulae in the MCELS images

Recently, Smith and the MCELS team (cf., Smith *et al.* 1999) published beautiful multi-colour images of the Magellanic Clouds emphasizing nebular emission of Hα (red color coded; R), O III λ5007 (blue; B), and S II $\lambda\lambda$6716, 31 (green; G). These images are not only very appealing (which explains why they have been shown in many talks during the conference!), but they also allow to readily discriminate between various ionization/excitation mechanisms at work.

These images easily allow to identify green and yellow ($y = R + G$) shock-ionized SNR, often superimposed on normal reddish H II regions, and contemplate the ionization structure of photoionized nebulae that often turn "yellowish" and "brownish" ($= y + R$) towards the edges where recombination towards lower-ionization species occurs.

Even a casual inspection of the LMC image in particular readily reveals a class of very blue nebulae (hereafter "BNe"); i.e. nebulae that appear to be dominated by the O III λ5007 blue hue of the image. I have verified that the colour balance is such that nebulae appear "very blue" when the intensity ratio O III λ5007/H$\alpha > 1.5$; i.e., being of very high excitation. I detect some 40 BNe in the LMC (not including the 30 Dor region which appears "burned-out") and about 10 BNe in the SMC, excluding NGC 346 and environment (cf. Fig. 3.)

What is the nature of the BNe? By employing the Strasbourg ALADIN SKY ATLAS one quickly realizes that a large fraction thereof represents photoionized nebulae around known early WN stars (WNE), several of which were not known before to excite ring nebulae around them (cf. the compilation by Dopita *et al.* (1994), and SMC WR update by Massey & Duffy (2001)). Late WN (WNL) and WC types appear to create less highly excited (violet-red coloured) nebulae more akin to more normal O star H II regions. The fact that such subtle differences become so strikingly apparent is indeed quite remarkable. It is also clear that many WNE BNe are much more extended than previously thought, and several such bubbles have O III λ5007-"skin" emission due to photoionized/incomplete shocks in the supersonically expanding bubbles (see, e.g., Dufour 1989).

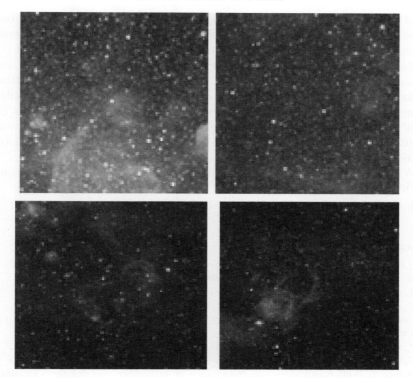

Figure 3. Examples of strongly O III λ5007 emitting blue nebulae in the Magellanic Clouds. These cut-outs each measure about $10' \times 10'$ corresponding to 150 pc in the Clouds. One needs to look at the colour version of the images in order to appreciate the unique "blueness" of the nebulae. From upper left, clockwise: (a) small round nebula North of the H II region N 19 (below) in the SMC, (b) previously unknown ring nebula around the faint WNE star SMC WR9, (c) a blue ring nebula in DEM 66 (= N 23A) in the LMC (d) the high excitation blue nebula NGC 1945 in the LMC located next to low exitation nebula NGC 1948 to the W (right)

We then might ask the question: What is the nature of the BNe not clearly associated with (known) WNEs? Remote possibilities include flash-ionized nebula around young O-rich supernova remnants (like in the case of SNR 0103−72.6), rare X-ray ionized nebulae or regions of incomplete shocks at the periphery of more easily discernable low-excitation SNRs. However, a more likely possibility is that we see here the fingerprints of hidden WN stars that have but weak stellar emission lines, i.e., being even fainer than those of the recently identified class of weak-lined SMC WRs. Such stars could have easily escaped detection by using slitless spectroscopy or narrow filter imaging techniques. Also, such objects would have weak winds and therefore be prone to be WRW. A possible example is the huge (150 pc diameter) newly detected faint ring nebula around SMC WR9 (a spectroscopic twin of WR 10) shown in Fig. 3.

4. A remark on the nature of He II λ4686 nebula N 44C

Not unexpectedly, He III region N 44C also appears as a prominent BN, in the MCELS image of the Large Cloud. However, its ionizing source has not been clearly identified even though an interpretation in terms of a fossil X-ray ionized nebula cannot be excluded. Here I propose an alternative scenario, not least inspired by the variable radial velocity of the central O7 star (Pakull 1989) which clearly points to a binary nature. Indeed, evolutionary calculations of massive binaries by Wellstein, Langer & Braun (2001) predict a sizable population of massive O stars with *low-mass*, (5–7 $M_\odot$), hot ($\sim 10^5$ K) Helium-star secondaries after mass transfer from the formerly more massive component to its companion has been completed. The He stars will be much fainter optically than the O star primaries and will therefore be quasi undetectable spectroscopically, readily explaining the apparent absence of such systems. However, the idea here is that such a star will act as a Warmer to excite a He III region that will in turn be detectable as a He II $\lambda 4686$ emitting nebula provided that the interstellar density is sufficiently high. Conceivably, the central (binary) star in N 44C is such a system.

Acknowledgements

I thank the editors for their patience with the delivery of my manuscript.

References

Crowther, P. A. & Hadfield, L. J. 2006, *A&A*, 449, 711

Dopita, M. A., Bell, J. F., Chu, Y. -H., & Lozinskaya, T. A. 1994, *ApJS*, 93, 455

Dufour, R. J. 1989, *Rev. Mexicana AyA*, 18, 87

Foellmi, C., Moffat, A. F. J., & Guerro, M. A. 2003, *MNRAS*, 338, 360

Hamann, W.-R. & Gräfener, G. 2004 *A&A*, 427, 697 — http://www.astro.physik.uni-potsdam. de/~wrh/PoWR/powrgrid1.html

Massey, P. & Duffy, A. S. 2001, *ApJ*, 550, 713

Niemela, V. S., Heathcote, S. A., & Weiller, W. C. 1991, in K.A. van der Hucht & B. Hidayat (eds.), *Wolf-Rayet stars and interrelations with other massive stars in galaxies*, IAU Conf.Proc. 143 (Dordrecht: Kluver), p. 425

Pakull, M. W. & Motch, C. 1989a, *Nature* 337, 337

Pakull, M. W. & Motch, C. 1989b, in E. J. A. Meurs & R. A. E. Fosbury (eds.), *ESO workshop on extranuclear activity in galaxies* (Garching: ESO), p. 285

Pakull, M. W. 1989, in K. S. de Boer, F. Spite, & G. Stasińska (eds.), *Recent developments in Magellanic Cloud research* (Meudon: Observatoire de Paris), p. 183

Pakull, M. W. 1991, in K. A. van der Hucht & B. Hidayat (eds.), *Wolf-Rayet stars and interrelations with other massive stars in galaxies*, IAU Conf.Proc. 143 (Dordrecht: Kluver), p. 391

Pakull, M. W. & Bianchi, L. 1991, in K. A. van der Hucht & B. Hidayat (eds.), *Wolf-Rayet stars and interrelations with other massive stars in galaxies*, IAU Conf.Proc. 143 (Dordrecht: Kluver), p. 260

Pakull, M. W & Angebault, L. P. 1986, *Nature*, 322, 511

Schmutz, W., Leitherer, C., & Gruenwald, R. 1992, *PASP*, 104, 1164

Smith, R. C. & the MCELS Team 1999, in Y. -H. Chu, N. Suntzeff, J. Hesser, & D. Bohlender (eds.), *New views of the Magellanic Clouds*, IAU Conf.Proc. 190, p. 28 — http://www.ctio.noao.edu/~mcels/

Smith, L. J., Norris, R. P. F., & Crowther, P. A. 2002, *MNRAS*, 337, 1309

Terlevich, R. & Melnick, J. 1985, *MNRAS*, 213, 841

Vink, J. S. & de Koter, A. 2005, *A&A*, 442, 587

Wellstein, S., Langer, N., & Braun, H. 2001, *A&A*, 369, 939

Yoon, S. -C. & Langer, N. 2005, *A&A*, 443, 643

The Magellanic System: Stars, Gas, and Galaxies
Proceedings IAU Symposium No. 256, 2008
Jacco Th. van Loon & Joana M. Oliveira, eds.

© 2009 International Astronomical Union
doi:10.1017/S1743921308028846

Supernova remnants in the Magellanic Clouds

Rosa N. M. Williams

Columbus State University, Coca-Cola Space Science Center, 701 Front Ave.,
Columbus, GA 31901 USA
email: rosanina@ccssc.org

Abstract. At the 1998 IAU Symposium on the Magellanic Clouds, Dr. Robert Petre observed that we were reaching a time where it was possible "to study the MC SNRs at a level of detail comparable with many Galactic remnants", while retaining the benefits of a global view in the MCs. Over the past decade, many researchers have taken advantage of these newly accessible populations. New MC-wide surveys at various wavelengths have enabled broader searches for SNR candidates, extending our census of MC SNRs to less prominent objects — older SNRs, SNRs in complex regions, et cetera. The use of light-echoes has provided a new avenue to probe young SNRs. Higher spatial and spectral resolutions in many wavelength regimes have enabled detailed studies of individual remnants, revealing progenitor types, pulsar-wind nebulae, expansion details, and environmental effects.

Perhaps the newest conceptual development is the increasing use of the MC SNRs to study physical problems of wider significance to many fields of astronomy. For example, researchers have examined the energy and hot gas inputs of MC SNRs to the ISM, including their collective effects within superbubbles, in order to evaluate their effects on stellar feedback cycles in a galaxy. Other scientists have investigated the fraction of SNR energies going to the acceleration of cosmic rays, which has significant implications for the role of SNRs in cosmic-ray production. Most recently, the onslaught of Spitzer data has led to new exploration of dust in MC SNRs, allowing us to probe dust creation, depletion, and destruction in the MC SNR populations. In summary, the study of SNRs in the MCs appears to have "come of age" over the past decade, becoming a mature field with rich potential for future scientific work.

Keywords. supernova remnants, Magellanic Clouds

1. Introduction

A supernova can result either from catastrophic core collapse and the resulting rebound in a massive star; or from "Type Ia" explosions in which a white dwarf accretes mass from a binary companion, until it reaches the Chandrasekhar limit and detonates and/or deflagrates. In either type of supernova, once the ejecta encounter interstellar matter, a shock wave proceeds outward into the interstellar medium (ISM). This encounter creates a reflected shock through which the ejecta pass, becoming shock-heated. As these shocks expand, and sweep up more interstellar material, they create a diffuse supernova remnant (SNR). In such a remnant, the outward ("blast wave") and reflected ("reverse") shocks generate emission throughout the electromagnetic spectrum. Synchrotron radiation, which dominates at radio wavelengths, is generated by the compression of magnetic fields and the acceleration of electrons by the shock. In the post-shock cooling region we can find substantial ionic and molecular line emission, which dominates at optical, ultraviolet, and (usually) X-ray wavelengths.

A galaxy's population of supernova remnants holds key clues to the overall structure and evolution of the ISM in that galaxy. SNRs are the primary source of energy, hot gas, and heavy elements injected into the ISM. Both individually and collectively as superbubbles, they provide a strong influence on large-scale ISM structure in a galaxy. The energy and heavy elements provided by SNRs will influence future generations of star formation, and thus are an important factor in the dynamical evolution of galaxies. SNRs clearly play a role in the cycle of dust formation and destruction in the ISM, although the nature and extent of their contribution is still not yet fully understood.

The past decade has seen remarkable advances in the study of supernova remnants outside the Milky Way Galaxy. In particular, the Large and Small Magellanic Clouds (LMC, SMC; collectively MCs) have proven to be a fertile field for SNR studies. A generation of new instruments, on the ground and in orbit, have enabled astronomers studying MC SNRs at a wide range of wavelengths to obtain resolutions and sensitivities previously only achievable for Galactic objects. At the same time, the relatively un-obscured view of these galaxies allows studies of the SNRs in the context of their surroundings, both individually and as galaxy-wide populations.

2. Data and demographics

A variety of new surveys of the Magellanic Clouds have been performed in numerous wavelength regimes. These include:

• The *ATCA* array and *Parkes* telescope have been used to complete a radio survey of both the Large and Small Magellanic Cloud at 3 and 6 cm (8640 and 4800 GHz). (See contribution by Dickel, this volume). There are also 21 cm maps of the LMC (Kim *et al.* 2003) and SMC (Stanimirović *et al.* 1999).

• The SAGE program (see contribution by Meixner, this volume) used *Spitzer* to survey the LMC and SMC in a series of narrow wavelength bands from 3 to 24 μm. The S^{3}MC program also performed narrow-band infrared imaging with *Spitzer* (see contribution by Sandstrom, this volume, on S^{3}MC and S^{4}MC). Additionally, the *AKARI* survey covered over half of the LMC (see contribution by Ita *et al.*, this volume).

• The Magellanic-Clouds Emission-Line Survey covered both MCs in Hα, [S II], and [O III] (Smith *et al.* 2005a).

• A *FUSE* Legacy project resulted in a large number of ultraviolet sightlines through SNRs in the MCs.

• There has not been a systematic study of the MCs with *XMM-Newton*, but the pointed observations do cover a significant fraction of the Clouds.

In addition, there have been a plethora of pointed observations in all of the wavelength regimes mentioned above, which have yielded detailed images and spectral data for many MC SNRs.

• X-ray images and spectra are available via pointed observations with *Chandra, XMM-Newton, ASCA*, and *ROSAT* for almost all of the MC SNRs (e.g., Filipović *et al.* 2008; van der Heyden *et al.* 2004; Williams *et al.* 1999).

• Ultraviolet images and spectra are available via *FUSE* observations for many of the brighter SNRs (e.g., Blair *et al.* 2006).

• Optical images, including some *Hubble Space Telescope* observations, and, e.g., echelle spectra are available for most of the MC SNR sample (e.g., Payne *et al.* 2007).

• Infrared images, spectra, and spectral maps are available via pointed observations with *Spitzer* for most of the LMC SNRs and perhaps half of the SMC SNRs (e.g., Williams, Borkowski *et al.* 2006; Borkowski, Williams *et al.* 2006; Williams *et al.* 2006; Stanimirović *et al.* 2005).

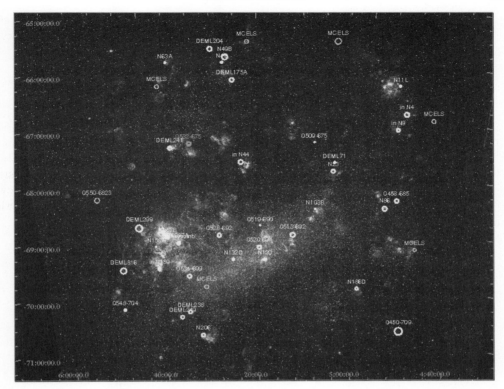

Figure 1. MCELS images of LMC with SNR known/confirmed/candidate locations overplotted. Color version: red, green, blue in image are Hα, [S II], and [O III] data respectively. White SNRs are known, cyan confirmed, yellow are candidates.

Table 1. Current SNR demographics.

	Total No.	Type Ia	Core-Collapse	Not Known	CCO/ PWN	O-rich
LMC	38	5	10	16	6	2
LMC unconfirmed [1]	7	5	2		1	
SMC	18	0	2	13	0	2
SMC unconfirmed [1]	2		4		1	

Notes: [1] Candidate remnants, unconfirmed type, etc.

- Radio images and polarization maps are available for many of the SNRs in both MCs (e.g., Filipović *et al.* 2005; contribution by Filipović *et al.*, this volume).

These new data provide a significant resource for the study of SNRs. For any given remnant, data is generally available in multiple wavelength regimes. In addition, such surveys greatly facilitate the discovery of faint, old SNRs, or SNRs in complex regions — populations which are themselves of significant interest. A number of new candidate remnants have indeed been suggested over the past decade (Table 1). However, only a few have actually been confirmed as SNRs; follow-up on the list of candidates has been somewhat desultory.

From Table 1, we can see that the number of "known SNRs" in the LMC, compared to the summary table of Petre (1999) has remained constant; two of the SNRs in previous

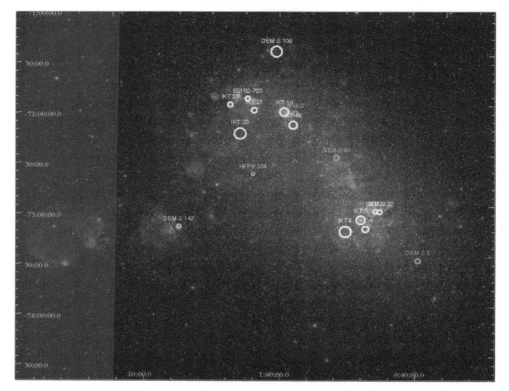

Figure 2. As Fig. 1, but now for the SMC.

list were re-identified as superbubbles, while two new discoveries joined the roster. The number of "confirmed SNRs" in the SMC has actually decreased, as a number of SNR candidates remain unconfirmed. The typing of SNRs tabulated by Petre (1999) rested partly on SNR characteristics (Balmer-line dominated SNRs typed as Ia, O-rich as CC) and partly on proximity of local OB populations (Chu & Kennicut 1988). Current typing of supernova remnants is based primarily on SNR characteristics, including, now, typing by abundance analysis of ejecta as discussed in §3.4 below. Pulsar-wind nebulae or other compact object phenomena have been identified for six objects in the LMC (and a seventh candidate suggested), doubling the previously known numbers; none have yet been confirmed within the SMC. These objects are discussed in §3.2.

3. Individual SNR studies

The wealth of new data, and of high-resolution, high-sensitivity instruments, has led to an explosion of research on Magellanic Cloud SNRs in the past decade. It is beyond the scope of this work to cover each paper in detail; instead, I summarize a few notable developments. These include recent developments in the study of SN 1987A; progress in uncovering and studying pulsar-wind nebulae (PWN) and other compact-object phenomena; observations of light-echoes from the youngest remnants; the use of X-ray spectra to determine abundances, allowing comparisons with model predictions; and observations of SNR interactions with their surroundings.

3.1. *SNR 1987A*

The most studied SNR in the MCs is of course SN/SNR 1987A; a brief search of the literature turns up as many papers about this object as for the rest of the MC SNRs together. A full review of this object is beyond the scope of this work; I will here only touch on a very select set of recent findings.

In 1995, the first "hot spot" was observed as an unresolved brightening along the equatorial ring surrounding SN 1987A. By 1999, there were many such spots, signaling the beginning of the transition from a supernova to a supernova remnant (Lawrence *et al.* 2000) as supernova ejecta encountered the circumstellar ring. Since then, emission interpreted as being from ejecta heated by the reverse shock has been observed (Michael *et al.* 2003; McCray 2007). The time-evolution of this SNR is being studied on a continuing basis at a wide range of wavelengths, with new observations every few months. This allows ongoing study of time-dependent effects in the evolution of a very young SNR. (e.g., Park *et al.* 2006; Heng *et al.* 2006; Smith *et al.* 2005b; Utrobin & Chugai 2005; Manchester *et al.* 2005).

The physical issues examined in SNR 1987A are many and varied. These include an inventory of the nucleosynthesis products, (e.g., Heng *et al.* 2008), and examinations of specific isotopes, (e.g., ^{55}Fe, Leising 2006). Exhaustive study has been made of shock structure, ionization stages, and internal shock velocities (Dewey *et al.* 2008; Gröningsson *et al.* 2008; Heng & McCray 2007; Reighard & Drake 2007; Haberl *et al.* 2006). Dust formation and destruction in SNRs is a subject of high current interest, and is studied in depth in SN 1987A (Dwek *et al.* 2008; Ercolano *et al.* 2007; Bouchet *et al.* 2006); the general case will be discussed in §4.3. Although efforts have also been bent toward finding a pulsar, the search has not yet resulted in a detection; see Manchester (2007) for a review. One should note that the upper limits for a detection are still relatively generous, so these non-detections are unsurprising.

3.2. *Compact-object phenomena*

The search for compact object phenomena in other SNRs has been more fruitful. Three examples had been known before the start of this decade. SNR 0540–693 in the LMC hosts the well-known pulsar (PSR) 0540–69, as well as a PWN bright in optical and X-rays. Morse *et al.* (2006) used *Hubble Space Telescope* data to show that the outer rim of the PWN had a "skin" of [O III] emission; Petre *et al.* (2007) followed up on this in a detailed study of the X-ray structure of the PWN, among other things showing that the X-ray extent was bounded by this [O III] feature.

In the SNR N157B, a long-suspected pulsar was detected by Marshall *et al.* (1998). Shortly afterward, Wang & Gotthelf (1998) identified X-ray emission from a PWN surrounding this pulsar. Further study of this region by Wang *et al.* (2001) showed it to have a "comet-like" head and tail structure, which they interpreted as a bow shock from the pulsar's supersonic motion through the SNR ejecta. (It should be noted that a variant interpretation has been given by van der Swaluw (2004), who argues that the motion of the pulsar is subsonic and that it is interacting with the reverse shock in the SNR.)

A soft gamma-ray repeater (SGR) 0526–66 is co-located on the sky with supernova remnant N49 in the LMC. However, whether the two objects were physically associated was considered questionable for some time (e.g., Kaplan *et al.* 2001). However, studies by Kulkarni *et al.* (2003) and Park *et al.* (2003) identified an X-ray counterpart to the SGR. Kulkarni *et al.* (2003) were able to use this counterpart to estimate the presumptive pulsar's spin-down rate, and calculate its age; that age is reasonably close to the estimated age of N49, which does suggest an association between the two.

In this decade, we have more than doubled the numbers of confirmed or probable compact-object phenomena. (Note that this also adds to the number of SNRs we can confidently classify as resulting from core-collapse SNe.) Gaensler *et al.* (2003) identified a small-diameter radio and X-ray source within SNR 0453–685, and determined it to be a relatively "Vela-like" PWN, probably subsonic. Klinger *et al.* (2002) noted a "peculiar linear feature" in radio maps of N206-SNR, and a subsequent X-ray examination by Williams *et al.* (2005) confirmed the presence of another "cometary" PWN resulting from a bow shock. Bamba *et al.* (2006) have more tentatively identified what may be a similar X-ray structure within the SNR DEM L 241. Finally, Hughes *et al.* (2006) and Hayato *et al.* (2006) uncovered and studied a point-like source in the SNR N23, whose properties are consistent with those of an object similar to the compact central object (CCO) recently discovered in the Galactic remnant Cas A.

3.3. *Light echoes*

It has long been known that the light of a supernova, as it propagates through the ISM, may be reflected at a delay toward us from various ISM features in a phenomenon known as a "light-echo". Such light-echoes had been noted to occur with SN 1987A (Crotts 1988; Rosa *et al.* 1988) in the Large Magellanic Cloud. In the current decade, however, the SuperMACHO collaboration discovered three more light-echoes within the LMC (Rest *et al.* 2005). These light-echoes were traced back to three of the LMC's youngest SNRs: N103B, SNR 0509–675, and SNR 0519–690. From the rate of travel, they were able to precisely determine ages for these SNRs; later studies of shock velocities from ultraviolet observations (Ghavamian *et al.* 2007) provided an independent check on these estimates. As the light from these echoes is a reflection of the supernova itself, Rest *et al.* (2005) were able to identify the progenitor type for each of the SNRs as Type Ia events; this is consistent with the Balmer-line dominated emission for which these remnants have been known (Tuohy *et al.* 1982; Smith *et al.* 1991). In addition, these light-echoes have the potential to be sensitive probes of the ISM through which they pass (e.g., Chevalier 1986; Crotts 1988).

3.4. *Spectral analysis of abundances*

Hughes *et al.* (1995) compared the X-ray spectra of some of the youngest, ejecta-dominated LMC SNRs to nucleosynthesis models for Type Ia versus core-collapse supernovae. The authors were able to distinguish between Type Ia and core-collapse progenitors for all of the SNRs in their sample. Later researchers found that in many cases, even when SNRs are dominated by swept-up ISM, there are still enough ejecta signatures in their X-ray spectra to identify the progenitor types with reasonable confidence. By this method, a number of older MC SNRs have recently been identified as resulting from Type Ia supernovae (e.g., Hendrick *et al.* 2003; Hughes *et al.* 2003; Lewis *et al.* 2003; Borkowski *et al.* 2006). In some instances, SNRs from core-collapse events have also been identified (e.g., Park *et al* 2003b,c; Williams *et al.* 2008).

An example of the application of this technique was provided by the two remnants of DEM L 316 in the LMC. The system has two shells which overlap along the line of sight. Williams *et al.* (1997) used various features of these remnants to argue that the two were coeval and actually colliding. However, Nishiuchi *et al.* (2001) used *ASCA* X-ray data to show that the abundances in one shell were most probably consistent with a Type Ia origin. Following up on this, Williams & Chu (2005) used *Chandra* X-ray spectra to confirm that result, and further identified the other shell as probably of core-collapse origin. The hypothesis that the two shells were coeval was clearly rendered improbable by these findings!

The increasing identification of remnants as Type Ia SNRs also provides a sample for more precise comparisons of SNR abundances with models. For example, Warren & Hughes (2004) used SNR 0509–675 to test nucleosynthesis models of Type Ia supernovae, and found that a delayed-detonation model best fit the observed abundances. Addressing a different issue, Badenes *et al.* (2007) investigated a number of known Type Ia SNRs to test a particular model in which accretion winds from the binary system of the progenitor white dwarf excavate a cavity in the surrounding ISM. However, the authors concluded that this "wind" model was inconsistent with the observed properties of ejecta in their sample of MC SNRs.

3.5. *SNR interactions with surroundings*

Another feature of the last decade has been the increase in studies examining MC SNRs in the context of their surroundings, and their interactions with the nearby ISM. The MCs offer rich samples, which have enabled research to study SNRs interacting with molecular clouds (Bilikova *et al.* 2007; Koo *et al.* 2007; Warren *et al.* 2003); with H II regions and complexes (Reid *et al.* 2006; Gorjian *et al.* 2004; Danforth *et al.* 2003; Nazé *et al.* 2002; Dickel *et al.* 2001; Williams *et al.* 2008); with (and within) superbubbles (Chen *et al.* 2006; Townsley *et al.* 2006); and even, in one case, with a single stellar-wind bubble (Velázquez *et al.* 2003). These studies address such questions as how the interaction with the denser material with a molecular cloud influences postshock cooling in an SNR; observational features of a superbubble that point to a recent interior SNR; the evolution of an SNR in the rarified environment of a superbubble; the formation and evolution of superbubbles themselves; and similar issues.

4. Using SNRs to address astrophysical issues

Possibly the most significant development in MC SNR research is the increasing use of Magellanic Cloud SNRs to address broader questions of interest to astronomy. Three examples of such work are discussed below: the impact of SNRs on the global ISM; the SNR role in cosmic-ray acceleration; and the creation and destruction of dust in SNRs.

4.1. *SNR inputs to global ISM*

SNRs are a key element of the stellar feedback cycle that is discussed extensively in these proceedings. Remnants affect the ISM individually, driving local conditions; and collectively as they combine with stellar wind to form the large-scale (> 100 pc) superbubbles. Thus, SNRs provide much of the energy that drives the stellar feedback processes in the ISM: thermal energy as hot gas; kinetic energy which goes to mechanical feedback (see contributions by Chu and by Oliveira, this volume); and nonthermal energy, some of which may go to cosmic-ray acceleration. Of course, SNRs also are the source of the nucleosynthesis mentioned above, providing heavy elements to the ISM.

A persistent question has been the "filling factor" of the hot gas phase (in the standard three-phase model of McKee & Ostriker (1977) for the ISM) in the Magellanic Clouds. In order to pin down this elusive quantity, two things are required; a fairly complete census of the sources of hot gas, i.e. SNRs and superbubbles; and knowledge of how long hot gas persists in SNR and superbubble interiors. The first element is addressed by uncovering faint, old remnants, and by trying to account for the number of SNRs within the rarefied cavities of superbubbles. The second element is addressed by studying the oldest SNRs; recent work has shown that hot gas persists deep in SNR interiors to late stages. In fact, these hot reservoirs are thought to result in "mixed-morphology" remnants, in which we see shell-like radio structure but X-ray filled interiors. Again, this line of inquiry must also account for the heating of gas inside superbubbles by SNR shocks.

4.2. *Cosmic-ray acceleration*

Energy from SNRs can go to another component: they are thought to be a primary source of GeV and TeV cosmic rays. Hughes *et al.* (2000) examined the SMC SNR E0102–723.

Based on the observed shock velocities, they inferred a high postshock ion temperature. Ions and electrons exchange energy through the Coulomb process, which links the electron temperature to the ion temperature. In this case, the electron temperature measured from X-ray spectra was considerably lower than expected from the shock velocity. The authors suggested that the "missing" energy may have largely gone to cosmic-ray acceleration.

Another question of interest is the maximum energy to which SNRs can accelerate cosmic rays. It is known that the nonthermal spectrum in SNRs steepens significantly from the extrapolated spectrum of radio synchrotron emission. Hendrick & Reynolds (2001) modeled the change in slope from radio to X-ray spectra, and fit this for 11 Magellanic Cloud SNRs. The "cutoff" in their model gives the maximum energy for cosmic-ray acceleration, and their fits indicated maximum cosmic-ray energies of about 80 TeV. While this is consistent with the overall expectation for the energy range SNR-produced cosmic rays, it shows that SNRs are unlikely to propel cosmic rays over the "knee" in the cosmic-ray spectrum above 100 TeV.

4.3. *Dust production and destruction*

Supernova ejecta are thought to be a major source of dust in the ISM, along with other primary sources such as the winds of evolved stars. Indeed, emission signatures of dust have been observed in supernovae. Warm (100 K) dust is expected to radiate strongly in infrared, but has proven difficult to detect in any but the youngest SNRs (e.g., Reach *et al.* 2006).

In observing SNR 1987A, Dwek *et al.* (2008) concluded that the infrared emission was primarily from swept-up ISM dust, not from SNR ejecta. Further, the dust-to-gas mass ratio in SNR 1987A has decreased over time — a sign of ongoing grain destruction. Tappe *et al.* (2006) also found infrared emission from dust in the young SNR N132D, but also inferred that the observed dust was from the swept-up ISM. They found a dust-to-gas mass ratio of about of tenth of the average ratio for the LMC. Notably, the authors found that the spectrum of the infrared emission was consistent with grain losses from dust destruction by sputtering. Borkowski, Williams *et al.* (2006) and Williams, Borkowski, *et al.* (2006), for LMC samples of Type Ia and core-collapse SNRs, respectively, likewise interpreted their *Spitzer* observations of infrared emission as being from swept-up dust. They, too, found low dust-to-gas mass ratios.

In addition, Williams *et al.* (2006) published a preliminary study, later confirmed by spectral mapping data, that showed that for certain specific remnants, ionic and molecular line emission from post-shocked gas forms a substantial fraction of the infrared emission. This leads to ever-decreasing dust estimates: not all SNRs are detected in infrared; when detected, some of the infrared emission may be line emission from the gas component; when dust is found it is often thought to be dust swept up from the ISM, rather than newly formed in ejecta; and even of the swept-up dust, often less is found in the SNR than is found in the typical MC ISM! SNRs may prove to be net destroyers of dust, rather than producers of it.

However, there is still room for substantial dust contribution from SNRs. An example is provided by SNR E0102–723. When this SNR was studied by Stanimirović *et al.* (2005), they estimated the proportion of line emission contribution to the infrared, and noted that the dust continuum appeared to originate from regions of the SNR associated with the reverse shock as well as the outer blast wave. This suggests that some emission is indeed from dust in the newly shock-heated ejecta. Sandstrom (see contribution by Sandstrom *et al.*, this volume) has been analyzing spectra from this SNR, and finds strong evidence for actual ejecta-formed dust. While there is still less dust found than models predict should be produced by SNRs, one must also consider that much of this dust may remain within the unshocked ejecta, and this "cold" dust may not be readily observable.

5. Summary

SNR studies in the Magellanic Clouds can now be considered a mature field of research. Researchers continue, of course, with the categorization, characterization, and comparison of Magellanic SNRs with Galactic SNRs. Work has been pushed to ever more sophisticated analyses of SNRs themselves, with MC SNRs being used to test various models of SNR origins, structure, and evolution. In addition, current studies are using MC SNRs to address key physical problems significant to a broad range of topics within astronomy.

Acknowledgements

The author gratefully acknowledges NASA Long-Term Space Astrophysics grant NNG-05GC97G, which has supported the author's research in this subject as well as travel to this conference.

References

Badenes, C., Hughes, J. P., Bravo, E., & Langer, N. 2007, *ApJ*, 662, 472
Bamba, A., Ueno, M., Nakajima, H., Mori, K., & Koyama, K. 2006, *A&A*, 450, 585
Bilikova, J., Williams, R. N. M., Chu, Y. -H., Gruendl, R. A., & Lundgren, B. F. 2007, *AJ*, 134, 2308
Blair, W. P., Ghavamian, P., Sankrit, R., & Danforth, C. W. 2006, *ApJS*, 165, 480
Borkowski, K. J., Hendrick, S. P., & Reynolds, S. P. 2006, *ApJ*, 652, 1259
Borkowski, K. J., Williams, B. J., *et al.* 2006, *ApJ*, 642, L141
Bouchet, P., Dwek, E., Danziger, J., *et al.* 2006, *ApJ*, 650, 212
Chen, Y., Wang, Q. D., Gotthelf, E. V., Jiang, B., Chu, Y. -H., & Gruendl, R. 2006, *ApJ*, 651, 237
Chevalier, R. A. 1986, *ApJ*, 308, 225
Chu, Y. -H. & Kennicutt, R. C., Jr. 1988, *AJ*, 96, 1874
Crotts, A. 1988, *ApJ*, 333, L51
Danforth, C. W., Sankrit, R., Blair, W. P., Howk, J. C., & Chu, Y. -H. 2003, *ApJ*, 586, 1179
Dewey, D., Zhekov, S. A., McCray, R., & Canizares, C. R. 2008, *ApJ*, 676, L131
Dickel, J. R., Williams, R. M., Carter, L. M., Milne, D. K., Petre, R., & Amy, S. W. 2001, *AJ*, 122, 849
Dwek, E., *et al.* 2008, *ApJ*, 676, 1029
Ercolano, B., Barlow, M. J., & Sugerman, B. E. K. 2007, *MNRAS*, 375, 753
Filipović, M. D., Haberl, F., Winkler, P. F., *et al.* 2008, *A&A*, 485, 63
Filipović, M. D., Payne, J. L., Reid, W., Danforth, C. W., Staveley-Smith, L., Jones, P. A., & White, G. L. 2005, *MNRAS*, 364, 217
Gaensler, B. M., Hendrick, S. P., Reynolds, S. P., & Borkowski, K.J. 2003, *ApJ*, 594, L111
Ghavamian, P., Blair, W. P., Sankrit, R., Raymond, J. C., & Hughes, J. P. 2007, *ApJ*, 664, 304
Gorjian, V., Werner, M. W., Mould, J. R., *et al.* 2004, *ApJS*, 154, 275
Gröningsson, P., Fransson, C., Lundqvist, P., *et al.* 2008, *A&A*, 479, 761
Haberl, F., Geppert, U., Aschenbach, B., & Hasinger, G. 2006, *A&A*, 460, 811
Hayato, A., Bamba, A., Tamagawa, T., & Kawabata, K. 2006, *ApJ*, 653, 280
Hendrick, S. P., Borkowski, K. J., & Reynolds, S. P. 2003, *ApJ*, 593, 370
Hendrick, S. P. & Reynolds, S. P. 2001, *ApJ*, 559, 903
Heng, K. & McCray, R. 2007, *ApJ*, 654, 923
Heng, K., Haberl, F., Aschenbach, B., & Hasinger, G. 2008, *ApJ*, 676, 361
Heng, K., McCray, R., Zhekov, S. A., *et al.* 2006, *ApJ*, 644, 959

Hughes, J. P., Rafelski, M., Warren, J. S., Rakowski, C., Slane, P., Burrows, D., & Nousek, J. 2006, *ApJ*, 645, L117

Hughes, J. P., Ghavamian, P., Rakowski, C. E., & Slane, P. O. 2003, *ApJ*, 582, L95

Hughes, J. P., Rakowski, C. E., & Decourchelle, A. 2000, *ApJ*, 543, L61

Hughes, J. P., Hayashi, I., Helfand, D., *et al.* 1995, *ApJ*, 444, L81

Kaplan, D. L., Kulkarni, S. R., van Kerkwijk, M. H., Rothschild, R. E., Lingenfelter, R. L., Marsden, D., Danner, R., & Murakami, T. 2001, *ApJ*, 556, 399

Kim, S., Staveley-Smith, L., Dopita, M. A., Sault, R. J., Freeman, K. C., Lee, Y., & Chu, Y. -H. 2003, *ApJS*, 148, 473

Klinger, R. J., Dickel, J. R., Fields, B. D., & Milne, D. K. 2002, *AJ*, 124, 2135

Koo, B. -C., Lee, H. -G., Moon, D. -S., *et al.* 2007, *PASJ*, 59, 455

Kulkarni, S. R., Kaplan, D. L., Marshall, H. L., Frail, D. A., Murakami, T., & Yonetoku, D. 2003, *ApJ*, 585, 948

Lawrence, S. S., Sugerman, B. E., Bouchet, P., *et al.* 2000, *ApJ*, 537, L123

Leising, M. D. 2006, *ApJ*, 651, 1019

Lewis, K. T., Burrows, D. N., Hughes, J. P., Slane, P. O., Garmire, G. P., & Nousek, J. A. 2003, *ApJ*, 582, 770

Manchester, R. N. 2007, in S. Immler, K. Weiler, & R. McCray (eds.), *Supernova 1987A: 20 Years After*, AIP Conf.Proc. 937 (New York: AIP), p. 134

Manchester, R. N., Gaensler, B. M., Staveley-Smith, L., Kesteven, M. J., & Tzioumis, A. K. 2005, *ApJ*, 628, L131

Marshall, F. E., Gotthelf, E. V., Zhang, W., Middleditch, J., & Wang, Q. D. 1998, *ApJ*, 499, L179

McCray, R. 2007, in S. Immler, K. Weiler, & R. McCray (eds.), *Supernova 1987A: 20 Years After*, AIP Conf.Proc. 937 (New York: AIP), p. 3

McKee, C. F. & Ostriker, J. P. 1977, *ApJ*, 218, 148

Michael, E., McCray, R., Chevalier, R., *et al.* 2003, *ApJ*, 593, 809

Morse, J. A., Smith, N., Blair, W. P., Kirshner, R. P., Winkler, P. F., & Hughes, J. P. 2006, *ApJ*, 644, 188

Nazé, Y., Hartwell, J. M., Stevens, I. R., Corcoran, M. F., Chu, Y. -H., Koenigsberger, G., Moffat, A. F. J., & Niemela, V. S. 2002, *ApJ*, 580, 225

Nishiuchi, M., Yokogawa, J., Koyama, K., & Hughes, J. P. 2001, *PASJ*, 53, 99

Park, S., Zhekov, S. A., Burrows, D. N., Garmire, G. P., Racusin, J. L., & McCray, R. 2006, *ApJ*, 646, 1001

Park, S., Burrows, D. N., Garmire, G. P., Nousek, J. A., Hughes, J. P., & Williams, R. M. 2003a, *ApJ*, 586, 210

Park, S., Hughes, J. P., Slane, P. O., Burrows, D. N., Warren, J. S., Garmire, G. P., & Nousek, J. A. 2003b, *ApJ*, 592, L41

Park, S., Hughes, J. P., Burrows, D. N., Slane, P. O., Nousek, J. A., & Garmire, G. P. 2003c, *ApJ*, 598, L95

Payne, J. L., White, G. L., Filipović, M. D., & Pannuti, T. G. 2007, *MNRAS*, 376, 1793

Petre, R., Hwang, U., Holt, S. S., Safi-Harb, S., & Williams, R. M. 2007, *ApJ*, 662, 988

Petre, R. 1999, in Y. H. Chu, N. Suntzeff, J. Hesser, & D. Bohlender (eds.), *New Views of the Magellanic Clouds*, IAU Conf.Proc. 190 (ASP), p. 74

Reach, W. T., Rho, J., Tappe, A., *et al.* 2006, *AJ*, 131, 1479

Reid, W. A., Payne, J. L., Filipović, M. D., Danforth, C. W., Jones, P. A., White, G. L., & Staveley-Smith, L. 2006, *MNRAS*, 367, 1379

Reighard, A. B. & Drake, R. P. 2007, *Ap&SS*, 307, 121

Rest, A., Suntzeff, N. B., Olsen, K., *et al.* 2005, *Nature*, 438, 1132

Rosa, M., Gouiffes, C., Ruiz, M., & Beresford, A. C. 1988, *IAU Circular*, 4564, 1

Smith, R. C., Points, S. D., Chu, Y. -H., Winkler, P. F., Aguilera, C., Leiton, R., & MCELS Team 2005a, *BAAS*, 37, 1200

Smith, N., Zhekov, S. A., Heng, K., McCray, R., Morse, J. A., & Gladders, M. 2005b, *ApJ*, 635, L41

Smith, R. C., Kirshner, R. P., Blair, W. P., & Winkler, P. F. 1991, *ApJ*, 375, 652

Stanimirović, S., Staveley-Smith, L., Dickey, J. M., Sault, R. J., & Snowden, S. L. 1999, *MNRAS*, 302, 417

Stanimirović, S., Bolatto, A. D., Sandstrom, K., Leroy, A. K., Simon, J. D., Gaensler, B. M., Shah, R. Y., & Jackson, J. M. 2005, *ApJ*, 632, L103

Tappe, A., Rho, J., & Reach, W. T. 2006, *ApJ*, 653, 267

Townsley, L. K., Broos, P. S., Feigelson, E. D., Brandl, B. R., Chu, Y. -H., Garmire, G. P., & Pavlov, G. G. 2006, *AJ*, 131, 2140

Tuohy, I. R., Dopita, M. A., Mathewson, D. S., Long, K. S., & Helfand, D. J. 1982, *ApJ*, 261, 473

Utrobin, V. P. & Chugai, N. N. 2005, *A&A*, 441, 271

van der Heyden, K. J., Bleeker, J. A. M., & Kaastra, J. S. 2004, *A&A*, 421, 1031

van der Swaluw, E. 2004, *Adv. Sp. Res.*, 33, 475

Velázquez, P. F., Koenigsberger, G., & Raga, A. C. 2003, *ApJ*, 584, 284

Wang, Q. D. & Gotthelf, E. V. 1998, *ApJ*, 494, 623

Wang, Q. D., Gotthelf, E. V., Chu, Y. -H., & Dickel, J. R. 2001, *ApJ*, 559, 275

Warren, J. S. & Hughes, J. P. 2004, *ApJ*, 608, 261

Warren, J. S., Hughes, J. P., & Slane, P. O. 2003, *ApJ*, 583, 260

Williams, B. J., Borkowski, K. J., *et al.* 2006, *ApJ*, 652, L33

Williams, R. M. & Chu, Y. -H. 2005, *ApJ*, 635, 1077

Williams, R. M., Chu, Y. -H., Dickel, J. R., Gruendl, R. A., Seward, F. D., Guerrero, M. A., & Hobbs, G. 2005, *ApJ*, 628, 704

Williams, R. M., *et al.* 2008, *ApJ*, submitted

Williams, R. M., Chu, Y. -H., & Gruendl, R. 2006, *AJ*, 132, 1877

Williams, R. M., Chu, Y. -H., Dickel, J. R., Gruendl, R. A., Seward, F. D., Guerrero, M. A., & Hobbs, G. 2005, *ApJ*, 628, 704

Williams, R. M., Chu, Y. -H., Dickel, J. R., Petre, R., Smith, R. C., & Tavarez, M. 1999, *ApJS*, 123, 467

Williams, R. M., Chu, Y. -H., Dickel, J. R., Beyer, R., Petre, R., Smith, R. C., & Milne, D. K. 1997, *ApJ*, 480, 618

The Magellanic System: Stars, Gas, and Galaxies
Proceedings IAU Symposium No. 256, 2008
Jacco Th. van Loon & Joana M. Oliveira, eds.

© 2009 International Astronomical Union
doi:10.1017/S1743921308028858

Resolving the dusty torus and the mystery surrounding LMC red supergiant WOH G64

Keiichi Ohnaka[1], Thomas Driebe[1], Karl-Heinz Hofmann[1], Gerd Weigelt[1] and Markus Wittkowski[2]

[1] Max-Planck-Institut für Radioastronomie, Auf dem Hügel 69, 53121 Bonn, Germany
email: kohnaka@mpifr-bonn.mpg.de

[2] European Southern Observatory, Karl-Schwarzschild-Str. 2, 85748 Garching, Germany

Abstract. We present mid-IR long-baseline interferometric observations of the red supergiant WOH G64 in the Large Magellanic Cloud with MIDI at the *ESO's Very Large Telescope Interferometer* (*VLTI*). Our MIDI observations of WOH G64 are the first *VLTI* observations to spatially resolve an individual stellar source in an extragalactic system. Our 2-D radiative transfer modeling reveals the presence of a geometrically and optically thick torus seen nearly pole-on. This model brings WOH G64 in much better agreement with the current evolutionary tracks for a 25 $M_\odot$ star — about a half of the previous estimate of 40 $M_\odot$ — and solves the serious discrepancy between theory and observation which existed for this object.

Keywords. techniques: interferometric, infrared: stars, circumstellar matter, stars: evolution, stars: late-type, stars: mass loss, supergiants, Magellanic Clouds

1. Introduction

Red supergiants (RSGs) in the Large and Small Magellanic Clouds (LMC and SMC, respectively) provide an excellent opportunity to observationally test the current stellar evolution theory for massive ($\geqslant 8$ $M_\odot$) stars. Another advantage of studying RSGs in the LMC and SMC is that we can probe metallicity effects on the mass loss.

WOH G64 is a luminous RSG in the LMC, surrounded by an optically thick dust envelope with the 10 μm silicate feature seen in self-absorption. The strength of the TiO bands suggests spectral types of M5–M7 (Elias *et al.* 1986; van Loon *et al.* 2005), which translate into $T_{\rm eff}$ = 3200–3400 K. The luminosities of 5–6$\times 10^5$ $L_\odot$ estimated by these authors assuming spherical shells correspond to an initial mass of $\sim$40 $M_\odot$. However, this $T_{\rm eff}$ is too low for the current evolutionary tracks for a 40 $M_\odot$ star (discrepancy in temperature $\approx$ 3000 K! — see the box in Fig. 1). The above authors note that the low obscuration in the near-IR/visible, despite the huge mid-/far-IR excess (Fig. 3a), may indicate the possible presence of a disk or torus. Such deviation from spherical symmetry affects the luminosity estimate. To examine this possibility, we carried out mid-IR high-spatial resolution observations of WOH G64 with *VLTI*/MIDI.

2. VLTI/MIDI Observations

MIDI is a 10 μm interferometric instrument which combines two beams from 8.2-m Unit Telescopes (UTs) or movable 1.8-m Auxiliary Telescopes (ATs). MIDI measures the "visibility amplitude" (= Fourier amplitude of the object's intensity distribution), which contains information about the object's size and shape. We observed WOH G64 in 2005 and 2007 with the UT3-UT4-62m baseline with a spatial resolution of $\sim$10 mas.

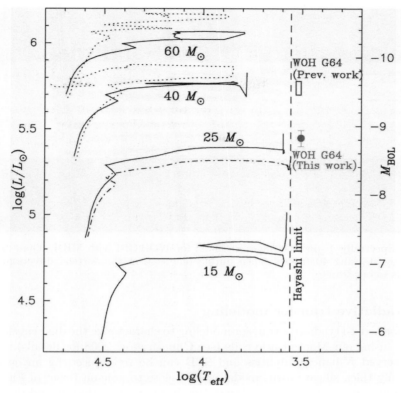

Figure 1. H-R diagram with theoretical evolutionary tracks with $Z = 0.008$ (solid lines: Schaerer *et al.* 1993; dotted lines: Meynet & Maeder 2005) and that newly calculated with $Z = 0.01$ by T. Driebe (dashed-dotted line). The box and circle represent the observationally derived locations of WOH G64. The dashed line represents the Hayashi limit.

We measured visibilities at four position angles differing by $\sim 60°$ but at almost the same baseline length (57–62 m). Interferometric fringes were spectrally dispersed with $\lambda/\Delta\lambda \approx 30$ between 8 and 13 μm (see Fig. 2). We found no significant temporal variation in the N-band spectra between the two epochs (also in agreement with the *Spitzer*/IRS data taken in 2005), and so we merged the data taken in 2005 and 2007.

3. Observational results: interferometry and spectroscopy

Figure 3c shows the N-band visibilities observed toward WOH G64. This is the first *VLTI* observations to spatially resolve an individual star in an extragalactic system. We fitted the observed visibilities with uniform-disks, and the resulting angular diameters are shown in Fig. 3d. The uniform-disk diameters increase from ~ 15 to 23 mas between 8 and 10 μm and is roughly constant above 10 μm. As Figs. 3c and 3d show, the visibilities and angular diameters measured at four position angles do not show a noticeable difference, suggesting that the object appears nearly centrosymmetric.

We also identified the H_2O absorption features at 2.7 and 6 μm in the spectra obtained with the Short Wavelength Spectrometer (SWS) and PHOT-S onboard the *Infrared Space Observatory* and the InfraRed Spectrometer (IRS) onboard the *Spitzer Space Telescope* (Fig. 3b). The 2.7 μm feature (ν_1 and ν_3 fundamental bands) originates in the photosphere and/or the extended molecular layers (so-called MOLsphere), while the 6 μm feature (ν_2 fundamental bands) is likely to be of circumstellar origin.

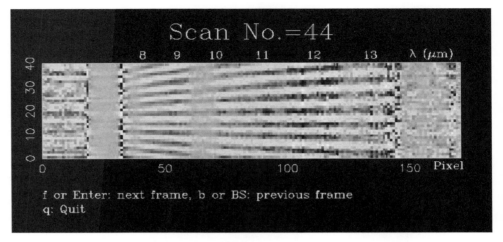

Figure 2. Spectrally dispersed fringes observed for WOH G64 with MIDI. These fringes were visualized by stacking 40 frames taken during one scan in the vertical direction. Each row corresponds to one frame.

4. 2-D radiative transfer modeling

We performed 2-D radiative transfer modeling to characterize the dust envelope around WOH G64, using our Monte Carlo code (see Ohnaka *et al.* 2008 for details).

The observed N-band visibilities and SED can be reproduced by an optically and geometrically thick silicate torus model viewed close to pole-on (inset of Fig. 3a). This pole-on model can explain the low obscuration in the near-IR/visible and the absence of position angle dependence of the visibilities (solid lines in Fig. 3). The derived dust torus parameters are as follows: $\tau_V = 30 \pm 5$ (optical depth at 0.55 μm in the equatorial plane), $\tau_V = 9 \pm 2$ (optical depth along the line of sight), inner boundary radius $= 15 \pm 5$ R$_\star$ ($R_\star = 1730$ R$_\odot$, $\rho \propto r^{-2}$ was assumed), torus half-opening angle $= 60 \pm 10°$. The derived luminosity, $\sim 2.8 \times 10^5$ L$_\odot$, is about a half of the previous estimates based on spherical models. This is because we look into the torus from nearly pole-on. Radiation escapes preferentially through the cavity (i.e., toward us), and the luminosity is overestimated when derived assuming spherical symmetry.

The new, lower luminosity brings the location of WOH G64 on the H-R diagram in much better agreement with theoretical evolutionary tracks for a 25 M$_\odot$ star (filled circle in Fig. 1). We also note that WOH G64 lies very close to or even beyond the Hayashi limit (dashed line in Fig. 1), which implies that this object may be experiencing unstable, violent mass loss. The derived total envelope mass, 3–9 M$_\odot$, represents a considerable fraction of its initial mass, which is consistent with this picture.

5. Conclusion

We have spatially resolved the circumstellar environment of the red supergiant WOH G64 in the LMC for the first time. Our 2-D radiative transfer modeling shows the presence of an optically and geometrically thick torus viewed close to pole-on and brings the location of WOH G64 in much better agreement with the current stellar evolution theory. Now, MIDI observations with shorter and longer baselines are indispensable for obtaining tighter constraints on the torus inner boundary radius, which is crucial for understanding metallicity effects on the dust formation and mass loss in RSGs.

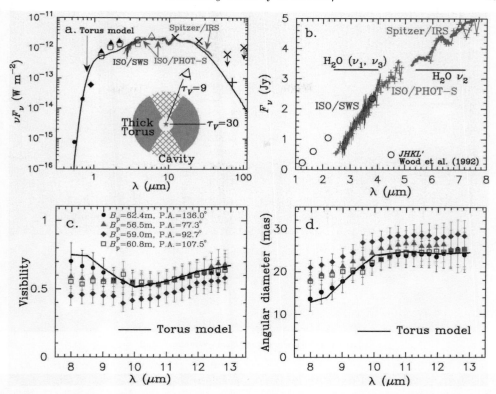

Figure 3. a: SED. filled circles: MACHO (Alcock *et al.* 2000), filled diamond: ASAS (Pojmański 2002, 2003; Pojmański & Maciejewski 2004, 2005; Pojmański *et al.* 2005), filled triangles: 2MASS (Cutri *et al.* 2003), open circles: Wood *et al.* (1992), open squares: Whitelock *et al.* (2003), open triangle: SAGE (Meixner *et al.* 2006), ×: IRAS, +: *Spitzer*/MIPS. **b:** H_2O absorption between 2 and 8 μm. **c:** N-band visibilities observed with MIDI and the torus model. The projected baseline lengths (B_p) and the position angles (P.A.) are also given. **d:** N-band angular diameters observed at four position angles and the torus model.

References

Alcock, C., Allsman, R., Alves, D., *et al.* 2000, *ApJ*, 542, 281

Cutri, R. M., Skrutskie, M. F., van Dyk, S., *et al.* 2003, *The IRSA 2MASS All-Sky Catalog of Point Sources*, NASA/IPAC Infrared Science Archive

Elias, J. H., Frogel, J. A., & Schwering, P. B. 1986, *ApJ*, 302, 675

Meynet, G. & Maeder, A. 2005, *A&A*, 429, 581

Meixner, M., Gordon, K. D., Indebetouw, R., *et al.* 2006, *AJ*, 132, 2268

Ohnaka, K., Driebe, T., Hofmann, K. -H., Weigelt, G., & Wittkowski, M. 2008, *A&A*, 484, 371

Pojmański G. 2002, *AcA*, 52, 397

Pojmański G. 2003, *AcA*, 53, 341

Pojmański G. & Maciejewski, G. 2004, *AcA*, 54, 153

Pojmański G. & Maciejewski, G. 2005, *AcA*, 55, 97

Pojmański G., Pilecki, B, & Szczygieł, D. 2005, *AcA*, 55, 275

Schaerer, D., Meynet, G., Maeder, A., & Schaller, G. 1993, *A&AS*, 98, 523

van Loon, J. Th., Cioni, M. -R. L., Zijlstra, A. A., & Loup, C. 2005, *A&A*, 438, 273

Whitelock, P., Feast, M. W., van Loon, J. Th., & Zijlstra, A. A. 2003, *MNRAS*, 342, 86

Wood, P. R., Whiteoak, J. B., Hughes, S. M. G., *et al.* 1992, *ApJ*, 397, 552

At the conference dinner inside Wrenbury Hall.

Getting quite lively...

Session VIII

Magellanic type systems as a class

The Magellanic System: Stars, Gas, and Galaxies
Proceedings IAU Symposium No. 256, 2008
Jacco Th. van Loon & Joana M. Oliveira, eds.

© 2009 International Astronomical Union
doi:10.1017/S1743921308028871

Magellanic type galaxies throughout the Universe

Eric M. Wilcots

Department of Astronomy, University of Wisconsin-Madison,
475 N. Charter St., Madison, WI 53706 USA
email: `ewilcots@astro.wisc.edu`

Abstract. The Magellanic Clouds are often characterized as "irregular" galaxies, a term that implies an overall lack of organized structure. While this may be a fitting description of the Small Cloud, the Large Magellanic Cloud, contrary to popular opinion, should not be considered an irregular galaxy. It is characterized by a distinctive morphology of having an offset stellar bar and single spiral arm. Such morphology is relatively common in galaxies of similar mass throughout the local Universe, although explaining the origin of these features has proven challenging. Through a number of recent studies we are beginning to get a better grasp on what it means to be a Magellanic spiral. One key result of these works is that we now recognize that the most unique aspect of the Magellanic Clouds is not their structure, but, rather, their proximity to a larger spiral such as the Milky Way.

Keywords. galaxies: dwarf, galaxies: evolution, galaxies: interactions, Magellanic Clouds, galaxies: structure

1. Introduction

While the Small Magellanic Cloud can properly be thought of as an irregular galaxy, the Large Magellanic Cloud shares a number of key morphological properties with a population of galaxies classified as Barred Magellanic Spirals (SBm). These properties include a stellar bar, the center of which may or may not be coincident with the dynamical center of the galaxy, a single, looping spiral arm, and often a large star-forming complex at one end of the bar. In the broadest of terms, Magellanic spirals are often simply referred to as being "asymmetric" or "lopsided", but they are not "irregular."

The earliest comprehensive look at Magellanic spirals was carried out by de Vaucouleurs & Freeman (1974) who noted the structural similarity between the LMC and a number of other nearby galaxies. Much of the subsequent work was aimed at understanding the origin of the lopsided structure that characterized Magellanic type galaxies. Odewahn (1994) carried out a photographic survey of Magellanic spirals and concluded that the vast majority of them had companions. The implication of this was that interactions were primarily responsible for the asymmetric properties of this class of galaxy. Theoretical work and some simulations could accurately account for the apparent lopsidedness of Magellanic spirals, but not the frequency of them.

Over the past decade or so we have also seen a proliferation of statistically significant samples of galaxies that show that objects sharing the basic properties of Magellanic spirals are common in both the local Universe and at intermediate redshift. It is becoming more and more apparent that the old adage of studying the Magellanic Clouds in order to learn more about the evolution of galaxies in general is quite true. The Clouds are the nearest examples of a broad population of galaxies that contributed and continue to contribute significantly to the global star formation rate and gas content of the Universe as a whole. In this contribution we present the results of a number of recent studies of the properties of other Magellanic spirals. What emerges is both a better understanding of Magellanic-type galaxies throughout the Universe and, perhaps more importantly, a recognition that the most unique aspect of the Magellanic Clouds is not their structure, but their proximity to the Milky Way.

2. In the Local Group

We begin our survey of Magellanic type galaxies throughout the Universe with a look at how the Magellanic Clouds fit into the Local Group. We have heard a great deal at this conference about the stellar content of the Clouds, so the focus here will be on the H I properties of galaxies in the Local Group. A casual glance at the morphological distribution of galaxies in the Local Group quickly reveals one of the most interesting aspects of the Magellanic Clouds. They are the only H I-rich companions of the large spirals, M 31 and the Milky Way; the low mass companions to the large spirals are almost exclusively either dwarf ellipticals or dwarf spheroidals. The LMC is the fourth most massive galaxy, by H I mass, in the Local Group behind only M 31, the Milky Way, and M 33. The Small Magellanic Cloud, on the other hand, is the most gas-rich of any of the Local Group irregulars, surpassing NGC 3109 and IC 10. The rest of the gas rich irregular galaxies tend to lie at large distances from either of the large spirals, suggesting a coarse morphology-density relation exists even in low mass galaxy groups such as our own. It is worth noting here that the distribution and kinematics of H I within the LMC and SMC is not unusual. The H I holes and extended H I distribution we see in the Clouds are also seen in almost every H I-rich galaxy, especially late-type ones, observed. In addition, galaxies with the H I mass of the Magellanic Clouds ($10^8 - 10^9$ M$_\odot$) represent the plurality of galaxies in similar groups. Freeland, Stilp, & Wilcots (2009) derived the H I mass function of galaxy groups shown in Figure 1. While the numbers are small, it is clear that mass function is relatively flat at low and intermediate masses; there is nothing uncommon about either the H I mass or the distribution of H I in the Magellanic Clouds.

Much attention has also been given to the Magellanic Stream, but it is interesting to note that M 33 and M 31 appear to be connected by an H I bridge of similar mass (Braun & Thilker 2004). While this tells us more about the complex dynamical history of the Local Group than it does about the Magellanic Clouds, it is yet another bit of evidence to show that the properties of the Magellanic Clouds (and Magellanic Stream) are reflective of conditions common within the local Universe. Indeed, the Magellanic Clouds are important because they are not uncommon objects, but rather are outstanding nearby examples of more distant objects.

3. A statistical approach

We can see how the Magellanic Clouds fit into the larger picture of the population of galaxies by making use of a number of different surveys of star-forming and gas-rich galaxies in the nearby Universe and at higher redshifts. We have already seen that the H I masses of the Clouds place them on the flat part of the H I mass function, slightly less massive than an M$_\star$ galaxy. H I mass functions, however, sample only the local Universe and, therefore, are not necessarily representative of the population of galaxies at intermediate and high redshift. We might ask the extent to which the Magellanic Clouds have counterparts in the more distant Universe. One population of interest is that of the luminous compact blue galaxies (LCBGs) which host a large fraction of the star formation at intermediate redshifts. In a study of the H I and optical properties of LCBGs, Garland et al. (2004) showed that 30% of the local counterparts to LCBGs are

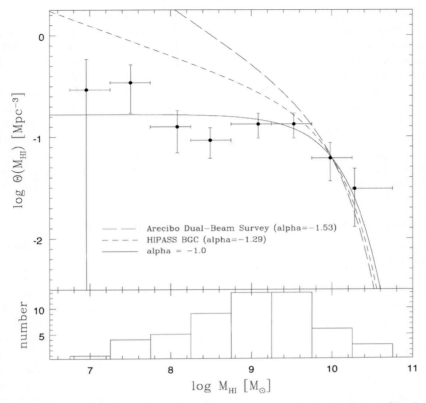

Figure 1. The H I mass function of galaxy groups similar to the Local Group (Freeland *et al.* 2009). The Magellanic Clouds reside on the flat part of the HIMF, somewhat less massive than an $M_\star$ galaxy. The comparison of the H I mass function shows that galaxies with masses similar to those of the LMC and SMC are common in the group environment

Magellanic type spirals and massive irregulars. This is consistent with the findings of Wolf *et al.* (2007) who showed that galaxies with masses similar to the Magellanic Clouds make up an increasingly large fraction of the star-forming galaxies at intermediate redshifts. In addition, galaxies like the LMC make up a strong plurality of the galactic contribution to the total H I density of the Universe and the total cross-section of the damped Lyα systems (Ryan-Weber *et al.* 2003). In the context of this presentation, these examples only serve to demonstrate that galaxies with the mass and luminosity of the Magellanic Clouds are common throughout the Universe and, as such, are important tracers of the overall evolution of the population of galaxies.

4. Magellanic type galaxies as companions

One of the themes to emerge from the body of work on Magellanic-type galaxies throughout the Universe is that the LMC and SMC are unique only in the fact that they are companions of significantly larger disk galaxies. Of the 75 Magellanic-type galaxies included in Odewahn (1994)'s survey, none are companions to a large spiral. Fundamentally this means that an ongoing encounter with a larger galaxy is not required to explain the morphological properties of Magellanic type galaxies. The intermediate H I mass galaxies in other groups are broadly distributed within the group and are not preferentially companions to the larger galaxies (Freeland *et al.* 2009). We can turn this argument around and ask how the Magellanic Clouds compare to the companions of other spirals. James, O'Neill, & Shane (2008) surveyed the star formation characteristics of the companions to 53 spiral galaxies. Of those 53 galaxies, only 9 had companions, and the LMC would be the most luminous of all of them. Zaritsky *et al.* (1997) also found that the LMC is among the brightest of all of the companions in their sample of disk galaxies. Comparing the Milky Way to a specific similar spiral, Pisano & Wilcots (2000) identified two H I-rich companions to NGC 6946, both of which were less massive than the Magellanic Clouds. SPH simulations reach a similar conclusion; Libeskind *et al.* (2007) found that companions of comparable luminosity and mass to the LMC were quite rare in the SPH simulations of galaxy and structure formation. Simply put, the Magellanic Clouds are not typical satellite galaxies; they are more luminous and more massive. At the same time the Magellanic Clouds are unique among galaxies of their type in the fact that they do reside in close proximity to a large galaxy.

5. Morphological characteristics of Magellanic spirals

What truly distinguishes Magellanic type galaxies from other classes is their characteristic optical morphology. Broadly speaking, Magellanic galaxies contain an optically visible stellar bar that is typically displaced from the apparent center of the galaxy. In fact, for most Magellanic type galaxies (and, indeed, late-type spirals in general) it is difficult to determine where the center of the galaxy actually resides. For example, while NGC 925 is not quite a Magellanic, it is a late type spiral for which the dynamical center as defined by the H I velocity field, the optical center of the bar, and the optical center of the outer isophotes do not coincide (Pisano, Wilcots, & Elmegreen 2000). For the LMC the dynamical center derived from carbon stars is not coincident with the dynamical center derived from H I observations (van der Marel *et al.* 2002). True Magellanic spirals such as the LMC often display a one-armed spiral morphology. An optical image of IC 1727 in Figure 2 shows this quite nicely, and NGC 3664 and NGC 4618 are other excellent examples of the characteristic bar and one-armed spiral morphology (Wilcots, Lehman, & Miller 1996).

In what was the first detailed review of the properties of Magellanic type galaxies, de Vaucouleurs & Freeman (1974) described a class of objects that, contrary to popular opinion, had a distinct morphology characterized by this strong optical asymmetry. Asymmetry, however, appears to run rampant amongst all disk galaxies. Approximately half of all spirals have asymmetric H I profiles (Richter & Sancisi 1994; Baldwin, Lynden-Bell, & Sancisi 1980) and this fraction increases to 75% for late-type spirals (Matthews, van Driel, & Gallagher 1998). Lopsidedness extends to the stellar distribution as well and Rix & Zaritsky (1995) found that nearly 30% of all spirals have asymmetric stellar distributions.

The prevalence of asymmetry among disk galaxies naturally led to a broad search for a cause. Baldwin *et al.* (1980) suggested that differential precession of the disks might account for the widespread observed lopsidedness. The short timescales of this process, however, are difficult to reconcile with the prevalence of assymetry. Interactions are more commonly identified as the cause of asymmetry in disk galaxies. Walker, Mihos, & Hernquist (1996) and Zaritsky & Rix (1997) both suggest that minor mergers could be responsible. Simulations show that the accretion of a small companion of $\sim 10\%$ of the mass of the primary can lead to a strongly asymmetric disk that persists for a few dynamical times (Walker *et al.* 1996). Using a combination of excess blue luminosity in lopsided galaxies and a quantitative measurement of the Fourier amplitudes of the asymmetry, Zaritsky & Rix (1997) derived an upper limit on the current rate of accretion of companions among field galaxies. Rudnick, Rix, & Kennicutt (2000) showed a similar

Figure 2. A V-band image of the classic Barred Magellanic spiral, IC 1727, showing the one-armed morphology and stellar bar.

correlation between lopsidedness and recent star formation that suggested that minor mergers might be the cause of both. In an analysis of photographic plates Odewahn (1994) suggested that essentially all Magellanic spiral type galaxies had companions, further leading credence to the notion that interactions were to blame for asymmetry. Wilcots *et al.* 1996) looked at the H I properties of a small sample of Magellanic spirals and suggested that minor mergers and accretion could be connected to the observed lopsidedness.

Interactions, and even minor mergers, are short-lived phenomena and the simulations indicate that the resulting asymmetry should only last for $\sim 10^9$ years (e.g., Walker *et al.* 1996) — too short to account for the prevalence of lopsidedness. A more compelling model was proposed by Levine & Sparke (1998) and Noordermeer, Sparke, & Levine (2001) that simply shows that a disk offset from the dynamical center of a dominant halo can result in long-lived asymmetry that is consistent with the observed velocity fields of lopsided late-type spirals. In other words, once a disk starts offset from the dynamical center, it is likely to remain offset. The observable effects of the Noordermeer *et al.* (2001) models include a velocity field in which the curvature of the isovelocity contours on the receding and approaching sides are different to the extent that one side of the rotation curve appears to have turned over while the other side continues to rise. The magnitude of the effect is a function of the fraction of the total mass in the disk, how far the disk is displaced from the dynamical center, and whether or not the disk orbits in a retrograde sense. Few true Magellanic spirals have been observed in enough detail to see the extent to which these models reflect reality.

Wilcots & Prescott (2004) took a systematic look at asymmetry in Magellanic type spirals by surveying the H I properties of a sample of 13 such galaxies. They found that only four of the 13 had actual companions, in stark contrast to the Odewahn (1994) study. In addition, the ongoing interactions are all quite weak and unlikely to do much to alter the morphology or dynamics of the primary galaxy. Most interestingly, the H I profiles of the Magellanic spirals with companions were no more or less asymmetric than the H I profiles of the galaxies that did not have a companion. Lastly, the H I profiles of the Magellanic spirals — a sample selected because of their optical asymmetry — were no more or less asymmetric than the H I profiles of typical galaxies in the field. It is extremely unlikely that current, or even recent, interactions have much to do with the optical asymmetry prevalent amongst Magellanic-type galaxies. Whatever the initial cause of lopsidedness amongst Magellanic spirals, it is clear that the asymmetry is long-lived and it is manifested largely in the stellar distribution but not so evident in the H I profiles or rotation curves.

6. Case study: NGC 4618 and NGC 4625

Up to this point we have concentrated on the properties of Magellanic spirals as a class of galaxy. The next step in building our understanding of Magellanic-type galaxies throughout the Universe is to investigate the detailed astrophysics of a small sample of such galaxies. NGC 4618 and NGC 4625 are probably the best nearby examples of Magellanic spirals outside of the Local Group and the targets of detailed study using both radio and optical techniques.

6.1. The effects of interactions

Both NGC 4618 and NGC 4625 are classified as Magellanic-type galaxies, they are interacting with one another, and are connected by a distinct H I bridge (Bush & Wilcots 2004). They are, in other words, very similar to the Magellanic Clouds, with the major exception that NGC 4618 and NGC 4625 are not companions to a larger galaxy. Bush & Wilcots (2004) calculated that the interaction began some 0.2–0.7 Gyr ago and, to date, has had little effect on the structure of the participating galaxies. The H I profiles and rotation curves of both galaxies remain remarkably symmetric, further raising doubts about the connection between interactions and the characteristic asymmetry of Magellanic-type galaxies.

The fact remains that a number of the most prominent Magellanic spirals such as the LMC and NGC 4618 *are* interacting. The effects of these interactions on the morphology and kinematics of the gas and stars in the galaxies involved are simply not well understood. There is obviously a wide range of variables that go into understanding the effect of a close passage between two galaxies on their morphologies. These include: the mass ratio of the galaxies, the fraction of the total mass in the disk, and the orientation of the interaction (retrograde vs. prograde). Bush & Wilcots (2004) completed a *VLA* study of the NGC 4618-NGC 4625 pair and their results show that one of the galaxies (NGC 4625) has a disk that accounts for only 6% of the total mass while the other (NGC 4618) has a disk that accounts for 45% of the total mass. This may be one the keys to understanding the effects of an interaction on Magellanic spirals. NGC 4618, the galaxy with a higher fraction of its mass in its disk, has suffered more tidal disruption than its more massive companion, NGC 4625, which apparently has a higher fraction of its mass in its halo. Whether or not the NGC 4618-NGC 4625 pair is characteristic of the class of Magellanic-type galaxies as a whole has yet to be seen.

Many of the ongoing interactions in which Magellanic-type galaxies are participating more typically symptomatic of minor mergers (Wilcots & Prescott 2004). NGC 4288 and

NGC 4861 are two excellent examples of this phenomenon. In both cases the smaller companion has less than 10% of the mass of the Magellanic-spiral and there is some suggestion that such accretion events are more reflective of the continuing growth of the individual galaxies (e.g., Wilcots *et al.* 1996).

6.2. *Stellar kinematics of a Magellanic spiral*

As one might expect, the only Magellanic spiral for which we have a good understanding of its internal structure and dynamics is the LMC. Van der Marel *et al.* (2002) used the kinematics of the carbon stars to measure the internal structure and stellar velocity dispersion of the LMC. Among other results, they found that the velocity dispersion of ~ 20 km s^{-1} implies that the disk of the LMC is remarkably thick. Whether these are unique properties of the LMC or more characteristics of the population of Magellanic spirals as a whole has yet to be determined and resolution of this particular issues requires a new suite of observations of other Magellanic-type galaxies. We have initiated just such a survey of the stellar kinematics of a sample of Magellanic spirals beginning with NGC 4618.

Because of its favorable inclination and relative proximity, NGC 4618 is probably the best Magellanic spiral beyond the LMC itself in which to study the structure and kinematics of these galaxies. Prescott *et al.* (in preparation) used the Sparsepak integral field unit (Bershady *et al.* 2005) and the bench spectrograph on the *WIYN* 3.5m telescope to measure the stellar kinematics in NGC 4618. The distribution of the Sparsepak fibers on NGC 4618 is shown in Figure 3 and we show the spectrum extracted from one of the central fibers in Figure 4. Information about the stellar kinematics was extracted using the cross-correlation of the observed spectrum of NGC 4618 with a series of stellar templates, and an example of the cross-correlation is shown in Figure 5.

Based on the analysis of the Mg I stellar absorption lines, Prescott *et al.* (in preparation) find the stellar velocity dispersion perpendicular to the disk to be ~ 23 km s^{-1}, comparable with the value found in the LMC. There is some variation of the velocity dispersion with radius; particularly, the velocity dispersion is 50% higher in the bar than it is in the rest of the disk. Assuming a scale height of ~ 300 pc, the disk mass is roughly 30% of the total dynamical mass of NGC 4618.

7. Are the bars real?

One of the enduring mysteries surrounding Magellanic-type galaxies is the nature of their stellar bars. Abraham & Merrifield (2000) found that the bar fraction decreases from early type spirals to Sc, but then *increases* for even later type disk galaxies. In other words, extreme late-type spirals are more likely to be barred than Sc-types. Noguchi (2001) saw a curious enhancement in his "concentration parameter" for extreme late-type galaxies, again suggesting that these galaxies are more likely to host central stellar concentrations like bars than Sc type galaxies. Perhaps one explanation is that an initially offset disk is more likely to be unstable to bar formation (Junqueira & Combes 1997). Even if extreme late-type galaxies are more susceptible to the formation of bars, it is not clear that the bars in Magellanic sprials are dynamically similar to those in earlier type galaxies. Strong stellar bars typically manifest themselves as S-shaped isovelocity contours in the observed velocity field of barred galaxies. The velocity field of NGC 4618, however, does not have that characteristic S-shape (Prescott *et al.*, in preparation); in short, the bar seems to be having only a modest effect on the gas kinematics in the central part of the galaxy. A similar situation is seen in IC 1727, another barred Magellanic spiral.

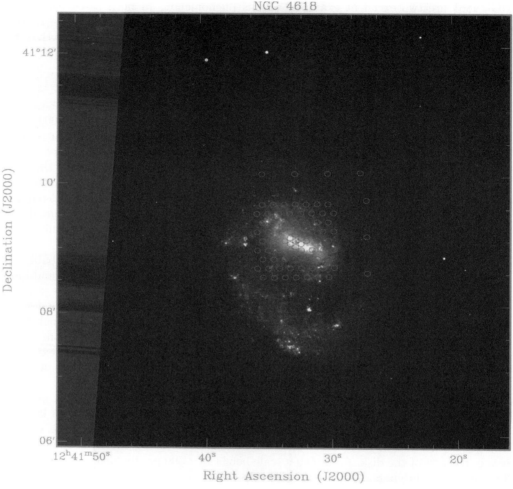

Figure 3. The small circles represent the individual Sparsepak fibers overlayed on an B-band image of NGC 4618.

While there has yet to be a true systematic study of the properties of bars in Magellanic type galaxies, the data obtained to date suggest that these bars are not "real."

8. Where do we go from here?

We have seen that the study of Magellanic-type galaxies now includes detailed observations of the structure and dynamics of a number of individual objects. In addition we now have surveys of other Magellanic-type galaxies at a range of wavelengths. What do we learn from these studies? First, the morphology and the kinematics of the Magellanic Clouds are not unique. The LMC and SMC are, very clearly, simply nearby examples of a population of galaxies that is well represented throughout the Universe. Second, Magellanic-type galaxies are not nearly as asymmetric or lopsided as they seem, particulary as measured by their H I profiles. Third, given the absence of evidence that interactions have played a large role in shaping the morphology of Magellanic spirals, we are still searching for a broadly applicable model for the origin of the asymmetry so

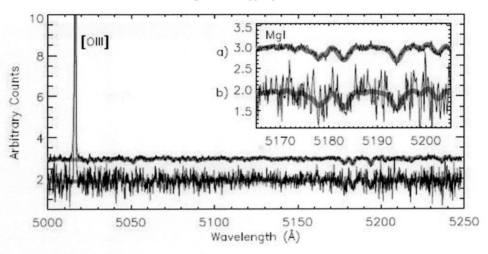

Figure 4. The spectrum of NGC 4618 extracted from one of the central Sparsepak fibers. The inset specifically shows the Mg I lines which formed the basis of much of the analysis.

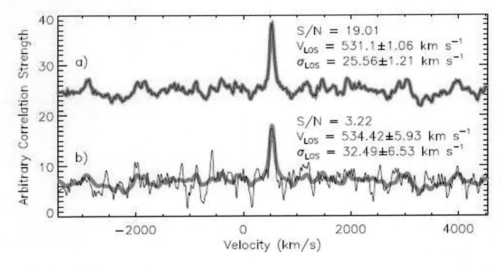

Figure 5. This figures shows the results of the cross-correlation of the observed spectrum of NGC 4618 with the spectrum of a stellar template. The centroid of the cross-correlation peak is the line of sight velocity while the width corresponds to the velocity dispersion in the galaxy.

common in these galaxies. Lastly, one can conclude that it is their proximity to the Milky Way that distinguishes the Magellanic Clouds from other similar galaxies.

Despite a number of significant advances in our understanding of Magellanic-type galaxies over the past few years, there remain a handful of key areas in which our ability to put the LMC and SMC in a proper context is limited. The first is the question of whether the bars in Magellanic spirals are "real." In the few cases studied to date there is little compelling kinematical data to suggest that the bars are "real." This is particularly the case with NGC 4618 Prescott *et al.* (in preparation) and IC 1727 as we have shown here. Clearly, we need more kinematical data for both the stars and gas on a much larger

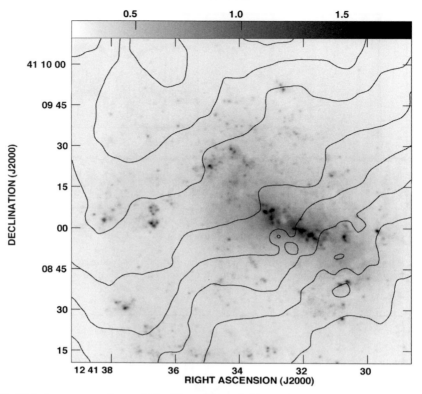

Figure 6. This is a comparison of the H I velocity field overlayed on an optical image of the SBm galaxy NGC 4618. The isovelocity contours show little evidence that the bar is strongly affecting the gas kinematics in the central part of the galaxy.

sample of Magellanic spirals. The obvious question would be that if bars in Magellanic spirals are not "real", then what are they?

Second on the list would be a better understanding of the impact of minor mergers on the structure and internal dynamics of late-type spirals. Wilcots & Prescott (2004) found that some Magellanic spirals do have small companions. We simply do not know what the effect of a minor merger might be on such late-type galaxies, especially if their disks are already off-set.

The third key area is to better understand the frequency with which Magellanic-type galaxies are found in other groups and what their place in those groups might be. This requires moderately deep and wide-field observations of a statistically significant sample of groups. We are in the midst of such a survey of the H I properties of galaxy groups and one example (GH 98) of the results is shown in the figure below. The contours correspond to the H I column density and they are overlayed over an image of the group obtained with the *WIYN* 0.9m telescope. Not surprisingly we find that groups are the sites of a number of interactions; in fact almost all of the H I detected galaxies are interacting (Freeland *et al.* 2009). Two important properties emerge from this work: it is not uncommon to find that H I galaxies are still "falling into" galaxy groups (i.e. groups are still growing) and that structures like the Magellanic stream are also not uncommon. Perhaps studying the Magellanic Clouds not only allows us to learn more about the evolution of galaxies, but also the evolution of galaxy groups. Given that our understanding of the origin of the Magellanic Clouds themselves continues to evolve, the future of the study of

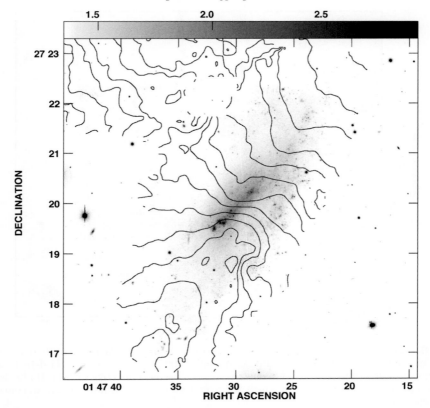

Figure 7. This is a comparison of the H I velocity field overlayed on an optical image of the central part of the SBm galaxy, IC 1727. While there is some evidence of streaming motions, the velocity field does not show the characteristic "S-shaped" isovelocity contours seen in strongly barred galaxies. The contours to the north and east of IC 1727 are the outer isovelocity contours of its companion NGC 672.

Magellanic-type galaxies throughout the Universe will likely focus on this aspect of their evolution. In other words the question is: what does the Magellanic system tell us about the evolution of the Local Group and what does the dynamical evolution of other galaxy groups tell us about the origin of the Magellanic Clouds?

References

Abraham, R. G. & Merrifield, M. R. 2000, *AJ*, 120, 2835

Baldwin, J. E., Lynden-Bell, D., & Sancisi, R. 1980, *MNRAS*, 193, 313

Bershady, M. A., Andersen, D. R., Verheijen, M. A. W., Westfall, K. B., Crawford, S. M., & Swaters, R. A. 2005, *ApJS*, 156, 311

Braun, R. & Thilker, D. A 2004, *A&A*, 417, 421

Bush, S. J. & Wilcots, E. M. 2004, *AJ*, 128, 2789

de Vaucouleurs, G. & Freeman, K. C. 1972, *Vistas in Astron.*, 14, 163

Freeland, E. E., Stilp, A. M., & Wilcots, E. M. 2009, *ApJ*, submitted

Garland, C. A., Pisano, D. J., Williams, J. P., Guzman, R., & Castander, F. J. 2004, *ApJ*, 615, 689

James, P. A., O'Neill, J., & Shane, N. S. 2008, *A&A*, 486, 131

Junqueira, S. & Combes, F. 1997, *ApL&C*, 36, 363

Levine, S. E. & Sparke, L. S. 1998, *ApJ*, 503, 125

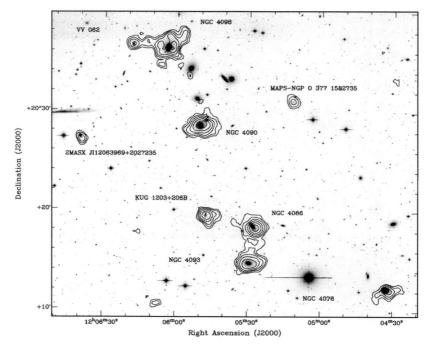

Figure 8. The contours represent the H I column density map overlayed on an R band image of the GH 98 galaxy group (Freeland, Stilp, & Wilcots 2009. Nearly all of the H I detections show clear signs of interactions and at least one pair of galaxies is connected with an H I bridge. This is an example of an environment in which the accretion of galaxies into the group continues in the present epoch.

Libeskind, N. I., Cole, S., Frenk, C. S., Okamoto, T., & Jenkins, A. 2007, *MNRAS*, 374, 16

Matthews, L. D., van Driel, W., & Gallagher, J. S. 1998, *AJ*, 116, 2196

Noguchi, M. 2001, *MNRAS*, 312, 194

Noordermeer, E., Sparke, L. S., & Levine, S. E. 2001, *MNRAS*, 328, 1064

Odewahn, S. C. 1994, *AJ*, 107, 1320

Pisano, D. J. & Wilcots, E. M. 2000, *MNRAS*, 319, 821

Pisano, D. J., Wilcots, E. M., & Elmegreen, B. G. 2000, *AJ*, 120, 763

Richter, O. -G. & Sancisi, R. 1994, *A&A*, 290, L9

Rix, H. -W. & Zaritsky, D. 1995, *ApJ*, 447, 82

Rudnick, G., Rix, H. -W., & Kennicutt, R. C. 2000, *ApJ*, 538, 569

Ryan-Weber, E. V., Webster, R. L., & Staveley-Smith, L. 2003, *MNRAS*, 343, 1195

van der Marel, R. P., Alves, D. R., Hardy, E., & Suntzeff, N. B. 2002, *AJ*, 124, 2639

Walker, I. R., Mihos, J. C., & Hernquist, L. 1996, *ApJ*, 460, 121

Wilcots, E. M., Lehman, C., & Miller, B. 1996, *AJ*, 111, 1575

Wilcots, E. M. & Prescott, M. K. M. 2004, *AJ*, 127, 1900

Wolf, C., Gray, M. E., Aragon-Salamanca, A., Lane, K. P., & Meisenheimer, K. 2007, *MNRAS*, 376, L1

Zaritsky, D. & Rix, H. -W. 1997, *ApJ*, 477, 118

Zaritsky, D., Smith, R., Frenk, C. S., & White, S. D. M. 1997, *ApJ*, 478, L53

The Magellanic System: Stars, Gas, and Galaxies
Proceedings IAU Symposium No. 256, 2008
Jacco Th. van Loon & Joana M. Oliveira, eds.

© 2009 International Astronomical Union
doi:10.1017/S1743921308028883

The Magellanic Group and the Seven Dwarfs

Elena D'Onghia[1][†] and George Lake[2]

[1]Institute for Theoretical Physik, University of Zurich,
Winterthurerstraße 190, 8057 Zurich, Switzerland
email:elena@physik.unizh.ch; lake@physik.unizh.ch

Abstract. The Magellanic Clouds were the largest members of a group of dwarf galaxies that entered the Milky Way (MW) halo at late times. This group, dominated by the LMC, contained $\sim 4\%$ of the mass of the Milky Way prior to its accretion and tidal disruption, but $\approx 70\%$ of the known dwarfs orbiting the MW. Our theory addresses many outstanding problems in galaxy formation associated with dwarf galaxies. First, it can explain the planar orbital configuration populated by some dSphs in the MW. Second, it provides a mechanism for lighting up a subset of dwarf galaxies to reproduce the cumulative circular velocity distribution of the satellites in the MW. Finally, our model predicts that most dwarfs will be found in association with other dwarfs. The recent discovery of Leo V (Belokurov *et al.* 2008), a dwarf spheroidal companion of Leo IV, and the nearby dwarf associations supports our hypothesis.

Keywords. Galaxy: halo, galaxies: clusters: general, galaxies: formation, galaxies: halos, Magellanic Clouds, cosmology: observations, dark matter

1. Introduction

In the cold dark matter (CDM) model, the dark halos of galaxies like the Milky Way build up hierarchically, through the accretion of less massive halos. When these subsystems avoid complete tidal disruption, they can survive in the form of satellite dwarf galaxies. However, the dwarf galaxies in the Local Group exhibit several puzzling features. Numerical simulations of CDM predict 10 to 30 times more satellites within 500 kpc of the Milky Way and M 31 than the modest observed population (e.g., Moore *et al.* 1999). This discrepancy between the expected and known numbers of dwarf galaxies has become known as the *missing dwarf problem*. The newly discovered population of ultra-faint dwarfs around the Milky Way and M 31 found in the Sloan Digital Sky Survey increases by a factor of two the number of known satellites (Simon & Geha 2007), but goes to even lower circular velocities where a comparable or even greater increase in the number of satellites is expected.

Another peculiarity is that many dwarf galaxies in the Local Group lie in the orbital plane of the Magellanic Clouds and Stream. These dwarfs have been associated with the Magellanic Clouds and termed the Magellanic Group (Lynden-Bell 1976; Fusi Pecci *et al.* 1995; Kroupa *et al.* 2005). In order to reproduce this planar configuration in the current scenario for structure formation, Libeskind *et al.* (2005) proposed that subhalos are anisotropically distributed in cosmological CDM simulations and that the most massive satellites tend to be aligned with filaments. Similarly, Zentner *et al.* (2005) suggested that the accretion of satellites along filaments in a triaxial potential leads to an anisotropic distribution of satellites.

Systems anisotropically distributed falling into the Galactic halo may not lie in a plane consistent with the orbital and spatial distribution of the MW satellites. For example, a

† Marie Curie fellow

theoretical bootstrap analysis of the spatial distribution of CDM satellites (taken from a set of CDM simulations) by Metz *et al.* (2008) finds that even if they are aligned along filaments, they will be consistent with being drawn randomly. This could mean that alignment of the satellites along filaments may not be sufficient to reproduce the observed planar structures.

As we propose here, the origin of planar distributions is facilitated by concentrating infalling satellites into groups.

Another issue is that the dSphs of the Local Group tend to cluster tightly around the giant spirals. Proximity to a large central galaxy might prevent dwarf irregulars from accreting material, turning off star formation, and they may then undergo tidal interactions to convert them into dwarf spheroidals. However, isolated dSphs like Tucana or Cetus found in the outskirts of the Local Group (Grebel *et al.* 2003) suggest that dSphs might also form at great distances from giant spirals prior to their being accreted. Clues to the questions raised by these observations may be contained in measurements of the metallicities of a large sample of stars in four nearby dwarf spheroidal galaxies: Sculptor, Sextans, Fornax, and Carina. Work by Helmi *et al.* (2006) shows that all four lack stars with low metallicity, implying that their metallicity distribution differs significantly from that of the Galactic halo, indicating a non-local origin for these systems.

2. Why do Magellanic Clouds need to be accreted in groups of dwarfs?

We propose that the Magellanic Clouds and seven of the eleven dwarf galaxies around the MW were accreted as a group that was then disrupted in the halo of our Galaxy. This is supported by observations indicating that dwarfs are often found in associations and by numerical simulations where subhalos are often accreted in small groups (e.g., Li & Helmi 2008). In particular, the LMC, SMC, and those dwarfs whose orbits are similar to those of the Magellanic Clouds may all have originally been part of such a group. This "LMC group" was dominated by the LMC and had a parent halo circular velocity of ~ 75 km s^{-1} with its brightest satellite, the SMC, having a rotation velocity of ~ 60 km s^{-1} as estimated from its H I distribution.

There is considerable evidence for tidal debris from the LMC group, supporting the proposal that it was tidally disrupted. The LMC and SMC have been modeled as a pair owing to their spatial proximity; as either a currently bound pair or one that became unbound on the last perigalacticon passage. The number of dwarfs assigned to the Magellanic Plane Group (Kunkel & Demers 1976) includes the following *candidates*: Sagittarius, Ursa Minor, Draco, Sextans and Leo II. Of the dwarfs known before the recent flurry of discoveries, 7 out of 10 within ~ 200 kpc might well be part of this group. The remaining three — Fornax, Sculptor and Carina — have been proposed to be part of a second grouping (Lynden-Bell 1982).

3. Evidence for nearby associations of dwarfs

CDM theory predicts that many dwarf galaxies should exist in the field. Numerical simulations show that the normalized mass function of subhalos is nearly scale-free. That is, when the circular velocity distribution function of the subhalos is normalized to the parent halo, it is nearly independent of the mass of the parent. Thus, groups of dwarf galaxies are a natural expectation of CDM models on small mass scales. However, like low mass satellites, these systems are difficult to observe.

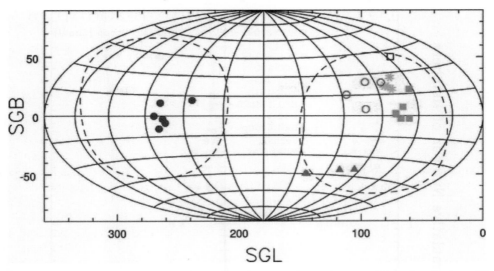

Figure 1. Distribution in supergalactic coordinates of associations of dwarfs galaxies with accurately known distances between 1.1 and 3.2 Mpc (Tully *et al.* 2006).

Tully *et al.* (2006) discovered a number of associations of dwarf galaxies within 5 Mpc of the MW. Figure 1 displays the distribution in supergalactic coordinates of these associations with accurately known distances between 1.1 and 3.2 Mpc. These groups have properties expected for bound systems with 1–10×10^{11} $M_{\odot}$, but are not dense enough to have virialized, and have little gas and few stars. Of the eight associations compiled by Tully *et al.* (2006), there are only three for which the two brightest galaxies differ by at least 1.5 magnitudes: NGC 3109, NGC 1313 and NGC 4214. In the other five, the two brightest galaxies are certain to merge if the associations collapse and virialize.

Figure 2 (left panel) shows the cumulative circular velocity distribution function inferred for the dwarf associations, the putative Magellanic Group (candidates listed previously), and the MW satellite galaxies. For each dwarf association we assume the largest dwarf galaxy circular velocity of the group to be the parent halo circular velocity. Magnitudes of member galaxies are converted to circular velocity assuming a Tully-Fisher relation in the B band (see D'Onghia & Lake (2008) for details). The MW data includes the newest dwarfs with a minimum $\sigma = 3.3$ km s^{-1} and a correction for incomplete sky coverage (Simon & Geha 2007).

Figure 2 shows that the nearby associations of dwarfs have a cumulative circular velocity distribution function similar to the MW, suggesting that such associations may be the progenitors of the brightest dwarf satellites in the MW. Thus, if these associations of dwarfs are accreted into larger galaxies, they can populate the bright end of the cumulative circular velocity distribution function of satellites. However, when normalized to the low mass of their parent, they have a far greater number of dwarfs.

4. Dwarfs in the LMC group can light up more efficiently

In our interpretation, the mass of the LMC group is $\sim 4\%$ of the Milky Way, yet most of the dwarfs known a decade ago are associated with it. There is a similar overabundance of dwarfs in the dwarf associations. Here, we suggest that dwarf galaxies formed in LMC-like groups will be luminous, while those that form by themselves in the halos of larger systems will be dark.

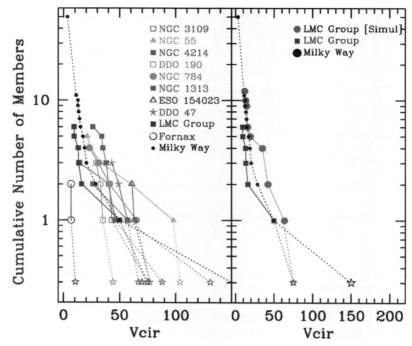

Figure 2. Cumulative circular velocity distribution of the satellites of the LMC group as compared to the nearby dwarf associations (left panel) and to the simulated LMC group in a ΛCDM model (right panel) (see D'Onghia & Lake 2008 for details).

It is generally assumed that galaxies with circular velocities ~ 30 km s^{-1} blow out their gas. When gas is blown out of a subhalo, it eventually thermalizes to the virial temperature of the parent halo, which is 2–5×10^6 K for bright galaxies such as the MW. At this temperature, the cooling times are long enough that there can be a considerable reservoir of hot gas and a subhalo with a velocity scale of 10–30 km s^{-1} will not reaccrete much gas, and it will be dark. However, in a small parent halo like the LMC, the virial temperature is only 2×10^5 K. This is at the peak of the cooling curve and the gas cools rapidly to 10^4 K. The low bulk motions in these halos might well permit reaccretion by some of the subhalos producing luminous dwarf galaxies. Note that our picture is consistent with the new proper motion measurements from Kallivayalil *et al.* (2006) and Kallivayalil, van der Marel & Alcock (2006) and orbit models from Besla *et al.* (2007). Prior to infall, the LMC group had a virial radius of ~ 75 kpc and a 3-D velocity dispersion of ~ 100 km s^{-1}. So, a thin plane would still be very unusual and a wide range of kinematics is expected for the disrupted satellites.

To investigate the plausibility of our model, we examined a catalog of high resolution galaxies in a cosmologically simulated volume to identify an analog to an LMC group with late infall into a MW galaxy. We note in this specific simulation that the LMC group is tidally disrupted before entering the virial radius of the MW, due to the specific mass distribution of this case. This could well be necessary to prevent the merger of the LMC and SMC prior to accretion. In Figure 2 (right panel), we display the cumulative peak circular velocity distribution of the satellites contributed by the simulated infalling group of dwarfs measured at $z = 0$ within the virial radius of the MW. This is compared to the corresponding quantity for dwarfs (filled squared symbols) in the MW which may have been part of an accreted group: LMC, SMC, Sagittarius, Ursa Minor, Draco, Sextans

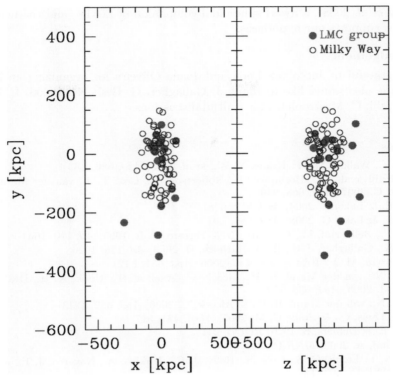

Figure 3. The spatial distribution of satellites within the virial radius of the Milky Way (blue open circles) as compared to the contributed subhalos from the break-up of the Magellanic Group at $z = 0$ within 600 kpc from the Milky Way center (magenta filled circles).

and Leo II. In Figure 2, only satellites that are accreted as part of the disrupted LMC group are displayed, because those are the dwarf galaxies that light up in our model. The remainder of the satellites that are not accreted in groups but are within the virial radius of the present-day MW are assumed to be dark.

We note that in this particular simulation, some satellites of the disrupted group are outside the MW radius at $z = 0$ and some are located inside. Figure 3 shows the spatial distribution of all the satellites within the virial radius of the Milky Way (blue filled circles) as compared to the subhalos of the disrupted Magellanic group at $z = 0$ (magenta stars). Despite the late infall, this particular group appears very well mixed, however almost half of the surviving subhalos of the group are at the present time located outside the virial radius of the final Milky Way. A few of them are in the outskirts of the Milky Way. These subhalos may reproduce the special cases like Tucana or Cetus that are located in low density regions of the Local Group.

5. Conclusion

We assume a model where the LMC was the largest member of a group of dwarf galaxies that was accreted into the MW halo. Our picture addresses several questions in galaxy formation: (*i*) It explains the association of some dwarf galaxies in the Local group with the LMC–SMC system. (*ii*) It provides a mechanism to light up dwarf galaxies. (*iii*) It predicts that other isolated dwarfs will have companions. The recent discovery of Leo V

(Belokurov *et al.* 2008), a dwarf spheroidal companion of Leo IV, and the nearby dwarf associations supports our hypothesis.

Acknowledgement

E.D. is grateful to Jacco van Loon and Joana Oliveira for organizing an interesting meeting. She also would like to thank J. Gallagher, G. Besla, K. Bekki, L. Hernquist, N. Kallivayalil, C. Mastropietro for fruitful discussions.

References

Belokurov, V., Walker, M. G., Evans, N. W., *et al.* 2008, *ApJ*, 686, L83

Besla, G., Kallivayalil, N., Hernquist, L., Robertson, B., Cox, T. J., van der Marel, R. P., & Alcock, C. 2007, *ApJ*, 668, 949

D'Onghia, E. & Lake, G. 2004, *ApJ*, 612, 628

D'Onghia, E. & Lake, G. 2008, *ApJ*, 686, L61

Fusi Pecci, F., Bellazzini, M., Cacciari, C., & Ferraro, F. R. 1995, *AJ*, 110, 1664

Grebel, E. K., Gallagher, J. S., III, & Harbeck, D. 2003, *AJ*, 125, 1926

Helmi, A., Irwin, M. J., Tolstoy, E., *et al.* 2006, *ApJ*, 651, L121

Kallivayalil, N., van der Marel, R. P., Alcock, C., Axelrod, T., Cook, K. H., Drake, A. J., & Geha, M. 2006, *ApJ*, 638, 772

Kallivayalil, N., van der Marel, R. P., & Alcock, C. 2006, *ApJ*, 652, 1213

Kroupa, P., Theis, C., & Boily, C. M. 2005, *A&A*, 431, 517

Kunkel, W. E. & Demers, S. 1976, *RGOB*, 241

Li, Y. & Helmi, A. 2008, *MNRAS*, 385, 1365

Libeskind, N. I., Frenk, C. S., Cole, S., Helly, J. C., Jenkins, A., Navarro, J. F., & Power, C. 2005, *MNRAS*, 363, 146

Lin, D. N. C. & Lynden-Bell, D. 1982, *MNRAS*, 198, 707

Lynden-Bell, D. 1976, *MNRAS*, 174, 695

Lynden-Bell, D. 1982, *Obs*, 102, 7L

Metz, M., Kroupa, P., & Libeskind, N. I. 2008, *ApJ*, 680, 287

Moore, B., Ghigna, S., Governato, F., Lake, G., Quinn, T., Stadel, J., & Tozzi, P. 1999, *ApJ*, 524, 19

Piatek, S., Pryor, C., Bristow, P., Olszewski, E. W., Harris, H. C., Mateo, M., Minniti, D., & Tinney, C. G. 2005, *AJ*, 130, 95

Simon, J. D. & Geha, M. 2007, *ApJ*, 670, 313

Tully, R. B., Rizzi, L., Dolphin, A. E., *et al.* 2006, *AJ*, 132, 729

Zentner, A. R., Kravtsov, A. V., Gnedin, O. Y., & Klypin, A. A. 2005, *ApJ*, 629, 219

The Magellanic System: Stars, Gas, and Galaxies
Proceedings IAU Symposium No. 256, 2008
Jacco Th. van Loon & Joana M. Oliveira, eds.

© 2009 International Astronomical Union
doi:10.1017/S1743921308028895

Evidence of Magellanic-like moderate redshift H I-rich galaxies

Brandon Lawton[1,2]**, Christopher W. Churchill**[2]**, Brian A. York**[3]**,
Sara L. Ellison**[3]**, Theodore P. Snow**[4]**, Rachel A. Johnson**[5]**,
Sean G. Ryan**[6] **and Chris R. Benn**[7]

[1]Space Telescope Science Institute, 3700 San Martin Drive, Baltimore, MD 21218, USA
email: lawton@stsci.edu

[2]Dept. of Astronomy, New Mexico State University, MSC 4500, P.O. Box 30001, Las Cruces,
NM 88003, USA
email: cwc@nmsu.edu

[3]Dept. of Physics and Astronomy, University of Victoria, 3800 Finnerty Rd., Victoria,
V8W 1A1, British Columbia, Canada

[4]Center for Astrophysics and Space Astronomy, University of Colorado at Boulder,
389 UCB, Boulder, CO 80309, USA

[5]Oxford Astrophysics, Denys Wilkinson Building, Keble Road, Oxford OX1 3RH, UK

[6]Centre for Astrophysics Research, University of Hertfordshire, College Lane,
Hatfield AL10 9AB, UK

[7]Isaac Newton Group, Apartado 321, E-38700 Santa Cruz de La Palma, Spain

Abstract. We present equivalent width measurements and limits of six diffuse interstellar bands (DIBs, $\lambda 4428$, $\lambda 5705$, $\lambda 5780$, $\lambda 5797$, $\lambda 6284$, and $\lambda 6613$) in seven damped Lyα absorbers (DLAs) over the redshift range $0.091 \leqslant z \leqslant 0.524$, sampling $20.3 \leqslant \log N(\text{H I}) \leqslant 21.7$. Based upon the Galactic DIB–$N(\text{H I})$ relation, the $\lambda 6284$ DIB equivalent width upper limits in four of the seven DLAs are a factor of $4-10$ times below the $\lambda 6284$ DIB equivalent widths observed in the Galaxy, but are not inconsistent with those present in the Magellanic Clouds. Assuming the Galactic DIB–$E(B - V)$ relation, we determine reddening upper limits for the DLAs in our sample. Based upon the $E(B - V)$ limits, the gas-to-dust ratios, $N(\text{H I})/E(B - V)$, of the four aforementioned DLAs are at least ~ 5 times higher than that of the Galactic ISM and are more consistent with the Large Magellanic Cloud. The ratios of two other DLAs are at least a factor of a few times higher. The best constraints on reddening derive from the upper limits for the $\lambda 5780$ and $\lambda 6284$ DIBs, which yield $E(B - V) \leqslant 0.08$ mag for four of the seven DLAs and are more consistent with the Magellanic Clouds rather than the Galaxy. Our results suggest that, in DLAs, quantities related to dust, such as reddening and metallicity, appear to have a greater impact on DIB strengths than does H I gas abundance. The molecules responsible for the DIBs in DLA selected sightlines are underabundant relative to sightlines in the Galaxy of similarly high $N(\text{H I})$. Using DIBs to study the ISM of DLAs provide evidence that at least some population of DLAs are more Magellanic-like than Galactic-like.

Keywords. astrochemistry, dust, extinction, galaxies: ISM, Magellanic Clouds, quasars: absorption lines

1. Introduction

Since their discovery in 1921 (Heger 1922), the diffuse intersteller bands (DIBs) have remained the longest known interstellar absorption features without a positive identification. There have been several hundred DIBs discovered to date (Jenniskens *et al.* 1994; Tuairisg *et al.* 2000; Weselak *et al.* 2000). The DIBs span the visible spectrum between

4000 and 13000 Å. Despite no positive identifications, several likely organic molecular candidates have emerged as the sources of the DIBs, including polycyclic aromatic
hydrocarbons (PAHs), fullerenes, long carbon chains, and polycyclic aromatic nitrogen
heterocycles (PANHs) (Herbig 1995; Snow 2001; Cox & Spaans 2006; Hudgins *et al.*
2005). The organic-molecular origin of the DIBs may give them an importance to as-
trobiology; they are now considered an important early constituent to the inventory of
organic compounds on Earth (Bada & Lazcano 2002).

Due to their relatively weak absorption strengths, the DIBs have been difficult to
detect in extragalactic sources. Aside from the hundreds of detections within the Galaxy
(Jenniskens *et al.* 1994; Tuairisg *et al.* 2000; Weselak *et al.* 2000), DIBs have been detected
in the Magellanic Clouds (Welty *et al.* 2006; Cox *et al.* 2006, 2007), seven starburst
galaxies (Heckman & Lehnert 2000), the active galaxy Centaurus A via supernova 1986G
(Rich 1987), spiral galaxy NGC 1448 via Supernovae 2001el and 2003hn (Sollerman *et al.*
2005), one damped Lyman-α galaxy (DLA) at $z = 0.524$ toward the quasar AO 0235+164
(Junkkarinen *et al.* 2004; York *et al.* 2006), and one galaxy selected by singly ionized
calcium (Ca II), J0013−0024, at $z = 0.157$ from the Sloan Digital Sky Survey (Ellison
et al. 2008).

There are several environmental factors that are known to enhance or inhibit DIB
strengths. Two important environmental factors that are often probed in galaxies are
the H I content, $N(\mathrm{H\,I})$, and the reddening, $E(B-V)$. $N(\mathrm{H\,I})$ is a measure of the gas
phase and $E(B-V)$ is a measure of the dust phase of the ISM in galaxies. $N(\mathrm{H\,I})$ in
DLAs is typically measured as a column density via the Lyman-α line in absorption, as
observed using a bright background source such as a quasar. DLAs are, by definition,
rich in H I gas, with a $\log N(\mathrm{H\,I}) > 20.3$ atoms cm^{-2}.

The ISM, and by extension the host galaxies, of DLAs are still relatively poorly understood (Ellison *et al.* 2005). We compare the DIB strengths in DLAs with the strengths
of DIBs in the Galaxy and the Magellanic Clouds, along sightlines with known H I abundances and reddening measurements, and find that the ISM of many DLAs are more
similar to the ISM of the Magellanic Clouds. DLAs, as a population, may represent unrecognized Magellanic-like galaxies. The results of further similar studies of DIB strengths
in galaxies may hold significant implications in our understanding of the H I and dust
abundances of galaxies with cosmic time.

For this study, fully explained in Lawton *et al.* (2008), we catalogue the strengths of
the $\lambda 4428$, $\lambda 5780$, $\lambda 5797$, $\lambda 6284$, and $\lambda 6613$ DIBs relative to the $E(B-V)$ and $N(\mathrm{H\,I})$
content of each of the seven DLAs in our sample. Observations were obtained, with
seven facilities, of seven DLAs toward six QSO sightlines. The facilities and instruments
used for this project are the *VLT*/FORS2, *VLT*/UVES, *APO*/DISIII, *Keck*/HIRES,
WHT/ISIS, and *Gemini-S*/GMOS.

2. Analysis and results

There are two detections included in this work, the $\lambda 5705$ and $\lambda 5780$ DIBs first reported by York *et al.* (2006), in the $z = 0.524$ DLA toward AO 0235+164. For all other
DLAs in our sample, we report upper limits on the $\lambda 4428$, $\lambda 5780$, $\lambda 5797$, $\lambda 6284$, and
$\lambda 6613$ DIB equivalent widths. We measured the equivalent width limits using a gener-
alized method of the Schneider *et al.* (1993) technique for finding lines and limits. We
compare our measured limits to the expected DIB equivalent widths from the known
Galactic DIB–$E(B-V)$ and DIB–$N(\mathrm{H\,I})$ relations (Welty *et al.* 2006). The $N(\mathrm{H\,I})$ quantities are known for the DLAs; however, Junkkarinen *et al.* (2004) published the only
reddening known for the DLA galaxies in our sample, AO 0235+164, with a measured

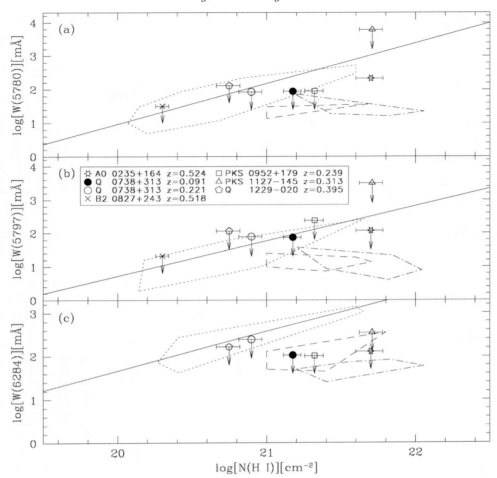

Figure 1. The DIB equivalent width–$N(\mathrm{H\,I})$ relations (Welty *et al.* 2006) with our DLAs added. —(a) $\lambda\,5780$ DIB. —(b) $\lambda\,5797$ DIB. —(c) $\lambda\,6284$ DIB. The solid lines are the best-fit weighted Galactic lines. The region enclosed by the dotted lines contain the Galactic data. The regions enclosed by the dashed lines contain the LMC data. The regions enclosed by the dot-dash lines contain the SMC data. Error bars are 1σ, and upper limits are marked with arrows. The vertical error bars for AO 0235+164 in panel (a) are smaller than the point size and all values for this DLA are from York *et al.* (2006).

$E(B-V) = 0.23$ mag. We estimate the upper limit to the reddening using our equivalent width limits and the Galactic DIB–$E(B-V)$ correlation. Our equivalent width limits are robust enough to constrain the upper reddening limits near the $E(B-V) < 0.04$ mag limit found by Ellison *et al.* (2005) for the highest redshift DLA galaxies.

The results from the $N(\mathrm{H\,I})$ model suggests that the organics that give rise to the DIBs in DLAs are underabundant relative to Galactic sightlines of the same hydrogen column density. Fig. 1 shows this by plotting the measured equivalent widths and upper equivalent width limits for the DLAs in our sample. The line is the best-fit to the Galactic data from Welty *et al.* (2006). The Galactic points are observed to lie within the dotted region while the Large Magellanic Cloud (LMC) sightlines are observed to lie within the dashed region. The Small Magellanic Cloud (SMC) sightlines are all within the dot-dashed region. The $\lambda\,6284$ DIB gives the best constraints and shows that this DIB is

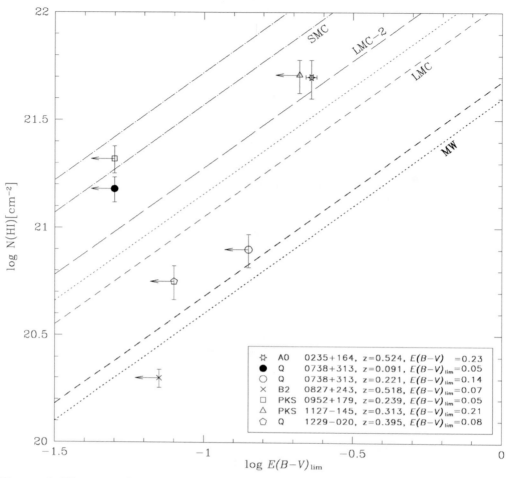

Figure 2. The gas-to-dust ratios of the DLAs in our sample relative to measured values in the Galaxy (MW), the LMC, and the SMC. The figure is modified from Cox *et al.* (2006). The plot measures the log column density [cm^{-2}] versus the upper limit to the log reddening for each DLA. The A0 0235+164 reddening measurement of 0.23 ± 0.01 is from Junkkarinen *et al.* (2004). The top two lines represent measured SMC gas-to-dust ratios while the middle three lines represent the LMC, and the bottom two lines represent the MW. The dot-dashed SMC lines are the upper and lower gas-to-dust ratios of Bouchet *et al.* (1985), with 52×10^{21} cm^{-2} mag^{-1} and 37×10^{21} cm^{-2} mag^{-1}, respectively. The long-dashed LMC line gives the gas-to-dust ratio of 19.2×10^{21} mag^{-1} from the LMC-2 data of Gordon *et al.* (2003). The dotted LMC line is a linear fit to the LMC data in Cox *et al.* (2006) and gives a gas-to-dust ratio of 14.3×10^{21} cm^{-2} mag^{-1}. The short-dashed line is the average LMC regions from Gordon *et al.* (2003) and has a gas-to-dust ratio of 11.1×10^{21} cm^{-2} mag^{-1}. The dashed MW line gives a gas-to-dust ratio of 4.8×10^{21} cm^{-2} mag^{-1} (Bohlin *et al.* 1978), and the dotted MW line is the fit to the Galactic data from Cox *et al.* (2006) which yields a ratio of 4.03×10^{21} cm^{-2} mag^{-1}. Four of the DLAs in our sample have gas-to-dust ratios more consistent with the LMC than the MW.

at least 4–10 times weaker in four of our DLAs compared to what is expected in the Galaxy. As is the case for the Magellanic Clouds, the Galactic DIB–$N(\text{H\,\textsc{i}})$ relation does not apply to DIBs in DLAs.

From our equivalent width limits we can estimate the upper limit to the reddening assuming the Galactic DIB–$E(B - V)$ relation holds for DLAs. There are little data on DIBs in DLAs; however, Ellison *et al.* (2008) create a fit to all known extragalactic points for the $\lambda 5780$ DIB. The Galactic $\lambda 5780$ DIB–$E(B-V)$ relation appears to remain valid when the extragalactic DIB equivalent width measurements are included. Our upper limits for $E(B-V)$ yield lower limits to the gas-to-dust ratios for our DLAs; these results are shown in Fig. 2. $E(B-V)_{lim}$ are the upper limits for the reddening determined by our best equivalent width limits. The best-fit lines for the Galaxy (MW), the LMC, and the SMC are given from the literature. Our limits are robust enough to constrain the gas-to-dust ratio for the DLAs as being at least ~ 5 times higher than the Galaxy for four of the DLAs. Those four DLAs have gas-to-dust ratios that are more consistent with the LMC or the SMC. The reddening limits allow us to extend the results of Ellison *et al.* (2005) to lower redshift DLAs. The $E(B-V) < 0.04$ mag from Ellison *et al.* (2005) appears to also be a relatively robust result for moderate redshift DLAs. Four of the seven DLAs have reddenings of $E(B-V) \leqslant 0.08$ mag.

3. Discussion

The LMC and SMC are sufficiently gas-rich that they would both be classified as DLAs if background QSOs probed through many of their intervening H I clouds. It is possible that DLAs may represent a population of gas-rich galaxies that are similar to the Magellanic Clouds. However, care should be taken when estimating global properties of a galaxy from a single line-of-sight. In the Galaxy, DIB strengths and ratios vary significantly based on the local environmental conditions. An example of this can be found within the SMC. The SMC Wing is the only location within the dwarf galaxy where DIBs are observed and is noticeably enshrouded in dust such that the molecules responsible for the DIBs are likely to be adequately protected from UV radiation (Ehrenfreund *et al.* 2002). The sightline Sk 143 toward the SMC Wing, along with containing DIB absorption, exhibits the 2175-Å dust bump (Ehrenfreund *et al.* 2002). This feature is unusual for the SMC but commonly found in the Galaxy. The only DLA with known DIB absorption, toward QSO AO 0235+164, is also the only DLA in our sample with the 2175-Å feature (Junkkarinen *et al.* 2004). Furthermore, the $\lambda 6284$ DIB observed along the SMC Wing sightline Sk 143 is uncommonly weak relative to the other observed DIBs, which is the same trait observed in the DLA toward QSO AO 0235+164 (York *et al.* 2006).

DLA metallicities are known to be low, on the order of ~ 0.1 solar metallicity, with the exception of AO 0235+164, which is metal abundant (Junkkarinen *et al.* 2004). Little is published about the radiation in moderate redshift DLA galaxies. Both the LMC and SMC have sub-solar metallicities with $Z \approx 0.4~Z_\odot$ and $Z \approx 0.2~Z_\odot$, respectively (Welty *et al.* 2006), relatively low reddening, and a large H I abundance. These are all similarities they share with the general DLA population. Further observations of DLAs will need to be done in order to understand the levels and sources of ionizing radiation, the dust grain populations, and the sizes and morphologies of their parent galaxies at all redshifts in order to put them into context with the Magellanic Clouds. Ongoing numerical simulations are also providing evidence that some population of DLAs may be gas-rich satellite or dwarf galaxies. Work done by Kacprzak *et al.* (2008), with ΛCDM cosmological simulations from Ceverino & Klypin (2007), show that lines-of-sight through an edge-on simulated galaxy, of approximately Galactic mass and at a redshift $z = 0.923$, probe many satellites of approximately LMC mass that would be classified as DLAs.

References

Bada, J. L. & Lazcano, A. 2002, *Science*, 296, 1982

Bohlin, R. C., Savage, B. D., & Drake, J. F. 1978, *ApJ*, 224, 132

Bouchet, P., Lequeux, J., Maurice, E., Prévot, L., & Prévot-Burnichon, M. L. 1985, *A&A*, 149, 330

Ceverino, D. & Klypin, A. 2007, *ApJ*, submitted [astroph/0712.3285]

Cox, N. L. J. & Spaans, M. 2006, *A&A*, 451, 973

Cox, N. L. J., Cordiner, M. A., Cami, J., Foing, B. H., Sarre, P. J., Kaper, L., & Ehrenfreund, P. 2006, *A&A*, 447, 991

Cox, N. L. J., Cordiner, M. A., Ehrenfreund, P., *et al.* 2007, *A&A*, 470, 941

Ehrenfreund, P., Cami, J., Jiménez-Vicente, J., *et al.* 2002, *ApJ*, 576, L117

Ellison, S. L., Hall, P. B., & Lira, P. 2005, *AJ*, 130, 1345

Ellison, S. L., York, B. A., Murphy, M. T., Zych, B. J., Smith, A. M., & Sarre, P. J. 2008, *MNRAS*, 383, L30

Gordon, K. D., Clayton, G. C., Misselt, K. A., Landolt, A. U., & Wolff, M. J. 2003, *ApJ*, 594, 279

Heckman, T. M. & Lehnert, M. D. 2000, *ApJ*, 537, 690

Heger, M. L. 1922, *Lick Observatory Bull. 10*, 337, 146

Herbig, G. H. 1995, *ARAA*, 33, 19

Hudgins, D. M., Bauschlicher, C. W., Jr., & Allamandola, L. J. 2005, *ApJ*, 632, 316

Jenniskens, P. & Désert, F. -X. 1994, *A&A*, 106, 39

Junkkarinen, V. T., Cohen, R. D., Beaver, E. A., Burbidge, E. M., & Lyons, R. W. 2004, *ApJ*, 614, 658

Kacprzak, G. G., Churchill, C. W., Ceverino, D., Steidel, C. C., Klypin, A., & Murphy, M. T. 2008, *ApJ*, submitted

Lawton, B., Churchill, C. W., York, B. A., Ellison, S. L., Snow, T. P., Johnson, R. A., Ryan, S. G., & Benn, C. R. 2008, *AJ*, 136, 994

Rich, R. M. 1987, *AJ*, 94, 651

Schneider, D. P., Hartig, G. F., Jannuzi, B. T., *et al.* 1993, *ApJS*, 87, 45

Snow, T. P. 2001, *Spectrochimica Acta Part A*, 57, 615

Sollerman, J., Cox, N., Mattila, S., Ehrenfreund, P., Kaper, L., Leibundgut, B., & Lundqvist, P. 2005, *A&A*, 429, 559

Tuairisg, S. O., Cami, J., Foing, B. H., Sonnentrucker, P., & Ehrenfreund, P. 2000, *A&A*, 142, 225

Welty, D. E., Federman, S. R., Gredel, R., Thorburn, J. A., & Lambert, D. L. 2006, *ApJS*, 165, 138

Weselak, T., Schmidt, M., & Krełowski, J. 2000, *A&A*, 142, 239

York, B. A., Ellison, S. L., Lawton, B., Churchill, C. W., Snow, T. P., Johnson, R. A., & Ryan, S. G. 2006, *ApJ*, 647, L29

At the conference dinner inside Wrenbury Hall.

Summary

The Magellanic System: Stars, Gas, and Galaxies
Proceedings IAU Symposium No. 256, 2008
Jacco Th. van Loon & Joana M. Oliveira, eds.

The Magellanic System:
Stars, Gas, and Galaxies

Joss Bland-Hawthorn[1] and Jay S. Gallagher[2]

[1] School of Physics, University of Sydney, Australia
email: jbh@physics.usyd.edu.ac

[2] Department of Astronomy, University of Wisconsin, USA
email: jsg@astro.wisc.edu

Abstract. We provide a brief overview of some key issues that came out of the IAUS 256 symposium on the Magellanic System (http://www.astro.keele.ac.uk/iaus256).

Keywords. history and philosophy of astronomy, sociology of astronomy, Magellanic Clouds

1. Introduction

We would like to start by thanking Jacco van Loon and the science organizing committee for proposing such a timely meeting, and for locating it in this serene corner of England. We suspect that only a tiny fraction of Britain is aware that Keele is a one-pub rural village in the heart of Staffordshire. The meeting was well organized and the facilities were excellent.

Meetings on the Magellanic System seem to come round once a decade, with the last held in 1998 in Canada. Over the past week, we have been treated to an extraordinary range of new observations over the Magellanic Clouds (MCs). The meeting has been well structured around putting the MCs into a cosmological context. The key questions that we were asked to address were:

- How does metal abundance influence star formation and stellar feedback?
- How does the structure of the multiphase ISM depend on the host?
- How has the interaction between the Galaxy and the MCs shaped their evolution?

After hearing a week's worth of lectures, we suggest, post facto, a couple of other questions that come to mind.

- To what extent have the *baryons* in the MCs evolved along different tracks relative to the Galaxy? To what extent are the MCs typical of sub-$L_\star$ galaxies?

A feature of this meeting has been the prevalence of detailed wide-field maps of the Magellanic System across almost the entire electromagnetic system. These are at comparable resolution and high sensitivity to the extent that we can consider a bolometric approach at each pixel over a very wide field. This enabled many of the speakers to show their specific area of study in a wider context, an approach that was highly effective. After all, we are studying entire galaxies that happen to be on our doorstep so why not take advantage of this fact wherever possible.

To achieve the detailed comparisons that we witnessed, several groups have organized themselves into large teams (e.g., SAGE: PI Margaret Meixner) that carries with it interesting sociological issues discussed below. It also is not at all clear that the theoretical community is ready for the data deluge that is almost upon us.

As expected, *Spitzer Space Telescope* (*SST*) infrared observations featured in many talks. IRAC observations, particularly at 6–8 μm, have successfully picked up PAH emission in metal-depleted gas within the MCs. Very small grains are prominent in the MIPS 24-μm band, and large grains in the longest MIPS bands. The MIPS bands are particularly sensitive to dust in high-starlight intensity environments, e.g., star forming regions. PAHs have yet to be detected in the Magellanic Bridge and Stream where metallicities are down to 1/10 solar; this may be consistent with the very low PAH abundance expected at these metallicities (e.g., Draine *et al.* 2007).

Several groups presented detailed star formation histories across the entire MC system. Just a decade ago, there were few YSOs known in the MC. Since Spitzer, we now see YSO populations across the entire disk of the LMC, while *HST* also is revealing hoards of pre-main sequence stars in young regions. It is encouraging the extent to which the multi-polyhedral CMD approach (q.v. Coimbra proceedings) is now widely used to analyze resolved stellar populations, and that different groups arrive at the same star formation histories. The stellar populations and gas/dust properties of the MC appear to be quite distinct from the Galaxy. This is evidenced by a range of different stellar components in the MC: a distinctive compact source (BH, XRB+Be) population, distributions of variable star types, and s-process chemical signatures in the RGB population, to name a few.

There is now clear evidence that stellar evolution is directly affected by the mean metallicity of the ISM. As is well known, there are distinct blue and red globular cluster populations, where the former are not seen in the Galaxy. Star cluster mass functions look distinct from the Galaxy in the sense that "infant mortality" seems much less prevalent. Moreover, a few of these get up to 10^4 stars, which seems extraordinary in some respects.

Some of us were left reeling at the news that the *HST* ACS proper motions (Kallivayalil *et al.* 2006) are now confirmed with a third epoch of observations, and by the analysis of an independent group (Piatek *et al.* 2008). The extraordinarily high orbital speed of the MC system ($\sim$ 380 km s^{-1}), a 50% increase on what was believed just a few years ago, was not anticipated by the plethora of published dynamical models since the 1970s. Several enterprising researchers have already exploited these results, but the "standard" orbit families will have to be re-addressed in light of this.

Interesting developments are the high baryon to dark matter ratio in both MC systems, and the possibility that these galaxies entered the outer Galaxy as a group. With the promise of future astrometric missions, we will soon be moving from kinematics to dynamics for much of the Local Group. This will provide tighter constraints for ΛCDM simulations that start from an inversion of the local density field, which routinely produce better galaxies than full-blown simulations.

2. Major developments since 1998

We start by listing what strike us as major developments since the last conference on the Magellanic System:

- All Local Group dwarfs contain ancient stars, including MCs
- MCs have distinct *baryons* (gas+stars) compared to the Galaxy
- MC star clusters survive long term relative to our Galaxy
- The Galaxy is *not* made up of today's dwarfs
- Remarkable change in orbit parameters for MC system
- MCs have very extended stellar distributions
- Growing complexity of dark matter for MC system?

- MCs possibly accreted as a group?
- MCs have high baryon/dark matter ratio: stripped outer halos?
- Stream, Bridge metal poor relative to MCs, remnants of large gas disks?
- Stream dissolving into Galactic halo
- Distinct magnetic field structure in MCs

We have stopped short of adding hundreds of references to the above list in order to achieve some measure of brevity for the proceedings. The interested reader can quickly track back from the body of contributions here, or contact us. Some of these lead naturally into "future questions" that we address below.

We were struck by the compelling case that was made by van der Marel & Cioni (2001) for the importance of angular projection effects in the LMC in particular. The offset bar looks much more central and the outer disk is found to be intrinsically elliptic once the 3D projection effects are taken into account.

A major aspect of the meeting is the extent to which the MCs have become profoundly important astrophysical laboratories, an issue that we pick up in the next section. Gary Da Costa asked the question in reverse: where has there been essentially no progress since the 1998 meeting? One topic that comes to mind perhaps is the thorniest of them all— what are the processes involved in massive star formation? But this is clearly one of the key research areas and we would be surprised if fundamental progress had not been made in this arena by the end of the next decade.

3. The Magellanic Clouds as astrophysical laboratories

While observing during the early 1830s from the Cape of Good Hope, Herschel resolved the MCs into myriads of stars, thereby laying the first foundations for extragalactic astronomy. This also set the stage for the MCs to serve as laboratories where key astrophysical processes could be readily observed over a wide range of scales. At this meeting, the tradition continues into a new century with the added advantage of increasingly high angular resolution, sensitivity and panchromatic coverage across the spectrum.

With the advent of FIR surveys from the *SST* and *AKARI*, connections between the ISM and stellar populations in the MC are becoming clearer. In this way we are extending our reach into the details of the central baryon lifecycle of IGM-ISM-stars-ISM-IGM-..., which are connected by radiation and gastrophysical processes that define much of the present day visible universe. Improving data on MC chemical abundances, also tracers of this cycle, reveal differences from the Milky Way. Some of these can be traced to the varying star formation histories, but some hint at surprisingly long term ISM–IGM interactions in the MCs. Our increasingly detailed snapshots of the MC ISM further emphasize the dynamic state of this system. Stars are pumping dust and new elements into the ISM, which in turn is modified by shocks, flows, and the stellar radiation fields. A qualitative modeling of these increasingly well-defined actions stands as a key step where MC studies contribute towards a full theory of galaxy evolution.

The traditional role of the MCs as test beds for ideas concerning stellar physics remains at full value. While the *HST* and *SST* combine to explore populations of young and dying stars, wide area surveys provide full samples and reveal rare and unexpected features of stellar populations. For example, the SMC seems to prefer Be X-ray binaries and eschews black hole systems. With their weak tidal fields, low stellar densities, and favorable viewing angles, the MC are a field of dreams for star cluster studies. They are taking on a leading role in understanding the birth and survival of star clusters as well their evolution as astrophysical systems that test our ability to model real N-body systems.

In recent years, the pendulum has swung in the direction of emphasizing differences between the MCs and the Galaxy. But one should not lose sight of the fact that, given that such differences are well established now, if there are populations that are essentially identical between both systems, this is of profound importance in its own right. While there are aspects of stellar astrophysics that are largely independent of their metallicity (cf. Russell-Vogt theorem), the same goes for gas/dust phases in both galaxies.

The ways in which the MCs respond to their multiply felt external perturbations from each other and the Milky Way offer increasingly sharp tests for models of galaxy dynamics. In the MCs, we have an interacting system of galaxies that is increasingly well mapped in phase space as the quality of distance and proper motions improve. That their behaviour doesn't quite match our current expectations (e.g., structure of the Magellanic Stream; stellar tidal debris) is cause for optimism that the MCs are yielding insights into hierarchical galaxy evolution. And doing so in a situation where microlensing surveys strongly constrain the nature and distribution of low luminosity baryonic matter.

4. Near field vs. far field: what are we learning?

We have a great deal to learn about the processes of galaxy formation and evolution from our nearest neighbours. Dwarf galaxies continue to challenge ΛCDM N-body simulators at all cosmic epochs. Are younger versions of the MCs quite typical of high redshift gas-rich galaxies? Are MCs typical of high-redshift GRB hosts? It is tempting to think so, but the quantitative case merits further development.

It seems plausible that the MCs were once part of a group that accreted onto the Local Group, and indeed such groups are being found now in the Local Universe (Tully *et al.* 2006). Thus, it is likely that dwarf galaxies were clustered in the early universe. It is interesting to speculate to what extent that the MC accretion event onto the Galaxy is typical of group accretion in general. We may ultimately find with detailed dynamical models that we can account for one or more of the other Galactic companions or streams as progenitor group companions.

At the present time, the most detailed baryonic simulations on the scale of individual galaxies come from semi-analytic models. Arguably, they have limited predictive power and are more of a consistency check. For example, no published simulations foresaw the extent to which the baryonic content (e.g., black hole / stellar population, gas / dust content) of the MCs evolved along a distinct track from the Galaxy.

The formation and evolution of the MCs must be complicated. The high baryon/dark matter ratio suggests that the once larger dark-matter halos have been tidally stripped, at least in part. Their evolution today is further complicated by their motion through the outer Galactic halo ("halos within halos"). The system is clearly losing gas to the Galactic halo, but somehow at least the LMC appears to have ongoing gas accretion.

A long held assumption is that Galactic accretion involves baryonic matter associated with dark matter. Indeed, the dark matter stabilizes the gas against disruption by the halo corona, at least for a while. This view is partly supported by the existence of the gas-rich MCs within 50 kpc of the Galaxy.

There is possible evidence for ram-pressure induced star formation along the leading edge of the LMC. For now, we can only speculate on galaxy-wide feedback processes that operate in a dynamic pair such as this. Are there large-scale galactic winds operating in either or both systems? These would have to compete with the inevitable cross wind due to the motion of the MCs through the halo. What are the expected kinematic signatures in such a configuration? Are we already seeing these signatures along QSO sight lines?

5. Present and future facilities

Here are just some of the present and future facilities that have the potential to make an impact on Magellanic System research. We have removed northern facilities like PanSTARRS that are not ideally situated for this field and cannot claim completeness; so this can be thought of as a minimal summary:

HST (serviced), *VLTI*, *SST* (warm), *AKARI* (warm)
VLT, *Magellan*, *SALT*, *Gemini* + Adaptive optics (MCAO, MOAO, ...)
VISTA, *VST*, *SkyMapper*
Chandra, *XMM*, *Suzaku*
INTEGRAL, *GLAST*, *Herschel*, *Planck*
WFMOS, HERMES, ...
ALMA, *ASKAP*, *meerKAT*, *NANTEN2*
JWST, *GAIA*, *SIM*, *JDEM* (SNAP, ADEPT, ...)
ELTs (*GMT*, *TMT*, *E-ELT*)
SKA

The world has invested in an amazing array of astronomical capabilities. Many of these can and should be applied to MC studies as part of our continued development of a quantitative understanding of the universe. The lively discussions and range of ideas presented here in Keele is the basis for optimism that we will progress along this path as the opportunities arise. The one point of concern is whether the resources will exist for theory to keep pace with the observations?

6. The sociology of large science teams

The MCs are complex systems whose properties are accessible in detail. Large data sets and associated sophisticated models in turn require diverse expertise to obtain full scientific value. While we empathize with White's (2007) lament on the trend of astronomers to organize themselves into oversized teams, the specialized requirements of multi-disciplinary research forces this upon many of us. We can expect more work in this vein; e.g., from the upcoming major ground-based surveys and new capabilities such as *Herschel* in space. The challenge then is for us to be aware of the concerns associated with team science while continuing to make the best of this approach.

MC research has the advantage of relatively well-defined goals (for astronomers!) that largely follow from the standard approach of comparing observations to models which in turn are based on theory. As we've seen and heard here, we now have the further advantage of working with wide and deep seas of data, which require the efforts of many people to produce, let alone analyze. At issue is our ability to maintain focus on key scientific goals without being overwhelmed by the many details, and to creatively engage a wide skill set in the process. Organizing our intellectual resources is of increasing importance in 21st century MC astronomy.

Large collaborations also raise a number of well-known practical issues, especially in times of tight funding. We seem to have landed in an era that is extraordinarily rich in facilities but perhaps somewhat less so in supporting the individual scientists. Obviously we want to make the best of this situation, and to a significant degree this will mean working in teams. Care needs to be taken to maintain a healthy level of competition; history is not kind to single behemoth models of human enterprise. Giving credit where credit is due also can become difficult, especially in the context of standard astronomical practices with regard to authorship on refereed papers, an issue that is receiving wide discussion in the bio-sciences and has led to very formal procedures in high energy physics. Wide access to data sets, especially in digested form, as well as results from modelling efforts can spread the possibility for creative work beyond the individual teams. Our MC programs with their clearly prescribed data sets and model parameters can benefit from and should support effective and wide access to information, e.g., through the Virtual Observatory effort. We've done quite well so far, and should plan to sustain this success into the future.

7. Relevant aside: how are we to approach complexity?

This is a fundamental question that is rarely discussed in polite company. After all, we are scientists and our job is to construct a well posed question (essentially a null hypothesis H_0 with a control sample) that allows for the possibility of a rejection of the hypothesis. But this question has certainly unsettled more than a few applied fields of science in recent years. John Horgan, in his provocative book "The End of Science", suggests that the era of reductionism is over and that the human race is facing fundamental barriers to the acquisition of new wisdom and knowledge. Robert Laughlin declares that "the central task of theoretical physics today is no longer to write down the ultimate equations but rather to catalogue and understand [complex] behaviour in its many guises..."

With regard to complex behaviour in astronomy, there are sociological issues to get over. At an Aspen meeting three decades ago, Ed Salpeter was confronted by a surly participant who quipped "but surely, that's just weather." "Aha!" he replied "someone who understands weather. Can you please stand and explain it to us?" The reader may be surprised to know that models for the phenomenon of rain have normalizing constants that differ by many orders of magnitude! And yet nobody would question the existence of rain, least of all a person from Staffordshire.

There is a wonderful quote by von Neumann from a speech he gave in Montréal in 1945: "Many branches of both pure and applied mathematics are in great need of computing instruments to break the present stalemate created by the failure of the purely analytical approach to non-linear problems." In fact, today, we realize that all fields of applied science (and mathematics for that matter) degenerate quickly into a wash of complexity. So now we are confronted by the prospect of acquiring vast amounts of information without receiving new wisdom. How are we to proceed?

Interestingly, we routinely make progress when there is a strong motivation to do so, e.g., medical science, environmental science, corporate finance. In most instances, we are teasing out weak signatures that are fundamental to the process at hand.

Lev Landau had an extraordinary capacity to tame complex problems. To paraphrase, consider a problem that is parametrized in the following way:

$$f = f(u_1, u_2, u_3, u_4, ...)$$

In other words, we write down all conceivable variables u_1, u_2, ... that might describe the process at hand. We leave nothing out. In the pre-supercomputer era, one is forced then to search long and hard for limiting and transitional cases at the boundaries of the polyhedral space that might be amenable to attack. Only along the interstices can you even consider constructing H_0 because here only a limited number of variables are operating. Once the interstices and vertices are defined (often with key discoveries along the way), we can now identify a large volume in f-space where one can certainly extract information, but little or no physical insight. One can think of this as finding the boundaries of a problem in our search for higher organizing principles.

Nowadays, we resort to massive computer calculations and physical algorithms with ever increasing sophistication. The f-space is then searched exhaustively. In this way, extraordinary progress has been made in understanding turbulent media under quite specific conditions.

We cannot help feeling that the gastrophysics/ISM community needs to embrace the large suite of mathematical tools that are available today. After all, cosmologists do this as a matter of course to extract physical parameters from the microwave background, and to compare numerical simulations with large galaxy surveys. This is precisely what the gastrophysics community needs to do in order to compare the multiphase density and velocity maps which already are available for the MCs with their complex simulations. It is well known that physical processes manifest themselves in the power spectrum of the density structure, even in the presence of 3D projection effects. For example, these mathematical tools are the mainstay of the many hundreds of Los Alamos PhDs that have studied complex fluid dynamics in the presence of an extreme impulsive event.

8. Where will we be in 10 years?

Predictions in any field are notoriously unreliable, but for what it's worth, here goes. Even while Moore's law shows evidence of breaking down, a moderately secure prediction is that computers will be considerably more powerful, and no doubt our gastrophysical and stellar atmospheric algorithms will continue to improve. Similarly we can hope to approach more closely to a complete theoretical description of stars, including their rotation and tendency for binarity, and from this build a more accurate set of evolutionary models.

We share the enthusiasm of this conference for the impending revolution that is likely to be brought on by improved astrometry, first with the *HST* servicing mission providing even longer timelines on the proper motion of the MCs. Astrometry will undergo a revolution in an era of *GAIA* and *SIM* in the next decade. We will steadily move from kinematics (geometry of motion) to 3D orbit and internal dynamics, all in full 3D perspective throughout the Milky Way system. One can envisage detailed and accurate simulations of dynamical friction, oval distortion, tidal shocking and disruption between the two galaxies, triggering star formation bursts that are consistent with the stellar record. It is not at all clear just how well this will work out. On the one hand, one of both of the MC may be accreting gas in a stochastic manner. But the new hydro simulations of the dissolving Stream, presented at this meeting, also suggest a replenishment rate for the Stream (i.e. a mass-loss rate from the MC system) of around 0.1–0.3 $M_\odot$ yr^{-1}.

We keenly await a time when a convergence of near-field and far-field cosmology is achieved, i.e. cradle to grave observations across cosmic time from the great *ALMA* and *JWST* experiments. This prospect was foreseen by Hoyle (1965) in his Princeton & Cambridge lectures:

It is not too much to say that the understanding of why there are different kinds of galaxy, of how galaxies originate, constitutes the biggest problem in present day astronomy. The properties of individual stars that make up the galaxies form the classical study of astrophysics, while the phenomena of galaxy formation touches on cosmology. In fact, the study of galaxies forms a bridge between conventional astronomy and astrophysics on the one hand, and cosmology on the other.

Stellar astrophysics is widely touted as one of the central pillars of modern astrophysics, but a vast amount remains to be understood. After all, just a few years ago, M.A. Asplund announced that the Sun does not in fact have canonical solar abundances (well, [Fe/H] needed to be downgraded by a hefty 0.2 dex). We know too little about non-LTE processes and accurate stellar ages continue to elude us, even with the best efforts of asteroseismologists. We were entranced by the prospect of the Eddington satellite measuring the mean molecular weight of the core of 100,000 dwarfs, but this seems to have gone the way of the dodo. Self-detonating supernova and stellar wind models seem a remote prospect but are clearly essential to future progress. Is it too much to expect major breakthroughs here?

We strongly encourage would-be theorists to give gastrophysics a serious look. In just a few years, we will be deluged with a staggering array of multiphase maps of gas, dust, particles and molecules spanning a wide range of ionization states and transitions. The stopgap divisions among ISM dust particles of aromatic features, very small and big grains looks set to subdivide further. This burgeoning field is crying out for an army of talented young recruits. Several groups around the world already treat multiphase turbulent plasmas, either with a view to studying the diffuse warm or hot media in galaxy clusters. The close association of H_2 and H^+ within cluster gas and in the M 82 wind reminds us of the importance of pressure in driving the formation of dust and molecules, even in the most unlikely environments. In recent years, a few teams have made impressive headway with studying the synergistic relationship of H_2 and H I, clearly an important step on the road to understanding the dynamical state of the interstellar medium. From here, it's "onward ho!" to a self-consistent model of star formation, with magnetic fields in tow. Good luck!

Who knows what awaits us with the commissioning of the *Large Hadron Collider* later this year. The speed with which the community accepted a non-zero value of Λ a decade ago was astonishing after two independent groups announced their high-redshift supernova results. Maybe the Dark Sector will invade our thinking again in just a few years with equal rapidity—time will tell.

9. Big future questions

It is abundantly clear to us that research on the Magellanic System is an extraordinarily rich area of study, at least for Southern observers. Here we have the opportunity to study two rather typical Magellanic-type galaxies in their entirety, their mutual interaction, and their interaction with the Galaxy. Dwarf galaxies continue to challenge ΛCDM predictions in the non-linear regime at all observable epochs. Here are some of the big questions we foresee in the next decade:

- How did the MCs form and how different were they from what we see today? Are there pre-ionization fossils within?
- Were the MCs much bigger at an earlier epoch? If so, where are the missing baryons and dark matter today?
- How did the MCs subsequently evolve? How is this affected by halos in halos?

- Did the MCs enter the Galactic sphere of influence as a group? Are there other identifiable dwarfs or streams that date back to this event?
- How far do their baryons and DM halos extend today?
- Do we have a complete baryon inventory?
- Where are the accreting baryons demanded by the recent SFH of the LMC?
- How did the bar form in the LMC?
- Do the MCs sustain large-scale winds?
- What are the eventual fate of the Bridge and Stream?
- How much can we learn of baryon evolution from MCs?
- How much can we learn of black hole evolution from MCs? Are there massive black holes ($> 10^2$ $M_\odot$) lurking there today?
- How did their magnetic fields seed and evolve?

10. Epilogue

At this meeting, we were graced by the presence of Mike Feast who has witnessed a few astronomical revolutions in his time—after all, his first Nature paper was published 60 years ago! Mike commented that the early days of MC research were very exciting because new things kept turning up all the time and there was no real framework to build upon. But today, he senses that we are arriving at something of a synthesis, i.e. a coherent picture is emerging at least to the baryon content of both galaxies.

It was said more than once at this meeting that the MC system may be a somewhat unusual accretion event since it is difficult to find a counterpart in the nearby universe. In fact, we would disagree with this point of view. In particular, the M 81 group is dominated by a large galaxy and at least three dwarfs caught up in a maelstrom of H I gas. To our view, this does not look so different from the Galaxy system. Both M 31 and the Galaxy appear to have their own gas streams which are clearly much more extensive at low column density than current observations may suggest.

It is worth noting that the Local Group, dominated by the two great rajahs encircled by courtiers, is in fact quite typical of the Universe today, e.g., as demonstrated by Tully's or Karaschentsev's group catalogues of the Local Volume. There is nothing "pathological" about the Local Group, the MC system or the galaxies within.

Let us conclude by saying that the pace of development in MC research appears to be quickening to the extent that we may not want to wait another decade to meet again. The conference chair has suggested that the next meeting be held in the Southern Hemisphere, with the faintest suggestion of South Africa as a fitting venue.

Heartfelt thanks to the Organizers of this IAU Symposium both for an excellent menu of science that was matched by the scenic venue. JSG thanks *NASA* for support of his Magellanic Cloud interests through funding for several *HST* GO and IDT research programs, and the University of Wisconsin-Madison for its investments in research support. JBH is supported by a Federation Fellowship from the Australian Research Council.

References

Draine, B. T., Dale, D. A., & Bendo, G., *et al.* 2007, *ApJ*, 663, 866

Hoyle, F. 1965, *Galaxies, Nuclei, and Quasars* (New York: Harper & Row), p. 10

Kallivayalil, N., van der Marel, R. P., Alcock, C., Axelrod, T., Cook, K. H., Drake, A. J., & Geha, M. 2006, *ApJ*, 638, 772

Piatek, S., Pryor, C., & Olszewski, E. W. 2008, *AJ*, 135, 1024

Tully, R. B., Rizzi, L., & Dolphin, A. E., *et al.* 2006, *AJ*, 132, 729

van der Marel, R. P. & Cioni, M.-R. 2001, *AJ*, 122, 1807

White, S. D. M. 2007, *Reports on Progress in Physics*, Vol. 70, p. 883

The *Magellanic Ale*, a beer brewed locally specially for the Magellanic System meeting.

Author Index

Object Index

Subject Index